ELEMENTS OF THE THEORY
OF MARKOV PROCESSES
AND THEIR APPLICATIONS

McGraw-Hill Series in Probability and Statistics

DAVID BLACKWELL, *Consulting Editor*

Elements of the Theory of Markov Processes and Their Applications

A. T. Bharucha-Reid

ASSISTANT PROFESSOR OF MATHEMATICS
UNIVERSITY OF OREGON

McGRAW-HILL BOOK COMPANY, INC.

New York   Toronto   London

1960

ELEMENTS OF THE THEORY OF MARKOV PROCESSES
AND THEIR APPLICATIONS

05156

Dedicated to

MY MOTHER

and

THE MEMORY OF MY FATHER

Preface

My purpose in this book is twofold: first, to present a nonmeasure-theoretic introduction to Markov processes, and second, to give a formal treatment of mathematical models based on this theory which have been employed in various fields. Since the main emphasis is on applications, this book is intended as a text and reference in applied probability theory.

The book is divided into three parts: Part I, Theory; Part II, Applications; and Appendixes. Part I consists of three chapters which are respectively devoted to processes discrete in space and time, processes discrete in space and continuous in time, and processes continuous in space and time (diffusion processes). In the first two chapters we restrict our attention to Markov chains with a denumerable number of states, with particular reference to branching stochastic processes. The reader interested in chains with a finite number of states should consult, for example, the books of Doob, Feller, Fréchet, Kemeny and Snell, and Romonovskiĭ.

While the main purpose of Part I is to present the elements of the theory required for the applications given in Part II, we hope that this material can also serve as a text for an introductory course (senior or first-year graduate level) on Markov processes for students of probability and mathematical statistics and research workers in applied fields. Since applications are given only in Part II, the instructor can select examples from the chapters devoted to applications. The prerequisites for Part I, and the rest of the book, are a knowledge of elementary probability theory (the first nine chapters of Feller, say), mathematical statistics (Mood level), and analysis (Rudin level). Some knowledge of matrices and differential equations is also required.

vii

Part II consists of six chapters devoted to applications in biology, physics, astronomy and astrophysics, chemistry, and operations research. An attempt has been made to consider in detail representative applications of the theory of Markov processes in the above areas, with particular emphasis on the assumptions on which the stochastic models are based and the properties of these models. We have restricted our attention to a formal treatment of the models involved and have, therefore, refrained from presenting any numerical results.

The fields we consider in Part II are not the only ones where Markov processes have found applications. We refer in particular to important studies in economics, psychology, sociology, and actuarial science. It is possible that applications in these areas will be treated in a subsequent volume, but for the present we have decided to restrict our attention to the biological and physical sciences and operations research.

The three appendixes are concerned with generating functions, integral transforms, and Monte Carlo methods. In the first two appendixes we have listed some properties of generating functions and Laplace and Mellin transforms that are required in the text. The third appendix is devoted, in the main, to references dealing with the use of Monte Carlo methods in the study of stochastic processes occurring in different applied fields.

A bibliography is given at the end of each chapter and appendix. In addition, a general bibliography of texts and monographs on stochastic processes is given at the end of the Introduction. With particular regard to Parts I and II, an attempt has been made to present bibliographies that are relatively complete and up to date. Reference in the text to items in the bibliography is denoted by the number of the item enclosed in brackets.

The author takes this opportunity to thank his colleagues and friends for their comments on preliminary drafts of this volume. In particular, he would like to express his gratitude to the following persons for reading all, or part, of the manuscript, and for their very helpful comments: S. Goldberg, J. Lopuszański, H. Messel, R. P. Pakshirajan, A. Rényi, H. E. Robbins, L. Takács, and K. Urbanik. The final draft of this book was prepared during the academic year 1958–1959 while the author was in residence at the Mathematical Institute of the Polish Academy of Sciences in Wroclaw, and he would like to thank E. Marczewski and H. Steinhaus for their hospitality during his stay in Wroclaw.

Finally, the author would like to express his thanks to Mrs. J. Rubalcava and Mrs. A. Kruczkiewicz for typing the manuscript and to the Sigma Xi–RESA Research Fund for a grant to facilitate preparation of the manuscript.

A. T. Bharucha-Reid

Contents

ix

APPENDIXES

Introduction

In the evolution of any scientific discipline there is a period in which attempts are made to develop mathematical theories in order to account for and explain the observations generated by the phenomena with which the discipline is concerned. During this period the qualitative and verbal theories are replaced, or supplemented, by quantitative and mathematical theories which express, in the form of some type of equation or equations, a postulated mechanism or model that can generate a theoretical set of observations. The theoretical and experimental observations can then be compared in order to see if the model is a reasonable one, i.e., if it is capable of accounting for the experimental observations obtained under the conditions stipulated by the model.

When formulating a mathematical model, we can select one of two approaches to the study of the phenomenon concerned. These two approaches, which are termed *deterministic* and *stochastic* (or *probabilistic*), reflect the causal nature of the postulated mechanism (or model) which we express in mathematical form. It is not of great interest to ask if the phenomenon with which we are concerned is deterministic or stochastic. We are only concerned with the formulation of a deterministic or stochastic model, the investigation of its properties, and its ability to account for experimental observations. We remark that the mathematical equations involved in the two approaches may be quite similar in form; however, the great difference in the two approaches is in the nature of the questions they pose and try to answer and in the interpretation of the results. To illustrate

1

these two approaches we shall formulate a simple deterministic model for bacterial growth and its stochastic analogue.

Let the function $x(t)$, real-valued and continuous, denote the number of bacteria in a population at time t. To describe the growth of the population, we must formulate a model based on some postulated mechanism for the manner in which the number of bacteria can change. We assume (1) that at time t there are x bacteria in the population and (2) that the population can only increase in size and that the increase in the interval $(t, t + \Delta t)$ is proportional to the number of bacteria present at time t. Hence, we have the relation

$$\Delta x(t) = \lambda x(t)\, \Delta t \qquad \lambda > 0$$

which leads to the differential equation

$$\frac{dx(t)}{dt} = \lambda x(t) \tag{0.1}$$

If we assume that $x(0) = x_0 > 0$, then the solution of Eq. (0.1) is

$$x(t) = x_0 e^{\lambda t} \tag{0.2}$$

In this simple model we have not made any assumption about the removal of bacteria from the population; hence it is clear on biological grounds, as well as from (0.2), that as $t \to \infty$ the population size will go from x_0 to ∞. The distinguishing feature of the deterministic solution (0.2) is that it tells us that, whenever the initial value x_0 is the same, the population size will always be the same for a given time $t > 0$.

We now consider the stochastic analogue of the above model. Let the integer-valued random variable $X(t)$ represent the number of bacteria in a population at time t, and let us assume that $X(0) = x_0 > 0$. In the stochastic approach we do not derive a functional equation for $X(t)$; instead, we attempt to find an expression for the probability that at time t the population size is equal to x. Hence, we seek $P_x(t) = \mathscr{P}\{X(t) = x\}$.

To formulate the stochastic model, we assume (1) that, if at time t there are $x > 0$ bacteria in the population, the probability that in the interval $(t, t + \Delta t)$ one bacterium will be added to the population is equal to $\lambda x\, \Delta t + o(\Delta t)$, $\lambda > 0$, and (2) that the probability of two or more bacteria being added to the population in $(t, t + \Delta t)$ is $o(\Delta t)$. These assumptions lead to the relation

$$P_x(t + \Delta t) = (1 - \lambda x\, \Delta t)P_x(t) + \lambda(x - 1)P_{x-1}(t) + o(\Delta t)$$

As $\Delta t \to 0$ we obtain the system of differential-difference equations

$$\frac{dP_x(t)}{dt} = -\lambda x P_x(t) + \lambda(x-1)P_{x-1}(t) \qquad x = x_0, x_0 + 1, \ldots \quad (0.3)$$

Since we have assumed that $X(0) = x_0$, Eq. (0.3) is to be solved with the initial conditions,

$$P_x(0) = 1 \qquad \text{for } x = x_0$$
$$= 0 \qquad \text{otherwise} \qquad\qquad (0.4)$$

The solution of Eq. (0.3) is given by

$$P_x(t) = \mathscr{P}\{X(t) = x\} = \binom{x-1}{x-x_0} e^{-\lambda x_0 t}(1 - e^{-\lambda t})^{x-x_0} \qquad (0.5)$$

for $x \geqslant x_0$.

To compare the two models, we first observe that in the deterministic approach the population size was represented by a real-valued and continuous function of time, while in the stochastic approach we start by assuming that the random variable denoting the population size is integer-valued. An examination of the deterministic solution (0.2) shows that, for λ and x_0 fixed, we have associated with every value of t a real number $x(t)$. From (0.5) we see that, for λ and x_0 fixed, and for every pair (x,t), $x \geqslant x_0$, $t \geqslant 0$, there exists a number $P_x(t)$, $0 \leqslant P_x(t) \leqslant 1$, which is the probability that the random variable will assume the value x at time t. It is of interest to note that the deterministic model is a special case of a stochastic model, in the sense that it yields results which hold with probability one.

We close this brief discussion of deterministic and stochastic models by considering the mean or expected population size. Let $m(t) = \mathscr{E}\{X(t)\}$. By definition

$$m(t) = \sum_{x=0}^{\infty} x P_x(t) = x_0 e^{\lambda t} \qquad (0.6)$$

Hence, we see that the expression for the *mean population size* (0.6) is the same as that for the *population size* (0.2) obtained from the deterministic model. In view of this correspondence, we can state that Eq. (0.1) describes the mean population size, while Eq. (0.3) takes into consideration random fluctuations. It is of interest to point out that this correspondence between the two models here considered does not hold in general; we shall encounter in the text several cases in which the deterministic solution is not the same as the stochastic mean.

The stochastic model considered above is an example of what is called a *stochastic process*. J. L. Doob has defined a stochastic process as *the mathematical abstraction of an empirical process whose development is governed by probabilistic laws.* It is important to note that the term "stochastic process" refers to the mathematical abstraction, model, or representation of the empirical process and *not* to the empirical process itself. In the example given above, the empirical process involved was the growth of a bacterial population, and the model was the system of differential-difference equations for the probabilities $P_x(t)$, which was derived on the basis of certain assumptions concerning the probabilistic manner in which the population could develop. The bacterial population in our example can be referred to as the *system*, and the possible population sizes $x_0, x_0 + 1, \ldots$ can be referred to as the *states* of the system. And we see that for any state x there is a probability $P_x(t)$, given by (0.5), that the system will be in that state at time t.

In the past twenty-five years the theory of stochastic processes has developed very rapidly and has found application in a large number of fields. In particular, a class of stochastic processes termed *Markov chains or processes*[1] has been investigated rather extensively; it is of great importance in many branches of science and engineering and in other fields. In this book we shall restrict our attention to this class of stochastic processes. For a discussion of other classes of stochastic processes we refer to the treatise of J. L. Doob, and to other books listed in the General Bibliography at the end of this Introduction.

Before defining a Markov chain, let us consider the stochastic process studied in classical probability theory, namely, a sequence of independent random variables. Let the random variables $X_1, X_2, \ldots,$ X_n (n finite or infinite) represent the results of n independent trials and let $E_1, E_2, \ldots$ represent the possible outcomes (states of the system) at each trial. For $i = 1, 2, \ldots, n$ let

$$p_j = \mathscr{P}\{X_i = E_j\} \qquad j = 1, 2, \ldots \tag{0.7}$$

Since the trials are independent, the probability of observing the sequence of outcomes $E_{j_1}, E_{j_2}, \ldots, E_{j_n}$ is simply

$$\mathscr{P}\{E_{j_1}, E_{j_2}, \ldots, E_{j_n}\} = p_{j_1}p_{j_2}\cdots p_{j_n} \tag{0.8}$$

That is, the probability of the sequence of outcomes is the product of the probabilities associated with the outcomes.

The classical situation was generalized by Markov by assuming that the outcome of any trial depends only on the outcome of the directly preceding trial. Hence, in this case we must introduce a *conditional*

[1] Named after the Russian mathematician A. A. Markov (1856–1922), who introduced the concept of chain dependence and did the basic pioneering work on this class of processes.

probability p_{ij}, associated with every pair of outcomes or states (E_i, E_j), which is defined as follows:

$$p_{ij} = \mathscr{P}\{X_{n+1} = E_j \mid X_n = E_i\} \qquad n = 0, 1, \ldots \qquad (0.9)$$

In addition to the probabilities p_{ij}, it is necessary to know the probabilities $q(j)$ which give the probability that the outcome of the initial trial was E_j. In this more general situation the probability of the sequence of outcomes $E_{j_1}, E_{j_2}, \ldots, E_{j_n}$ is given by

$$\mathscr{P}\{E_{j_1}, E_{j_2}, \ldots, E_{j_n}\} = q(j_1)p_{j_1 j_2}p_{j_2 j_3} \cdots p_{j_{n-1}, j_n} \qquad (0.10)$$

The random variables of a Markov chain are clearly not independent, but *dependent*, with the dependence extending over one unit of time. Hence, we can conclude that the concept of a Markov chain is obtained by abstraction from an empirical process associated with systems whose state changes with time n, according to some probability law, in such a manner that the probability of the system going from a given state E_i at time n_0 to a state E_j at time $n_0 + 1$ depends only on the state E_i at time n_0 and is independent of the states of the system at times prior to n_0.

General Bibliography

1 Arley, N.: "On the Theory of Stochastic Processes and Their Application to the Theory of Cosmic Radiation," John Wiley & Sons, Inc., New York, 1948.
2 Bartlett, M. S.: "An Introduction to Stochastic Processes," Cambridge University Press, New York, 1955.
3 Blanc-Lapierre, A., and R. Fortet: "Théorie des fonctions aléatoires," Masson et Cie, Paris, 1953.
4 Bochner, S.: "Harmonic Analysis and the Theory of Probability," University of California Press, Berkeley, 1955.
5 Chung, K. L.: "Markov Chains with Stationary Transition Probabilities," Springer-Verlag, Berlin, Vienna, *in press*.
6 Doob, J. L.: "Stochastic Processes," John Wiley & Sons, Inc., New York, 1953.
7 Dynkin, E. B.: "Foundations of the Theory of Markov Process" (in Russian), Gosudarstv. Izdat. Fiz.-Mat. Lit., Moscow, 1959.
8 Feller, W.: "An Introduction to Probability Theory and Its Applications," vol. 1, 2d ed., John Wiley & Sons, Inc., New York, 1957.
9 Fisz, M.: "Calculus of Probabilities and Mathematical Statistics" (in Polish), Państwowe Wydawnictwo Naukowe, Warsaw, 1958.
10 Fortet, R.: "Calcul des probabilités," Centre National de la Recherche Scientifique, Paris, 1950.
11 Fréchet, M.: "Recherches théoriques modernes sur le calcul des probabilités: II. Méthode des fonctions arbitraires. Théorie des événements en chaine dans le cas d'un nombre fini d'états possibles," 2d ed., Gauthier-Villars, Paris, 1952.

12   Gnedenko, B. V.: "Course in the Theory of Probability" (in Russian), Gosudarstv. Izdat. Tehn.-Teor. Lit., Moscow, 1954.
13   Halmos, P. R.: "Measure Theory," D. Van Nostrand Company, Inc., Princeton, N.J., 1950.
14   Harris, T. E.: "Branching Processes," Springer-Verlag, Berlin, Vienna, *in press.*
15   Hille, E., and R. S. Phillips: "Functional Analysis and Semi-groups," American Mathematical Society, New York, 1957.
16   Itô, K.: "Theory of Probability" (in Japanese), Iwanami Shôten, Tokyo, 1953.
17   Kawada, Y.: "The Theory of Probability" (in Japanese), Kyôritsusha, Tokyo, 1952.
18   Kemeny, J. G., and J. L. Snell: "Finite Markov Chains," D. Van Nostrand Company, Inc., Princeton, N.J., 1960.
19   Khintchine, A.: "Asymptotische Gesetze der Wahrscheinlichkeitsrechnung," Springer-Verlag, Berlin, Vienna, 1933.
20   Kolmogorov, A. N.: "Grundbegriffe der Wahrscheinlichkeitsrechnung," Springer-Verlag, Berlin, Vienna, 1933.
21   Kunisawa, K.: "Limit Theorems in Probability Theory" (in Japanese), Chûbunkan, Tokyo, 1949.
22   Lévy, P.: "Théorie de l'addition des variables aléatoires," Gauthier-Villars, Paris, 1937.
23   Lévy, P.: "Processus stochastiques et mouvement Brownien," Gauthier-Villars, Paris, 1948.
24   Loève, M.: "Probability Theory," D. Van Nostrand Company, Inc., Princeton, N.J., 1955.
25   Lundberg, O.: "On Random Processes and Their Application to Sickness and Accident Statistics," Almqvist and Wiksells, Uppsala, 1940.
26   Mann, H. B.: "Introduction to the Theory of Stochastic Processes Depending on a Continuous Parameter," *Natl. Bur. Standards, Appl. Math. Ser., No.* 24, 1953.
27   Onicescu, O., G. Mihoc, and C. T. Ionescu Tulcea: "The Calculus of Probability and Its Applications" (in Roumanian), Editura Academiei Republicii Populare Romine, Bucharest, 1956.
28   Rényi, A.: "The Calculus of Probabilities" (in Hungarian), Tankónyvkiado, Budapest, 1954.
29   Romanovskiĭ, V. I.: "Discrete Markov Chains" (in Russian), Gosudarstv. Izdat. Tehn.-Teor. Lit., Moscow, 1949.
30   Sarymsakov, T. A.: "Elements of the Theory of Markov Processes" (in Russian), Gosudarstv. Izdat. Tehn.-Teor. Lit., Moscow, 1954.

PART I

Theory

syādasti nāsti ca avaktavyacsa
(may be, it is and it is not and also indeterminate)
From the Indian-Jaina dialectic of predication.

1

Processes Discrete in Space and Time

1.1 Introduction

In this chapter the elements of the theory of discrete Markov chains with a denumerable number of states will be presented. This theory will be developed with special reference to discrete branching processes. The mathematical model of a discrete branching process can be thought of as a representation of the generation-by-generation growth of a population. The integer-valued random variable X_n, $n = 0$, $1, \ldots$, studied is taken to represent the number of individuals in the population at time n, or in the nth generation.

In Sec. 1.2 we shall consider the study of discrete branching processes as discrete Markov chains with a denumerable number of states. The fundamental definitions and properties of this class of stochastic processes, together with their interpretation for branching processes, are given. In Sec. 1.3 the moments and cumulants of the fundamental random variable are studied.

A problem of great interest in the theory of branching stochastic processes is to determine conditions for which the random variable X_n will assume the value zero for some n. This problem, which leads to the fundamental theorem concerning branching processes, is considered in Sec. 1.4. A related problem, namely, that of obtaining the probability distribution of the number of generations before the population becomes extinct, is discussed in Sec. 1.5.

Section 1.6 is devoted to the study of limit theorems. We consider some general theorems for discrete Markov chains with a denumerable number of states, as well as some special limit theorems for branching processes. In Sec. 1.7 we consider a simple random-walk process on the nonnegative integers and discuss the relationship between random-walk processes and branching processes. Finally, in Sec. 1.8, we consider N-dimensional processes, i.e., processes in which the population is made up of N different types of individuals.

1.2 Fundamental Definitions and Properties

A. Discrete Branching Processes and Markov Chains. We consider a stochastic process $\{X_n,\ n = 0,\ 1,\ 2,\ \ldots\}$, that is, a family of random variables, defined on the space $\mathfrak{X}$ of all possible values that the random variables can assume. The space $\mathfrak{X}$ is called the *state space* of the process, and the elements $x \in \mathfrak{X}$, the different values that X_n can assume, are called the *states*. In this chapter the state space $\mathfrak{X}$ will be taken to be the set of nonnegative integers $x = 0, 1, 2, \ldots$.   Hence the processes we consider in this chapter are discrete with respect to both the state variable x and the time variable n.

The process $\{X_n,\ n = 0, 1, 2, \ldots\}$ will be said to represent a *simple discrete branching process*[1] if the following conditions are satisfied:

(*a*) $X_0 = x_0 = 1$.

(*b*) $p(x) = \mathscr{P}\{X_1 = x\}$, with $\sum\limits_{x=0}^{\infty} p(x) = 1$.

(*c*) The conditional distribution of X_{n+1}, given $X_n = j$, is the same as the sum of j independent random variables, each having the same probability distribution as X_1. Hence, if we denote by $p_{n+1}(k)$ the probability that there are k individuals in the population in the $(n + 1)$st generation, then

$$p_{n+1}(k) = \sum_{j=0} p_n(j)[p(k)]^{*j} \tag{1.1}$$

that is, the conditional distribution is the j-fold convolution of $p(k)$ with itself.[2]

As stated in Sec. 1.1, the mathematical model for a discrete branching process can be thought of as representing the generation-by-generation development or growth of a population, the integer-valued random variable X_n representing the number of individuals in the population in the nth generation. In order to study the development of the population, i.e., the sequence $X_0, X_1, X_2, \ldots$, it is necessary to obtain an expression for the probability that the population size in the $(n + 1)$st generation, say, assumes a given value when the history

---

[1] For abstract formulations of branching processes we refer to Otter [49] and Urbanik [58].

[2] Cf. Appendix A for the definition of the convolution operation.

of the population, i.e., the population sizes in previous generations, is known. Hence we seek the conditional probability

$$\mathscr{P}\{X_{n+1} = x_{n+1} \mid X_n = x_n, X_{n-1} = x_{n-1}, \ldots, X_0 = x_0 = 1\} \quad (1.2)$$

If the structure of the stochastic process $\{X_n, n = 0, 1, 2, \ldots\}$ is such that the conditional probability distribution of X_{n+1} depends only on the value of X_n and is *independent of all previous values*, we say that the process has the *Markov property* and call it a *Markov chain*. More precisely,

$$\mathscr{P}\{X_{n+1} = x_{n+1} \mid X_n = x_n, \ldots, X_0 = x_0 = 1\} = \mathscr{P}\{X_{n+1} = x_{n+1} \mid X_n = x_n\}$$

$$(1.3)$$

Let us now write

$$p_{ij} = \mathscr{P}\{X_{n+1} = j \mid X_n = i\} \qquad i, j = 0, 1, 2, \ldots \quad (1.4)$$

Since the p_{ij} are conditional probabilities, they satisfy the conditions

$$p_{ij} \geqslant 0 \qquad \text{for all } i \text{ and } j \quad (1.5)$$

and

$$\sum_{j=0}^{\infty} p_{ij} = 1 \qquad i = 0, 1, 2, \ldots \quad (1.6)$$

For i and j fixed these probabilities can be interpreted as the conditional probability that at time $n + 1$ the system is in state j, given that at time n the system was in state i. Should the p_{ij} be such that they depend only on the states i and j and not on the time n, we say that the conditional probabilities are *constant* or *stationary*. In this chapter we shall restrict our attention to chains with stationary probabilities p_{ij}. If the p_{ij} are known for all i and j, all we need in order to completely define the process is the probability distribution $q(x)$, say, where

$$q(x) = \mathscr{P}\{X_0 = x\} \qquad x = 0, 1, \ldots \quad (1.7)$$

that is, $q(x)$ is the probability that at time zero the system is in the state x. Clearly,

$$0 \leqslant q(x) \leqslant 1 \qquad \text{and} \qquad \sum_{x=0}^{\infty} q(x) = 1$$

From the above discussion we see that a discrete branching process $\{X_n, n = 0, 1, 2, \ldots\}$ is a Markov chain. Since we have assumed that the state space $\mathfrak{X}$ is the set of nonnegative integers, we can be more precise and say that the discrete branching process is a special case of a *Markov chain with a denumerable·number of states*. This class of Markov chains was first considered by Kolmogorov [43][1] in his fundamental paper on the subject.

[1] Numbers in brackets are those of the bibliography given at the end of the chapter.

B.   Transition Probabilities and Markov Matrices. In order
to conform with the terminology of Markov chain theory, we call the
conditional probability p_{ij} the *probability of a transition* from the
state i to the state j, and call $P = (p_{ij})$ the *matrix of transition
probabilities*:

$$P = \begin{bmatrix} p_{00} & p_{01} & p_{02} & \cdots \\ p_{10} & p_{11} & p_{12} & \cdots \\ p_{20} & p_{21} & p_{22} & \cdots \\ \cdot & \cdot & \cdot & \cdot & \cdot & \cdot & \cdot & \cdot & \cdot & \cdot & \cdot \end{bmatrix}$$

Clearly P is a *square matrix* (of infinite order since the chain has a
denumerable number of states) with *nonnegative elements*, since $p_{ij} \geqslant 0$
for all i and j, and with *row sums equal to unity*, since $\sum\limits_{j=0}^{\infty} p_{ij} = 1$
for all i. A matrix satisfying the above conditions is called a *stochastic*,
or *Markov*, *matrix*.[1] A Markov chain is completely defined by a matrix
of transition probabilities P and a column vector, say $Q = (q(0),
q(1), \dots)$, which gives the probability distribution for the states $x =
0, 1, 2, \dots$ at time zero. In the case of a simple discrete branching
process we have $Q = (0, 1, 0, \dots)$, since we assume $X_0 = 1$.

In addition to the so-called *one-step* transition probabilities p_{ij}, it is
of interest to consider the *higher*, or *n-step*, transition probabilities,
written $p_{ij}^{(n)}$. These express the probability of a transition from the
state i to the state j in $n \, (>1)$ steps, or a change in the population
size from i to j in n generations. The n-step transition probabilities
can be defined recursively by

$$p_{ij}^{(1)} = p_{ij} \qquad p_{ij}^{(n+1)} = \sum_{v=0} p_{iv}^{(n)} p_{vj} \tag{1.8}$$

We now wish to show that for all k

$$p_{ij}^{(k)} = P\{X_{m+k} = j \mid X_m = i\} \tag{1.9}$$

For $k = 1$ the above is true by (1.4). Let us now assume that (1.9) is
true for $k = n + 1$. We have

$$\mathscr{P}\{X_{m+n+1} = j \mid X_m = i\} = \sum_{v=0}^{\infty} \mathscr{P}\{X_{m+n} = v \mid X_m = i\}$$
$$\times \, \mathscr{P}\{X_{m+m+1} = j \mid X_{m+n} = v\}$$
$$= \sum_{v=0}^{\infty} p_{iv}^{(n)} p_{vj}$$
$$= p_{ij}^{(n+1)}$$

Hence, by induction, we have shown that (1.9) is true for all k.

[1] If the column sums also equal unity the matrix is called *doubly stochastic*.

In general, we have the following relation:

$$p_{ij}^{(m+n)} = \sum_{v=0}^{\infty} p_{iv}^{(m)} p_{vj}^{(n)} \tag{1.10}$$

For $m = 1$ (1.10) is (1.8). That (1.10) is true for all m can also be shown by induction. Equation (1.10) is a statement of the fact that in passing from state i to j in $m + n$ steps the first m steps take the system from i to some intermediate state v and the last n steps then take the system from v to j. In order to obtain all possible ways that the transition from i to j can take place, it is necessary to sum over all possible intermediate states. If we now denote by P^n the matrix of n-step transition probabilities, i.e., $P^n = (p_{ij}^{(n)})$, the above equation in matrix form becomes

$$P^{m+n} = P^m P^n \tag{1.11}$$

Equation (1.11) is the matrix form of the *Chapman-Kolmogorov functional equation*. This functional equation, which characterizes Markov chains, is of fundamental importance in the theory of Markov chains, and it is this equation which establishes the connection between Markov chains and the theory of semigroups of operators.[1]

C. Absolute Probabilities. If when $n = 0$ the probability of the system being in the state i is $q(i)$, i.e., $q(i) = \mathscr{P}\{X_0 = i\}$, then the *unconditional* or *absolute probability of the system being in the state j at time n* is given by

$$q^{(n)}(j) = \sum_{i=0}^{\infty} q(i) p_{ij}^{(n)} \tag{1.12}$$

Hence, given the $q(i)$ and the n-step transition probabilities $p_{ij}^{(n)}$, the $q^{(n)}(j)$ can be calculated. A problem of special interest is the behavior of $q^{(n)}(j)$ as $n \to \infty$. In Sec. 1.6 we shall study the asymptotic behavior of the $p_{ij}^{(n)}$, which in turn enables us to study the asymptotic behavior of the absolute probabilities.

D. Classification of States. Given a Markov chain $\{X_n, n = 0, 1, \ldots\}$, its states can be classified in a number of ways. The classification is of great importance in the study of the asymptotic behavior of the n-step transition probabilities $p_{ij}^{(n)}$ and also in the physical interpretation of the states of the chain. In this section we shall consider the different classes of states, with emphasis on definitions and relations which are fundamental in the theory.

---

[1] For a discussion of Markov chains from the point of view of semigroup theory we refer to the treatise of Hille and Phillips [33] and to references given there.

Definition: A state i is called *essential* if the existence of a positive integer n and a state j ($\neq i$) such that $p_{ij}^{(n)} > 0$ implies the existence of a positive integer m such that $p_{ji}^{(m)} > 0$. A state which is not essential is called *inessential*.

Definition: Let i and j be two states of a Markov chain; then i and j are said to *communicate* if there exist an $n \geqslant 1$ and an $m \geqslant 1$ such that $p_{ij}^{(n)} > 0$ and $p_{ji}^{(m)} > 0$. If i and j communicate, we write $i \to j$. It is easy to see that the following relations hold for communicating states:

    (a) $i \to j$ implies $j \to i$.
    (b) $i \to j$ and $j \to k$ implies $i \to k$.
    (c) If $i \to j$ for some j, then $i \to i$.

Definition: A set of states $S \in \mathfrak{X}$ is called *closed* if no one-step transition is possible from any state in S to any state in $\mathfrak{X} - S$, the complement of the set S. Hence, $p_{ij} = 0$ for $i \in S$ and $j \in \mathfrak{X} - S$. If the set S contains only one state, this state is called an *absorbing state*. It is clear that a necessary and sufficient condition for a state i to be an absorbing state is that $p_{ii} = 1$. If the state space $\mathfrak{X}$ contains two or more closed sets, the chain is called *decomposable* or *reducible*.

The Markov matrix associated with a decomposable chain can be written in the form of a partitioned matrix; for example,

$$P = \begin{bmatrix} P_1 & 0 \\ 0 & P_2 \end{bmatrix}$$

In the above, P_1 and P_2 represent Markov matrices which describe the transitions within the two closed sets of states. A chain, or matrix, which is not decomposable is called *indecomposable* or *irreducible*, and a chain is indecomposable if and only if every state can be reached from every other state.[1]

Definition: If $i \to i$, the greatest common divisor of the set of integers n such that $p_{ii}^{(n)} > 0$ is called the *period* of the state i. We denote the period of i by $\omega(i)$ and say that i is *periodic* with period $\omega(i)$. If $i \nrightarrow i$, we define $\omega(i) = 0$. A state that is not periodic is called *aperiodic*.

We now introduce and define several probabilities that will be useful in the sequel. Let

$$K_{ij}^{(n)} = \mathscr{P}\{X_n = j \mid X_0 = i, X_h \neq j, i \leqslant h < n\} \qquad (1.13)$$

---

[1] These matrix properties can be discussed in terms of the theory of graphs in topology; cf. Solow [55] and Veblen [60].

Hence $K_{ij}^{(n)}$ is the conditional probability of the system being in the state j at time n under the conditions that at time zero it was in the state i and has not been in j before time n. Similarly, we put

$$L_{ij} = \mathscr{P}\{X_h = j \text{ for at least one } h \geqslant 1 \mid X_0 = i\} \tag{1.14}$$

From (1.13) we see that

$$L_{ij} = \sum_{n=1}^{\infty} K_{ij}^{(n)} \tag{1.15}$$

Finally, we define

$$Q_{ij} = \mathscr{P}\{X_h = j \text{ infinitely often} \mid X_0 = i\}$$

$$= \mathscr{P}\{X_h = j \text{ for an infinite number of values of } h \mid X_0 = i\} \tag{1.16}$$

We now utilize the above probabilities in classifying the states of a Markov chain.

Definition: A state i is called *recurrent* if $Q_{ii} = 1$ or $L_{ii} = 1$ and *nonrecurrent* (*transient*) if $Q_{ii} = 0$ or $L_{ii} < 1$.

In order that the above definition be well defined, it is necessary to prove the equivalence of the conditions on Q_{ii} and L_{ii}. This can be done as follows: Let $F_n = \mathscr{P}\{X_h = i \text{ for at least } n \text{ values of } h \mid X_0 = i\}$. Then from (1.14) we have $F_1 = L_{ii}$, $F_2 = L_{ii}F_1, \ldots,$ $F_n = L_{ii}F_{n-1}$. Hence $F_n = (L_{ii})^n$. Therefore

$$\begin{aligned} Q_{ii} = \lim_{n \to \infty} F_n &= \lim_{n \to \infty} (L_{ii})^n = 0 \qquad \text{if } L_{ii} < 1 \\ &= 1 \qquad \text{if } L_{ii} = 1 \end{aligned}$$

In order to prepare ourselves for the study of the asymptotic behavior of $p_{ij}^{(n)}$, we state and prove a fundamental theorem which establishes a relationship between the $p_{ij}^{(n)}$ and Q_{ij}.

Theorem 1.1: Consider the series

$$\sum_{n=1}^{\infty} p_{ij}^{(n)} \tag{1.17}$$

If (1.17) converges, then $Q_{ij} = 0$. (This is always true if the state j is nonrecurrent.) And if (1.17) diverges, then $Q_{ij} > 0$ and $Q_{jj} = 1$. If $i \to j$, we have $Q_{ij} = Q_{ji} = 1$.

Proof: Consider the sequence of events $\{E_n\}$, $n = 1, 2, \ldots,$ where we denote by E_n the occurrence of the state j at time $m + n$ when at time m the system was in state i. Let $\mathscr{P}\{X_m = i\} > 0$. Therefore,

$$\mathscr{P}\{E_n\} = \mathscr{P}\{X_m = i\}p_{ij}^{(n)}$$

Now if (1.17) converges, we have, by an application of the Borel-Cantelli lemma (cf. [17]), that

$$\mathscr{P}\{E_n \text{ infinitely often}\} = 0$$

Therefore $Q_{ij} = 0$.

From the definition of the $p_{ij}^{(n)}$ and $K_{ij}^{(n)}$ we have, for every i and j,

$$p_{ij}^{(n)} = \sum_{v=1}^{n} K_{ij}^{(v)} p_{jj}^{(n-v)} \tag{1.18}$$

with $p_{jj}^{(0)} = 1$. If we now sum both sides of (1.18) over n, we obtain

$$\sum_{n=1}^{N} p_{ij}^{(n)} = \sum_{n=1}^{N} \sum_{v=1}^{n} K_{ij}^{(v)} p_{jj}^{(n-v)}$$

$$= \sum_{v=1}^{N} K_{ij}^{(v)} \sum_{n=v}^{N} p_{jj}^{(n-v)} \leqslant \sum_{v=1}^{N} K_{ij}^{(v)} \left(1 + \sum_{n=1}^{N} p_{jj}^{(n)}\right) \tag{1.19}$$

If we now let $N \to \infty$ and use (1.15), we have

$$\sum_{n=1}^{\infty} p_{ij}^{(n)} \leqslant L_{ij} \left(1 + \sum_{n=1}^{\infty} p_{jj}^{(n)}\right) \tag{1.20}$$

If we put $j = i$ in (1.19) and divide both sides by $1 + \sum_{n=1}^{N} p_{ii}^{(n)}$, we obtain

$$\frac{\sum_{n=1}^{N} p_{ii}^{(n)}}{1 + \sum_{n=1}^{N} p_{ii}^{(n)}} \leqslant \sum_{v=1}^{N} K_{ii}^{(v)}$$

Therefore if (1.17) diverges, it follows from (1.20) that $\sum_{n=1}^{\infty} p_{jj}$ diverges, and we have

$$1 \leqslant \sum_{v=1}^{\infty} K_{ii}^{(v)} = L_{ii}$$

Hence $L_{ii} = 1$ and $Q_{ii} = 1$, since

$$Q_{ii} = 0 \qquad \text{if } L_{ii} < 1$$
$$= 1 \qquad \text{if } L_{ii} = 1$$

Thus we have shown that

$$Q_{ii} = 0 \qquad \text{if (1.17) converges}$$
$$= 1 \qquad \text{if (1.17) diverges} \tag{1.21}$$

In general, the divergence of (1.17) implies, by (1.20), the divergence of $\sum_{n=1}^{\infty} p_{jj}^{(n)}$. Hence $Q_{jj} = 1$. Should (1.17) diverge, this implies the existence of infinitely many $p_{ij}^{(n)} > 0$, hence $L_{ij} > 0$, and it follows that $Q_{ij} = L_{ij}Q_{jj} > 0$.

If i and j communicate, we have $L_{ij} > 0$, $L_{ji} > 0$. It can be shown that, if $Q_{ii} = 1$ and $L_{ij} > 0$, then $Q_{ij} = 1$. Hence $Q_{jj} = 1$ implies $Q_{ji} = 1$. If the state j is nonrecurrent, $Q_{ji} = 0$ by (1.16). Hence (1.17) must converge.

Additional concepts we now introduce are those of *recurrence* and *first-passage times* for a given state. Suppose a state i is recurrent, and $\mathscr{P}\{X_n = i\} > 0$. Given that $X_n = i$, we introduce a random variable T_i defined as follows: $T_i = m$ if $X_{n+k} \neq i$ for $1 \leqslant k \leqslant m$ and $X_{n+m} = i$. From (1.13) we see that

$$\mathscr{P}\{T_i = n\} = K_{ii}^{(n)}$$

The random variable T_i is called the *recurrence time* for the state i. It is clear that T_i can assume only positive integer values. The expected value of T_i, called the *mean recurrence time* for i, is given by

$$M_i = \mathscr{E}\{T_i\} = \sum_{n=1}^{\infty} n\mathscr{P}\{T_i = n\} = \sum_{n=1}^{\infty} nK_{ii}^{(n)} \leqslant \infty \tag{1.22}$$

Definition: A recurrent state i is called *positive* if $M_i < \infty$, or *null* if $M_i = \infty$. A recurrent state which is neither null nor periodic is called *ergodic*.

From (1.13) we see that $K_{ij}^{(n)}$ is the probability of passing from i to j in exactly n steps, and not before. If $L_{ij} = \sum_{n=1}^{\infty} K_{ij}^{(n)} = 1$, we can define the *mean first-passage time* from the state i to the state j as

$$M_{ij} = \sum_{n=1} nK_{ij}^{(n)} \tag{1.23}$$

For some theorems which provide criteria for determining whether a chain is ergodic, transient, or recurrent, we refer to the theorems of F. G. Foster given in Sec. 9.2B.

E. Absorption Probabilities. We now consider the problem of determining the probability that a system starting from some transient state i will eventually pass into a closed set of states S.

Denote by $\alpha_i^{(n)}$ the probability that a system starting from a transient state i will at time n, and not before, enter a closed set S. That is,

$$\alpha_i^{(n)} = \mathscr{P}\{X_n \in S \mid X_0 = i, X_h \neq S, i \leqslant h < n\}$$

Now, denote by α_i the probability that the system starting from a transient state i will eventually enter and remain in a closed set S. Hence

$$\alpha_i = \sum_{n=1} \alpha_i^{(n)} \tag{1.24}$$

Clearly

$$\alpha_i^{(1)} = \sum_{j \in S} p_{ij} \tag{1.25}$$

the summation being taken over all states in the closed set S. It is also clear that

$$\alpha_i^{(n+1)} = \sum_{v \in T} p_{iv} \alpha_v^{(n)} \tag{1.26}$$

the summation being taken over the set T of all transient states. The $\alpha_i^{(n)}$ are uniquely determined by the above recurrence relations. If we sum (1.26) over n, we see that the α_i are given by the solutions of the system of linear equations

$$\alpha_i = \sum_{v \in T} p_{iv} \alpha_v + \alpha_i^{(1)} \tag{1.27}$$

For i fixed, the α_i can be interpreted as the probability that the system will eventually be absorbed in the closed set S when at time zero the system was in state i. Similarly, $1 - \alpha_i$ can be interpreted as the probability of absorption in other closed sets, or remaining indefinitely in the set of transient states. In the nomenclature of branching processes, α_i is the probability that the population will eventually die out, i.e., reach the absorbing state $x = 0$, when the initial population size was $x = i$, $0 < i < \infty$. Similarly, $1 - \alpha_i$ is the probability that the population will not become extinct.

F. Transition Probabilities of a Discrete Branching Process. Generating Functions. It is difficult to obtain explicit expressions for the transition probabilities associated with a discrete branching process; however, the use of generating functions[1] simplifies matters considerably and in many cases enables us to study certain properties of the process without recourse to the transition probabilities. Let

$$F(s) = \sum_{x=0}^{\infty} p(x)s^x \qquad |s| \leqslant 1 \tag{1.28}$$

be the *generating function* of the probabilities $p(x)$. In Sec. 1.2A $p(x)$ was defined as the probability that in the first generation there are x ($x = 0, 1, 2, \ldots$) individuals in the population. If we now assume that the individuals in the population do not interact, i.e., are

---

[1] Cf. Appendix A.

statistically independent, it follows from the definition of the generating function that

$$p_{ij} = \text{coefficient of } s^j \text{ in } [F(s)]^i \qquad (1.29)$$

In order to determine the n-step transition probabilities, we need the following theorem.

Theorem 1.2: The generating function of the number of individuals in the nth generation, given that $X_0 = 1$, is

$$F_n(s) = F[F_{n-1}(s)] \qquad (1.30)$$

where $F_1(s) = F(s), \ldots, F_{n+1}(s) = F[F_n(s)] = F_n[F(s)]$; that is, $F_n(s)$ is the nth functional iterate of $F(s)$. Also, if $X_0 = x_0 > 1$, the generating function of the population size in the nth generation is $[F_n(s)]^{x_0}$.

Proof: From the conditions defining a discrete branching process $X_0 = 1$, and X_1 has the probability distribution $p(x)$. Also, the number of offsprings of *each* of the X_1 individuals of the first generation is a random variable S_i, say, with probability distribution $p(x)$. Let $X_2 = \sum_{i=1}^{X_1} S_i$. Since we assume that the offsprings do not interact, the S_i are mutually independent. Now $F(s)$ is the generating function of the S_i and X_1. Hence, by using Theorem A.4 we see that the generating function of X is $F_2(s) = F[F(s)]$. Similarly, the number of individuals in the third generation X_3 are second-removed offsprings of the X_1 individuals in the first generation. Therefore, X_3 is the sum of X_1 mutually independent random variables, each of which has the same probability distribution as X_2. Hence, the generating function of X_3 is $F_3(s) = F[F[F(s)]] = F[F_2(s)]$. By proceeding in this way, we see that the generating function $F_n(s) = F[F_{n-1}(s)]$.

Now let $X_0 = x_0 > 1$. Because of the assumption of statistical independence, the x_0 individuals in the (zero)th generation give rise to x_0 independent populations or processes *each* of which has the generating function $F_n(s)$ for the number of individuals in the nth generation. By applying Theorem A.3, we obtain $[F_n(s)]^{x_0}$ as the generating function of the number of individuals in the nth generation when $X_0 = x_0 > 1$.

From the above theorem, it is now clear that for fixed n the n-step transition probability

$$p_{ij}^{(n)} = \text{coefficient of } s^j \text{ in } [F_n(s)]^i \qquad (1.31)$$

1.3   Calculation of Moments and Cumulants

In this section we shall study the moments and cumulants of the random variables X_n, $n = 1, 2, \ldots$, when $\{X_n, \ n = 0, 1, \ldots\}$ is a simple branching process. We first consider the moments of X_1. Let us denote by m and σ^2 the mean and variance of X_1, respectively. It is well known that

$$m = \mathscr{E}\{X_1\} = \sum_{x=0}^{\infty} xp(x) \tag{1.32}$$

and

$$\sigma^2 = \mathscr{D}^2\{X_1\} = \sum_{x=0}^{\infty} x^2 p(x) - m^2 \tag{1.33}$$

Let us assume that m and σ^2 are finite. By using the method of generating functions, it is relatively simple to determine the moments and cumulants of X_1 and X_n, $n > 1$.

First, we see that differentiation of the generating function $F(s)$ with respect to s yields, upon putting $s = 1$,

$$F'(1) = \sum_{x=0}^{\infty} xp(x)s^{x-1} \bigg]_{s=1} = \sum_{x=0}^{\infty} xp(x)$$

$$= \mathscr{E}\{X_1\} = m \tag{1.34}$$

In order to obtain the mean of X_n, we note that differentiation of $F_n(s) = F[F_{n-1} s)]$ with respect to s yields

$$F'_n(s) = F'[F_{n-1}(s)]F'_{n-1}(s)$$

Therefore $\qquad\qquad F'_n(1) = F'(1)F'_{n-1}(1)$

since $F_{n-1}(1) = 1$. Now $F'(1) = m$, hence

$$F'_n(1) = mF'_{n-1}(1)$$
$$= m^2 F'_{n-2}(1)$$
$$= m^3 F'_{n-3}(1)$$
$$= m^{n-1}F'_1(1)$$
$$= m^n$$

Therefore
$$\mathscr{E}\{X_n\} = m^n \tag{1.35}$$

From (1.33) we see that the variance of X_1, in terms of generating functions, is given by

$$\sigma^2 = \mathscr{D}^2\{X_1\} = F''(1) + F'(1) - [F'(1)]^2 \tag{1.36}$$

Similarly, we have

$$\mathscr{D}^2\{X_n\} = F_n''(1) + F_n'(1) - [F_n'(1)]^2 \tag{1.37}$$

In order to obtain an explicit expression for $\mathscr{D}^2\{X_n\}$, we proceed as follows: Differentiation of $F_{n+1}(s) = F_n[F(s)]$ with respect to s and putting $s = 1$ yields

$$F_{n+1}'(1) = F_n'[F(1)]F'(1) = m^{n+1}$$

On differentiating again, we obtain

$$F_{n+1}''(1) = F''(1)[F_n'(1)]^2 + F'(1)F_n''(1) \tag{1.38}$$

Similarly, twofold differentiation of $F_{n+1}(s) = F[F_n(s)]$ gives

$$F_{n+1}''(1) = F''(1)F_n'(1) + [F'(1)]^2 F_n(1) \tag{1.39}$$

If $m = F'(1) \neq 1$, we can equate (1.38) and (1.39) and solve for $F_n''(1)$. We have

$$F_n''(1) = \sigma^2 m^n \frac{m^n - 1}{m^2 - m} + m^n(m^n - 1)$$

By using the above in (1.37), we obtain

$$\mathscr{D}^2\{X_n\} = \sigma^2 m^n \frac{m^2 - 1}{m^2 - m} \qquad m \neq 1 \tag{1.40}$$

When $m = 1$, an application of L'Hôpital's rule yields

$$\mathscr{D}^2\{X_n\} = n\sigma^2 \tag{1.41}$$

The higher moments of X_n can be found by similar calculations.

In addition to the moments, it is of interest in many statistical applications to consider the *cumulants* (or *semi-invariants*) of X_n. The cumulant generating function is defined by the relation

$$K(u) = \log F(s) \qquad s = e^u \tag{1.42}$$

Hence, we see that the kth cumulant of X_1, say i_k is given by

$$i_k = \frac{d^k}{du^k}\left[\log\left(\sum_{x=0}^{\infty} p(x)e^{ux}\right)\right]_{u=0}$$

$$= \frac{d^k K(u)}{du^k}\bigg]_{u=0} \tag{1.43}$$

that is, the kth cumulant of X_1 is given by the kth derivative of $K(u)$ evaluated at $u = 0$.

If μ_k is the kth moment of X_1 and i_k is the kth cumulant, the following relations between μ_k and i_k are well known:[1]

$$\mu_1 = i_1$$

$$\mu_2 = i_1^2 + i_2 \qquad (1.44)$$

$$\mu_3 = i_1^2 + 3i_1 i_2 + i_3$$

$$\cdot$$
$$\cdot$$
$$\cdot$$

and

$$i_1 = \mu_1$$

$$i_2 = \mu_2 - \mu_1^2 \qquad (1.45)$$

$$i_3 = \mu_3 - 3\mu_1\mu_2 + 2\mu_1^2$$

$$\cdot$$
$$\cdot$$
$$\cdot$$

The relations in (1.44) are obtained by differentiation of $F(u) = e^{K(u)}$, while the relations in (1.45) are obtained by differentiation of $K(u) = \log F(u)$.

Now let $i_{k,n}$ denote the kth cumulant of X_n. To obtain the $i_{k,n}$ we need the following theorem.[2]

Theorem 1.3: The cumulant generating function of X_{n+1}, given that $X_0 = 1$, is

$$K_{n+1}(u) = K[K_n(u)] \qquad (1.46)$$

Proof: Let $p_{n+1}(k)$ denote the probability that there are k individuals in the $(n + 1)$st generation. From (1.1) we have

$$p_{n+1}(k) = \sum_{j=0}^{\infty} p_n(j)[p(k)]^{*j} \qquad (1.47)$$

If we multiply both sides of (1.47) by e^{uk} and sum over k, we obtain

$$F_{n+1}(u) = \sum_{j=0}^{\infty} p_n(j) \sum_{k=0}^{\infty} [p(k)]^{*j} e^{uk} \qquad (1.48)$$

---

[1] Cf. Cramér [11] and Gnedenko and Kolmogorov [22].
[2] Woodward [62].

From the definition of the cumulant generating function, (1.48) becomes

$$F_{n+1}(u) = \sum_{j=0}^{\infty} p_n(j)[F(u)]^j$$

$$= \sum_{j=0}^{\infty} p_n(j)e^{jK(u)}$$

$$= F_n[K(u)]$$

On taking logarithms of both sides, we have

$$K_{n+1}(u) = K_n[K(u)] = K[K_n(u)]$$

The solution of $K_n(u) = K[K_{n-1}(u)]$ is formally given by

$$K_n(u) = K_0[K[K[\cdots K(u)\cdots]]] \tag{1.49}$$

Hence, the $i_{k,n}$ are given by

$$i_{k,n} = \frac{d^k}{du^k}\{K_0[K[K[\cdots K(u)\cdots]]]\}_{u=0} \tag{1.50}$$

As an example, we have

$$i_{1,n} = i_{1,0}(i_1)^n \tag{1.51}$$

$$i_{2,n} = i_{2,0}(i_1)^{2n} + i_{1,0}i_2(i_1)^{n-1}\left[\frac{(i_1)^n - 1}{i_1 - 1}\right] \qquad i \neq 1 \tag{1.52}$$

1.4 The Fundamental Theorem Concerning Branching Processes

A. History and Statement of Theorem. When there is a nonzero probability that individuals in the population can die or be removed from the population, it is of interest to determine the probability that the population will eventually die out and under what conditions this event occurs with probability one. If we assume $X_0 = 1$, what we seek are necessary and sufficient conditions for

$$\lim_{n\to\infty} p_{10}^{(n)} = 1$$

where $p_{10}^{(n)}$ is the probability of passing from the state $x = 1$ to the absorbing state $x = 0$ in time n, or n generations.

This problem was first raised, in 1873, by Francis Galton in connection with the extinction of family surnames. Galton's problem can be briefly stated as follows: If each male in a population has a family with x sons, where the random variable x has a distribution with generating function $F(s)$, and so on for the next generation, what is the probability of any particular male line dying out? The first solution of this problem was given, in 1874, by H. W. Watson.[1] In 1930, Steffensen [56,57] gave the complete solution of the problem. Since that time the problem of determining the extinction probability by various methods has attracted the attention of many individuals.

The *fundamental theorem* can be stated as follows:

Theorem 1.4: Let $p(0) > 0$ and let $\omega = \mathscr{P}\{X_n = 0\}$ for some n. If $m = \lim\limits_{s \to 1} F'(s) = 1$, $\omega = 1$, while if $m > 1$, ω is the smallest nonnegative root of the functional equation $F(s) = s$.

In this section we shall present one proof of this theorem. The proof we give is based on the fixed points of the function $F(s)$; hence, we shall first give a brief discussion of functional iterates and fixed points.

B. Remarks on Functional Iteration and Fixed Points. Let $\varphi(s)$ be a monotone increasing function and consider the functional iterates of $\varphi(s)$ defined by

$$\varphi_1(s) = \varphi(s), \ \varphi_2(s) = \varphi[\varphi(s)], \ \ldots, \ \varphi_n(s) = \varphi[\varphi_{n-1}(s)]$$

For s fixed, let

$$\xi = \lim_{n \to \infty} \varphi_n(s)$$

If $\varphi(s)$ is continuous, ξ satisfies the equation $\varphi(\xi) = \xi$, and if ξ satisfies the above equation, it is called a *fixed* or *double point* of $\varphi(s)$.

Now, the set of all fixed points of a continuous function is closed, i.e., the values of s which are not fixed points form a collection of disjoint intervals whose end points are fixed points. Hence, if we have a sequence of functions $\varphi_n(s)$, for s fixed, we obtain a sequence of fixed points ξ_n with $\xi_n \to \xi$. This fixed point forms the end point of the interval containing s.

For s fixed we have either $\varphi(s) < s$ or $\varphi(s) > s$, and since $\varphi(s)$ is monotone, it is clear that $\varphi_n(s) < \varphi_{n-1}(s)$ or $\varphi_n(s) > \varphi_{n-1}(s)$. This relation obtains for all n. Also, limit points for these sequences will exist unless they tend to $+\infty$ or $-\infty$. If $\xi = \lim\limits_{n \to \infty} \varphi_n(s)$, we must have $\varphi(\xi) = \xi$. The fixed point ξ is the next fixed point to s; its position depends on whether $\varphi(s) > s$ or $\varphi(s) < s$. The above follows from the fact that, if $\varphi(s)$ is monotone and $\varphi(\xi) = \xi$ and $\varphi(\xi^*) = \xi^*$, then for all $s \in (\xi, \xi^*)$ we have

$$\xi = \varphi(\xi) < \varphi(s) < \varphi(\xi^*) = \xi^*$$

For additional properties of functional iterates and fixed points we **refer** to Hadamard [26], Picard [50], and Schröder [52].

---

[1] Galton and Watson [21].

C. Proof of the Fundamental Theorem (fixed-point method).[1]

The generating function $F(s) = \sum\limits_{x=0}^{\infty} p(x)s^x$ is a power series in s whose coefficients, being probabilities, are all nonnegative. In addition, $F(0) = p(0) > 0$ and $F(1) = 1$; hence, $F(s)$ is monotone increasing in the interval $[0,1]$. Now let ω be the first nonnegative fixed point of

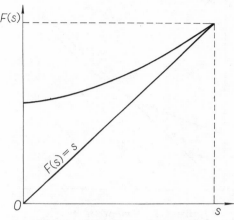

Figure 1.1 Case when $m \leqslant .1.$.

$F(s)$. Since the set of fixed points of $F(s)$ is closed, we have established the existence of ω. From the above conditions it follows that

$$\omega = \lim_{n \to \infty} F_n(0)$$

Now
$$p_{10}^{(n)} = F_n(s)\bigg]_{s=0}$$

Therefore, ω gives the limit of the probability of extinction, and $1 - \omega$ is the probability that the population will not die out.

Having established the existence of a limiting extinction probability, we will now show that $\omega = 1$ if, and only if, $m \leqslant 1$. Assume $\omega = 1$. Since $p(0) > 0$ and $F(s)$ is convex in the interval $[0,1)$, we have that $[F(s) - 1]/(s - 1)$ is bounded by 1 and is monotone increasing with s. Hence, $F'(1)$ exists and is $\leqslant 1$. Conversely, if $F'(1) \leqslant 1$, $F'(s)$ is either constant in $[0,1)$ or strictly increasing with s. Hence, $F'(s) < 1$ in either case. Application of the mean-value theorem gives $F(s) > s$ in $[0,1)$; hence, $\omega = 1$.

Let us now consider the case $m > 1$. Consider the equation $F(s) - s = 0$. This equation is nonnegative at $s = 0$ and vanishes when $s = 1$;

[1] Hawkins and Ulam [32].

its second derivative is nonnegative for all $s \in [0,1)$. Hence, it cannot have more than one positive root less than 1. In addition, its derivative evaluated at $s = 1$ is $m - 1$, so there is a positive root less than 1 when $m > 1$.

The above relations are shown in Figs. 1.1 and 1.2.

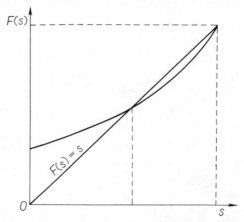

Figure 1.2   Case when $m > 1$.

1.5   Remarks on the Number of Generations to Extinction

In the preceding section we showed that, when the expected number of individuals in the first generation is less than or equal to unity, the population will become extinct with probability one, i.e., $\mathscr{P}\{X_n = 0\} = 1$ for some n when $m \leqslant 1$. In the terminology of Markov chains theory, the time n is called the first-passage time for the absorbing state $x = 0$. The problem of determining the moment generating function of the number of generations to extinction has been considered by Harris [27]; however, the methods employed are rather involved. In this section we shall consider the problem within the framework of the theory of Markov chains and obtain a system of equations for the mean first-passage time or the mean number of generations to extinction. For another approach to this problem we refer to the paper of Castoldi [7].

At time zero, i.e., $n = 0$, let the initial state, or population size, be $x = i, i = 1, 2, \ldots$. Let N_i be the smallest integer n such that $X_{n+1} = 0$, given that $X_0 = i$. Hence $N_i - 1$ is the number of generations preceding extinction. Throughout this discussion we assume $m \leqslant 1$. Let T be the set of transient states. We define

$$\rho_i^n = \mathscr{P}\{X_n \in T \mid X_0 = i\} \tag{1.53}$$

Clearly ρ_i^n satisfies the equations

$$\rho_i^{(1)} = \sum_{v \in T} p_{iv} \qquad \rho_i^{(n+1)} = \sum_{v \in T} p_{iv} \rho_v^{(n)} \tag{1.54}$$

From (1.54) we see that $\rho_i^{(1)} \leqslant 1$, and $\rho_i^{(2)} \leqslant \rho_i^{(1)}$; hence, in general, $\rho_i^{(n+1)} \leqslant \rho_i^{(n)}$. Since the $\rho_i^{(n)}$ form a decreasing sequence, we have

$$\rho_i = \lim_{n \to \infty} \rho_i^{(n)}$$

$$= \mathscr{P}\{X_n \in T \text{ for all } n > 0 \mid X_0 = i\} \tag{1.55}$$

From (1.54) we have

$$\rho_i = \sum_{v \in T} p_{iv} \rho_v \tag{1.56}$$

Now,

$$\mathscr{P}\{N_i = n\} = \rho_i^{(n-1)} - \rho_i^{(n)} \tag{1.57}$$

These probabilities will add to unity if and only if $\rho_i = 0$, that is, whenever the probability of extinction is one.

Now let M_i denote the mean number of generations to extinction. Then

$$M_i = \sum_{n=0}^{\infty} n \mathscr{P}\{N_i = n\} = \sum_{n=0}^{\infty} \rho_i^{(n)} \qquad 1 = 1, 2, \ldots \tag{1.58}$$

From (1.54) it follows that M_i is the solution of the system

$$M_i = \sum_{v \in T} p_{iv} M_v + 1 \qquad i = 1, 2, \ldots \tag{1.59}$$

1.6 Limit Theorems

A. Introduction. The subject of limit theorems occupies a very prominent place in probability theory. As Gnedenko and Kolmogorov [22] have remarked, ". . . the epistemological value of the theory of probability is revealed only by limit theorems." Limit theorems are also of importance in elucidating the physical significance of results obtained by studying the asymptotic behavior of sequences of random variables associated with stochastic models in various applications.

In this section we shall study some limit theorems which are concerned with the following:

1. The n-step transition probabilities $p_{ij}^{(n)}$
2. The conditional distribution of X_n
3. The normed random variable $Y_n = X_n/m^n$
4. The total number of individuals produced in all generations

The first four limit theorems we consider (Theorems 1.5 to 1.8) are for arbitrary Markov chains with a denumerable number of states, while the remaining theorems are for branching processes.

For other limit theorems for Markov chains with a denumerable number of states, such as the weak and strong laws of large numbers, the central limit theorem, and the law of the iterated logarithm, we refer to the paper of Chung [10].

B.   Limiting Behavior of the n-step Transition Probabilities $p_{ij}^{(n)}$. In Sec. 1.2 we defined $p_{ij}^{(n)}$ as the probability that in n steps the system would pass from the state i into the state j. It is of interest in many applications to investigate the behavior of $p_{ij}^{(n)}$ as n approaches infinity. In this section we shall state and prove several theorems concerning the limiting behavior of $p_{ij}^{(n)}$.

Definition: A probability distribution $\{\pi_k\}$ is called a *stationary* (or *equilibrium*) distribution if[1]

$$\pi_j = \sum_{i=0}^{\infty} \pi_i p_{ij} \tag{1.60}$$

For indecomposable aperiodic chains the relationship between the different classes of states and the existence of a stationary distribution is given by the following theorems.

Theorem 1.5: If the state $j\,(j \neq 0)$ is either transient or recurrent null, then

$$\lim p_{ij}^{(n)} = 0$$

for every i.

Proof: From Theorem 1.1 we have $\sum_{n=1}^{\infty} p_{ij}^{(n)} < \infty$ for j non-recurrent, hence $p_{ij}^{(n)} \to 0$.

The above theorem is immediately applicable to discrete branching processes, since for these processes all states except 0 are nonrecurrent; hence, the probability of passing from any state in the set T (the set of all transient or nonrecurrent states) to another state in T in an infinite amount of time is zero.

The next theorem we consider is concerned with the important case in which a nonzero limit exists.

Theorem 1.6: If the states of the chain are neither transient nor recurrent null (i.e., they are ergodic), then for every pair of states i and j

$$\lim_{n \to \infty} p_{ij}^{(n)} = \pi_j \tag{1.61}$$

---

[1] We refer to Blackwell [6] for a discussion of the adjoint system $\pi_i^* = \sum_{j=0}^{\infty} p_{ij}\pi_j^*$ and to Feller [16] for a discussion of the relation between the two systems.

exists and is independent of i, and $\pi_j = M_j^{-1}$, where M_j is the mean recurrence time of the state j. The π_j satisfy the equations

$$\pi_j = \sum_{i=0}^{\infty} \pi_i p_{ij} \qquad [1.60]$$

with $\sum_{j=0}^{\infty} \pi_j = 1$, and the distribution $\{\pi_j\}$ is uniquely determined.

Proof: By hypothesis, the states j have finite mean recurrence times M_j. Let $\pi_j = M_j^{-1}$. Now $p_{jj}^{(n)} \to \pi_j$; hence, (1.61) holds for $i = j$. For any fixed i we have $\sum_{n=1}^{\infty} K_{ij}^{(n)} = 1$, since if this condition were not satisfied, there would be a positive probability of the system passing from i to j and not returning. This, however, contradicts the hypothesis that j is a recurrent state. Hence, for every $\epsilon > 0$ we can select an N such that $\sum_{n=1}^{N} K_{ij}^{(n)} = 1 - \epsilon$. Now, the last N terms in the series $p_{ij}^{(n)} = \sum_{\nu=1}^{n} K_{ij}^{(\nu)} p_{jj}^{(n-\nu)}$ differ arbitrarily little from $\pi_j \sum_{n=1}^{N} K_{ij}^{(n)}$, and hence from π_j. Finally, the sum of the first $n - N$ terms is less than the Nth remainder in the series $\sum_{n=1}^{\infty} K_{ij}^{(n)}$, and hence less than ϵ. This proves (1.61).

To show that the π_j satisfy (1.60), we first observe that $\sum_{j=0}^{\infty} \pi_j \leqslant 1$. In order to show that this inequality obtains, we have only to note that for fixed i and n the probabilities $p_{ij}^{(n)}$ $(j = 1, \dots)$ add to unity; hence, $\sum_{j=1}^{N} \pi_j \leqslant 1$ for every N. Consider now the fundamental equation $p_{ij}^{(m+n)} = \sum_{k=1}^{\infty} p_{ik}^{(m)} p_{kj}^{(n)}$. If we put $n = 1$ and let $m \to \infty$, we observe that $p_{ij}^{(m+n)} \to \pi_j$, and the right-hand side tends to $\pi_k p_{kj}$. Addition of an arbitrary, but finite, number of terms yields the inequalities

$$\pi_j \geqslant \sum_{k=0}^{\infty} \pi_k p_{kj} \qquad (1.62)$$

If we sum the above inequalities over all k, we obtain on each side the finite sum $\Sigma \pi_j$. Hence the inequality in (1.62) is impossible. This proves (1.60).

To show that the distribution $\{\pi_j\}$ is uniquely determined, let $\{\rho_j\}$ be a sequence of positive quantities satisfying the conditions

$$\rho_j = \sum_{k=0}^{\infty} \rho_k p_{kj} \qquad \sum_{j=0}^{\infty} |\rho_j| < \infty \qquad (1.63)$$

If we multiply the equation in (1.63) by p_{js} and sum over j, we obtain

$$\rho_s = \sum_{k=0}^{\infty} \rho_k p_{ks}^{(2)}$$

On repeating this operation, we have that for every n

$$\rho_s = \sum_{k=0}^{\infty} \rho_k p_{ks}^{(n)}$$

Since we have assumed $\sum_{j=0}^{\infty} |\rho_j| < \infty$, passage to the limit in the above system yields

$$\rho_s = \pi_s \sum_{k=0}^{\infty} \rho_k$$

Since the sum is a constant independent of s, the ratio ρ_s/π_s is constant. Hence, by putting $\rho_j = \pi_j$ we obtain

$$\sum_{j=0}^{\infty} \pi_j = 1$$

We now give a result, due to Derman [12], which proves the existence and uniqueness of a set of numbers $\{v_j\}$ which satisfy (1.60). We have the following theorem.

Theorem 1.7: If all of the states of an indecomposable chain are recurrent, then there exists one and only one set of positive numbers $\{v_j\}$, with $v_0 = 1$, which satisfy (1.60), and

$$v_j = \lim_{n \to \infty} \frac{\sum\limits_{r=0}^{n} p_{ij}^{(r)}}{\sum\limits_{r=0}^{n} p_{00}^{(r)}} \qquad (1.64)$$

Proof: Let

$$_jp_{ik}^{(n)} = \mathscr{P}\{X_n = k, X_n \neq j \text{ for } 1 \leqslant m \leqslant n \mid X_0 = i\} \qquad n = 1, 2, \ldots$$

Chung [9] has shown that $\sum\limits_{n=1}^{\infty} {_ip_{ik}^{(n)}} = {_ip_{ik}^*}, 0 < {_ip_{ik}^*} < \infty$, for all i and k. To prove that (1.64) satisfies (1.60), we consider the following relations:

$$\sum_{k=0}^{\infty} {_0p_{0k}^*} p_{kj} = {_0p_{00}^*} p_{0j} + \sum_{k=1}^{\infty} {_0p_{0k}^*} p_{kj}$$

$$= p_{0j} + \sum_{n=1}^{\infty} \sum_{k=1}^{\infty} {_0p_{0k}^{(n)}} p_{kj}$$

$$= p_{0j} + \sum_{n=1}^{\infty} {_0p_{0j}^{(n+1)}}$$

$$= {_0p_{0j}^{(1)}} + \sum_{n=2}^{\infty} p_{0j}^{(n)} = {_0p_{0j}^*} \qquad j = 0, 1, \ldots$$

In the second equality we have put $_0p_{00}^* = 1$, since all of the states are recurrent. Chung has also shown that

$$\lim_{n \to \infty} \frac{\sum_{r=0}^{n} p_{jj}^{(r)}}{\sum_{r=0}^{n} p_{ii}^{(r)}} = {_i}p_{ij}^*$$

Hence, by putting $i = 0$, we have shown that the v_j defined by (1.64) are positive solutions to (1.60).

To show uniqueness, let us suppose there exists a set of positive numbers $\{a_j\}$, with $a_0 = 1$, that satisfy (1.60). We introduce the *inverse probabilities*.[1]

$$q_{ik}^{(n)} = \frac{a_k}{a_i} p_{ki}^{(n)} \qquad (1.65)$$

We then have

$$q_{ik}^{(n)} \geqslant 0$$

$$\sum_{k=0}^{\infty} q_{ik}^{(n)} = 1 \qquad (1.66)$$

$$q_{ik}^{(n+1)} = \sum_{j=0}^{\infty} q_{ij} q_{jk}^{(n)}$$

Hence the matrix of inverse transition probabilities $Q = (q_{ik})$, i, $k = 0, 1, \ldots$, is a Markov matrix.

We can now write

$$\frac{\sum_{r=1}^{n} q_{i0}^{(r)}}{\sum_{r=0}^{n} q_{00}^{(r)}} = \frac{1}{a_i} \frac{\sum_{r=1}^{n} p_{0i}^{(r)}}{\sum_{r=0}^{n} p_{00}^{(r)}} \qquad (1.67)$$

Chung has shown that the left-hand side of (1.67) approaches unity as $n \to \infty$. Hence, we have that a_i must be unique. This completes the proof of the theorem.

The importance of Derman's theorem lies in the fact that the equations for stationary distributions admit of a unique solution in the case when the mean recurrence times are infinite. In this case the a_i are not probabilities, since their sum is unbounded; however, Eqs. (1.60) are satisfied.

The next results we consider are concerned with the limiting behavior of normed sums of the n-step transition probabilities. The

[1] Cf. Feller [17].

first theorem we consider is due to Kolmogorov [43], who showed that the Cesàro limits

$$\pi_{ij} = \lim_{n \to \infty} \frac{1}{n} \sum_{v=1}^{n} p_{ij}^{(v)}$$

always exist. The proof follows Yosida and Kakutani [64].[1]

Theorem 1.8 (*Mean Ergodic Theorem*): Consider an arbitrary Markov chain with a denumerable number of states; then the limits

$$\pi_{ij} = \lim_{n \to \infty} \frac{1}{n} \sum_{v=1}^{n} p_{ij}^{(v)} \tag{1.68}$$

exist for any pair of states i and j, and satisfy

$$\sum_{k=1}^{\infty} \pi_{ik} p_{kj}^{(n)} = \sum_{k=1}^{\infty} p_{ik}^{(n)} \pi_{kj} = \sum_{k=1}^{\infty} \pi_{ik} \pi_{kj} = \pi_{ij} \tag{1.69}$$

$i, j, n = 1, 2, \ldots,$ and

$$\sum_{j=1}^{\infty} \pi_{ij} \leqslant 1 \qquad i = 1, 2, \ldots \tag{1.70}$$

Proof: Let $q_{ij}^{(n)} = \dfrac{1}{n} \sum_{v=1}^{n} p_{ij}^{(r)}$, so that

$$0 \leqslant q_{ij}^{(n)} \leqslant 1 \tag{1.71}$$

and

$$\sum_{j=1}^{\infty} q_{ij}^{(n)} = 1 \tag{1.72}$$

Because of (1.71) there exists, by the Cantor diagonal process,[2] an increasing sequence of positive integers $\{n_v\}$, $v = 1, 2, \ldots,$ such that $\lim_{v \to \infty} q_{ij}^{(n_v)} = \pi_{ij}$ exists for any i and j. We note that

$$\left| \sum_{k=1}^{\infty} p_{ik} q_{kj}^{(n)} - q_{ij}^{(n)} \right| = \left| \frac{1}{n} (p_{ij}^{(n+1)} - p_{ij}) \right| \leqslant \frac{1}{n}$$

hence

$$\lim_{v \to \infty} \sum_{k=1}^{\infty} p_{ik} q_{kj}^{(n_v)} = \lim_{v \to \infty} q_{ij}^{(n_v)} = \pi_{ij}$$

Also, since $\sum_{j=1}^{\infty} p_{ij} = 1$ is absolutely convergent, we have

$$\lim_{v \to \infty} \sum_{k=1}^{\infty} p_{ik} q_{kj}^{(n_v)} = \sum_{k=1}^{\infty} p_{ik} (\lim_{v \to \infty} q_{kj}^{(n_v)}) = \sum_{k=1}^{\infty} p_{ik} \pi_{kj}$$

---

[1] Cf. [65] for an operator-theoretical treatment of the mean ergodic theorem.
[2] Cf. Rudin [51].

Therefore

$$\sum_{k=1}^{\infty} p_{ik}\pi_{kj} = \pi_{ij}$$

and consequently

$$\sum_{k=1}^{\infty} p_{ik}^{(n)}\pi_{kj} = \pi_{ij} \tag{1.73}$$

$$\sum_{k=1}^{\infty} q_{ik}^{(n)}\pi_{kj} = \pi_{ij} \tag{1.74}$$

Similarly, we have

$$\left| \sum_{k=1}^{\infty} q_{ik}^{(n)}p_{kj} - q_{ij}^{(n)} \right| = \left| \frac{1}{n}(p_{ij}^{(n+1)} - p_{ij}) \right| \leqslant \frac{1}{n}$$

hence

$$\lim_{\nu\to\infty} \sum_{k=1}^{\infty} q_{ik}^{(n_\nu)}p_{kj} = \lim_{\nu\to\infty} q_{ij}^{(n_\nu)} = \pi_{ij}$$

Since $q_{ik}^{(n)}$ and p_{kj} are nonnegative, we have, as $\nu \to \infty$,

$$\pi_{ij} \geqslant \sum_{k=1}^{\infty} \pi_{ik}p_{kj} \tag{1.75}$$

Now, if for a pair of integers i' and j' the above inequality is strict, then by interchanging the summations, we have

$$\sum_{k=1}^{\infty} \pi_{i'k} = \sum_{k=1}^{\infty}\sum_{j=1}^{\infty} \pi_{i'k}p_{kj} \leqslant \sum_{j=1}^{\infty} \pi_{i'j}$$

which is a contradiction. Hence we have

$$\sum_{k=1}^{\infty} \pi_{ik}p_{kj} = \pi_{ij}$$

and consequently

$$\sum_{k=1}^{\infty} \pi_{ik}p_{kj}^{(n)} = \pi_{ij} \tag{1.76}$$

$$\sum_{k=1}^{\infty} \pi_{ik}q_{kj}^{(n)} = \pi_{ij} \tag{1.77}$$

If we now put $n = n_\nu$ in (1.77) and let $\nu \to \infty$, we have, since $\sum_{k=1}^{\infty} \pi_{ik} \leqslant 1$ is absolutely convergent,

$$\sum_{k=1}^{\infty} \pi_{ik}\pi_{kj} = \pi_{ij} \tag{1.78}$$

Hence (1.69) is proved by (1.73), (1.76), and (1.78).

In order to prove (1.68), we assume the existence of a pair of integers i', j' such that $\lim_{n\to\infty} q_{i',j'}^{(n)}$ does not exist. This implies the existence of another increasing sequence of positive integers $\{m_\nu\}$, $\nu = 1, 2, \ldots$, such that

$$\lim_{\nu \to \infty} q_{ij}^{(m_\nu)} = Q_{ij}$$

exists for any i and j and such that $\pi_{i'j'} \neq Q_{i'j'}$. The Q_{ij} satisfy the same conditions as π_{ij}. By putting $n = m_\nu$ in (1.74) and (1.77) and letting $\nu \to \infty$, we have

$$\sum_{k=1}^{\infty} Q_{ik}\pi_{kj} \leqslant \pi_{ij} \tag{1.79}$$

and

$$\sum_{k=1}^{\infty} \pi_{ik}Q_{kj} = \pi_{ij} \tag{1.80}$$

If we now interchange π_{ij} and Q_{ij} in (1.79) and (1.80), we have

$$\sum_{k=1}^{\infty} \pi_{ik}Q_{kj} \leqslant Q_{ij} \tag{1.81}$$

$$\sum_{k=1}^{\infty} Q_{ik}\pi_{kj} = Q_{ij} \tag{1.82}$$

From (1.79) and (1.82) we have $Q_{ij} \leqslant \pi_{ij}$, and from (1.80) and (1.81) we have $Q_{i'j} \geqslant \pi_{ij}$. The above is a contradiction; hence, the limit (1.68) must exist for any i and j, and this completes the proof.

Theorem 1.8 states that $\sum_{j=1}^{\infty} \pi_{ij} \leqslant 1$. Foster [20] has given a classification of chains based on the properties of the π_{ij}. A chain is termed *dissipative* if $\pi_{ij} = 0$ for all i and j; it is termed *semidissipative* if $\pi_{ij} = 0$ for some i and j but $\sum_{j=1}^{\infty} \pi_{ij} < 1$ for some i; and it is termed *nondissipative* if $\sum_{j=1}^{\infty} \pi_{ij} = 1$ for all i.

The structure of the limit matrix $\Pi = (\pi_{ij})$ can be described by classifying a state j as *positive* when $\pi_{jj} > 0$ and *dissipative* when $\pi_{jj} = 0$. The set of all positive states, if not empty, can be divided into disjoint subsets, with two states i and j being in the same class if and only if $\pi_{ij} > 0$. Let j be a positive state; then we can write $\pi_j = \pi_{jj}$ and the limit elements π_{ij} can be expressed in terms of the π_j and a set of numbers $\alpha(i,C)$, defined for each state i and each positive class C, where $0 \leqslant \alpha(i,C) \leqslant 1$. The $\alpha(i,C)$ give the probability that the system starting at state i will ultimately enter the set of states C, and remain in C. The positive states are those states of the chain which are recurrent with finite mean recurrence time M_j, given by $M_j = 1/\pi_j$; the dissipative states are those states which are either recurrent with infinite mean recurrence time or nonrecurrent. The properties of the π_{ij}, π_j, and $\alpha(i,C)$ can be summarized as follows:[1]

(a) $\pi_{ij} = 0$ for all i, if j is dissipative

(b) $\pi_{ij} = \pi_j\alpha(i,C)$ for all i, if $j \in C$

(c) $\sum_{j \in C} \pi_j = 1$ for each positive class C

(d) $\alpha(i,C) = 1$ for $i \in C$
           0 for $i \notin C$

where i is a positive state and C is a positive class.

(e) If i is dissipative and $\{C^\gamma : \gamma = 1, 2, \ldots \}$ are positive classes, then

$$\sum_\gamma \alpha(i,C^\gamma) \leqslant 1$$

(f) If C is a positive class, then

$$\sum_{k \in C} \pi_k\pi_{kj} = \pi_j \quad \text{for } j \in C$$

$$= 0 \quad \text{for } j \notin C$$

and

$$\sum_{k=0}^{\infty} p_{ik}\alpha(k,C) = \alpha(i,C) \quad \text{for all } i$$

(g)

$$\sum_{k=0}^{\infty} \pi_{ik}\pi_{kj} = \sum_{k=0}^{\infty} \pi_{ik}p_{kj} = \sum_{k=0}^{\infty} p_{ik}\pi_{kj} = \pi_{ij}$$

for all i and j.

The problem of determining the limit matrix Π when the matrix of transition probabilities P is given has been considered by many investigators using different methods (cf. [17]). For the application of semigroup theory to this problem we refer to the paper of Kendall and Reuter [42].

We now give two limit theorems, due to Foster, associated with what has been termed a *generalized discrete branching process*. This process is described by a Markov matrix $P = (p_{ij})$ with the following properties:

(a) $p_{ij} \geqslant 0$ $\sum_{j=0}^{\infty} p_{ij} = 1$ $i = 0, 1, 2, \ldots$

(b) $p_{00} = 1$

(c) $p_{i0} > 0$ for each i

Condition (a) is the normal one for the elements of a Markov matrix. From Sec. 1.2D we see that $p_{00} = 1$ is the necessary and sufficient condition that the state $x = 0$ be an absorbing state. Condition (c) means that there is always a positive probability that a population of i individuals does not reproduce.

---

[1] Cf. Kendall and Reuter [42].

Theorem 1.9: Let P be a Markov matrix defining a generalized branching process; then the limit

$$\pi_{ij} = \lim_{n \to \infty} \frac{1}{n} \sum_{\nu=1}^{n} p_{ij}^{(\nu)}$$

exists such that

$$\pi_{ij} = 0 \qquad i = 0, \ldots, j \neq 0$$

while

$$\lim_{n \to \infty} p_{i0}^{(n)} = \pi_{i0} > 0 \qquad i = 1, 2, \ldots$$

Proof: For any given state i, the row vector π_{ij}, if not null, is a proper vector of the matrix P; also $\pi_{ij} \geq 0$ for each j. Consider now any proper vector $\mathbf{x}_j$ of P with components $x_j \geq 0$. Since $p_{ij}^{(n)} \geq 0$ and $\sum_{j=1}^{\infty} p_{ij}^{(n)} = 1$, we have the linear system

$$\sum_{i=0}^{\infty} p_{i0}x_i = x_0 \qquad p_{00} \neq 1 \qquad p_{i0} > 0$$

Hence the only possibility for the vector $\mathbf{x}_j$ is $(x_0, 0, 0, \ldots, 0)$, where $x_0 \geq 0$. Therefore, we have $\pi_{i0} \geq 0$, while $\pi_{ij} = 0$, $j = 0$. To show that $\pi_{i0} = \lim_{n \to \infty} p_{i0}^{(n)} > 0$, we note that

$$p_{i0}^{(n)} = \sum_{k=0}^{\infty} p_{ik}p_{k0} \geq p_{00}p_{i0} = p_{i0} > 0$$

and, by induction, we see that, in general,

$$0 \leq p_{i0} \leq p_{i0}^{(n)} \leq \cdots \leq 1$$

Hence $\lim_{n \to \infty} p_{i0}^{(n)} = \lambda \leq 1$, where λ is a constant. But this limit must be π_{i0} (> 0), and this completes the proof.

In terms of branching processes, Theorem 1.9 states that there is always a nonzero probability of the population becoming extinct regardless of how large the population may be in the (zero)th generation.

We now consider the limit of $p_{i0}^{(n)}$ for a nondissipative chain whose transition probabilities satisfy conditions (a) to (c). We have the following theorem.

Theorem 1.10: If the Markov matrix P satisfies conditions (a) to (c) and, in addition, the chain is nondissipative, then

$$\lim_{n \to \infty} p_{i0}^{(n)} = 1$$

Proof: From Theorem 1.9 the above limit π_{i0} exists and is nonzero. Moreover, $\pi_{ij} = 0$, $j \neq 0$. By the definition of nondissipative chains, $\sum_j \pi_{ij} = 1$ for each i. Since π_{i0} is the only nonzero element of the above series, we have $\pi_{i0} = 1$.

Whereas Theorem 1.9 gives conditions for the existence of a nonzero extinction probability, Theorem 1.10 gives conditions such that the extinction probability will always equal unity.

For additional limit theorems for Markov chains and their application in the theory of queues we refer to Sec. 9.2B.

C. Limiting Behavior of the Conditional Distribution of X_n. In Sec. 1.4 we showed that $\lim_{n \to \infty} X_n = 0$ when the mean $m \leqslant 1$. In view of this result, we might be led to believe that, whenever $m \leqslant 1$, we would have no interesting asymptotic theory associated with X_n. However, Yaglom [63] has shown that, if we consider the *conditional distribution* of X_n, given that $X_n \neq 0$, then we obtain nontrivial limiting distributions. Yaglom's results are given by Theorems 1.11 to 1.13.

Theorem 1.11: Let $p_n(0) = \mathscr{P}\{X_n = 0\}$ for some n, and let $q_n = 1 - p_n(0)$. Now let

$$p_n^*(x) = \frac{p_n(x)}{q_n} = \mathscr{P}\{X_n = x \mid X_n \neq 0\} \qquad (1.83)$$

Let $F(s)$ and $G(s)$ be the generating functions of $p(x)$ and $p_n^*(x)$, respectively. For $m < 1$ and $F''(1) < \infty$

$$\lim_{n \to \infty} p_n^*(x) = p^*(x) \qquad \sum_{x=1}^{\infty} p^*(x) = 1$$

In addition
$$\lim_{n \to \infty} G_n(s) = G(s)$$

where $\qquad G[F(s)] = mG(s) + (1-m)K^{-1} \qquad |s| \leqslant 1 \qquad (1.84)$

with $G(1) = 1$, $G'(1-) = K^{-1}$, and $K = \lim_{n \to \infty} [1 - F_n(0)]/m^n$.

Proof: We first obtain the generating function of $p_n^*(x)$. Clearly

$$G_n(s) = \sum_{x=1}^{\infty} \mathscr{P}\{X_n = x \mid X_n \neq 0\} s^x \qquad (1.85)$$

$$= \sum_{x-1}^{\infty} \frac{\mathscr{P}\{X_n = x, X_n \neq 0\}}{\mathscr{P}\{X_n \neq 0\}} s^x = \sum_{x=1}^{\infty} \frac{\mathscr{P}\{X_n = x\}}{\mathscr{P}\{X_n \neq 0\}} s^x$$

$$= \frac{\sum_{x=1}^{\infty} \mathscr{P}\{X_n = x\} s^x}{1 - F_n(0)} = \frac{F_n(s) - F_n(0)}{1 - F_n(0)}$$

$$= 1 + \frac{F_n(s) - 1}{1 - F_n(0)}$$

From the above we see that $G_n(s)$ is completely known when $F(s)$ is given.

In order to establish the functional equation satisfied by $G(s)$, we utilize a classical result due to Königs [48]:[1] If the function $\theta(s)$ is analytic in a given domain D with $\theta(s_0) = s_0$ and $|\theta'(s_0)| = a < 1$ for some $s_0 \in D$, then in this domain $[\theta_n(s) - s_0]/a^n$ converges uniformly (as $n \to \infty$) to a continuous function $\xi(s)$ which satisfies Schröder's functional equation $\xi[\theta(s)] = a\xi(s)$ and the conditions $\xi(s_0) = 0$, $\xi'(s_0) = 1$.

In applying Königs' theorem, we note that $\theta(s) = F(s)$, which is analytic at least in $D = \{s \colon -1 \leqslant s \leqslant 1\}$. We remark that, when considering the continuity and differentiability of $F(s)$ at $s = 1$ (or on the unit circle), we approach the point from inside the circle.

Now, taking $s_0 = 1$, we have $F(s_0) = s_0$. Also, $F'(1) = m < 1$, by hypothesis. Therefore, $[F_n(s) - 1]/m^n$ converges uniformly to an analytic function $H(s)$, where $H(s)$ satisfies

$$H[F(s)] = mH(s)$$

$$H(1) = 0 \qquad H'(1) = 1$$

Therefore,

$$\lim_{n \to \infty} G_n(s) = 1 + \frac{H(s)}{K} = G(s)$$

say. Note that the existence of $\lim_{n \to \infty} [1 - F_n(0)]/m^n$ also follows from Königs' result. To complete the proof, we see that the conditions on $H(s)$ can be written as

$$G[F(s)] = mG(s) + (1 - m)K^{-1}$$

$$G(1) = 1 + \frac{H(1)}{K} = 1$$

$$G'(1-) = \frac{H'(1)}{K} = K^{-1}$$

[1] Bellman [2] has considered a generalization of the theorem of Königs which involves, in place of Schröder's functional equation, the functional equation

$$\frac{\alpha_1}{p_1} \xi[\theta_1(s)] + \frac{\alpha_2}{q_2} \xi[\theta_2(s)] = \xi(s)$$

where $\xi(s)$ is a continuous function and $\theta_1(s) = \sum_{i=1}^{\infty} p_i s^i$ and $\theta_2(s) = \sum_{i=1}^{\infty} q_i s^i$ are generating functions with $|p_1| > 0$; $|q_1| < 1$; $\alpha_1, \alpha_2 \geqslant 0$; and $\alpha_1 + \alpha_2 = 1$. The above equation is associated with a branching process where the generation time is deterministic and where the choice of a branching distribution is stochastic.

It is clear from the above that the limiting distribution $p^*(x)$ will depend on the particular generating function $F(s)$ and will assume a number of forms depending on the expected value m. A special case was studied by Fisher [18], who considered a generating function of the form $F(s) = e^{s-1}$. We now state two other theorems of Yaglom concerning the limiting distribution. It is of interest to note that a common limiting distribution is obtained when $m = 1$.

Theorem 1.12: Let $p_n(0) = \mathscr{P}\{X_n = 0\}$ and let $q_n = 1 - p_n(0)$. Define the random variable Y_n by

$$Y_n = \left(\frac{q_n}{m^n}\right) X_n \tag{1.86}$$

and let

$$S_n(y) = \mathscr{P}\{Y_n < y \mid Y_n > 0\} \tag{1.87}$$

Then, if $m = 1$, $F''(1) \neq 0$, and $F'''(1) < \infty$, we have

$$\lim_{n \to \infty} S_n(y) = S(y) = 1 - e^{-y} \qquad \text{for } y \geqslant 0$$
$$= 0 \qquad \text{for } y < 0 \tag{1.88}$$

Theorem 1.13: If $m > 1$ and $F''(1) \neq m^2 - m < \infty$, then the characteristic function of the limiting distribution

$$\psi(\tau) = \int_{-\infty}^{\infty} e^{i\tau y}\, dS(y)$$

satisfies the functional equation

$$F[(1 - \xi)\psi(\tau) + \xi] = (1 - \xi)\psi(m\tau) + \xi \tag{1.89}$$

where $\xi \in [0,1)$ is the root of $F(s) = s$ and $\psi(0) = i$.

D. Limiting Behavior of the Normed Random Variables $Y_n = X_n/m^n$. We now consider the limiting behavior of the process $\{Y_n = X_n/m^n,\ n = 0, 1, 2, \ldots\}$. Hence, Y_n is defined as the population size in the nth generation normed by the expected population size in the nth generation.

Before considering the convergence properties of Y_n, we note several interesting properties. First, we note that

$$\mathscr{E}\{Y_n\} = \frac{1}{m^n}\, \mathscr{E}\{X_n\} = 1 \tag{1.90}$$

Second, we have

$$\mathscr{E}\{Y_n^2\} = \frac{1}{m^{2n}}\, \mathscr{E}\{X_n^2\} = \frac{1}{m^{2n}}\left[\sigma^2 m^2 \frac{m-1}{m^2-m} + m^n(m^n - 1) + m^n\right]$$
$$= 1 + \frac{\sigma^2}{m^2 - m}\left(1 - \frac{1}{m^n}\right) \qquad m \neq 1$$

Hence
$$\mathcal{D}^2\{Y_n\} = \frac{\sigma^2}{m^2 - m}\left(1 - \frac{1}{m^n}\right) \tag{1.91}$$

Third, if we consider the conditional expectation of X_{n+i}, given X_n, we have

$$\mathcal{E}\{X_{n+i} \mid X_n\} = \sum_{x=1}^{\infty} x \text{ (coefficient of } s^x \text{ in } [F_i(s)]^{X_n})$$

$$= \frac{d}{ds}[F_i(s)]^{X_n}\bigg]_{s=1}$$

$$= X_n[F_i(1)]^{X_n}F_i'(1)$$

$$= m^i X_n \tag{1.92}$$

Consequently,
$$\mathcal{E}\{Y_{n+i} \mid Y_n\} = Y_n \tag{1.93}$$

By virtue of (1.93), the sequence Y_n forms a *discrete parameter martingale*, i.e., a stochastic process $\{Y_n, n = 0, 1, 2, \ldots\}$ such that

(a) $\mathcal{E}\{|Y_n|\} < \infty$ for all n

(b) $\mathcal{E}\{Y_{n+1} \mid Y_n, Y_{n-1}, \ldots, Y_0\} = Y_n$

with probability one. We now state and prove the following theorem.[1]

Theorem 1.14: If $m > 1$, the sequence $\{Y_n\}$ converges to a random variable Y^* with probability one.

Proof: The proof is based on the following *martingale convergence theorem* (cf. the treatise of Doob): Let $\{Y_n, n \geqslant 0\}$ be a discrete parameter martingale. Then $\mathcal{E}\{|Y_0|\} \leqslant \mathcal{E}\{|Y_1|\} \leqslant \cdots$. If $\lim_{n\to\infty} \mathcal{E}\{|Y_n|\} = K < \infty$, then $\lim_{n\to\infty} Y_n = Y^*$ exists with probability one and $\mathcal{E}\{|Y^*|\} < K$. In particular, $K < \infty$ if the Y_n are all real and nonnegative or all real and nonpositive.

In our case we have $Y_n \geqslant 0$ for all n; also, $\mathcal{E}\{|Y_n|\} = \mathcal{E}\{Y_n\} = 1$. Since the quantities $\mathcal{E}\{|Y_n|\}$ are uniformly bounded, it follows from the above theorem that $\lim Y_n = Y^*$ with probability one.

We remark that, from the fundamental theorem, when $m \leqslant 1$, we have $X_n \to 0$ with probability one; hence, $Y_n \to Y^* = 0$. Also, when $m > 1$, $X_n \to 0$ with probability ω, $0 < \omega < 1$.

---

[1] Harris [27] has proved the following theorem: If $m > 1$, the sequence $\{Y_n\}$ converges in quadratic mean to a random variable Y. This result, however, is not as strong as Theorem 1.14, since convergence in quadratic mean implies only convergence in probability.

Having shown that the limit of the sequence $\{Y_n\}$ is a random variable, it is of interest to consider the probability distribution of the limiting value. Let

$$H_n(y) = \mathscr{P}\{Y_n \leqslant y\} \tag{1.94}$$

and

$$\varphi_n(s) = \mathscr{E}\{\exp\{Y_n s\}\} = \int_0^\infty e^{sy}\, dH_n(y) \tag{1.95}$$

be the moment generating function of Y_n. In addition, when $m > 1$, we put

$$H(y) = \mathscr{P}\{Y^* \leqslant y\} \tag{1.96}$$

and

$$\varphi(s) = \mathscr{E}\{\exp\{Y^* s\}\} = \int_0^\infty e^{sy}\, dH(y) \tag{1.97}$$

Now $\varphi_n(s) = F_n[\exp\{s/m^n\}]$; hence, by using the relation $F_{n+1}(s) = F_n[F(s)]$, we have that

$$\varphi_{n+1}(ms) = F[\varphi_n(s)] \tag{1.98}$$

The result of Theorem 1.14 implies that

$$\lim_{n\to\infty} H_n(s) = H(s) \qquad \text{and} \qquad \lim_{n\to\infty} \varphi_n(s) = \varphi(s)$$

for $m > 1$ and $\mathscr{R}(s) \leqslant 0$. Hence, the moment generating function $\varphi(s)$ satisfies the functional equation

$$\varphi(ms) = F[\varphi(s)] \tag{1.99}$$

Harris [27] has studied the distribution function $H(y)$ and its properties. In particular, for the process defined by the generating function $F(s) = 0.4s + 0.6s^2$, $H(y)$ has been computed for certain values of y. Bartlett [1] has used a χ^2 distribution to approximate $H(y)$.

Before closing this section, we remark that Theorem 1.14 is related to the following general theorem in the theory of Markov chains due to Doeblin [13].

Theorem 1.15: There exists a sequence $\{\lambda_n\}$, with $\lambda_n \to \infty$, such that

$$\lim_{n\to\infty} \frac{p_{ij}^{(n)}}{\lambda_n} = \alpha_j > 0$$

where the limit α_j is independent of i.

In the case of discrete branching processes, we can define $\lambda_n = [F'(1)]^n = m^n$.

E. Limiting Population Size. In addition to the random variable X_n, it is of interest to study the random variable Z defined as

$$Z = \sum_{n=0}^{\infty} X_n \qquad (1.100)$$

That is, Z is the *cumulative population size*, or the total number of individuals in all generations. We have already shown (Sec. 1.4) that, when $m \leqslant 1$, the random variable Z is finite with probability one. Let

$$q(z) = \mathscr{P}\{Z = z\}$$

and let
$$H(s) = \sum_{z=1}^{\infty} q(z)s^z \qquad |s| \leqslant 1 \qquad (1.101)$$

be the generating function of $q(z)$. We have the following theorem.[1]

Theorem 1.16: If $m \leqslant 1$, $H(s)$ satisfies the functional equation

$$H(s) = sF[H(s)] \qquad (1.102)$$

where $F(s) = \sum_{x=0}^{\infty} p(x)s^x$.

Proof: We consider the partial sums

$$Z_k = 1 + X_1 + X_2 + \cdots + X_k$$

We know that $X_n = 0$ for some n with probability one when $m \leqslant 1$. Therefore, for $m \leqslant 1$, Z_k converges to some random variable Z with probability one. Now, Z_k as defined above represents the total number of individuals in the $(k + 1)$st generation starting from the (zero)th. Put $Z_k = 1 + Y_k$, so $Y_k = X_1 + \cdots + X_k$; that is, Y_k represents the total number of individuals in the k succeeding generations if we have $X_0 = 1$.

Let
$$G_k(s) = \sum_{n=0}^{\infty} q_{k,n} s^n$$

be the generating function of Y_k, where $q_{k,0} = p_0$ and $q_{k,n} = \mathscr{P}\{Y_k = n\}$. The event $Y_k = n$ can be divided up into n mutually exclusive events with probabilities

$$\mathscr{P}\{X_1 = i\}\mathscr{P} \{ \text{of producing a total of } n - i \text{ individuals in the following}$$
$$k - 1 \text{ generations}\} \qquad i = 1, 2, \ldots, n$$

Now, if we start with one individual at time zero, the probability of producing a total of n individuals in the succeeding k generations is the coefficient of s^n in $G_k(s)$. Therefore, if $X_0 = i$, where the i

---

[1] Hawkins and Ulam [32] and Otter [49].

individuals are identical and statistically independent, then the probability of producing a total of n individuals in the succeeding k generations is the coefficient of s^n in $[G_k(s)]^i$. Thus

$$q_{k,n} = \sum_{i=1}^{n} \mathscr{P}\{X_1 = i\} \mathscr{P} \{\text{of producing a total of } n - i \text{ individuals in the}$$

following $k - 1$ generations$\}$

$$= \sum_{i=1}^{n} p_i \,(\text{coefficient of } s^{n-i} \text{ in } [G_{k-1}(s)]^i)$$

$$= \sum_{i=1}^{n} p_i \,(\text{coefficient of } s^{n-i} \text{ in } [sG_{k-1}(s)]^i)$$

$$= \text{coefficient of } s^n \text{ in } \left(\sum_{i=1}^{n} p_i[sG_{k-1}(s)]^i \right)$$

$$= \text{coefficient of } s^n \text{ in } \left(\sum_{i=0}^{\infty} p_i[sG_{k-1}(s)]^i \right)$$

$$= \text{coefficient of } s^n \text{ in } F[sG_{k-1}(s)]$$

Therefore
$$G_k(s) = \sum_{n=0}^{\infty} q_{k,n} s^n = F[sG_{k-1}(s)]$$

Now if H_k is the generating function of the Z_k, then it is easily seen that $H_k(s) = sG_k(s)$. Hence

$$H_k(s) = sF[H_{k-1}(s)]$$

On passing to the limit, we see that the generating function $H(s)$ of Z satisfies

$$H(s) = sF[H(s)]$$

Otter [49] has shown that the above equation has a unique analytic solution. He also gives asymptotic expressions for the probabilities $q(z)$.

1.7   Representation as Random-walk Processes

A.   Introduction. In many applied problems the generating-function representation of a discrete branching process leads to mathematical problems which are difficult, if not impossible, to treat in a satisfactory manner. In these instances it is often possible to study some of the properties of the stochastic process by resorting to some other type of representation. In this section we shall discuss the representation of discrete branching processes as random-walk processes.

In Chap. 3 the theory of diffusion processes will be considered, and the representation of discrete branching processes as diffusion processes will be presented at that time.

B.   Random-walk and Branching Processes.   A *random walk* is the motion along a line executed by a particle which, at each unit time, can move one unit to the right, move one unit to the left, or stand still, the probabilities of these transitions depending in general on the position of the particle, but being independent of time. We can, therefore, identify the possible positions or states of the particle with a finite or infinite set of integers. As before, we denote by $p_{ij}^{(n)}$ the probability that a particle initially at position i will be at position j after n units of time. If the Markov matrix $P = (p_{ij})$ is given, we can study the properties of the motion of the particle.[1]

The interpretation of a simple discrete branching process as a random walk on the nonnegative integers 0, 1, 2, . . . is as follows: At $t = 0$ the "particle," which we take to symbolize the population size, is at position $i, 0 < i < \infty$. At $t = k$ ($k = 1, 2, \ldots$) the particle has probability p independent of position of moving one step to the right and probability $q = 1 - p$ of moving one step to the left.[2]

The position of the particle at time $k \geqslant 0$ represents the population size in the kth generation. A unit step to the right can be interpreted as a "birth," which results in the addition of a new individual to the population; while a unit step to the left can be interpreted as a "death," which results in the removal of an individual from the population. The motion executed by the particle represents the evolution of the population. We also remark that the position 0 is an absorbing barrier (or state); hence, once the particle reaches 0 the random walk is terminated, this being tantamount to the extinction of the population.

In view of the above remarks, we can conclude that the random walk described above represents a twofold discrete branching process.

It is also possible to consider a "generalized" random-walk process in which steps greater than 1 are permitted at each unit of time. Random-walk processes of this type can be used to represent branching processes in which a single individual in each generation can either die or give birth to $n > 1$ individuals.

The relationship between certain formulas in the theory of branching

---

[1] For a very elegant approach to random-walk problems we refer to the paper of Karlin and McGregor [39]. In this approach an integral representation of the Markov matrix P is obtained; through it the probabilistic structure of the process may be analyzed. Integral representations of P have also been obtained by Kac [35].

[2] For a simple random walk on the nonnegative integers the matrix P is a Jacobi matrix, that is, $p_{ij} = 0$ for $|i - j| > 1$.

processes and random walks was first pointed out by Good [24], and Harris [29] has given a detailed discussion of this correspondence.

C. Gambler's Ruin and the Problem of Extinction of a Population. We now consider the relationship between the classical gambler's ruin problem and the problem of extinction of a population after n generations. The gambler's ruin problem is equivalent to a random walk on the finite set of integers $x = 0, 1, 2, \ldots, a$, where 0 and a are absorbing barriers. Let $\pi_{i,n}$ denote the probability that the random walk will terminate with the nth step at the barrier $x = 0$ (gambler's ruin at the nth trial), when the initial position is $x = i$, $0 < i < a - i$.

To obtain an equation for the $\pi_{i,n}$, we first note that, after the first step, the position is either $i + 1$ or $i - 1$, with probabilities p and $q = 1 - p$, respectively. Hence, for $1 < i < a - 1$ and $n \geqslant 1$ the usual reasoning leads to the difference equation

$$\pi_{i,n+1} = p\pi_{i+1,n} + q\pi_{i-1,n} \tag{1.103}$$

This equation is to be solved with the boundary conditions

$$\pi_{0,n} = \pi_{a,n} = 0 \qquad \text{for } n > 1$$
$$\pi_{0,0} = 1 \qquad \pi_{i,0} = 0 \qquad \text{for } i > 0 \tag{1.104}$$

Equation (1.104) is now valid for all i with $0 < i < a$ and $n \geqslant 0$.

Let us now introduce the generating function

$$\Pi_i(s) = \sum_{n=0}^{\infty} \pi_{i,n} s^n \qquad |s| \leqslant 1 \tag{1.105}$$

Equation (1.103) can now be written as

$$\Pi_i(s) = ps\,\Pi_{i+1}(s) + qs\,\Pi_{i-1}(s)$$
$$= s\{p\Pi_{i+1}(s) + q\Pi_{i-1}(s)\} \tag{1.106}$$

with boundary conditions

$$\Pi_0(s) = 1 \qquad \Pi_a(s) = 0 \tag{1.107}$$

To solve (1.106), we assume a solution of the form $\Pi_i(s) = \beta^i(s)$. By substituting this expression into (1.106), we find that $\beta(s)$ must satisfy the characteristic equation

$$ps\beta^2(s) - \beta(s) + qs = 0 \tag{1.108}$$

The roots of (1.108) are

$$\beta_{1,2}(s) = \frac{1 \pm \sqrt{1 - 4pqs^2}}{2ps} \tag{1.109}$$

Hence

$$\Pi_i(s) = c_1(s)\beta_1^{\,i}(s) + c_2(s)\beta_2^{\,i}(s) \tag{1.110}$$

is the solution of (1.106), where $c_1(s)$ and $c_2(s)$ are arbitrary functions of s. To determine $c_1(s)$ and $c_2(s)$, we see that the boundary conditions require that

$$c_1(s) + c_2(s) = 1$$
$$\beta_1^a(s)c_1(s) + \beta_2^a(s)c_2(s) = 0 \qquad (1.111)$$

From this system we obtain

$$c_1(s) = \frac{\beta_2^a(s)}{\beta_2^a(s) - \beta_1^a(s)} \qquad c_2(s) = \frac{-\beta_1^a(s)}{\beta_2^a(s) - \beta_1^a(s)} \qquad (1.112)$$

The solution can now be written

$$\Pi_i(s) = \frac{\beta_1^a(s)\beta_2^i(s) - \beta_2^a(s)\beta_1^i(s)}{\beta_1^a(s) - \beta_2^a(s)} \qquad (1.113)$$

Now, (1.113) is the generating function of the probability of absorption of $x = 0$ (ruin) at the nth trial. Hence the probabilities $\pi_{i,n}$ can be obtained by expanding $\Pi_i(s)$ as a power series in s. These probabilities are given by

$$\pi_{i,n} = a^{-1}2^n p^{(n-i)/2}q^{(n+i)/2}\sum_{k=1}^{a-1}\cos^{n-1}\frac{\pi k}{a}\sin\frac{\pi k}{a}\sin\frac{\pi i k}{a} \qquad (1.114)$$

In order to establish a connection between a "birth-and-death" type branching process and the gambler's ruin problem, it is necessary to consider the case when $a = \infty$, that is, the case in which $x = 0$ is the only absorbing barrier (or state). In this case the boundary conditions (1.107) are replaced by the single condition $\Pi_0(s) = 1$. From (1.109) we see that $\beta_1 > 1$ and $\beta_2 < 1$ for $|s| < 1$; hence, in order for $\Pi_i(s)$ to be bounded, we must have $c_1(s) = 0$. Hence, the generating function in the case is

$$\Pi_i(s) = \beta_2^i(s) = \left(\frac{1 - \sqrt{1 - 4pqs^2}}{2ps}\right)^i \qquad (1.115)$$

Therefore, for fixed i the probability $\pi_{i,n}$ is given by the coefficient of s^n in the power series expansion of $\beta_2^i(s)$. In this case we have, by letting $a \to \infty$ in (1.114),

$$\pi_{i,n} = 2^n p^{(n-i)/2}q^{(n+i)/2}\int_0^1 (\cos^{n-1}\pi x \sin \pi x \sin \pi x i)\,dx \qquad (1.116)$$

Consider now the probability of eventual ruin or absorption. This is given by

$$Q = \sum_{n=0}^{\infty}\pi_{i,n} = \Pi_i(1) = \beta_2^i(1) = \left(\frac{1 - \sqrt{1 - 4pq}}{2p}\right)^i \qquad (1.117)$$

Now, $1 - 4pq = (p - q)^2$; hence, from (1.115) we see that $\beta_2(1) = q/p$ for $p \geqslant q$, and $\beta_2(1) = 1$ for $q \geqslant p$. Hence

$$Q = \left(\frac{q}{p}\right)^i \qquad \text{for } p \geqslant q$$

$$= 1 \qquad \text{for } p \leqslant q \qquad (1.118)$$

The gambler's ruin problem represents the following branching process: The population size in the zero generation is i, $0 < i < \infty$. Since the above random walk is equivalent to a twofold branching, the associated generating function is

$$F(s) = q + ps^2 \qquad (1.119)$$

that is, an individual has probability p $(0 \leqslant p \leqslant 1)$ of producing a new individual and probability $q = 1 = p$ of dying. From the theory developed in Sec. 1.4 we know that the probability that the population will become extinct after n generations when the initial number of individuals was i is given by $p_{i0}^{(n)}$, where $p_{i0}^{(n)}$ is the coefficient of s^0 in the power series expansion of $[F_n(s)]^i$.

Now let Q^* denote the probability of eventual extinction. From the fundamental theorem we know that Q is given by the smallest non-negative root of $F(s) = s$. Hence

$$Q^* = \frac{1 - \sqrt{1 - 4pq}}{2p} \qquad (1.120)$$

For $X_0 = i$ the probability of eventual extinction is $[Q^*]^i = Q$, as given by (1.118).

1.8 N-dimensional Branching Processes

A. Introduction. In Secs. 1.2 to 1.7 we have been concerned with the study of mathematical models for discrete branching processes which represent the growth of populations when each member has probability $p(x)$ of producing in each generation $x \geqslant 0$ individuals of the same type. These processes, which involve only one type of individual, are called *one-* (or *uni-*) *dimensional* processes. In many problems arising in biology and physics it is of interest to consider a more general scheme in which an individual when transformed can give rise to individuals of different types. In these cases we encounter N- (or *multi-*) *dimensional processes*, where N is the number of types present in the population.

There now exists a rather complete theory of N-dimensional branching processes which has been developed by Soviet and American mathematicians (cf. Everett and Ulam [14], Harris [28], Kolmogorov and Sevastyanov [47], Sevastyanov [53, 54]). This section will be devoted to the characterization of these processes and a discussion of some of their elementary properties.

B. Generating Functions and Moments. Let the random vector $\mathbf{X}_n = (X_{1n}, X_{2n}, \ldots, X_{Nn})$ represent the number of individuals in the population in the nth generation. The component scalar random variables X_{in} $(i = 1, 2, \ldots, N)$ represent the number of individuals of the ith type in the nth generation. In the one-dimensional case the random variables X_n assumed values in the state space $\mathfrak{X}$ which consisted of the nonnegative integers; in the N-dimensional case, however, the state space $\mathfrak{X}$ is the N-dimensional Euclidean space. Since the component random variables X_{in} can assume only integer values, the space $\mathfrak{X}$ has a lattice structure, with the points $\mathbf{x} = (x_1, x_2, \ldots, x_N)$ representing the possible states of the system.

We now assume that an individual of type i $(i = 1, 2, \ldots, N)$ in the nth generation has probability $p_i(\alpha_1, \alpha_2, \ldots, \alpha_N) \geqslant 0$, where $\sum_{\alpha_i \geqslant 0} p_i(\alpha_1, \alpha_2, \ldots, \alpha_N) = 1$, of producing in the next generation α_1 individuals of type 1, α_2 individuals of type 2, $\ldots$, and α_N individuals of type N. We also assume that this probability is independent of the development of the ith individual and the $N - 1$ other individuals in the population. Let the initial random vector $\mathbf{X}_0 = (X_{10}, X_{20}, \ldots, X_{N0})$, for example, $\mathbf{X}_0 = (1, 1, \ldots, 1)$, together with the probabilities $p_i(\alpha_1, \alpha_2, \ldots, \alpha_N)$, $i = 1, 2, \ldots, N$, be given. These quantities determine the probability law for the process $\{\mathbf{X}_n\}$. We assume that the Markov property holds; hence $\{\mathbf{X}_n, n = 0, 1, \ldots\}$ is a vector-Markov process with a denumerable number of states.

We now introduce the generating functions $F_i(\mathbf{s})$ and $F_{in}(\mathbf{s})$ defined by

$$F_i(\mathbf{s}) = F_{i1}(\mathbf{s}) = \sum_{\alpha_i=0} p_i(\alpha_1, \alpha_2, \ldots, \alpha_N) s_1^{\alpha_1} s_2^{\alpha_2} \cdots s_N^{\alpha_N} \qquad (1.121)$$

$$F_{in}(\mathbf{s}) = \sum_{\alpha_i=0} \mathscr{P}\{X_{1n} = \alpha_1, X_{2n} = \alpha_2, \ldots, X_{Nn} = \alpha_N\} s_1^{\alpha_1} s_2^{\alpha_2} \cdots s_N^{\alpha_N}$$

where $\mathbf{s} = (s_1, s_2, \ldots, s_N)$, $i = 1, 2, \ldots, N$, and $n = 1, 2, \ldots$. The fundamental functional relation in this case is

$$F_{i,n+1}(\mathbf{s}) = F_i[F_{1n}(\mathbf{s}), F_{2n}(\mathbf{s}), \ldots, F_{Nn}(\mathbf{s})]$$

$$= [F_i(\mathbf{s})]^{n+1} \qquad (1.122)$$

where $i = 1, 2, \ldots, N$ and $n = 1, 2, \ldots$. $F_{i,n+1}(\mathbf{s})$ is the generating function for the $(n + 1)$st generation of offspring from one individual of type i at time zero.

As in the one-dimensional case, we can obtain the moments of $\mathbf{X}_n$ by differentiation of the generating function. In this case we define

$$m_{ij}^{(n)} = \frac{\partial F_{in}}{\partial s_j}\bigg]_{s_1 = s_2 = \cdots = s_N = 1} \tag{1.123}$$

as the expected number of individuals of type j $(j = 1, 2, \ldots, N)$ in the nth generation from one individual to type i $(i = 1, 2, \ldots, N)$. It can also be shown that the relation

$$m_{ij}^{(n)} = [m_{ij}^{(1)}]^n \tag{1.124}$$

exists between the first moments of the nth generation and those of the first generation. If we let $M = (m_{ij}^{(1)})$ denote the first-moment matrix, then differentiation of (1.122) with respect to s and evaluation at $s_1 = s_2 = \cdots = s_N = 1$ gives

$$\mathscr{E}\{\mathbf{X}_n\} = \mathbf{X}_0 M^n \tag{1.125}$$

If we now let $v_{ij}^{(n)}$ represent the second moment of the individuals of type j in the nth generation from one individual of type i, we have the relation

$$\frac{\partial^2 F_{in}}{\partial s_j^2}\bigg]_{s_1 = s_2 = \cdots = s_N = 1} = v_{ij}^{(n)} - m_{ij}^{(n)}$$

or

$$v_{ij}^{(n)} = \frac{\partial^2 F_{in}}{\partial s_j^2}\bigg]_{s_1 = s_2 = \cdots = s_N = 1} + m_{ij}^{(n)} \tag{1.126}$$

A detailed treatment of the first and second moments based on Jacobian and Hessian matrices is given in [14].

C. **Fundamental Theorem in the N-dimensional Case.** We now give a result, due to Sevastyanov, which establishes a relationship between the first-moment matrix M and the probability that the population will eventually become extinct.

Let $p_i^{(n)}(0, 0, \ldots, 0)$ denote the probability that the population becomes extinct in the nth generation when initially one individual of type i is present and let

$$\rho_i = \lim_{n \to \infty} p_i^{(n)}(0, 0, \ldots, 0) = \lim_{n \to \infty} F_{in}(0, 0, \ldots, 0)$$

$$= \lim_{n \to \infty} F_i^n(0, 0, \ldots, 0) \tag{1.127}$$

denote the probability that the population will eventually become extinct when initially one individual of type i is present. If $\rho_i = 1$ for all i, the process $\{\mathbf{X}_n\}$ is called *degenerate*. As in the one-dimensional case, the ρ_i are determined by the system of equations

$$s_i = F_i(s_1, s_2, \ldots, s_N) \qquad i = 1, 2, \ldots, N \tag{1.128}$$

From the theory of absorption probabilities it is easy to verify that the s_i are determined by the system of equations

$$s_i(n+1) = \sum_{j=1}^{N} m_{ij} s_j(n) \qquad i = 1, 2, \ldots, N \qquad (1.129)$$

with initial conditions

$$\begin{aligned} s_i(0) = \delta_{ik} &= 1 \qquad \text{for } i = k \\ &= 0 \qquad \text{for } i \neq k \end{aligned} \qquad (1.130)$$

The solution of (1.129) is of the form

$$s_i(n) = \sum_{j=1}^{N} \varphi_{ij}^{(n)}(\lambda_j)^n \qquad i = 1, 2, \ldots, N \qquad (1.131)$$

where the λ_j are the characteristic roots of M, of multiplicity k, and $\varphi_{ij}(n)$ is a polynomial in n of degree not greater than $k-1$.

Before stating Sevastyanov's theorem, we remark that the progeny of individuals in N-dimensional process can be classified in a manner analogous to the classification of states given in Sec. 1.2D. For example: a *closed class* S is a set of types each of which can produce (after some number of generations) any other type in S. A *final class* is a closed class S each type of which, with probability one, produces in the next generation just *one* individual in S; it may, however, produce individuals of types not contained in S.

Theorem 1.17: Let λ be the characteristic root of M which is at least as large in absolute value as any other root. In order that the process $\{\mathbf{X}_n\}$ be degenerate, it is necessary and sufficient that (1) $|\lambda| \leqslant 1$ and (2) there are no final classes.

D. A Limit Theorem for N-dimensional Processes. In this section we state and prove the N-dimensional analogue of Theorem 1.14. This theorem is due to Harris [28].

Theorem 1.18: Suppose that $\lambda > 1$, that λ is simple and larger in magnitude than any other characteristic root of M, and that for every $i, j = 1, 2, \ldots, N$ there is a positive probability that an individual of type i will have in some generation of its offspring an individual of type j. Consider the random variable

$$S_m = \sum_{i=1}^{N} \frac{X_{im}}{\lambda^m} \qquad (1.132)$$

Then

$$\lim_{m \to \infty} S_m = S^* \qquad (1.133)$$

with probability one.

Proof: Analogous with the one-dimensional case, we have

$$\mathscr{E}\{\mathbf{X}_{m+n} \mid \mathbf{X}_m\} = \mathbf{X}_m M^n$$

If we divide both sides of the above expression by λ^{m+n} and post-multiply both sides by the column vector $\mathbf{e}$ (the right eigenvector of λ), we have

$$\mathscr{E}\left\{\frac{\mathbf{X}_{m+n}\mathbf{e}}{\lambda^n} \,\middle|\, \mathbf{X}_m\right\} = \frac{\mathbf{X}_m M^n \mathbf{e}}{\lambda^{m+n}} = \frac{\mathbf{X}_m \mathbf{e}}{\lambda^m} \qquad (1.134)$$

Now let

$$\beta_m = \frac{\mathbf{X}_m \mathbf{e}}{\lambda^m}$$

Then (1.134) gives

$$\mathscr{E}\{\beta_{m+n} \mid \beta_m\} = \beta_m$$

Application of the martingale convergence theorem to $\{\beta_m\}$ yields

$$\lim_{m \to \infty} \beta_m = \beta^*$$

with probability one. Everett and Ulam [14] have shown that the direction of the vector $\mathbf{X}_m$ (if it does not eventually vanish) approaches a limit. Hence, this result, together with the convergence of $\{\beta_m\}$, proves that each component of $\mathbf{X}_m/\lambda^m$ converges, and the statement of the theorem follows.

We refer to the paper of Jiřina [34], in which asymptotic behavior of the N-dimensional branching process is studied in the case when $\lambda < 1$.

Problems

1.1 Let $\{X_n, n = 0, 1, \dots\}$ be a sequence of independent random variables, each of which assumes nonnegative integral values. Define the sequence of partial sums

$$S_n = \sum_{i=0}^{n} X_i \qquad n = 0, 1, \dots$$

Show that $\{S_n, n = 0, 1, \dots\}$ is a Markov chain.

1.2 Prove the following propositions: Two states i and j of a Markov chain communicate if and only if $L_{ij} > 0$ and $L_{ji} > 0$.

1.3 Classify the states of the Markov chains whose transition probabilities are given by:

(a) $p_{02} = 1$; $\quad p_{11} = 1$; $\quad p_{i,i-1} = p_{i,i+1} = \frac{1}{2}$ $\quad$ for $i = 2, 3, \dots$
(b) $p_{00} = \frac{1}{2}$; $\quad p_{01} = \frac{1}{2}$; $\quad p_{i,i-1} = p_{i,i+1} = \frac{1}{2}$ $\quad$ for $i = 1, 2, \dots$
(c) $p_{00} = \frac{1}{3}$; $\quad p_{01} = \frac{2}{3}$; $\quad p_{i,i-1} = \frac{1}{3}, p_{i,i+1} = \frac{2}{3}$ $\quad$ for $i = 1, 2, \dots$

Assume that at time zero the system is in state 0. The following result is required: A class of states S of a Markov chain is recurrent if and only if for some $i \in S$ we have $Q_{ii} = 1$. In this case $Q_{ij} = L_{ij} = 1$ for every $i, j \in S$. If S is a nonrecurrent class, $Q_{ij} = 0$ and $L_{ii} < 1$ for all $i, j \in S$; however, it is possible that $L_{ij} = 1$ for some $i, j \in S$. An inessential class of states is nonrecurrent, but an essential class may be either recurrent or nonrecurrent.

1.4 Calculate the higher moments and cumulants of X_n, where $\{X_n,\ n = 0, 1, \ldots\}$ is a simple branching process.

1.5 Let $\mathfrak{X}$ be an arbitrary metric space with distance function $\rho(x,y)$. A mapping T of $\mathfrak{X}$ into itself is said to be a *contraction* if there exists a number $\alpha < 1$ such that $\rho(Tx,Ty) \leqslant \alpha\rho(x,y)$ for any two points $x, y \in \mathfrak{X}$. It is well known (*principle of contraction mappings*[1]) that every contraction mapping defined on a complete metric space $\mathfrak{X}$ has one and only one fixed point; i.e., the equation $Tx = x$ has one and only one solution. Show that the generating function $F(s)$ defined in $[0,1]$ and satisfying the Lipschitz condition $|F(s_2) - F(s_1)| \leq K|s_2 - s_1|$, with $K < 1$, is a contraction mapping; and therefore, the sequence $s_0,\ s_1 = F(s_0),\ s_2 = F(s_1),\ \ldots$ converges to the single root of the equation $F(s) = s$.

1.6 Prove the fundamental theorem concerning branching processes by utilizing the theory of absorption probabilities.[2] *Hint:* Let $\alpha_n,\ n = 1, 2, \ldots,$ be the probability that the population will eventually die out, given that $X_0 = n$. Then the α_n satisfy the system $\alpha_n = \sum_{i=1}^{\infty} p_{ni}\alpha_i + p_{n0},\ n = 1, 2, \ldots$. Now utilize the relationship between the transition probabilities and the iterates of the generating function $F(s)$.

1.7 Let $\mathfrak{Y}$ be a partially ordered space, i.e., a linear space in which one of the relations $y_1 > y_2, y_1 < y_2,$ or $y_1 = y_2$ is defined for a certain pair of its elements. $\mathfrak{Y}$ is said to be of type $\mathscr{K}$ if the following axioms are satisfied: (1) ϕ (the null element of $\mathfrak{Y}$) is not greater than 0; (2) $y_1 > \phi$ and $y_2 > \phi$ implies $y_1 + y_2 > \phi$; (3) for any $y \in \mathfrak{Y}$ there exists a $y' \geqslant \phi$ such that $y' - y \geqslant \phi$; (4) if $y > \phi, y \in \mathfrak{Y}$, and λ is a positive real number, then $\lambda y > \phi$; and (5) any bounded set $E \in \mathfrak{Y}$ possesses a least upper bound.

Kantorovič [36,37] has proved the following *theorem:* Let $\mathfrak{Y}$ be a space of type $\mathscr{K}$ and let $y' > 0$ be a certain fixed element of $\mathfrak{Y}$. Consider the functional equation $G(y) = y$. Assume (1) that $G(y)$ is defined for every $y \in \mathfrak{Y}$ such that $y \in [0,y']$ and that for any such y we have $G(y) \in \mathfrak{Y}$; (2) if $0 \leqslant y_1 \leqslant \cdots \leqslant y'$ and $\lim_{n\to\infty} y_n = y$, then $\lim_{n\to\infty} G(y_n) = G(y)$; (3) $G(y)$ is monotone increasing for $y \in [0,y']$; (4) $G(0) \geqslant 0$; (5) $G(y') \leqslant y'$. If all these conditions are satisfied, then $G(y) = y$ has a solution y^* such that $y^* \in [0,y']$ and this solution can be found by the method of successive approximations.

Use the above theorem to establish the existence of a solution of the functional equation $F(s) = s$, which occurs in the fundamental theorem concerning branching processes.

1.8 A Markov chain with transition matrix P is said to be *periodic* with period ω ($\omega = 1, 2, \ldots$) if $P^{n+\omega} = P^n$ and ω is the smallest positive integer with this property. Determine the limit matrix Π for periodic chains.[3]

1.9[4] Let $\{X_n, n = 0, 1, \ldots\}$ be a branching process with generating function $F(s) = \sum_{x=0}^{\infty} p(x)s^x,\ 0 < p_0 < 1,$ and with transition probabilities p_{ij}. Show that there exist numbers $\pi_i \geqslant 0$, not all zero, that satisfy the system of equations

[1] Cf. Kolmogorov and Fomin [46] and Vulikh [61].
[2] Cf. Feller [15].
[3] Cf. Bharucha-Reid [5].
[4] Due to T. E. Harris, unpublished.

$\pi_i = \sum_{j=1}^{\infty} \pi_j p_{ji}$, $i = 1, 2, \ldots$, with $\sum_{i=1}^{\infty} \pi_i = \infty$, where the generating function
$\Pi(s) = \sum_{i=1}^{\infty} \pi_i s^i$ is analytic for $|s|$ less than the probability of extinction and
satisfies the functional equation $\Pi[F(s)] = \Pi(s) + \Pi(p_0)$.

1.10 Consider a random-walk process in which a moving particle can occupy
any of the points $i = a + 1, a + 2, \ldots, b - 1$ on the line segment $[a,b]$. If the
particle is at position i at time t, then with probability p_i it will be at $i + 1$ at
time $t + 1$ and with probability q_i it will be at $i - 1$ at time $t + 1$, where $p_i +
q_i = 1$ for $i = a + 1, \ldots, b - 1$. Determine the probability, say $f(a;x)$,
$x \in (a,b)$, that a particle starting at position x at $t = 0$ will land at the position a
before landing at b. Consider x and b to be fixed, and a variable.[1]

Bibliography

1 Bartlett, M. S.: "An Introduction to Stochastic Processes," Cambridge University Press, New York, 1955.
2 Bellman, R.: On Limit Theorems for Non-commutative Operations: II. A Generalization of a Result of Königs, RAND Research Paper 485, 1954.
3 Bellman, R., and R. Kalaba: Random Walk, Scattering, and Invariant Imbedding: I. One-dimensional Discrete Case, *Proc. Natl. Acad. Sci. U.S.*, vol. 43, pp. 930–933, 1957.
4 Bharucha-Reid, A. T.: Note on the Fundamental Theorem Concerning Branching Processes (Abstract), *Bull. Am. Math. Soc.*, vol. 62, p. 414, 1956.
5 Bharucha-Reid, A. T.: Ergodic Projections for Semi-groups of Periodic Operators, *Studia Math.*, vol. 17, pp. 189–197, 1958.
6 Blackwell, D.: On Transient Markov Processes with a Countable Number of States and Stationary Transition Probabilities, *Ann. Math. Statist.*, vol. 26, pp. 654–658, 1955.
7 Castoldi, L.: Sulla distribuzione dei tempi di estinzione nelle discendenze biologiche, *Boll. un. mat. ital.*, ser. 3, vol. 11, pp. 152–167, 1956.
8 Chung, K. L.: "Notes on Markov Chains," Graduate Mathematical Statistics Society, Columbia University, New York, 1951.
9 Chung, K. L.: Contributions to the Theory of Markov Chains: I, *J. Research Natl. Bur. Standards*, vol. 50, pp. 203–208, 1953.
10 Chung, K. L.: Contributions to the Theory of Markov Chains: II, *Trans. Am. Math. Soc.*, vol. 76, pp. 397–419, 1954.
11 Cramér, H.: "Mathematical Methods of Statistics," Princeton University Press, Princeton, N.J., 1946.
12 Derman, C.: A Solution to a Set of Fundamental Equations in Markov Chains, *Proc. Am. Math. Soc.*, vol. 5, pp. 332–334, 1954.
13 Doeblin, W.: Sur deux problèmes de M. Kolmogoroff concernant les chaines dénombrables, *Bull. soc. math. France*, vol. 66, pp. 210–220, 1938.
14 Everett, C. J., and S. Ulam: Multiplicative Systems in Several Variables: I, II, III, *Los Alamos Scientific Laboratory Declassified Documents* LADC-534 (AECD-2164), LADC-533 (AECD-2165), LA-707, 1948.
15 Feller, W.: Diffusion Processes in Genetics, *Proc. Second Berkeley Symposium on Math. Statistics and Probability*, pp. 227–246, 1951.

[1] This inhomogeneous version of the gambler's ruin problem is due to Bellman and Kalaba [3].

16   Feller, W.: Boundaries Induced by Positive Matrices, *Trans. Am. Math. Soc.*, vol. 83, pp. 19–54, 1956.

17   Feller, W.: "An Introduction to Probability Theory and Its Applications," vol. 1, 2d ed., John Wiley & Sons, Inc., New York, 1957.

18   Fisher, R. A.: "The Genetical Theory of Natural Selection," Oxford University Press, New York, 1930.

19   Fortet, R.: Les processus stochastiques en cascades, *Trabajos Estatist.*, vol. 1, pp. 11–34, 1953.

20   Foster, F. G.: Markov Chains with an Enumerable Number of States and a Class of Cascade Processes, *Proc. Cambridge Phil. Soc.*, vol. 48, pp. 587–591, 1952.

21   Galton, F., and H. W. Watson: On the Probability of the Extinction of Families, *J. Anthrop. Inst.*, vol. 4, pp. 138–144, 1874.

22   Gnedenko, B. V., and A. N. Kolmogorov: "Limit Distributions for Sums of Independent Random Variables," Addison-Wesley Publishing Company, Reading, Mass., 1954. (Translated from the Russian by K. L. Chung.)

23   Good, I. J.: The Number of Individuals in a Cascade Process, *Proc. Cambridge Phil. Soc.*, vol. 45, pp. 360–363, 1949.

24   Good, I. J.: Random Motion on a Finite Abelian Group, *Proc. Cambridge Phil. Soc.*, vol. 47, pp. 756–762, 1951.

25   Good, I. J.: The Joint Distribution for the Sizes of the Generation in a Cascade Process, *Proc. Cambridge Phil. Soc.*, vol. 51, pp. 240–242, 1955.

26   Hadamard, J.: Two Works on Iteration and Related Questions, *Bull. Am. Math. Soc.*, vol. 50, pp. 67–75, 1944.

27   Harris, T. E.: Branching Processes, *Ann. Math. Statist.*, vol. 19, pp. 474–494, 1948.

28   Harris, T. E.: Some Mathematical Models for Branching Processes, *Proc. Second Berkeley Symposium on Math. Statistics and Probability*, pp. 305–328, 1951.

29   Harris, T. E.: First Passage and Recurrence Distributions, *Trans. Am. Math. Soc.*, vol. 73, pp. 471–486, 1952.

30   Harris, T. E.: The Existence of Stationary Measures for Certain Markov Processes, *Proc. Third Berkeley Symposium on Math. Statistics and Probability*, vol. 2, pp. 113–124, 1956.

31   Harris, T. E.: Transient Markov Chains With Stationary Measures, *Proc. Am. Math. Soc.*, vol. 8, pp. 937–942, 1957.

32   Hawkins, D., and S. Ulam: Theory of Multiplicative Processes: I, *Los Alamos Scientific Laboratory Declassified Document* 265, 1944.

33   Hille, E., and R. S. Phillips: "Functional Analysis and Semi-groups," American Mathematical Society, New York, 1957.

34   Jiřina, M.: The Asymptotic Behavior of Branching Stochastic Processes (in Russian), *Czechoslovak Math. J.*, vol. 7, pp. 130–153, 1957.

35   Kac, M.: Random Walk and the Theory of Brownian Motion, *Am. Math. Monthly*, vol. 54, pp. 369–391, 1947.

36   Kantorovič, L.: The Method of Successive Approximations for Functional Equations, *Acta Math.*, vol. 71, pp. 63–97, 1939.

37   Kantorovič, L., B. Z. Vulikh, and A. G. Pinsker: "Functional Analysis in Partially Ordered Spaces" (in Russian), Gosudarstv. Izdat. Techn.-Teor. Lit., Moscow, 1950.

38   Karlin, S., and J. McGregor: Representation of a Class of Stochastic Processes, *Proc. Natl. Acad. Sci. U.S.*, vol. 41, pp. 387–391, 1955.

DISCRETE TIME PROCESSES

39 Karlin, S., and J. McGregor: Random Walks, *Illinois J. Math.*, vol. 3, pp. 66–81, 1959.

40 Kendall, D. G.: Stochastic Processes and Population Growth, *J. Roy. Statist. Soc.*, ser. B, vol. 11, pp. 230–264, 1949.

41 Kendall, D. G.: On Non-dissipative Markov Chains with an Enumerable Infinity of States, *Proc. Cambridge Phil. Soc.*, vol. 48, pp. 587–591, 1952.

42 Kendall, D. G., and G. E. H. Reuter: The Calculation of the Ergodic Projection for Markov Chains and Processes with a Countable Infinity of States, *Acta Math.*, vol. 97, pp. 103–144, 1957.

43 Kolmogorov, A. N.: Anfangsgründe der Markoffschen Ketten mit unendlich vielen möglichen Zuständen, *Rec. math. Moscou (Mat. Sbornik)*, vol. 1, no. 43, pp. 607–610, 1936.

44 Kolmogorov, A. N.: Zur Lösung einer biologischen Aufgabe, *Comm. Math. Mech. Chebychev Univ., Tomsk*, vol. 2, pp. 1–6, 1938.

45 Kolmogorov, A. N., and N. A. Dmitriev: Branching Stochastic Processes, *Compt. rend. acad. sci. U.R.S.S.*, n.s., vol. 56, pp. 5–8, 1947.

46 Kolmogorov, A. N., and S. V. Fomin: "Elements of the Theory of Functions and Functional Analysis," vol. 1: "Metric and Normed Spaces" (in Russian; English translation, Graylock Press, Rochester, N.Y., 1957), Izdat. Moscow Univ., Moscow, 1954.

47 Kolmogorov, A. N., and B. A. Sevastyanov: The Calculation of Final Probabilities for Branching Stochastic Processes (in Russian), *Doklady Akad. Nauk S.S.S.R.*, n.s., vol. 59, pp. 783–786, 1947.

48 Königs, G.: Nouvelles recherches sur les intégrales des certaines équations fonctionnelles, *Ann. sci. école norm. super. de Paris*, ser. 3, vol. 1, pp. 3–41, 1884.

49 Otter, R.: The Multiplicative Process, *Ann. Math. Statist.*, vol. 20, pp. 206–224, 1949.

50 Picard, E.: "Leçons sur quelques équations fonctionnelles," Gauthier-Villars, Paris, 1950.

51 Rudin, W.: "Principles of Mathematical Analysis," McGraw-Hill Book Company, Inc., New York, 1953.

52 Schröder, E.: Über iterirte Funktionen, *Math. Ann.*, vol. 3, pp. 296–322, 1871.

53 Sevastyanov, B. A.: On the Theory of Branching Stochastic Processes (in Russian), *Doklady Akad. Nauk S.S.S.R.*, n.s., vol. 59, pp. 1407–1410, 1947.

54 Sevastyanov, B. A.: Theory of Branching Stochastic Processes (in Russian), *Uspehi Mat. Nauk*, n.s., vol. 6, no. 6, pp. 47–99, 1951.

55 Solow, R.: On the Structure of Linear Models, *Econometrica*, vol. 21, pp. 29–46, 1952.

56 Steffensen, J. F.: Om Sandyligheden for at Afkommet Uddør, *Mat. Tidssker.*, ser. B., vol. 19, pp. 19–23, 1930.

57 Steffensen, J. F.: Deux problèmes du calcul des probabilités, *Ann. inst. H. Poincaré*, vol. 3, pp. 331–344, 1933.

58 Urbanik, K.: On a Stochastic Model of a Cascade, *Bull. acad. polon. sci. Classe III*, vol. 3, pp. 349–351, 1955.

59 Urbanik, K.: Remarks on the Equation of Branching Stochastic Processes (in Polish), *Zeszyty Nauk. Uniw. Wroclaw. im. B. Bieruta*, ser. B, no. 1, pp. 17–26, 1956.

60 Veblen, O.: "Analysis Situs," American Mathematical Society, New York, 1922.

61 Vulikh, B. Z.: "Introduction to Functional Analysis" (in Russian), Gosudarstv. Izdat. Fiz.-Mat. Lit., Moscow, 1958.
62 Woodward, P. M.: A Statistical Theory of Cascade Multiplication, *Proc. Cambridge Phil. Soc.*, vol. 44, pp. 404–412, 1948.
63 Yaglom, A. M.: Certain Limit Theorems of the Theory of Branching Stochastic Processes (in Russian), *Doklady Akad. Nauk S.S.S.R.*, n.s., vol. 56, pp. 795–798, 1947.
64 Yosida, K., and S. Kakutani: Markoff Process with an Enumerable Infinite Number of Possible States, *Japan. J. Math.*, vol. 16, pp. 47–55, 1939.
65 Yosida, K., and S. Kakutani: Operator-theoretical Treatment of Markoff's Process and Mean Ergodic Theorem, *Ann. Math.*, vol. 42, pp. 188–228, 1941.

2

Processes Discrete in Space and Continuous in Time

2.1 Introduction

In Chap. 1 we considered processes (or chains) discrete in space and time. We shall now consider processes discrete in space and continuous in time; they are referred to as *discontinuous Markov processes*.[1] In Sec. 2.2 we derive the fundamental integrodifferential equations of Markov processes of this type. These equations are specialized in Sec. 2.3, where we study the Kolmogorov differential equations. The Kolmogorov equations, which are differential-difference equations, characterize (in most cases) Markov processes that are discrete in space and continuous in time. In Sec. 2.4 we consider some particular Markov processes. Here the emphasis is on the methods of solving particular cases of the Kolmogorov equations and studying the properties of the solutions. Section 2.5 is devoted to an introduction to the Bellman-Harris theory of age-dependent processes. These processes, which in general are non-Markovian, are of great importance in many fields of applications. In Sec. 2.6 we consider some limit theorems associated with processes discrete in space and continuous in time. Finally, in Sec. 2.7, we consider N-dimensional processes.

2.2 Fundamental Equations of Discontinuous Markov Processes

In this section we shall derive the fundamental functional equations of discontinuous Markov processes. Denote by $\mathfrak{X}$ the state space

---

[1] For the theory of more general discontinuous Markov processes which include the processes considered in this chapter as a special case, we refer to the paper of Moyal [65].

associated with the process $\{X(t),\, t \geqslant 0\}$, that is, $\mathfrak{X}$ is the space of all possible values x which the random variable $X(t)$ can assume. Denote by E an event, which is a subset of $\mathfrak{X}$, that is, $E \subset X$. The probability of the event E is a set function $P\{E\}$, termed the *probability function*, which satisfies the following conditions:

$$0 \leqslant P\{E\} \leqslant 1 \tag{2.1}$$

$$P\{\mathfrak{X}\} = 1 \tag{2.2}$$

and
$$P\left\{\sum_{i=1}^{\infty} E_i\right\} = \sum_{i=1}^{\infty} P\{E_i\} \tag{2.3}$$

if the E_i are disjoint, i.e., they have no common points. If for some $t > 0$ the random variable $X(t)$ assumes a value $\xi \in E$, we say that that event E has occurred. If the stochastic process $\{X(t),\, t \geqslant 0\}$ is a Markov process, the *conditional probability function*

$$P(E,t;\xi,\tau) = \mathscr{P}\{X(t) \in E \mid X(\tau) = \xi\} \qquad t > \tau \tag{2.4}$$

where $\mathscr{P}\{X(\tau) = \xi\} > 0$, is uniquely determined. For the continuous time parameter case, we say that $\{X(t),\, t \geqslant 0\}$ is a *Markov process* if

$$P(E,t;\xi,\tau) = \mathscr{P}\{X(t) \in E \mid X(\eta),\, 0 < \eta \leqslant \tau\}$$
$$= \mathscr{P}\{X(t) \in E \mid X(\tau) = \xi\}$$

i.e., the probability distribution of $X(t)$ is completely determined for all $t > \tau$ by the knowledge of the value assumed by $X(\tau)$ and in particular is independent of the history of the process for all $t < \tau$. Since $P(E,t;\xi,\tau)$ gives the conditional probability of $X(t) \in E$ under the hypothesis that at a fixed time $\tau < t$, $X(\tau) = \xi$, we can call the functions $P(E,t;\xi,\tau)$ the *transition probabilities of the Markov process* $\{X(t),\, t \geqslant 0\}$. Analytically, a Markov process is completely determined by its transition probabilities. In addition, if $X(t)$ has a given initial value, we can define the absolute probability of the event $X(t) \in E$, and, therefore, an *absolute probability function*

$$P(E,t) = \mathscr{P}\{X(t) \in E\} \tag{2.5}$$

If $P(E,t)$ is defined for $t = \tau$, and if $\{X(t),\, t \geqslant 0\}$ is Markovian, we have

$$P(E,t) = \int_{\mathfrak{X}} P(E,t;\omega,\tau)\, d_{\Omega} P(\Omega,\tau) \tag{2.6}$$

This result follows from the general composition rule for probabilities. The above integral, which is a Lebesgue-Stieltjes integral, is taken over the state space $\mathfrak{X}$. The differential in (2.6) refers to the set Ω, and ω denotes an element corresponding to Ω.

From the above discussion it is clear that for all $t \geqslant \tau$ the transition probabilities $P(E,t;\xi,\tau)$ must satisfy the following conditions:

$$0 \leqslant P(E,t;\xi,\tau) \leqslant 1 \tag{2.7}$$

and

$$\int_{\mathfrak{X}} d_E P(E,t;\xi,\tau) = 1 \tag{2.8}$$

In addition, if we subdivide the interval (τ,t) by selecting a point $s \in [\tau,t]$ and consider the probability of the transition from ξ into E through all points $s \in (\tau,t)$, we obtain the identity

$$P(E,t;\xi,\tau) = \int_{\mathfrak{X}} P(E,t;\omega,s) \, d_\Omega P(\Omega,s;\xi,\tau) \tag{2.9}$$

for all $t \geqslant \tau$ and $s \in [\tau,t]$; Eq. (2.9) is the *Chapman-Kolmogorov functional equation*, and it is the analogue of Eq. (1.10) of Chap. 1. For $t = \tau$ the set function $P(E,t;\xi,\tau)$ is reduced to the unit distribution (or characteristic function of the set E)

$$\chi(E) = 1 \qquad \text{for } \xi \in E$$
$$= 0 \qquad \text{for } \xi \in \mathfrak{X} - E \tag{2.10}$$

If we now assume that $P(E,t;\xi,\tau)$ is a continuous function of τ and t, for $t \geqslant \tau$, and Borel measurable with respect to ξ, we have

$$\lim_{t \to \tau} P(E,t;\xi,\tau) = \lim_{\tau \to t} P(E,t;\xi,\tau) = \chi(E) \tag{2.11}$$

We now introduce two new functions: (1) the *intensity function* $q(\xi,t)$ and (2) the *relative transition probability function* $Q(E;\xi,t)$. The intensity function has the following interpretation: $q(\xi,t) \, \Delta t + o(\Delta t)$ is the probability that $X(t)$ will undergo a random change in the interval $(t, t + \Delta t)$ when $X(t) = \xi$; hence, $1 - q(\xi,t) \, \Delta t + o(\Delta t)$ is the probability that no change will take place. Similarly, $Q(E;\xi,t)$ is the conditional probability of $X(t)$ assuming a value in E at time $t + \Delta t$, that is, $X(t + \Delta t) \in E$, when it is given that $X(t) = \xi$ and has undergone a change in the interval $(t, t + \Delta t)$. It is clear that these two functions satisfy the following conditions:

$$q(\xi,t) \geqslant 0 \qquad \text{for all } \xi \text{ and } t \tag{2.12}$$

$$0 \leqslant Q(E;\xi,t) \leqslant 1 \qquad \text{for all } E, \xi, \text{ and } t \geqslant \tau \tag{2.13}$$

$$Q(E;\xi,t) = 0 \qquad \text{for all } \xi \in E, t \geqslant \tau \tag{2.14}$$

$$\int_{\mathfrak{X}} d_\Omega Q(\Omega;\xi,t) = 1 \qquad \text{for all } \xi, t \geqslant \tau \tag{2.15}$$

A final condition is that for small values of Δt

$$P(E, t + \Delta t; \xi, t) = [1 - q(\xi,t) \, \Delta t]\chi(E) + Q(E;\xi,t)q(\xi,t) \, \Delta t + o(\Delta t) \tag{2.16}$$

We shall now derive the fundamental functional equations that describe discontinuous Markov processes. These equations will be referred to as the *Feller integrodifferential equations*, since they were obtained by W. Feller in his fundamental paper [23] (cf. also [25]). To derive these equations, we shall utilize the Chapman-Kolmogorov equation and condition (2.16).

In the Chapman-Kolmogorov equation (2.9) we put $t = t + \Delta t$ and $s = t$; then by using (2.16), we have

$$P(E, t + \Delta t; \xi, \tau) = \int_{\mathfrak{x}} P(E, t + \Delta t; \omega, t) \, d_{\Omega} P(\Omega, t; \xi, \tau)$$

$$= \int_{\mathfrak{x}} [1 - q(\omega, t) \, \Delta t] \chi(E) \, d_{\Omega} P(\Omega, t; \xi, \tau)$$

$$+ \Delta t \int_{\mathfrak{x}} Q(E; \omega, t) q(\omega, t) \, d_{\Omega} P(\Omega, t; \xi, \tau) + o(\Delta t) \quad (2.17)$$

Hence

$$P(E, t + \Delta t; \xi, \tau) = P(E, t; \xi, \tau) - \Delta t \int_{E} q(\omega, t) \, d_{\Omega} P(\Omega, t; \xi, \tau)$$

$$+ \Delta t \int_{\mathfrak{x}} Q(E; \omega, t) q(\omega, t) \, d_{\Omega} P(\Omega, t; \xi, \tau) + o(\Delta t) \quad (2.18)$$

By transposing the first term on the right and dividing by Δt, we have

$$\frac{P(E, t + \Delta t; \xi, \tau) - P(E, t; \xi, \tau)}{\Delta t} = - \int_{E} q(\omega, t) \, d_{\Omega} P(\Omega, t; \xi, \tau)$$

$$+ \int_{\mathfrak{x}} Q(E; \omega, t) q(\omega, t) \, d_{\Omega} P(\Omega, t; \xi, \tau) + \frac{o(\Delta t)}{\Delta t}$$

Hence, in the limit as $\Delta t \to 0$ we obtain

$$\frac{\partial P(E, t; \xi, \tau)}{\partial t} = - \int_{E} q(\omega, t) \, d_{\Omega} P(\Omega, t; \xi, \tau)$$

$$+ \int_{\mathfrak{x}} Q(E; \omega, t) q(\omega, t) \, d_{\Omega} P(\Omega, t; \xi, \tau) \quad (2.19)$$

If we now put $s = \tau + \Delta \tau$ in (2.9) and use (2.16), we can show, as in the above case, that the first partial derivative of $P(E, t; \xi, \tau)$ with respect to τ exists, and therefore obtain an equation similar to (2.19). We have

$$\frac{\partial P(E, t; \xi, \tau)}{\partial \tau} = q(\xi, \tau) P(E, t, \xi, \tau) - q(\xi, \tau) \int_{\mathfrak{x}} P(E, t; \omega, \tau) \, d_{\Omega} Q(\Omega, \xi; \tau) \quad (2.20)$$

The above integrodifferential equations are the *Feller equations describing discontinuous Markov processes*. The first equation, (2.19), is called the *forward equation*, since it involves differentiation with respect to the later time t, while the second equation, (2.20), is called the *backward equation*, since it involves differentiation with respect to the earlier time τ.

2.3 Infinite Systems of Stochastic Differential Equations

A. The Kolgomorov Differential Equations. We now turn from the general form of Eqs. (2.19) and (2.20) and consider a specialized form which is of interest in many applications. We consider the case in which the random variable $X(t)$ can only assume a denumerable number of values, which we will denote by the nonnegative integers $x = 0, 1, 2, \ldots$. In this case the process $\{X(t), t \geqslant 0\}$ is said to be a *Markov chain with continuous time parameter*. For these integer-valued processes the transition probability set function $P(E,t;\xi,\tau)$ defining the Markov process is reduced to a point function $P(j,t;i,\tau)$, which we will write as $P_{ij}(\tau,t)$. The subscripts i and j are referred to as the *initial* and *final* states, respectively. Similarly, we rewrite $Q(E;\xi,t)$ as $Q_{ij}(t)$. This reduction from set to point functions reduces the Lebesgue-Stieltjes integrals, Eq. (2.6), to infinite sums (or finite sums if the number of states is finite). The Feller integrodifferential equations, (2.19) and (2.20), reduce to the system of differential equations

$$\frac{\partial P_{ij}(\tau,t)}{\partial t} = -q_j(t)P_{ij}(\tau,t) + \sum_{k=0}^{\infty} q_k(t)Q_{kj}(t)P_{ik}(\tau,t) \qquad (2.21)$$

and

$$\frac{\partial P_{ij}(\tau,t)}{\partial \tau} = q_i(\tau)\left\{ P_{ij}(\tau,t) - \sum_{k=0}^{\infty} Q_{ik}(\tau)P_{kj}(\tau,t) \right\} \qquad (2.22)$$

$i,j = 0, 1, \ldots$. The first system of equations is called the *forward system of differential equations*, and the second system is called the *backward system of differential equations*. These two infinite systems of differential equations were first derived by Kolmogorov [56] in a fundamental paper in which he laid the foundations of the theory of Markov processes. Hence, the above equations are called the *Kolmogorov differential equations*. It is of interest to note that the Kolmogorov equations form a system of *adjoint* equations.

In place of conditions (2.7) and (2.8) we have

$$0 \leqslant P_{ij}(\tau,t) \leqslant 1 \qquad \text{for all } i,j, \text{ and } \tau,t \qquad (2.23)$$

$$\sum_{j=0}^{\infty} P_{ij}(\tau,t) \leqslant 1 \qquad \text{for all } i,\tau, \text{ and } t \qquad (2.24)$$

The Chapman-Kolmogorov equation is now written as

$$P_{ij}(\tau,t) = \sum_{k=0}^{\infty} P_{ik}(\tau,s)P_{kj}(s,t) \qquad (2.25)$$

for all $t \geqslant \tau$, and $s \in [\tau, t]$. The initial conditions, for both systems of equations, are

$$P_{ij}(t,t) = \delta_{ij} = 1 \qquad \text{for } i = j$$
$$= 0 \qquad \text{for } i \neq j \qquad (2.26)$$

Finally, in place of the regularity condition (2.16), we now have

$$P_{ij}(t, t + \Delta t) = [1 - q_i(t) \, \Delta t]\delta_{ij} + q_i(t)Q_{ij}(t) \, \Delta t + o(\Delta t) \qquad (2.27)$$

As given above, Eqs. (2.21) and (2.22) are not time-homogeneous, since they depend explicitly on both τ and t. We now restrict our attention to the *time-homogeneous* or *stationary* case. Hence, we assume

$$P_{ij}(\tau, t) = P_{ij}(t - \tau) \qquad (2.28)$$

that is, the transition probabilities depend only on the duration of the time interval, and not on the initial time. Hence, in this case, we can put $\tau = 0$, and restrict our attention to the transition probabilities $P_{ij}(t)$.

From Sec. 2.2 we have that $q_i(t) \, \Delta t + o(\Delta t)$ is the probability that $X(t)$ will undergo a random change in the interval $(t, t + \Delta t)$ when $X(t) = i$. Similarly, $Q_{ij}(t)$ is the conditional probability that $X(t + \Delta t) = j$, given that $X(t) = i$ and a random change has taken place in the interval $(t, t + \Delta t)$.

In order to make the above more precise, we assume that to every state $i \in \mathfrak{X}$ there corresponds a continuous function $q_i(t) \geqslant 0$ such that

$$\lim_{\Delta t \to 0} \frac{1 - P_{ii}(t, t + \Delta t)}{\Delta t} = q_i(t) \qquad (2.29)$$

Analytically, (2.29) requires that $\lim_{\tau \to t} P_{ii}(t,\tau) = 1$ and that $P_{ii}(t,\tau)$ has at $\tau = t$ a derivative with respect to τ. The states of a Markov process can be classified on the basis of the intensity functions $q_i(t)$, the state i being called *stable* if for every fixed t $q_i(t) < \infty$, *instantaneous* if $q_i(t) = \infty$, or *final* if $q_i(t) = 0$.[1]

Similarly, we assume that for every pair of states (i,j), with $i \neq j$, there exist relative transition probabilities $Q_{ij}(t)$, continuous in t, such that

$$\lim_{\Delta t \to 0} \frac{P_{ij}(t, t + \Delta t)}{\Delta t} = q_i(t)Q_{ij}(t) \qquad (2.30)$$

and for every fixed t and initial state i

$$\sum_{j=0}^{\infty} Q_{ij}(t) = 1$$

with $Q_{ii}(t) = 0$.

[1] Cf. Chung [17] and Lévy [62].

Let us now consider the Kolmogorov equations in matrix form. Put

$$q_i(t) = -a_{ii}(t)$$
$$q_i(t)Q_{ij}(t) = a_{ij}(t) \qquad i \neq j \tag{2.31}$$

and let $A(t) = (a_{ij}(t))$. The elements $a_{ij}(t)$ are continuous functions of time with

$$a_{ij}(t) \geqslant 0 \qquad i \neq j \tag{2.32}$$
$$a_{ii}(t) \leqslant 0 \tag{2.33}$$

and
$$\sum_{j=0}^{\infty} a_{ij}(t) \leqslant 0 \qquad \text{for all } i \tag{2.34}$$

In view of the relationship between the $a_{ij}(t)$ and the $q_i(t)$ and $Q_{ij}(t)$, we call $a_{ij}(t) \, \Delta t + o(\Delta t)$ the *infinitesimal transition probabilities* of the Markov process $\{X(t), t \geqslant 0\}$.[1]

If we now put $P(t) = (P_{ij}(t))$, the Kolmogorov equations can be written in the form

$$\frac{dP(t)}{dt} = P(t)A(t) \tag{2.35}$$

$$\frac{dP(t)}{dt} = A(t)P(t) \tag{2.36}$$

with
$$P(0) = I \qquad \text{(the identity matrix)} \tag{2.37}$$

The Chapman-Kolmogorov equation in matrix form is

$$P(\tau + t) = P(\tau)P(t) \qquad \tau, t \geqslant 0 \tag{2.38}$$

We now assume that the elements $a_{ij}(t)$ are independent of time; hence, $A(t) = A$, a constant matrix. The above equations become

$$\frac{dP(t)}{dt} = P(t)A \tag{2.39}$$

$$\frac{dP(t)}{dt} = AP(t) \tag{2.40}$$

---

[1] Throughout this chapter we shall study the transition probabilities from the point of view of classical analysis, i.e., we shall be concerned with the study of the Kolmogorov equations the solutions of which give the transition probabilities $P_{ij}(t)$. Another approach to the study of the transition probabilities is based on the properties of the *sample functions* or *realizations* of the Markov process. For a Markov process $\{X(t), t \geqslant 0\}$ and a given state space $\mathfrak{X}$ we denote by Ω the collection of functions $\omega(t)$ taking values in $\mathfrak{X}$. For discontinuous processes the functions $\omega(t)$, as functions of time, are integer-valued step functions, with the discontinuities of the sample functions $\omega(t)$ representing the succession of transitions of the process.

For a rigorous discussion of the sample-function approach we refer to the treatise of Doob and [17,22,62]. Also, for a discussion of the analytical properties of the transition probabilities $P_{ij}(t)$ we refer to [4,5,18,52,57,88].

Let us now consider the backward equation (2.40). It is well known that the scalar differential equation

$$\frac{dz(t)}{dt} = az(t)$$

has the solution $z(t) = z(0)e^{at}$. This suggests that the solution of (2.40) can be written in the form

$$P(t) = P(0)e^{At} = e^{At} \tag{2.41}$$

If we introduce the matrix series

$$e^{At} = \sum_{n=0}^{\infty} \frac{A^n t^n}{n!} \qquad A^0 = I$$

then
$$P(t) = \sum_{n=0}^{\infty} \frac{A^n t^n}{n!} \tag{2.42}$$

In the case when A is a finite matrix (i.e., the Markov chain has a finite number of states) the above solution is valid. However, in the infinite case the above series may not converge, especially when the matrix elements a_{ij} are unbounded. In either case, it is known from the theory of semigroups that $P(t)$ admits an exponential representation.[1]

B. Existence and Uniqueness of Solutions of the Kolmogorov Differential Equations. In this section we consider the existence and uniqueness theory for the Kolmogorov equations. While several workers, utilizing different methods, have contributed to this theory, we shall follow Reuter and Ledermann [74], since their methods permit an elementary derivation of Feller's important results.[2] Before developing the existence and uniqueness theory, we present a brief summary of Feller's results:

1. Given a set of coefficients $a_{ij}(t)$ satisfying (2.32) to (2.34), there exists a set of functions $P_{ij}(\tau,t) = F_{ij}(\tau,t)$ that satisfy (2.21) to (2.25).

2. The inequality (2.24) cannot, in general, be replaced by equality. Necessary and sufficient conditions are given such that equality holds for the solution $P_{ij}(\tau,t) = F_{ij}(\tau,t)$.

3. The solutions $F_{ij}(\tau,t)$ are minimal solutions of (2.21) and (2.22), i.e., any other solution $P_{ij}(\tau,t) = G_{ij}(\tau,t)$ satisfies $G_{ij}(\tau,t) \geqslant F_{ij}(\tau,t)$.

4. The solution of (2.21) and (2.22) is unique if equality in (2.24) obtains for $P_{ij}(\tau,t) = F_{ij}(\tau,t)$.

The method employed by Reuter and Ledermann is based on the

---

[1] Cf. Hille and Phillips [37].
[2] Cf. Feller [25].

construction of an infinite matrix $F(t) = (F_{ij}(t))$ which satisfies the Kolmogorov equations.

$$\frac{dF(t)}{dt} = F(t)A \tag{2.43}$$

$$\frac{dF(t)}{dt} = AF(t) \tag{2.44}$$

with $F(0) = I$. The method of construction[1] utilizes the nth section of A, that is, an $n \times n$ matrix $A^{(n)}$, to find a solution $F^{(n)}(t)$ of the *finite* Kolgomorov equations, showing that

$$\lim_{n \to \infty} F^{(n)}t = F(t)$$

satisfies (2.43) and (2.44). For the matrix $A^{(n)} = (a_{ij}^{(n)})$ the above system becomes

$$\frac{dF^{(n)}(t)}{dt} = F^{(n)}(t)A^{(n)} \tag{2.45}$$

$$\frac{dF^{(n)}(t)}{dt} = A^{(n)}F^{(n)}(t) \tag{2.46}$$

with $F^{(n)}(0) = I^{(n)}$.

Before stating and proving the existence theorem, we state three lemmas that will be used in the sequel.

Lemma 2.1: The solution $F^{(n)}(t) = e^{A^{(n)}(t)}$ has the following properties:

$$\frac{dF^{(n)}(t)}{dt} = F^{(n)}(t)A^{(n)} \tag{2.45}$$

$$\frac{dF^{(n)}(t)}{dt} = A^{(n)}F^{(n)}(t) \tag{2.46}$$

$$F^{(n)}(0) = I^{(n)} \tag{2.47}$$

$$F^{(n)}(\tau + t) = F^{(n)}(\tau)F^{(n)}(t) \tag{2.48}$$

Also, the matrix elements $a_{ij}^{(n)}$ satisfy

$$a_{ij}^{(n)} \geqslant 0 \quad \text{for } i \neq j \qquad a_{ii}^{(n)} \leqslant 0 \tag{2.49}$$

$$\sum_{j=i}^{n} a_{ij}^{(n)} = \sum_{j=1}^{\infty} a_{ij} - \sum_{j=n+1}^{\infty} a_{ij} \leqslant 0 \tag{2.50}$$

For a statement and discussion of these properties we refer to Bellman [7] and Bourbaki [15].

---

[1] This method is also applicable to the case where A is a function of time.

Lemma 2.2: The solutions $F_{ij}^{(n)}(t)$ are nonnegative for all $t \geqslant 0$.

Proof: Let

$$\min_{1 \leqslant i \leqslant n} a_{ii} = \mu \leqslant 0$$

and introduce the substitution

$$Y_{ij}(t) = e^{-\mu t} F_{ij}^{(n)}(t)$$

Then $Y(t) = Y_{ij}(t)$ satisfies

$$\frac{dY(t)}{dt} = Y(t)B \qquad Y(0) = I^{(n)} \tag{2.51}$$

where $b_{ij} = a_{ij} - \mu \delta_{ij} \geqslant 0$. Since B is a finite matrix, (2.51) has the solution

$$Y(t) = e^{Bt} = \sum_{n=0}^{\infty} \frac{B^n t^n}{n!}$$

Hence the elements $Y_{ij}(t) \geqslant 0$, and $F_{ij}(t) \geqslant 0$.

Lemma 2.3:

$$\sum_{j=1}^{n} F_{ij}^{(n)}(t) \leqslant 1 \qquad \text{for all } i \text{ and all } t \geqslant 0$$

Proof: Consider the equation

$$\frac{dF_{ij}^{(n)}(t)}{dt} = \sum_{k=1}^{n} F_{ik}^{(n)}(t) a_{kj}^{(n)}$$

On summing both sides over j, we have

$$\frac{d}{dt}\left(\sum_{j=1}^{n} F_{ij}^{(n)}(t) \right) = \sum_{j=1}^{n} \sum_{k=1}^{n} F_{ik}^{(n)} a_{kj}^{(n)}$$

$$= \sum_{k=1}^{n} \left(\sum_{j=1}^{n} a_{kj}^{(n)} \right) F_{ik}^{(n)}$$

Now, from Lemma 2.2 $F_{ik}^{(n)} \geqslant 0$, and from (2.50) $\sum_{j=1}^{n} a_{ij}^{(n)} \leqslant 0$; hence, the right-hand side is less than or equal to zero. Therefore, $\sum_{j=1}^{n} F_{ij}^{(n)}(t)$ is a decreasing function of t, and the statement of the lemma follows from the fact that

$$\sum_{j=1}^{n} F_{ij}^{(n)}(0) = \sum_{j=1}^{n} \delta_{ij} = 1$$

We now state and prove the following theorem.

Theorem 2.1 (*Existence Theorem*):

$$\lim_{n \to \infty} F_{ij}^{(n)}(t) = F_{ij}(t)$$

where $F_{ij}(t)$ is a continuous function of t with a continuous first derivative, and

$$\frac{dF_{ij}(t)}{dt} = \sum_{k=1}^{\infty} F_{ik}(t) a_{kj} \tag{2.52}$$

$$\frac{dF_{ij}(t)}{dt} = \sum_{k=1}^{\infty} a_{ik} F_{kj}(t) \tag{2.53}$$

$$F_{ij}(t) \geqslant 0 \tag{2.54}$$

$$F_{ij}(0) = \delta_{ij} \tag{2.55}$$

$$\sum_{j=1}^{\infty} F_{ij}(t) \leqslant 1 \tag{2.56}$$

$$F_{ij}(\tau + t) = \sum_{k=1}^{\infty} F_{ik}(\tau) F_{kj}(t) \tag{2.57}$$

Proof: To show $\lim_{n \to \infty} F_{ij}^{(n)}(t) = F_{ij}(t)$, we make use of the following lemma.

Lemma 2.4: For all $t \geqslant 0$

$$F_{ij}^{(n+1)}(t) \geqslant F_{ij}^{(n)}(t) \tag{2.58}$$

Proof: We consider (2.45) with n replaced by $n + 1$; hence,

$$\frac{dF_{ij}^{(n+1)}(t)}{dt} = \sum_{k=1}^{n+1} F_{ik}^{(n+1)}(t) a_{kj}$$

$$= \sum_{k=1}^{n} F_{ik}^{(n+1)}(t) a_{kj}^{(n)} + F_{i,n+1}^{(n+1)}(t) a_{n+1,j} \tag{2.59}$$

Let us now denote by $Z(t)$ the nth section of $F^{(n+1)}(t)$, i.e., $Z_{ij}(t) = F_{ij}^{n+1}(t)$, for $i, j, = 1, 2, \ldots, n$, and let

$$C_{ij}(t) = F_{i,n+1}^{(n+1)}(t) a_{n+1,j}$$

We can now write (2.59) (in matrix form) as

$$\frac{dZ(t)}{dt} = Z(t) A^{(n)} + C(t) \tag{2.60}$$

The solution[1] of (2.60), since $Z(0) = I^n$, is given by

$$Z(t) = F^{(n)}(t) + \int_0^t F^{(n)}(t - \tau) C(\tau) \, d\tau \tag{2.61}$$

[1] Cf. Bellman [7].

By Lemma 2.2 the elements of $F^{(n)}(t - \tau)$ are nonnegative for $\tau \in [0,t]$. The elements of $C(\tau)$ are also nonnegative, since $C_{ij}(t) = F_{i,n+1}^{(n+1)}(t)a_{n+1,j}$. Hence, from (2.61) $Z_{ij}(t) = F_{ij}^{(n+1)}(t) \geqslant F_{ji}^{(n)}(t)$.

Now, the sequence $F_{ij}^{(n)}(t)$ is monotone by the above lemma. Its boundedness is implied by Lemmas 2.2 and 2.3, since they imply $0 \leqslant F_{ji}^{(n)}(t) \leqslant 1$ for all t. From the above it follows that $\lim\limits_{n\to\infty} F_{ij}^{(n)}(t) = F_{ij}(t)$ exists, and $0 \leqslant F_{ij}(t) \leqslant 1$. In addition, it is clear that $F_{ij}(0) = \delta_{ij}$, so that $F_{ij}(t)$ satisfies (2.52) and (2.55). In order to prove the remaining properties, i.e., (2.52), (2.53), (2.56), and (2.57), we must justify the passage to the limit in (2.45), (2.46), (2.48), and the result of Lemma 2.3. The following lemma, which we state without proof, will be used to justify the passage to the limit.

Lemma 2.5: For i fixed, let $\{c_i(n)\}$ be an increasing sequence with $\lim\limits_{n\to\infty} c_i(n) = c_i$; then

$$\sum_{i=1}^{\infty} c_i = \lim_{n\to\infty} \left(\sum_{i=1}^{\infty} c_i(n) \right)$$

If we now assume that $F_{ij}^{(n)}(t) = 0$ for $i > n$ or $j > n$, we can write

$$\sum_{j=1}^{\infty} F_{ij}^{(n)}t = 1 \tag{2.62}$$

$$F_{ij}^{(n)}(\tau + t) = \sum_{k=1}^{\infty} F_{ik}(\tau)F_{kj}(t) \tag{2.63}$$

Because of Lemmas 2.2 and 2.4, we can use Lemma 2.5 to deduce (2.56) and (2.57) from (2.62) and (2.63), respectively.

We now consider Eqs. (2.52) and (2.53). Now all terms on the right in these equations are positive except for the term with $k = i$ arising from the diagonal terms of A. This suggests separating the diagonal terms from the rest. Hence we write

$$\alpha_{ij} = a_{ij} \qquad \text{for } i \neq j$$

$$= 0 \qquad \text{otherwise}$$

In view of the above we can rewrite Eq. (2.46) as

$$\frac{dF_{ij}^{(n)}(t)}{dt} = \sum_{k=1}^{\infty} a_{ik}F_{kj}^{(n)}(t)$$

$$= a_{ii}F_{ij}^{(n)}(t) + \sum_{k=1}^{\infty} \alpha_{ik}F_{kj}^{(n)}(t) \qquad n \geqslant i,\, n \geqslant j \tag{2.64}$$

On integrating the above equation from 0 to t, we have

$$F_{ij}^{(n)}(t) = \delta_{ij} + a_{ii}\int_0^t F_{ij}^{(n)}(\tau)\,d\tau + \int_0^t \left[\sum_{k=1}^\infty \alpha_{ik}F_{kj}^{(n)}(\tau) \right] d\tau \qquad (2.65)$$

By Lemmas 2.2, 2.3, and 2.5,

$$F_{ij}^{(n)}(\tau) \to F_{ij}(\tau) \qquad \text{and} \qquad \sum_{k=1}^\infty \alpha_{ik}F_{kj}^{(n)}(\tau) \to \sum_{k=1}^\infty \alpha_{ik}F_{kj}(\tau)$$

By the convergence theorem for monotone sequences,[1] we may pass to the limit under the integral signs in (2.65) to obtain

$$F_{ij}(t) = \delta_{ij} + a_{ii}\int_0^t F_{ij}(\tau)\,d\tau + \int_0^t \left(\sum_{k=1}^\infty \alpha_{ik}F_{kj}(\tau) \right) d\tau$$

Now the second integral is bounded, since $\sum_{k=1}^\infty \alpha_{ik} < \infty$ and $0 \leqslant F_{ij}(\tau) \leqslant 1$. This equation shows that $F_{ij}(t)$ is continuous for all $t \geqslant 0$; hence, we can rewrite it as

$$F_{ij}(t) = \delta_{ij} + \int_0^t \left(\sum_{k=1}^\infty a_{ik}F_{kj}(\tau) \right) d\tau \qquad (2.66)$$

Since $\sum_{k=1}^\infty |a_{ik}| < \infty$ and $0 \leqslant F_{kj}(\tau) \leqslant 1$, the series in (2.66) is uniformly convergent for all $\tau \geqslant 0$. Hence $\sum_{k=1}^\infty a_{ik}F_{kj}(\tau)$ is continuous for all $\tau \geqslant 0$. Differentiation of (2.66) proves (2.53) and also shows that $F_{ij}(t)$ is continuous for all $t \geqslant 0$.

If we now compare the right-hand members of (2.52) and (2.53), we obtain

$$\sum_{k=1}^\infty F_{ik}^{(n)}(t)a_{kj} = \sum_{k=1}^\infty a_{ik}F_{kj}^{(n)}(t)$$

which, in view of (2.64), can be written

$$F_{ij}^{(n)}(t)a_{jj} + \sum_{k=1}^\infty F_{ik}^{(n)}(t)\alpha_{kj} = a_{ii}F_{ij}^{(n)}(t) + \sum_{k=1}^\infty \alpha_{ik}F_{kj}^{(n)}(t) \qquad (2.67)$$

By Lemma 2.5, (2.67) becomes

$$F_{ij}(t)a_{jj} + \sum_{k=1}^\infty F_{ik}(t)\alpha_{kj} = a_{ii}F_{ij}(t) + \sum_{k=1}^\infty \alpha_{ik}F_{kj}(t)$$

hence

$$\sum_{k=1}^\infty F_{ik}(t)a_{kj} = \sum_{k=1}^\infty \alpha_{ik}F_{kj}(t) \qquad (2.68)$$

Now (2.68) shows that (2.52) follows from (2.53), but the latter has been proved; so that (2.52) is proved also. This completes the proof of Theorem 2.1.

[1] Cf. Halmos [32].

Having established the existence of solutions of the Kolmogorov equations, we now consider the minimal property and uniqueness of these solutions.

Theorem 2.2 (*Minimal Property*): 1. If the row vector $(P_{i1}(t), P_{i2}(t), \ldots)$ for fixed i satisfies

$$\frac{dP_{ij}(t)}{dt} = \sum_{k=1}^{\infty} P_{ik}(t)a_{kj} \tag{2.69}$$

and

$$\frac{dP_{ij}(t)}{dt} = \sum_{k=1}^{\infty} a_{ik}P_{kj}(t) \tag{2.70}$$

with $P_{ij}(0) = \delta_{ij}$, then

$$P_{ij}(t) \geqslant F_{ij}(t) \tag{2.71}$$

2. If the column vector $(P_{1j}(t), P_{2j}(t), \ldots)$, for j fixed, satisfies (2.69) and (2.70), then $P_{ij}(t) \geqslant F_{ij}(t)$.

Proof: We only consider the proof of part 1, since the proof for part 2 follows similar lines. Let $P_{ij}(t) = Z_j(t)$ for i fixed. Then (2.69) becomes

$$\frac{dZ_j(t)}{dt} = \sum_{k=1}^{\infty} Z_k(t)a_{kj} \qquad Z_j(0) = \delta_{ij} \tag{2.72}$$

Now, denote by $Z^{(n)}(t)$ the row vector $(Z_1(t), \ldots, Z_2(t), \ldots, Z_n(t))$. Then (2.72) can be written as

$$\frac{dZ^{(n)}(t)}{dt} = \sum_{k=1}^{n} Z_k^{(n)}(t)a_{kj} + \sum_{k=n+1}^{\infty} Z_k(t)a_{kj}$$

$$= \sum_{k=1}^{n} Z_k^{(n)}(t)a_{kj}^{(n)} + B_j(t) \tag{2.73}$$

where $B_j(t) \geqslant 0$. In vector form the above becomes

$$\frac{dZ^{(n)}(t)}{dt} = Z^{(n)}(t)A^{(n)} + B(t) \tag{2.74}$$

We then have[1]

$$Z^{(n)}(t) = Z^{(n)}(0)F^{(n)}(t) + \int_0^t F^{(n)}(t - \tau)B(\tau)\,d\tau \tag{2.75}$$

or $$Z_j(t) = F_{ij}^{(n)}(t) + \int_0^t \left(\sum_{k=1}^{n} B_k(\tau)F_{kj}^{(n)}(t - \tau) \right) d\tau \geqslant F_{ij}^{(n)}(t) \tag{2.76}$$

for $j = 1, 2, \ldots, n$ and $n \geqslant i$. On passing to the limit, we obtain

$$Z_j(t) = P_{ij}(t) \geqslant F_{ij}(t)$$

[1] Cf. Bellman, *loc. cit.*

The use of Theorem 2.2 enables us to obtain sufficient conditions for uniqueness. We now consider the following:

Theorem 2.3 (*Uniqueness Theorem*): 1. If

$$\sum_{j=1}^{\infty} F_{ij}(t) = 1 \tag{2.77}$$

for some fixed i and if $P_{ij}(t)$ satisfies (2.69), $P_{ij}(t) \geqslant 0$ and

$$\sum_{j=1}^{\infty} P_{ij}(t) \leqslant 1 \tag{2.78}$$

then $P_{ij}(t) = F_{ij}(t)$.

2. If (2.77) holds for all i and $P_{ij}(t)$ satisfies (2.70), $P_{ij}(t) \geqslant 0$, and (2.78) for all i, then $P_{ij}(t) = F_{ij}(t)$.[1]

Proof: Again we consider only part 1. By Theorem 2.2 (part 1) we have $P_{ij}(t) \geqslant F_{ij}(t)$ for i fixed and $j = 1, 2, \ldots$. This result, together with (2.77) and (2.78), implies $P_{ij}(t) = F_{ij}(t)$.

It is of importance to note that the forward Kolmogorov equations may have a unique solution even if (2.77) does not hold. However, (2.77) is necessary in order that the backward Kolmogorov equations have a unique solution.

When condition (2.77) is not satisfied, we have the following theorem.

Theorem 2.4: If $\sum_{j=1}^{\infty} F_{ij}(t) < 1$ for some $i = i_0$ and $t = t_0 > 0$, then there exist solutions $P_{ij}(t)$ that satisfy the backward Kolmogorov equations, are different from $F_{ij}(t)$, are nonnegative, and are such that $\sum_{j=1}^{\infty} P_{ij}(t) \leqslant 1$. For some of these solutions the above condition can be replaced by equality, i.e., $\sum_{j=1}^{\infty} P_{ij}(t) = 1$.

Proof: Let $G_i(t) = \sum_{j=1}^{\infty} F_{ij}(t)$. For j fixed, $F_{ij}(t)$ satisfies the backward equations; hence, $G_i(t)$ satisfies

$$\frac{dG_i(t)}{dt} = \sum_{k=1}^{\infty} a_{ik} G_k(t) \tag{2.79}$$

with $G_i(0) = \sum_{j=1}^{\infty} F_{ij}(0) = \sum_{j=1}^{\infty} \delta_{ij} = 1$, $0 \leqslant G_i(t) \leqslant 1$, and $G_i(t_0) < 1$.

---

[1] A process satisfying the usual conditions for Markov processes, but with $\sum_{j=1}^{\infty} a_{ij} \leqslant 0$ and $\sum_{j=1}^{\infty} P_{ij}(t) \leqslant 1$, has been called a *quasi-process* (or *dishonest process*); cf. Jensen [38].

If we now put $H_i(t) = 1 - \sum_{j=1}^{\infty} F_{ij}(t) = 1 - G_i(t)$, then $H_i(t)$ satisfies

$$\frac{dH_i(t)}{dt} = - \sum_{k=1}^{\infty} a_{ik}(1 - H_k(t)) = \sum_{k=1}^{\infty} a_{ik}H_k(t) \qquad (2.80)$$

since $\sum_{k=1}^{\infty} a_{ik} = 0$.   Also, $H_i(0) = 1 - G_i(0) = 0$, $0 \leqslant H_i(t) \leqslant 1$, $H_{i_0}(t_0) > 0$.

Now let C_j be a set of nonnegative constants such that $0 < \sum_{j=1}^{\infty} C_j \leqslant 1$, and put

$$P_{ij}(t) = F_{ij}(t) + C_j H_i(t) \qquad (2.81)$$

Then the $P_{ij}(t)$ satisfy

$$\frac{dP_{ij}(t)}{dt} = \sum_{k=1}^{\infty} a_{ik}F_{kj}(t) + C_j \sum_{k=1}^{\infty} a_{ik}H_k(t)$$

$$= \sum_{k=1}^{\infty} a_{ik}P_{kj}(t) \qquad (2.82)$$

with $P_{ij}(0) = \delta_{ij}$ and $P_{ij}(t) \geqslant 0$. Also

$$\sum_{j=1}^{\infty} P_{ij}(t) = \sum_{j=1}^{\infty} F_{ij}(t) + \left(\sum_{j=1}^{\infty} C_j \right) H_i(t)$$

$$= 1 = \left(1 - \sum_{j=1}^{\infty} C_j \right) H_i(t) \leqslant 1$$

Hence, $P_{ij}(t)$ satisfies the backward equations with $P_{ij}(t) \geqslant 0$, and $\sum_{j=1}^{\infty} P_{ij}(t) \leqslant 1$. In addition, $P_{i_0 j}(t_0) \neq F_{i_0 j}(t_0)$ for any j with $C_j > 0$, so that $P(t)$ is different from $F(t)$. In conclusion we note that, by selecting the C_j such that $\sum_{j=1}^{\infty} C_j = 1$, we have $\sum_{j=1}^{\infty} P_{ij}(t) = 1$ for all i.

For additional contributions to the existence and uniqueness theory of the Kolmogorov equations, as well as the analysis of certain pathological cases, we refer to Doob [22], Feller [27], Hille [36], Hille and Phillips [37], Kato [48], Kendall [53], Kendall and Reuter [54], and Reuter [73].

2.4   Some Discontinuous Markov Processes and Their Properties

A.   Introduction. In this section we consider several discontinuous Markov processes; in particular, we consider the Poisson, simple birth, Pólya, simple death, and simple birth-and-death processes. In each case we derive the differential-difference equation describing the probability law of the process and obtain the solution of the equation and discuss its properties. For the birth and birth-and-death processes the problems of uniqueness are considered. In some cases we use several methods to obtain a solution to the differential-difference equation,[1] our main purpose being to illustrate the various methods that are available for treating stochastic differential-difference equations.

B. The Poisson Process. The Poisson process is the simplest of the discontinuous Markov processes. This process occupies a unique position in the theory of probability and has found many applications in fields such as biology, physics, and telephone engineering. We shall first consider the derivation of the differential equation for the probability distribution associated with the Poisson process. Later, we consider the Kolmogorov equations describing the Poisson process.

In order to derive the differential equation, it is necessary to state the assumptions which specify the manner in which the Poisson process develops.

Assumptions for the Poisson Process. (1) The probability of a change in the interval $(t, t + \Delta t)$ is $\lambda \Delta t + o(\Delta t)$, where λ is a positive constant. (2) The probability of more than one change in the interval $(t, t + \Delta t)$ is $o(\Delta t)$. (3) The probability of no change in $(t, t + \Delta t)$ is $1 - \lambda \Delta t + o(\Delta t)$. The above probabilities are independent of the state of the system.

Let
$$P_x(t) = \mathscr{P}\{X(t) = x\} \qquad x = 0, 1, 2, \dots$$

In view of the above assumptions, we are led to the following relation for $P_x(t + \Delta t)$:

$$P_x(t + \Delta t) = (1 - \lambda \Delta t)P_x(t) + \lambda P_{x-1}(t) \Delta t + o(\Delta t) \qquad (2.83)$$

This relation states that the only ways, in view of the assumptions, the system can be in the state $x \geq 0$ at time $t + \Delta t$ are that the system at time t was in x and did not change in the interval $(t, t + \Delta t)$ or the system was in state $x - 1$ at time t and experienced a positive unit change in the interval $(t, t + \Delta t)$. The probability of no change is $1 - \lambda \Delta t + o(\Delta t)$; this gives rise to the first term on the right-hand side of (2.83). Similarly, the probability of a unit change in $(t, t + \Delta t)$ is $\lambda \Delta t + o(\Delta t)$. This gives rise to the second term.

If we transpose the term $P_x(t)$ on the right-hand side, divide by Δt, and let $\Delta t \to 0$, we obtain the differential equation

$$\frac{dP_x(t)}{dt} = -\lambda P_x(t) + \lambda P_{x-1}(t) \qquad x \geq 1 \qquad (2.84)$$

[1] For other methods of solving differential-difference equations we refer to Pinney [68].

When $x = 0$, $P_{x-1}(t) = 0$; hence we have

$$\frac{dP_0(t)}{dt} = -\lambda P_0(t) \qquad (2.85)$$

These two equations characterize the Poisson process; they are to be solved with the initial condition

$$P_x(0) = 1 \qquad \text{for } x = 0$$
$$= 0 \qquad \text{for } x = 1, 2, \ldots \qquad (2.86)$$

From (2.85) and (2.86) we have

$$P_0(t) = e^{-\lambda t} \qquad (2.87)$$

Having obtained $P_0(t)$, we can now use (2.84) and write

$$\frac{dP_1(t)}{dt} = -\lambda P_1(t) + e^{-\lambda t}$$

hence $$P_1(t) = \lambda t e^{-\lambda t}$$

By continuing in this way, we have

$$P_x(t) = e^{-\lambda t}\left[\lambda \int_0^t P_{x-1}(\tau)e^{\lambda \tau}\, d\tau + C_x\right] \qquad x = 1, 2, \ldots \qquad (2.88)$$

From (2.86) we have $C_x = 0$ for $x = 1, 2, \ldots$. From (2.88) we obtain by induction

$$P_x(t) = \frac{(\lambda t)^x}{x!} e^{-\lambda t} \qquad (2.89)$$

The above expression, which is called the *Poisson distribution*, gives the probability that at time $t \geqslant 0$ the system is in state x ($x = 0, 1, 2, \ldots$). Another interpretation is that $P_x(t)$ is the probability of x changes during a time interval of length t.

Before considering the properties of the Poisson process, we will demonstrate the application of the method of the Laplace transformation[1] to the system of equations describing the Poisson process. Let $P_x(s) = \mathscr{L}\{P_x(t)\}$ denote the Laplace transform of $P_x(t)$. Application of the Laplace transformation to (2.85) yields

$$sP_0(s) - 1 = -\lambda P_0(s) \qquad \text{or} \qquad P_0(s) = \frac{1}{s + \lambda} \qquad (2.90)$$

Inversion gives

$$P_0(t) = e^{-\lambda t}$$

---

[1] Cf. Appendix B.

Now, by applying the transformation to (2.84), we have

$$sP_x(s) = -\lambda P_x(s) + \lambda P_{x-1}(s) \qquad \text{or} \qquad P_x(s) = \frac{\lambda P_{x-1}(s)}{s + \lambda} \qquad (2.91)$$

From (2.90), we have

$$P_1(s) = \frac{\lambda}{(s + \lambda)^2} \qquad P_2(s) = \frac{\lambda^2}{(s + \lambda)^3}$$

and, in general,

$$P_x(s) = \frac{\lambda^x}{(s + \lambda)^{x+1}} \qquad (2.92)$$

By inverting, we obtain the Poisson distribution

$$P_x(t) = \frac{(\lambda t)^x}{x!} e^{-\lambda t}$$

To turn now to the properties of the Poisson process, we first show that $P_x(t)$ is an honest probability distribution, i.e., $\sum_{x=1}^{\infty} P_x(t) = 1$. We have

$$\sum_{x=0}^{\infty} P_x(t) = e^{-\lambda t} \sum_{x=0}^{\infty} \frac{(\lambda t)^x}{x!} = e^{-\lambda t} e^{\lambda t} = 1 \qquad (2.93)$$

We also note that $P_x(t)$, as a function of x, is monotone decreasing for $\lambda t < 1$. In addition, $P_x(t)$ has a maximum in the neighborhood of $x \sim \lambda t$ for $\lambda t > 1$.

Let us now consider the expression for the mean and variance of the Poisson process. By definition

$$m(t) = \mathscr{E}\{X(t)\} = \sum_{x=0}^{\infty} x P_x(t)$$

hence

$$m(t) = e^{-\lambda t} \sum_{x=0}^{\infty} x \frac{(\lambda t)^x}{x!} = \lambda t e^{-\lambda t} \sum_{x=1}^{\infty} \frac{(\lambda t)^{x-1}}{(x-1)!} = \lambda t e^{-\lambda t} e^{\lambda t} = \lambda t \qquad (2.94)$$

Hence, the mean number of events in the interval of length t is proportional to t, the constant of proportionality being λ. To obtain the variance, we first find the second moment. By definition

$$\sum_{x=0}^{\infty} x^2 P_x(t) = \lambda t e^{-\lambda t} \sum_{x=1}^{\infty} x \frac{(\lambda t)^{x-1}}{(x-1)!}$$

$$= \lambda t e^{-\lambda t} \sum_{x=1}^{\infty} [(x-1) + 1] \frac{(\lambda t)^{x-1}}{(x-1)!}$$

$$= \lambda t e^{-\lambda t} \left[\lambda t \sum_{x=2}^{\infty} \frac{(\lambda t)^{x-2}}{(x-2)!} + \sum_{x=1}^{\infty} \frac{(\lambda t)^{x-1}}{(x-1)!} \right]$$

$$= \lambda t e^{-\lambda t} (\lambda t e^{\lambda t} + e^{\lambda t})$$

$$= (\lambda t)^2 + \lambda t = m^2(t) + m(t)$$

Hence $\qquad\qquad \mathscr{D}^2\{X(t)\} = (\lambda t)^2 + \lambda t - (\lambda t)^2 = \lambda t \qquad\qquad (2.95)$

We find, then, that for the Poisson process the mean and variance are equal.

Thus far we have considered only the function $P_x(t)$, which has been interpreted as the probability that exactly x changes have taken place in an interval of length t. Let us now consider the transition probabilities $P_{ij}(t)$ associated with the Poisson process. From the theory of Sec. 2.2 and the assumptions for the Poisson process, we see that the intensities q_i and the relative transition probabilities Q_{ij} are given by

$$
\begin{aligned}
q_i &= \lambda && \text{for } i = 0, 1, \ldots \\
Q_{ij} &= 1 && \text{for } j = i + 1 \\
&= 0 && \text{otherwise}
\end{aligned}
\qquad (2.96)
$$

Hence the matrix of infinitesimal transition probabilities is given by

$$
A = \begin{bmatrix}
-\lambda & \lambda & 0 & 0 & 0 & 0 & \cdots & 0 & 0 \\
0 & -\lambda & \lambda & 0 & 0 & 0 & \cdots & 0 & 0 \\
0 & 0 & -\lambda & \lambda & 0 & 0 & \cdots & 0 & 0 \\
\cdot & \cdot & \cdot & \cdot & \cdot & \cdot & \cdots & \cdot & \cdot \\
\cdot & \cdot & \cdot & \cdot & \cdot & \cdot & \cdots & \cdot & \cdot \\
\cdot & \cdot & \cdot & \cdot & \cdot & \cdot & \cdots & \cdot & \cdot \\
\cdot & \cdot & \cdot & \cdot & \cdot & \cdot & \cdots & \cdot & \cdot \\
0 & 0 & 0 & 0 & 0 & 0 & \cdots & 0 & 0
\end{bmatrix}
$$

that is,

$$
\begin{aligned}
a_{ij} &= \lambda && \text{for } j = i + 1, i = 0, 1, \ldots \\
&= 0 && \text{otherwise} \\
a_{ii} &= -\lambda
\end{aligned}
\qquad (2.97)
$$

From the above we can immediately write the *Kolmogorov equations for the Poisson process*:

$$
\frac{dP_{ij}(t)}{dt} = -\lambda P_{ij}(t) + \lambda P_{i,j-1}(t) \qquad\qquad (2.98)
$$

$$
\frac{dP_{ij}(t)}{dt} = -\lambda P_{ij}(t) + \lambda P_{i+1,j}(t) \qquad\qquad (2.99)
$$

It is easy to verify that

$$
\begin{aligned}
P_{ij}(t) &= \frac{(\lambda t)^{j-i}}{(j-i)!}\, e^{-\lambda t} && \text{for } j \geqslant i \\
&= 0 && \text{for } j < i
\end{aligned}
\qquad (2.100)
$$

satisfies the above system of equations, with the initial condition $P_{ij}(0) = \delta_{ij}$.

For additional properties of Poisson processes, in particular non-homogenous and composed Poisson processes, we refer to the papers of Fisz and Urbanik [29], Florek et al. [30], Marczewski [63], Prékopa [69], Rényi [71,72], and Ryll-Nardzewski [75,76].

C. **The Pure Birth Process.** We now consider the pure birth process. This process is similar to the Poisson process; however, for the pure birth process the probability of a change in the interval $(t, t + \Delta t)$ is a function of the state that the system is in at time t. We have the following.

Assumptions for the Pure Birth Process. (1) If at time t the system is in state x ($x = 0, 1, 2, \ldots$), the probability of the transition $x \to x + 1$ in the interval $(t, t + \Delta t)$ is $\lambda_x \Delta t + o(\Delta t)$. (2) The probability of a transition from x to a state different from $x + 1$ is $o(\Delta t)$. (3) The probability of no change is $1 - \lambda_x \Delta t + o(\Delta t)$.

In view of the above assumptions we have the following relation:

$$P_x(t + \Delta t) = (1 - \lambda_x \Delta t)P_x(t) + \lambda_{x-1}P_{x-1}(t) \Delta t + o(\Delta t) \qquad (2.101)$$

By transposing the term $P_x(t)$, dividing by Δt, and passing to the limit, we obtain the system of differential equations

$$\frac{dP_0(t)}{dt} = -\lambda_0 P_0(t) \qquad (2.102)$$

$$\frac{dP_x(t)}{dt} = -\lambda_x P_x(t) + \lambda_{x-1}P_{x-1}(t) \qquad x = 1, 2, \ldots \qquad (2.103)$$

This system is to be solved with the initial condition

$$P_x(0) = \delta_{x,i} = 1 \qquad \text{for } x = i$$
$$= 0 \qquad \text{for } x \neq i \qquad (2.104)$$

We now consider in detail what is called a *simple birth process* or a *linear birth process.*[1] In this case we put

$$\lambda_x = \lambda x \qquad x > 1, \lambda > 0 \qquad (2.105)$$

i.e., the probability of a transition in $(t, t + \Delta t)$ is proportional to x, the constant of proportionality being λ. In this case (2.103) becomes

$$\frac{dP_x(t)}{dt} = -\lambda x P_x(t) + \lambda(x - 1)P_{x-1}(t) \qquad x \geqslant 1 \qquad (2.106)$$

---

[1] Nonlinear birth processes have been considered by John [42].

We note that when $x = 1$,

$$\frac{dP_1(t)}{dt} = -\lambda P_1(t)$$

hence
$$P_1(t) = C_1 e^{-\lambda t}$$

As in the case of the Poisson process, we can now write the recursion relation

$$P_x(t) = e^{-\lambda x t}\left[\lambda(x-1)\int_0^t P_{x-1}(\tau)e^{x\tau}\,d\tau + C_x\right] \tag{2.107}$$

From the initial condition (2.104) we see that $C_1 = 1$ for $x = 1$ and $C_x = 0$ for $x \neq 1$. By induction, we obtain the *Yule-Furry distribution*

$$P_x(t) = e^{-\lambda t}(1 - e^{-\lambda t})^{x-1} \qquad \text{for } x = 1, 2, \ldots$$

$$= 0 \qquad \text{for } x = 0 \tag{2.108}$$

We now consider some of the properties of the Yule-Furry process. It is easily verified that (2.108) represents an honest probability distribution, since $P_x(t) \geqslant 0$ for $t > 0$, and

$$\sum_{x=0}^{\infty} P_x(t) = e^{-\lambda t}\sum_{x=1}^{\infty}(1 - e^{-\lambda t})^{x-1}$$

$$= e^{-\lambda t}[1 - (1 - e^{-\lambda t})]^{-1} = 1 \tag{2.109}$$

It is also of interest to note that $P_x(t)$ is a monotone decreasing function of x and that $P_x(t)$ is a distribution of geometric form, the common ratio approaching zero as $t \to \infty$.

To determine the mean of $X(t)$, we have

$$m(t) = \mathscr{E}\{X(t)\} = \sum_{x=0}^{\infty} x P_x(t) = e^{-\lambda t}\sum_{x=1}^{\infty} x(1 - e^{-\lambda t})^{x-1}$$

If we now put $z = 1 - e^{-\lambda t}$, the above becomes

$$m(t) = e^{-\lambda t}\sum_{x=1}^{\infty} x z^{x-1} = e^{-\lambda t}\frac{d}{dz}\left(\sum_{x=1}^{\infty} z^x\right) = e^{-\lambda t}\frac{d}{dz}\left(\frac{1}{1-z}\right)$$

$$= e^{-\lambda t}\frac{1}{(1-z)^2} = e^{-\lambda t}\left[\frac{1}{1 - (1 - e^{-\lambda t})}\right]^2 = e^{\lambda t} \tag{2.110}$$

Similarly, we find that the second moment is given by $2m^2(t) - m(t)$; hence, the variance is

$$\mathscr{D}^2\{X(t)\} = e^{\lambda t}(e^{\lambda t} - 1) \tag{2.111}$$

Let us now use the method of generating functions to find the distribution $P_x(t)$. As before, we put

$$F(s,t) = \sum_{x=0}^{\infty} P_x(t)s^x$$

Then it is easily verified that

$$\frac{\partial F}{\partial t} = \lambda s(s-1)\frac{\partial F}{\partial s} \tag{2.112}$$

The general solution of (2.112) (cf. Sneddon [81]) is

$$F(s,t) = f\left(\left(1 - \frac{1}{s}\right)e^{\lambda t}\right)$$

where $f(\cdot)$ is an arbitrary function. Since at time $t = 0$ $X(0) = 1$, we have $F(s,0) = s$, so that

$$s = f\left(1 - \frac{1}{s}\right)$$

Hence we seek a function $f(\xi)$ such that the above holds. We see that $f(\xi) = 1/(1-\xi)$ satisfies the above condition. Therefore,

$$F(s,t) = \frac{se^{-\lambda t}}{1 - (1 - e^{-\lambda t})s} \tag{2.113}$$

is the solution of (2.112). We now put $A = e^{-\lambda t}$ and $B = 1 - e^{-\lambda t}$. Then (2.113) becomes

$$F(s,t) = \frac{As}{1 - Bs}$$

For $|Bs| < 1$, the above can be written as

$$F(s,t) = As \sum_{x=0}^{\infty} B^x s^x$$

Therefore $\qquad P_x(t) = AB^{x-1} = e^{-\lambda t}(1 - e^{-\lambda t})^{x-1}$

So far we have considered the probability of being in the state $x \geqslant 1$ at time t when at time zero the system was in the state $x = 1$. We now consider the Kolmogorov equations for the birth process. From the assumptions of the linear birth process, we have

$$q_i = \lambda i \qquad \text{for } i = 1, 2, \ldots$$
$$Q_{ij} = 1 \qquad \text{for } j = i+1, i = 1, 2, \ldots \tag{2.114}$$
$$= 0 \qquad \text{otherwise}$$

Therefore, the matrix of infinitesimal transition probabilities is given by

$$A = \begin{bmatrix} 0 & 0 & 0 & 0 & 0 & 0 & \cdots \\ 0 & -\lambda & \lambda & 0 & 0 & 0 & \cdots \\ 0 & 0 & -2\lambda & 2\lambda & 0 & 0 & \cdots \\ 0 & 0 & 0 & -3\lambda & 3\lambda & 0 & \cdots \\ \cdot & \cdot & \cdot & \cdot & \cdot & \cdot & \cdot \cdot \cdot \cdot \cdot \cdot \cdot \cdot \end{bmatrix}$$

Hence the *Kolmogorov equations for the birth process* are given by

$$\frac{dP_{ij}(t)}{dt} = -\lambda_j P_{ij}(t) + \lambda_{j-1} P_{i,j-1}(t)$$

$$\frac{dP_{ij}(t)}{dt} = -\lambda_i P_{ij}(t) + \lambda_i P_{i+1,j}(t) \tag{2.115}$$

Now, from the relationship between transition probabilities and generating functions we know that $P_{ij}(t)$ is given by the coefficient of s^j in the series expansion of $[F(s,t)]^i$. Hence

$$[F(s,t)]^i = \left(\frac{As}{1 - Bs}\right)^i$$

$$= A^i s^i \sum_{j=0}^{\infty} \binom{i+j-1}{j} B^j s^j$$

Therefore

$$P_{ij}(t) = \binom{j-1}{j-i} e^{-i\lambda t}(1 - e^{-\lambda t})^{j-i} \qquad \text{for } j \geqslant i$$

$$= 0 \qquad\qquad\qquad\qquad \text{otherwise} \tag{2.116}$$

These transition probabilities give the probability of being in the state $x = j$ at time t when at time zero $x = i$. It is easily verified that the mean and variance in this case are given by

$$m(t) = \mathscr{E}\{X(t)\} = ie^{\lambda t} \tag{2.117}$$

and
$$\mathscr{D}^2\{X(t)\} = ie^{\lambda t}(e^{\lambda t} - 1) \tag{2.118}$$

Before ending our discussion of the birth process, we consider an interesting problem which is encountered for the first time in connection with the birth process. We refer to the problem of determining under what conditions the distribution $P_x(t)$ is an honest probability distribution, i.e., $\sum_{x=0}^{\infty} P_x(t) = 1$. It is clear that $\{P_x(t)\}$ is uniquely determined, since the solution of (2.103), satisfying the initial condition (2.104), can be calculated by induction.

In the general birth process the birth rate depends on the state x that the system is in at time t; hence, it is possible for a rapid increase in the λ_x to lead to the condition $\sum_{x=0}^{\infty} P_x(t) < 1$, i.e., the associated birth process is a *dishonest process*. The interpretation of this condition is as follows: Should the λ_x increase too rapidly, the random variable $X(t)$ would, with positive probability, assume the value $x = \infty$ in a *finite* period of time. Since we do not have any "probability mass" being created or absorbed, the total probability must always equal unity. Hence, the above sum, representing the probability for all finite values of time, would be less than unity.

The condition that must be satisfied in order for $\sum_{x=0}^{\infty} P_x(t) = 1$ is given by the following theorem.

Theorem 2.5 (*Feller-Lundberg*): For a pure birth process with parameters λ_x

$$\sum_{x=0}^{\infty} P_x(t) = 1$$

for all t, if and only if

$$\sum_{x=0}^{\infty} \frac{1}{\lambda_x} = \infty$$

Proof:[1] Let

$$S_n(t) = \sum_{k=0}^{n} P_k(t) \qquad 0 \leqslant S_n(t) \leqslant 1$$

From the equations for the general birth process we obtain

$$\frac{dS_0(t)}{dt} = -\lambda_0 S_0(t)$$

$$\frac{dS_n(t)}{dt} = -\lambda_n [S_n(t) - S_{n-1}(t)] \qquad n > 0 \tag{2.119}$$

Since $P_k(0) = \delta_{xk}$, we have

$$S_n(0) = \sum_{x=0}^{n} \delta_{xk}$$

Now let

$$\sum_{k=0}^{\infty} \frac{1}{\lambda_k} = C \qquad 0 < C < \infty$$

If we denote by $S_n(z)$ the Laplace transform of $S_n(t)$, we have, from (2.119),

$$\mathscr{L}\{S_n(t)\} = S_n(z) = \frac{S_n(0) + \lambda_n S_{n-1}(z)}{\lambda_n + z} \tag{2.120}$$

[1] Due to Morgenstern [64].

For $n = 0$, we have

$$S_0(z) = \frac{\delta_{x0}}{\lambda_0 + z} < \frac{1}{\lambda_0}$$

In general, we obtain from (2.120)

$$S_n(z) < \frac{1 + \lambda_n S_{n-1}(z)}{\lambda_n} = \frac{1}{\lambda_n} + S_{n-1}(z)$$

By induction we have

$$S_n(z) < \sum_{k=0}^{n} \frac{1}{\lambda_k} \leqslant C$$

Hence, in the limit as $n \to \infty$, we have

$$\lim_{n \to \infty} S_n(z) \leqslant \sum_{k=0}^{\infty} \frac{1}{\lambda_k} \leqslant C \qquad z > 0$$

The condition $\sum_{k=0}^{\infty} P_k(t) = 1$ requires that

$$\lim_{n \to \infty} S_n(t) = 1$$

for all t, or in terms of the Laplace transform

$$\lim_{n \to \infty} S_n(z) = \frac{1}{z}$$

for all $z > 0$. For $z = z_0 < 1/C$ this leads to a contradiction. This completes the proof.

D.   The Pólya Process. The processes considered so far have all been homogeneous, i.e., the intensity functions were independent of time. We now consider the *Pólya process*, which is a nonhomogeneous pure birth process. The assumptions for the Pólya process are the same as for the birth process, except that we now have that the probability of a change in state in the interval $(t, t + \Delta t)$ is

$$\lambda \frac{1 + \alpha x}{1 + \alpha \lambda t} \Delta t + o(\Delta t) \tag{2.121}$$

where α and λ are nonnegative constants. By proceeding as before, we see that the Pólya process is described by the following differential equations:

$$\frac{dP_0(t)}{dt} = -\frac{\lambda}{1 + \alpha \lambda t} P_0(t) \tag{2.122}$$

$$\frac{dP_x(t)}{dt} = \lambda \frac{1 + \alpha(x - 1)}{1 + \alpha \lambda t} P_{x-1}(t) - \lambda \frac{1 + \alpha x}{1 + \alpha \lambda t} P_x(t) \qquad x = 1, 2, \ldots \tag{2.123}$$

Similarly, we have the initial condition

$$P_x(0) = \delta_{x0} \qquad (2.124)$$

From (2.122) we have

$$P_0(t) = (1 + \alpha\lambda t)^{-1/\alpha} \qquad (2.125)$$

The solution of (2.123) can be obtained by the same successive method used for the Poisson and birth processes. We have

$$P_x(t) = \frac{(\lambda t)^x}{x!} (1 + \alpha\lambda t)^{-x-(1/\alpha)} \prod_{i=1}^{x-1}(1 + \alpha i) \qquad x = 1, 2, \ldots \quad (2.126)$$

To begin the study of the properties of the Pólya process, we first show that $\sum_{x=0}^{\infty} P_x(t) = 1$. We have

$$\sum_{x=0}^{\infty} P_x(t) = (1 + \alpha\lambda t)^{-1/\alpha} \sum_{x=0}^{\infty} \frac{(\lambda t)^x}{x!} (1 + \alpha\lambda t)^{-x} \prod_{i=1}^{x-1}(1 + \alpha i)$$

$$= (1 + \alpha\lambda t)^{-1/\alpha} \sum_{x=0}^{\infty} (-\alpha)^x \binom{1/\alpha}{x} \left(\frac{\lambda t}{1 + \alpha\lambda t}\right)^x$$

$$= (1 + \alpha\lambda t)^{-1/\alpha} \sum_{x=0}^{\infty} \binom{-1/\alpha}{x} \left(\frac{-\alpha\lambda t}{1 + \alpha\lambda t}\right)^x$$

$$= (1 + \alpha\lambda t)^{-1/\alpha} \left(1 - \frac{\alpha\lambda t}{1 + \alpha\lambda t}\right)^{-1/\alpha} = 1$$

In the above calculations we have used the result

$$\frac{1}{x!} \prod_{i=1}^{x-1}(1 + \alpha i) = (-\alpha)^x \binom{-1/\alpha}{x}$$

To determine the mean of $X(t)$, we have

$$m(t) = \mathscr{E}\{X(t)\} = \sum_{x=0}^{\infty} x P_x(t) = (1 + \alpha\lambda t)^{-1/\alpha} \sum_{x=0}^{\infty} x \binom{-1/\alpha}{x} \left(\frac{-\alpha\lambda t}{1 + \alpha\lambda t}\right)^x$$

$$= (1 + \alpha\lambda t)^{-1/\alpha} \left(-\frac{1}{\alpha}\right) \left(\frac{-\alpha\lambda t}{1 + \alpha\lambda t}\right) \sum_{x=0}^{\infty} \binom{-(1 + 1/\alpha)}{x - 1} \left(\frac{-\alpha\lambda t}{1 + \alpha\lambda t}\right)^{x-1}$$

$$= \lambda t(1 + \alpha\lambda t)^{-1-1/\alpha} \left(1 - \frac{\alpha\lambda t}{1 + \alpha\lambda t}\right)^{-1-1/\alpha} = \lambda t \qquad (2.127)$$

In the above calculations we have used the result

$$x\binom{-1/\alpha}{x} = \left(-\frac{1}{\alpha}\right)\binom{-(1 + 1/\alpha)}{x - 1}$$

Similar calculations for the second moment give

$$(1 + \alpha)m^2(t) + m(t) = (1 + \alpha)(\lambda t)^2 + \lambda t$$

hence $$\mathscr{D}^2\{X(t)\} = \lambda t(1 + \alpha\lambda t) \tag{2.128}$$

Hence, for the Pólya process the mean is a linear function of time and the variance is a quadratic function of time.

We now wish to show the relationship between the Pólya process and the Poisson and Yule-Furry processes. Let us first consider the behavior of the Pólya distribution as the parameter approaches zero. From (2.125) and (2.126) we have

$$\lim_{\alpha \to 0} (1 + \alpha\lambda t)^{-1/\alpha} = e^{-\lambda t}$$

$$\lim_{\alpha \to 0} \frac{(\lambda t)^x}{x!} (1 + \alpha\lambda t)^{x-1/\alpha} \prod_{i=1}^{x-1}(1 + \alpha i) = \frac{(\lambda t)^x}{x!} e^{-\lambda t} \qquad x = 1, 2, \ldots \tag{2.129}$$

From the above we see that the Poisson process is a special case of the Pólya process, obtained in the limit as α approaches zero.

Similarly, if we put $\alpha = 1$, we have

$$P_0(t) = \frac{1}{1 + \lambda t} \tag{2.130}$$

$$P_x(t) = (\lambda t)^x(1 + \lambda t)^{-x-1} = [1 - (1 + \lambda t)^{-1}]^x(1 + \lambda t)^{-1} \qquad x = 1, 2, \ldots$$

If we now introduce a new time parameter $e^{\lambda\tau} = 1 + \lambda t$, we see that (2.126) gives the probability distribution for the Yule-Furry process.

We close this discussion of the Pólya process by obtaining the Kolmogorov equations that describe this process. For the Pólya process the intensity functions depend on the state variable x and time; hence

$$q_i(t) = \lambda \left(\frac{1 + \alpha i}{1 + \alpha\lambda t} \right) \qquad \text{for } i = 0, 1, \ldots$$

$$Q_{ij} = 1 \qquad \qquad \text{for } j = i + 1, i = 0, 1, \ldots \tag{2.131}$$

$$= 0 \qquad \qquad \text{otherwise}$$

The matrix of infinitesimal transition probabilities is also a function of time; thus

$$A = A(t) = \begin{bmatrix} -\lambda\dfrac{1}{1 + \alpha\lambda t} & \lambda\dfrac{1}{1 + \alpha\lambda t} & 0 & 0 & \cdots 0 \\[2ex] 0 & -\lambda\dfrac{1 + \alpha}{1 + \alpha\lambda t} & \lambda\dfrac{1 + \alpha}{1 + \alpha\lambda t} & 0 & \cdots 0 \\[2ex] 0 & 0 & -\lambda\dfrac{1 + 2\alpha}{1 + \alpha\lambda t} & \lambda\dfrac{1 + 2\alpha}{1 + \alpha\lambda t} & \cdots 0 \\[1ex] \cdot & \cdot & \cdot & \cdot & \cdots \end{bmatrix}$$

Hence the *Kolmogorov equations for the Pólya process are*

$$\frac{dP_{ij}(t)}{dt} = -\lambda \frac{1 + \alpha j}{1 + \alpha \lambda t} P_{ij}(t) + \lambda \frac{1 + \alpha(j-1)}{1 + \alpha \lambda t} P_{i,j-1}(t) \qquad (2.132)$$

$$\frac{dP_{ij}(t)}{dt} = -\lambda \frac{1 + \alpha i}{1 + \alpha \lambda t} P_{ij}(t) + \lambda \frac{1 + \alpha(i+1)}{1 + \alpha \lambda t} P_{i+1,j}(t) \qquad (2.133)$$

E. The Simple Death Process. The processes considered thus far all come under the general heading of "birth processes," i.e., the random variable $X(t)$ is a strictly increasing function of time. We now define and investigate a process which can be termed a *death process.* In this case the random variable $X(t)$ is a strictly decreasing function of time.

Assumptions for the Death Process. (1) At time zero the system is in a state $x = x_0$, i.e., $X(0) = x_0 \geqslant 1$. (2) If at time t the system is in the state x ($x = 1, 2, \ldots$), then the probability of the transition $x \to x - 1$ in the interval $(t, t + \Delta t)$ is $\mu_x \Delta t + o(\Delta t)$. (3) The probability of a transition from $x \to x - i$, where $i > 1$, is $o(\Delta t)$. (4) The probability of no change is $1 - \mu_x \Delta t + o(\Delta t)$. The above assumptions lead to the relation

$$P_x(t + \Delta t) = (1 - \mu_x \Delta t)P_x(t) + \mu_x P_{x+1}(t) \Delta t + o(\Delta t) \qquad (2.134)$$

This relation leads to the differential equation

$$\frac{dP_x(t)}{dt} = -\mu_x P_x(t) + \mu_{x+1} P_{x+1}(t) \qquad (2.135)$$

As in the case of the birth process, we assume that the death rate is linear. i.e., $\mu_x = \mu x$, $\mu > 0$. With this assumption we obtain the equation for the *simple death process or linear death process:*[1]

$$\frac{dP_x(t)}{dt} = -\mu x P_x(t) + \mu(x + 1)P_{x+1}(t) \qquad (2.136)$$

This equation is to be solved with the initial condition

$$P_x(0) = \delta_{xx_0} = 1 \qquad \text{for } x = x_0 > 1$$
$$= 0 \qquad \text{otherwise} \qquad (2.137)$$

The solution of (2.136) which satisfies the above initial condition is

$$P_x(t) = \binom{x_0}{x} e^{-x_0 \mu t}(e^{\mu t} - 1)^{x_0 - x} \qquad 0 \leqslant x \leqslant x_0 \qquad (2.138)$$

[1] In Sec. 4.4B we consider a nonlinear death process due to N. T. J. Bailey.

It is easy to verify that the mean and variance of the simple death process are given by

$$m(t) = \mathcal{E}\{X(t)\} = x_0 e^{-\mu t} \tag{2.139}$$

and

$$\mathcal{D}^2\{X(t)\} = x_0 e^{-\mu t}(1 - e^{-\mu t}) \tag{2.140}$$

To obtain the Kolmogorov equations, we note that

$$
\begin{aligned}
q_i &= \mu i &&\text{for } i = 1, 2, \ldots \\
Q_{ij} &= 1 &&\text{for } j = i - 1, i = 1, 2, \ldots \\
&= 0 &&\text{otherwise}
\end{aligned}
\tag{2.141}
$$

Hence the matrix of infinitesimal transformation probabilities is given by

$$
A = \begin{bmatrix}
0 & 0 & 0 & 0 & \cdots & 0 & 0 \\
\mu & -\mu & 0 & 0 & \cdots & 0 & 0 \\
0 & 2\mu & -2\mu & 0 & \cdots & 0 & 0 \\
0 & 0 & 3\mu & -3\mu & \cdots & 0 & 0 \\
\cdot & \cdot & \cdot & \cdot & \cdot & & \cdot
\end{bmatrix}
$$

It is of interest to note that in the case of the death process the nonzero elements of A are those on the diagonal and lower diagonal. In the previous cases, all birth-type processes, the nonzero elements were on the diagonal and upper diagonal. The *Kolmogorov equations for the death process* are

$$\frac{dP_{ij}(t)}{dt} = \mu_{j+1} P_{i,j+1}(t) - \mu_j P_{ij}(t) \tag{2.142}$$

$$\frac{dP_{ij}(t)}{dt} = \mu_i P_{i-1,j}(t) - \mu_i P_{ij}(t) \tag{2.143}$$

F. The Birth-and-Death Process. In this section we consider a process which combines the features of the simple birth and simple death processes; i.e., we consider a process such that the random variable $X(t)$ can experience positive as well as negative jumps. This process, called the *birth-and-death process*, is of considerable theoretical interest; and birth-and-death type processes, as we shall see, are encountered in many fields of application.

Assumptions for the Birth-and-Death Process. (1) If at time t the system is in the state x ($x = 1, 2, \ldots$), the probability of the transition $x \to x + 1$ in the interval $(t, t + \Delta t)$ is $\lambda_x \Delta t + o(\Delta t)$. (2) If at time t the system is in the state x ($x = 1, 2, \ldots$), the probability of the transition $x \to x - 1$ in the interval $(t, t + \Delta t)$ is $\mu_x \Delta t + o(\Delta t)$. (3) The probability of a transition to a state other than a neighboring state is $o(\Delta t)$. (4) The probability of no change is $1 - (\lambda_x + \mu_x)\,\Delta t + o(\Delta t)$. (5) The state $x = 0$ is an absorbing state.

These assumptions lead to the relation

$$P_x(t + \Delta t) = \lambda_{x-1}P_{x-1}(t)\,\Delta t + [1 - (\lambda_x + \mu_x)\,\Delta t]P_x(t)$$
$$+ \mu_{x+1}P_{x+1}(t)\,\Delta t + o(\Delta t) \quad (2.144)$$

The above relation leads in the limit to the differential equation

$$\frac{dP_x(t)}{dt} = \lambda_{x-1}P_{x-1}(t) - (\lambda_x + \mu_x)P_x(t) + \mu_{x+1}P_{x+1}(t) \quad (2.145)$$

This equation holds for $x = 1, 2, \ldots$. For $x = 0$ we have

$$\frac{dP_0(t)}{dt} = \mu_1 P_1(t) \quad (2.146)$$

since $\lambda_{-1} = \lambda_0 = \mu_0 = 0.$[1] If at time zero the system is in the state $x = x_0,\ 0 < x_0 < \infty$, the initial conditions are

$$P_x(0) = \delta_{xx_0} = 1 \qquad \text{for } x = x_0$$
$$= 0 \qquad \text{otherwise} \quad (2.147)$$

As in the case of the birth process and the death process, the coefficients λ_x and μ_x are arbitrary functions of x. For the processes considered earlier the differential equations were of a form which permitted us to obtain the probabilities $P_x(t)$ by recurrence methods. The equations for the birth-and-death process are not of this form; hence, it is necessary to determine the $P_x(t)$ simultaneously.

We now obtain the solution of the birth-and-death equations in the case of a linear birth-and-death process,[2] i.e., we assume $\lambda_x = \lambda x$ and $\mu_x = \mu x;\ \lambda,\ \mu > 0$. When we introduce the generating function $F(s,t) = \sum_{x=0}^{\infty} P_x(t)s^x$, Eqs. (2.145) and (2.146) become

$$\frac{\partial F(s,t)}{\partial t} = [\lambda s^2 - (\lambda + \mu)s + \mu]\frac{\partial F(s,t)}{\partial s} \quad (2.148)$$

[1] In the general case Eq. (2.146) should read

$$\frac{dP_0(t)}{dt} = -\lambda_0 P_0(t) + \mu_1 P_1(t)$$

since $x = 0$ may not be an absorbing state.

[2] A detailed study of linear birth-and-death processes with $\lambda_x = \lambda x^2 + \alpha$ and $\mu_x = \mu x^2 + \beta$ is presented in the paper by Karlin and McGregor [47]. The properties of a quadratic birth-and-death process with $\lambda_x = \lambda(x^2 + \alpha x)$ and $\mu_x = \mu(x^2 + \alpha x)$, where $\lambda, \mu > 0,\ \alpha \geqslant 0$, and $x = 1$ when $t = 0$, have been studied by John [41].

The general solution of (2.148) is

$$F(s,t) = f\left(\frac{\mu - \lambda s}{1 - s} e^{-(\lambda-\mu)t}\right) \qquad (2.149)$$

where $f(\cdot)$ is an arbitrary function. If we now assume that $X(0) = x_0 = 1$, then $F(s,0) = s$, so that

$$s = f\left(\frac{\mu - \lambda s}{1 - s}\right)$$

hence

$$f(\xi) = \frac{\mu - \xi}{\lambda - \xi}$$

Therefore,

$$F(s,t) = \frac{\mu(1 - e^{(\lambda-\mu)t}) - (\lambda - \mu e^{(\lambda-\mu)t})s}{\mu - \lambda e^{(\lambda-\mu)t} - \lambda(1 - e^{(\lambda-\mu)t})s} \qquad (2.150)$$

If we expand (2.150) as a power series in s, the coefficients of s^x give

$$P_x(t) = [1 - \alpha(t)][1 - \beta(t)][\beta(t)]^{x-1} \qquad x = 1, 2, \ldots \qquad (2.151)$$
$$P_0(t) = \alpha(t)$$

where

$$\alpha(t) = \frac{\mu(e^{(\lambda-\mu)t} - 1)}{\lambda e^{(\lambda-\mu)t} - \mu} \qquad (2.152)$$

$$\beta(t) = \frac{\lambda(e^{(\lambda-\mu)t} - 1)}{\lambda e^{(\lambda-\mu)t} - \mu} \qquad (2.153)$$

To obtain expressions for the mean and variance of $X(t)$, we differentiate the generating function (2.150) and proceed as before to obtain

$$m(t) = \mathscr{E}\{X(t)\} = e^{(\lambda-\mu)t} \qquad (2.154)$$

$$\mathscr{D}^2\{X(t)\} = \frac{\lambda + \mu}{\lambda - \mu} e^{(\lambda-\mu)t}(e^{(\lambda-\mu)t} - 1) \qquad (2.155)$$

In the processes studied earlier, the behavior of the mean as a function of time was not very interesting. In the case of the birth-and-death process, however, the asymptotic behavior of the mean depends on the relationship that obtains between the parameters λ and μ. From (2.154) we see that

$$\lim_{t \to \infty} m(t) = 0 \qquad \text{for } \lambda < \mu$$
$$= 1 \qquad \text{for } \lambda = \mu$$
$$= \infty \qquad \text{for } \lambda > \mu \qquad (2.156)$$

From the above, we see that, when $\lambda = \mu$, the expected rate of growth is zero and the mean population size is stationary.

In the special case $\lambda = \mu$ we have from (2.154) and (2.155)

$$m(t) = 1 \qquad \text{when } X(0) = 1 \tag{2.157}$$

$$\mathscr{D}^2\{X(t)\} = 2\lambda t \tag{2.158}$$

Associated with processes of the birth-and-death type is the extinction probability $P_0(t)$ that the population will die out by time t. From (2.151) and (2.152) we have

$$P_0(t) = \frac{\mu(e^{(\lambda-\mu)t} - 1)}{\lambda e^{(\lambda-\mu)t} - \mu} \tag{2.159}$$

The probability that the population will eventually die out is obtained by letting $t \to \infty$ in (2.159). Hence

$$\lim_{t\to\infty} P_0(t) = 1 \qquad \text{for } \lambda < \mu$$

$$= \frac{\mu}{\lambda} \qquad \text{for } \lambda > \mu \tag{2.160}$$

From the above result we see that the population dies out with probability one when the death rate is greater than the birth rate, but when the birth rate is greater than the death rate the probability of eventual extinction is equal to the ratio of the rates. It is interesting to compare this result with the probability of absorption obtained in Sec. 1.7 when a random-walk model was used to represent a birth-and-death process.

A more general (nonhomogeneous) birth-and-death process has been studied by Kendall [49] in which the birth-and death rates are arbitrary functions of time, and not linear functions of the state variable as we have assumed above.[1] In this general case the solution given by (2.151) still holds, but we now have

$$\alpha(t) = 1 - \frac{e^{-\gamma(t)}}{w(t)} \tag{2.161}$$

$$\beta(t) = 1 - \frac{1}{w(t)} \tag{2.162}$$

where

$$\gamma(t) = \int_0^t [\mu(\tau) - \lambda(\tau)]\, d\tau \tag{2.163}$$

$$w(t) = e^{-\gamma(t)}\left[1 + \int_0^t \mu(\tau)e^{\gamma(\tau)}\, d\tau\right] \tag{2.164}$$

[1] We refer also to the paper of Lamens [59] and Lamens and Consael [60].

For this nonhomogeneous process, the mean and variance of $X(t)$ are given by

$$m(t) = \mathscr{E}\{X(t)\} = e^{-\gamma(t)} \tag{2.165}$$

$$\mathscr{D}^2\{X(t)\} = e^{-2\gamma(t)} \int_0^t [\lambda(\tau) + \mu(\tau)]e^{\gamma(\tau)} \, d\tau \tag{2.166}$$

To obtain the probability of extinction, we use (2.151) and (2.161) to get

$$P_0(t) = \frac{\displaystyle\int_0^t \mu(\tau)e^{\gamma(\tau)} \, d\tau}{1 + \displaystyle\int_0^t \mu(\tau)e^{\gamma(\tau)} \, d\tau} \tag{2.167}$$

Here we see that the necessary and sufficient condition for $\lim\limits_{t\to\infty} P_0(t) = 1$ is that the integral in the above expression diverge. Since the integrand is nonnegative, it follows that the integral must either diverge to plus infinity or remain bounded.

In order to obtain the transition probabilities for the simple linear birth-and-death process, we first note that the intensity function and relative transition probabilities are given by

$$\begin{aligned}
q_i &= (\lambda + \mu)i && \text{for } i = 1, 2, \ldots \\
Q_{ij} &= \frac{\lambda}{\lambda + \mu} && \text{for } j = i + 1, i = 1, 2, \ldots \\
&= \frac{\mu}{\lambda + \mu} && \text{for } j = i - 1 \\
&= 0 && \text{for } |i - j| > 1
\end{aligned} \tag{2.168}$$

Hence the matrix of infinitesimal transition probabilities is given by

$$A = \begin{bmatrix}
0 & 0 & 0 & 0 & 0 & \cdots & 0 & 0 \\
\mu & -(\lambda + \mu) & \lambda & 0 & 0 & \cdots & 0 & 0 \\
0 & 2\mu & -2(\lambda + \mu) & 2\lambda & 0 & \cdots & 0 & 0 \\
0 & 0 & 3\mu & -3(\lambda + \mu) & 3\lambda & \cdots & 0 & 0 \\
\cdot & \cdot & \cdot & \cdot & \cdot & \cdot & \cdot & \cdot
\end{bmatrix}$$

The Kolmogorov equations for the general birth-and-death process are

$$\frac{dP_{ij}(t)}{dt} = \lambda_{j-1}P_{i,j-1}(t) - (\lambda_j + \mu_j)P_{ij}(t) + \mu_{j+1}P_{i,j+1}(t) \tag{2.169}$$

$$\frac{dP_{ij}(t)}{dt} = \lambda_i P_{i+1,j}(t) - (\lambda_i + \mu_i)P_{ij}(t) + \mu_i P_{i-1,j}(t) \tag{2.170}$$

To obtain explicit expressions for the transition probabilities which satisfy the Kolmogorov equations, we can utilize the expression for the generating function. On using (2.150), (2.152), and (2.153), we have

$$[F(s,t)]^i = \left[\frac{\alpha(t) + [1 - \alpha(t) - \beta(t)]s}{1 - \beta(t)s} \right]^i \tag{2.171}$$

Now

$$(\alpha(t) + [1 - \alpha(t) - \beta(t)]s)^i = \sum_{n=0}^{\infty} \binom{i}{n} [\alpha(t)]^i [1 - \alpha(t) - \beta(t)]^i s^i$$

and

$$[1 - \beta(t)s]^{-i} = \sum_{n=0}^{\infty} \binom{i+n-1}{n} [\beta(t)]^i s^i \qquad |\beta(t)s| < 1$$

We have, therefore,

$$P_{ij}(t) = \sum_{n=0}^{i} \binom{i}{n} \binom{i+j-n-1}{i-1} [\alpha(t)]^{i-n} [\beta(t)]^{j-n}$$
$$\times [1 - \alpha(t) - \beta(t)]^n \qquad i \geqslant j \tag{2.172}$$

In the case $X(0) = x_0 > 1$, the expressions for the mean and variance of $X(t)$ are obtained by multiplying (2.154) and (2.155), or (2.165) and (2.166), by x_0. The asymptotic behavior of the mean remains the same, except that, when $\lambda = \mu$, the mean $m(t) = x_0$ rather than unity. Similarly, the probability of ultimate extinction is unity when $\lambda < \mu$ but equal to $(\mu/\lambda)^{x_0}$ when $\lambda > \mu$. For the general case, when λ and μ are functions of time, the probability of extinction is obtained by raising (2.167) to the (x_0)th power.

We now state some results concerning the uniqueness problem for birth-and-death processes. In particular, we state three theorems due to Reuter and Ledermann which generalize the Feller-Lundberg theorem.[1] For proofs we refer to the paper of Reuter and Ledermann [74]. These theorems give sufficient conditions and necessary conditions for $\sum_{j=0}^{\infty} F_{ij}(t) = 1$, for i fixed, and sufficient conditions that the forward Kolmogorov equation (2.169) have a unique solution.

Theorem 2.6: Any one of the following conditions is sufficient in order that $\sum_{j=0}^{\infty} F_{ij}(t) = 1$, for i fixed, and that the forward Kolmogorov equation have the unique solution $P_{ij}(t) = F_{ij}(t)$:

(a) $\lambda_n = 0$ for some $n \geqslant 1$

(b) $\lambda_0 = 0$ for $n \geqslant i$ and $\sum_{n=i}^{\infty} w_n = \infty$

[1] Cf. also John [40].

where

$$w_n = \frac{1}{\lambda_n} + \frac{\mu_n}{\lambda_n \lambda_{n-1}} + \cdots + \frac{\mu_n \cdots \mu_{i+1}}{\lambda_n \cdots \lambda_i} + \frac{\mu_n \cdots \mu_i}{\lambda_n \cdots \lambda_i} \qquad n \geqslant i$$

(c) $\quad \lambda_n > 0 \quad$ for $n \geqslant i \quad$ and $\quad \sum_{n=i}^{\infty} \frac{1}{\lambda_n} = \infty$

(d) $\quad \lambda_n > 0 \quad$ for $n \geqslant i \quad$ and $\quad \sum_{n=i}^{\infty} \frac{\mu_n \mu_{n-1} \cdots \mu_i}{\lambda_n \lambda_{n-1} \cdots \lambda_i} = \infty$

From the above, we see that (c), by restricting the magnitude of the λ_n, means that the population will not become infinite in a finite period of time. Also, (d) ensures the finiteness of the population by making the μ_n of comparable magnitude to the λ_n, so that the birth and death rates tend to balance. Similarly, (b) covers the cases in which the above two factors are combined.

Theorem 2.7: If $\lambda_n > 0$, for $n \geqslant i$, either of the following conditions is sufficient in order that $\sum_{j=0}^{\infty} F_{ij}(t) < 1$ for some $t > 0$:

(a) $\quad \sum_{n=i}^{\infty} w_n < \infty \quad$ and $\quad \mu_{n+1} w_n = 0(1)$

(b) $\quad \dfrac{\mu_{n+1}}{\lambda_n} \leqslant k < 1 \quad$ for $n \geqslant m \geqslant i \quad$ and $\quad \sum_{n=1}^{\infty} \frac{1}{\lambda_n} < \infty$

This theorem states that every solution $P_{ij}(t)$ of the forward Kolmogorov equation is dishonest, e.g., $\sum_{j=0}^{\infty} P_{ij}(t) < 1$, for some t.

Theorem 2.8: If $\lambda_n > 0$, for $n \geqslant i$, either of the following conditions is sufficient in order that the forward equation should have the unique solution $P_{ij}(t) = F_{ij}(t)$:

(a) $\qquad\qquad \mu_n = 0 \qquad$ for infinitely many n

(b) $\qquad\qquad \mu_n > 0 \qquad$ for $n \geqslant m \geqslant i$

and $\qquad \dfrac{\mu_n \mu_{n-1} \cdots \mu_m}{\lambda_n \lambda_{n-1} \cdots \lambda_n} = 0(1) \qquad$ as $n \to 0$

For other studies[1] on the uniqueness problem for the Kolmogorov equations associated with birth-and-death processes we refer to the papers of Breiman [16], Dobrushin [20], John [40], Karlin and McGregor [44,45], and Koopman [58].

---

[1] This list supplements the references given in Sec. 2.3B.

Of the above studies, we refer in particular to those of Karlin and McGregor and Koopman. The method used by Karlin and McGregor is concerned with obtaining integral representations for the transition probabilities $P_{ij}(t)$ in terms of a system of orthogonal functions, these orthogonal functions appearing as the eigenfunctions of the matrix of infinitesimal transition probabilities associated with the Markov process. Koopman uses the analytic theory of continued fractions in his approach to the uniqueness problem. This theory is applicable because the matrix of infinitesimal transition probabilities is a continuant,[1] and the relationship between continuants and continued fractions is well known.[2] It is shown that the necessary and sufficient condition for uniqueness of the forward equation is that the triangular series

$$\sum_{i=0}^{\infty} \sum_{j=1}^{\infty} \frac{\lambda_j \lambda_{j+1} \cdots \lambda_{j+i}}{\mu_j \mu_{j+1} \cdots \mu_{j+i} \mu_{j+i+1}} = \infty$$

For the backward equation

$$\sum_{i=0}^{\infty} \sum_{j=0}^{\infty} \frac{\mu_{j+1} \cdots \mu_{j+i}}{\lambda_j \lambda_{j+1} \cdots \lambda_{j+i}} = \infty$$

We close this section on birth-and-death processes by stating some results, due to Karlin and McGregor [46], on the classification of processes of this type. Let

$$H_{ij}(t) = \mathscr{P}\{X(\tau) = j \text{ for some } \tau \in (0,t] \mid X(0) = i\} \qquad i \neq j$$

and

$$H_{ii}(t) = \mathscr{P}\{X(\tau_1) \neq i, X(\tau_2) = i \text{ for some } \tau_1, \tau_2, 0 < \tau_1 < \tau_2 \leqslant t \mid X(0) = i\}$$

Hence, $H_{ij}(t)$ and $H_{ii}(t)$ are the first-passage time and recurrence time distributions, respectively. The integral

$$I = \int_0^{\infty} dH_{ii}(t)$$

is the probability that, if the system starts in state i, it leaves i and then returns to i in a finite period of time.

We can now define the states as follows:

Definition: The ith state is *recurrent* if $I = 1$ and *transient* if $I \neq 1$. If i is a recurrent state, it can be classified as *ergodic* or

---

[1] A *continuant* is a matrix with nonzero elements on the lower diagonal, diagonal, and upper diagonal only.

[2] Cf. Hellinger and Wall [34].

recurrent null according as its mean recurrence time $\int_0^\infty t\,dH_{ii}(t)$ is finite or infinite.

Similarly, for processes we have the following.

Definition: A process is called recurrent, ergodic, recurrent null, or transient if every one of its states has the corresponding property.

Karlin and McGregor have given the following criteria for the classification of birth-and-death processes:
1. A birth-and-death process is *recurrent* if and only if

$$\sum_{n=0}^{\infty} \frac{1}{\lambda_n \pi_n} = \infty$$

2a. A birth-and-death process is *ergodic* if and only if

$$\sum_{n=0}^{\infty} \pi_n < \infty \qquad \text{and} \qquad \sum_{n=0}^{\infty} \frac{1}{\lambda_n \pi_n} = \infty$$

b. A birth-and-death process is *recurrent null* if and only if

$$\sum_{n=0}^{\infty} \pi_n = \infty \qquad \text{and} \qquad \sum_{n=0}^{\infty} \frac{1}{\lambda_n \pi_n} = \infty$$

3. From (2), a birth-and-death process is *transient* if and only if

$$\sum_{n=0}^{\infty} \pi_n = \infty \qquad \text{and} \qquad \sum_{n=0}^{\infty} \frac{1}{\lambda_n \pi_n} < \infty$$

In the above

$$\pi_0 = 1 \qquad \pi_n = \frac{\lambda_0 \lambda_1 \cdots \lambda_{n-1}}{\mu_1 \mu_2 \cdots \mu_n} \qquad n \geqslant 1$$

2.5   Age-dependent Branching Stochastic Processes

A.   Introduction.   The Fundamental Functional Equation. The stochastic processes considered thus far have all been of the Markov type. These processes can be characterized by the property that the interval of time, say of length τ, between transitions or changes of state is a random variable with a negative exponential distribution. In this section we consider a non-Markovian process in which τ can have a general distribution $G(\tau)$, so that $dG(\tau)$ is the probability that an individual born at time τ is transformed in the interval $(\tau, \tau + d\tau)$.

This process, described by Bellman and Harris [9], is called *age-dependent*.[1] The Bellman-Harris process is of considerable importance in many fields of application, since, as we shall see, it incorporates several features which enable the construction of more realistic stochastic models.

The Bellman-Harris process is formulated as follows: Let $X(t)$ be an integer-valued random variable representing the number of individuals in the population at time t and let $P_x(t) = \mathscr{P}\{X(t) = x\}, x = 0, 1, 2, \ldots$. Let

$$F(s,t) = \sum_{x=0}^{\infty} P_x(t)s^x \qquad |s| \leqslant 1 \tag{2.173}$$

be the generating function for the probabilities $P_x(t)$. As we remarked earlier, the generation time τ from the birth of an individual until he is transformed is a random variable with distribution $G(\tau)$, $0 < \tau < \infty$. In addition, when the individual is transformed, he has probability q_n, $n \geqslant 0$, of being transformed into n individuals, each new individual having the same distribution $G(\tau)$ for its generation time as the parent. As usual, we assume the individuals are statistically independent.

The generating function (2.173) has been shown to satisfy the Stieltjes functional equation

$$F(s,t) = \int_0^t h[F(s, t-\tau)]\, dG(\tau) + s[1 - G(t)] \tag{2.174}$$

where

$$h(s) = \sum_{n=0}^{\infty} q_n s^n \tag{2.175}$$

that is, $h(s)$ is the generating function of the transformation probabilities q_n.

The equation for the generating function (2.174) can be derived as follows: By definition

$$P_x(t) = \mathscr{P}\{X(t) = x\} = \int_0^t \mathscr{P}\{X(t) = x \mid \tau\}\, dG(\tau)$$

where $\mathscr{P}\{X(t) = x \mid \tau\}$ is the probability of having x individuals at

---

[1] Age-dependent processes belong to a class of stochastic processes known as *regenerative processes*, which can be described as follows: if a process $\{X(t), t \geqslant 0\}$ is such that for some particular time T and for all $t > T$ the conditional probability distribution of $X(t)$, given $X(T)$, is equal to the conditional probability distribution of $X(t)$, given $X(\tau)$, for all $\tau < T$, the time T (or the event by which it can be identified) is called a *regeneration point* for the process. A regenerative process is any process which has such points. It is of interest to note that Markov processes can be characterized as stochastic processes for which every time t is a regeneration point.

time t from one individual at time zero who is known to have branched or undergone a transformation at $t = \tau$. Now

$$\mathscr{P}\{X(t) = x \mid \tau\} = \sum_{n=0}^{\infty} q_n \left\{ \sum_{i_1 + \cdots + i_n = x} P_{i_1}(t - \tau) \cdots P_{i_n}(t - \tau) \right\}$$

where the term in braces is the coefficient of s^x in the expansion of

$$\left[\sum_{x=0}^{\infty} P_x(t - \tau)s^x \right]^n = F^n(s, t - \tau)$$

(The reasoning used above is the same as that used in the theory of compound probability distributions; cf. Appendix A, especially Theorem A.4.) Hence, we have

$$P_x(t) = \int_0^t \left[\sum_{n=0}^{\infty} q_n \left\{ \sum_{i_1 + \cdots i_n = x} P_{i_1}(t - \tau) \cdots P_{i_n}(t - \tau) \right\} \right] dG(\tau)$$

$$= \int_0^t \left[\sum_{n=0}^{\infty} q_n F^n(s, t - \tau) \right] dG(\tau)$$

By multiplying $P_x(t)$ by s^x and summing over x, we obtain, after adding the term for $P_1(t) = 1 - G(t)$ [i.e., $X(0) = 1$, and the individual is not transformed], Eq. (2.174).

If $G(t)$ has a density function $g(t)$ of bounded total variation, i.e.,

$$G(t) = \int_0^t g(\tau) \, d\tau \qquad \int_0^{\infty} |dg(\tau)| < \infty$$

we can rewrite (2.174) as

$$F(s,t) = \int_0^t \left[\sum_{n=0}^{\infty} q_n F^n(s, t - \tau) \right] g(\tau) \, d\tau + s[1 - G(t)] \qquad (2.176)$$

From the above we see that the Bellman-Harris process is completely determined by the distribution $G(\tau)$ and the generating function $h(s)$. As stated earlier, the arbitrary nature of G enables us to consider branching processes that do not have the Markov property. In addition, the transformation probabilities can be defined in a manner which permits the representation of death processes, when $q_0 \neq 0$, or birth processes in which multiple births can occur.

B. Integral Equations for the Mean and Variance. We now consider the equations for the mean and variance of $X(t)$. As before, we can obtain the moments by differentiation of the generating function $F(s,t)$. Let $m_1(t)$ and $m_2(t)$ denote the first and second derivatives of $F(s,t)$ with respect to s, evaluated at $s = 1$. Hence, differentiation of (2.174) yields the integral equation

$$m_1(t) = 1 - G(t) + K_1 \int_0^t m_1(t - \tau)q(\tau) \, d\tau \qquad (2.177)$$

where $K_1 = \sum_{n=1}^{\infty} n q_n$. Since $m_1(t) = m(t) = \mathscr{E}\{X(t)\}$, the expected value of $X(t)$ is obtained by solving Eq. (2.177). Similarly, if we differentiate (2.174) twice, we obtain the integral equation

$$m_2(t) = K_2 \int_0^t [m_1(t-\tau)]^2 q(\tau)\,d\tau + K_1 \int_0^t m_2(t-\tau)q(\tau)\,d\tau \qquad (2.178)$$

where $K_2 = \sum_{n=2}^{\infty} n(n-1)q_n$. Since $m_1(t)$ is a known function [obtained by solving Eq. (2.177)], the first integral in (2.178) can be evaluated without difficulty. After solving integral equation (2.178) for $m_2(t)$, the variance of $X(t)$ can be obtained, since $\mathscr{D}^2\{X(t)\} = m_2(t) + m_1(t) - [m_1(t)]^2$.

It is of interest to note that the equations for the moments are integral equations of the *renewal type*, so called because of their occurrence in renewal theory.[1] Equations of this type have been investigated by Feller [26] and Täcklind [82].

While we have considered only the first two moments, Bellman and Harris have shown that the kth moments

$$\mu_k(t) = \sum_{x=0}^{\infty} x^k P_x(t) \qquad k = 1, 2, \ldots \qquad (2.179)$$

exist for $t \in [0, \infty)$, provided the moments $\sum_{n=0}^{\infty} n^k q_n$ exist for $k = 1, 2, \ldots$. The asymptotic theory developed by Feller has been used by Bellman and Harris to discuss the limiting behavior of the moments μ_k. In particular, it has been shown that, under the following assumptions:

(a) $dG(\tau) \geqslant 0 \qquad G(0) = 1 \qquad G(\infty) = 1$

(b) $G(t) = \int_0^t g(\tau)\,d\tau$

(c) $\int_0^{\infty} e^{-\alpha\tau}|dg(\tau)| < \infty \qquad$ for some $\alpha > 0$ whose value depends on K_1

(d) $1 < K_1 < \infty$

the asymptotic relations

$$\mu_k \sim \beta_k e^{k\beta t} \qquad k = 1, 2, \ldots \qquad (2.180)$$

obtain as $t \to \infty$, where β is defined as the positive root of

$$1 = K_1 \int_0^{\infty} e^{-\beta\tau}\,dG(\tau)$$

provided that $\sum_{n=0}^{\infty} n^k q_n < \infty, k = 1, 2, \ldots$.

[1] For an excellent discussion of renewal equations we refer to Bellman [8].

C. Probability of Extinction. In Sec. 1.4 we showed that, if the expected number of individuals in the first generation was less than or equal to unity (when $X_0 = 1$), the population would die out with probability one; and if the expected number was greater than unity, the probability of extinction would be given by the unique nonnegative root, less than unity, of the functional equation $F(s) = s$. In the case of age-dependent branching processes Bellman and Harris have shown that there is a positive probability that $X(t)$ never assumes the value zero if and only if $\sum_{n=0}^{\infty} nq_n > 1$. In this case, the probability that $X(t) = 0$, for some time t, is given by the unique nonnegative root, less than unity, of the functional equation

$$h(s) = s \qquad (2.181)$$

where $h(s)$ is defined by (2.175).

D. Some Branching Models Based on Specific $h(s)$ and $G(\tau)$. We now construct several branching models by assuming specific forms for $h(s)$ and $G(\tau)$ and inserting them in the functional equation for the generating function. Some of the models we consider have been studied in Sec. 2.4, where the Kolmogorov equations were used; hence, we will only show that these processes can be obtained from the Bellman-Harris formulation.

1. *A Deterministic Model.* The *Galton-Watson process* is a binary birth process with birth rate $\lambda > 0$ in which the population size doubles at intervals of time equal to $1/\lambda$. Since the process is binary, $h(s) = s^2$, i.e., $q_2 = 1$, while the other transformation probabilities are zero. $G(\tau)$ in this case is the step function

$$G(\tau) = 1 \qquad \text{for } \tau \geqslant \frac{1}{\lambda}$$

$$= 0 \qquad \text{for } \tau < \frac{1}{\lambda} \qquad (2.182)$$

With $h(s)$ and $G(\tau)$ thus defined, the equation for the generating function becomes

$$F\left(s, \frac{k+1}{\lambda}\right) = F^2\left(s, \frac{k}{\lambda}\right) \qquad (2.183)$$

where $F(s,0) = s$ if $X(0) = 1$.

2. *The Simple Birth Process.* To obtain the simple birth process, which is of the Markov type, we put $h(s) = s^2$ and $G(t) = 1 - e^{-\lambda t}$, the negative exponential distribution. The generating function in this case satisfies the integral equation

$$F(s,t) = \int_0^t F^2(s, t - \tau)e^{-\lambda \tau}\, d\tau + se^{-\lambda t} \qquad (2.184)$$

Whenever the distribution is a negative exponential, i.e., the process is of the Markov type, the integral equation for the generating function can be expressed as a differential equation. If we multiply (2.184) by $e^{\lambda t}$ and introduce the new variable $\xi = t - \tau$, we obtain

$$F(s,t)e^{\lambda t} = \int_0^t F^2(s,\xi)\lambda e^{\lambda \xi}\, d\xi + s$$

By differentiating with respect to t, we obtain the differential equation

$$\frac{\partial F}{\partial t} = \lambda F(F - 1) \tag{2.185}$$

Solution of this equation with the initial condition $F(s,0) = s$ gives the same solution obtained earlier, i.e., (2.113).

3. *The Simple Birth-and-Death Process.* To represent this process, we put $G(t) = 1 - e^{-(\lambda+\mu)t}$, $q_0 = \dfrac{\mu}{\lambda + \mu}$, and $q_2 = \dfrac{\lambda}{\lambda + \mu}$. Hence, the generating function satisfies

$$F(s,t) = \int_0^t \{\mu + \lambda F^2(s,t)\}e^{-(\lambda+\mu)\tau}\, d\tau + se^{-(\lambda+\mu)t} \tag{2.186}$$

By proceeding as before, we obtain the differential equation

$$\frac{\partial F}{\partial t} = \lambda F^2 = (\lambda + \mu)F + \mu \tag{2.187}$$

Solving this equation with initial condition $F(s,0) = s$ yields (2.150).

4. *A Non-Markovian Birth Process.* Kendall [50,51] has considered a non-Markovian binary birth process which can be described as a multiple-phase birth process. This process can be described as follows: When an individual is born, it passes through a series of distinct phases, say k in number, and only after it has completed the kth phase does division into two individuals take place. The times spent in the phases are independent random variables, each having the distribution function $1 - e^{-k\lambda t}$. Hence the distribution function of the generation time τ is the k-fold convolution of the distribution function for each phase, that is,

$$dG(\tau) = \frac{k^k \lambda^k}{\Gamma(k)}e^{-k\lambda\tau}\tau^{k-1}\, d\tau \qquad 0 < \tau < \infty \tag{2.188}$$

It is of interest to note that, when $k = 1$, the multiple-phase process is the same as the simple birth process. Also, if k tends to infinity, the multiple-phase process becomes the deterministic, or Galton-Watson, process which is characterized by a fixed generation time $\tau = 1/\lambda$.

5. *Other Age-dependent Processes.* In subsequent chapters we shall consider other age-dependent processes that arise in biology and physics.

E. The Waugh Model. Transformation Probabilities as Functions of Time. The age-dependent processes considered by Bellman and Harris were assumed to be binary fission processes with constant transformation probabilities. Waugh [89] has considered a generalization of the Bellman-Harris process in which the transformation probabilities are functions of time, that is, $q_n = q_n(t)$. In this case the generating function of the transformation probabilities becomes

$$h(s,t) = \sum_{n=0}^{\infty} q_n(t)s^n \qquad (2.189)$$

and the generating function of the probabilities $P_x(t)$ satisfies

$$F(s,t) = \int_0^t h[F(s, t-\tau), \tau]g(\tau)\,d\tau + s[1 - G(t)] \qquad (2.190)$$

Similarly, the integral equations for the first two derivatives of $F(s,t)$, evaluated at $s = 1$, are

$$m_1(t) = 1 - G(t) + \int_0^t K_1(\tau)m_1(t-\tau)g(\tau)\,d\tau \qquad (2.191)$$

$$m_2(t) = \int_0^t K_2(\tau)[m_1(t-\tau)]^2 g(\tau)\,d\tau + \int_0^t K_1(\tau)m_2(t-\tau)g(\tau)\,d\tau \qquad (2.192)$$

respectively, where $K_1(\tau) = \sum_{n=1}^{\infty} nq_n(\tau)$ and $K_2(\tau) = \sum_{n=2}^{\infty} n(n-1)q_n(\tau)$. Waugh has given a detailed analysis of a non-Markovian birth-and-death process in which the distribution of the generation time is of the convolution form considered by Kendall. We refer to Waugh's paper for details.

We now consider a model of the Waugh type.[1] This model, which is of the birth-and-death type, is based on the following assumptions:

$$G(\tau) = 1 - e^{-\lambda\tau} \qquad \lambda > 0 \qquad (2.193)$$

and $\qquad q_0(\tau) = 1 - \beta e^{-\alpha\tau} \qquad \alpha > 0, 0 \leqslant \beta \leqslant 1$

$$(2.194)$$

$$q_1(\tau) = 1 - [q_0(\tau) + q_2(\tau)] = (\beta - \gamma)e^{-\alpha\tau} \qquad q_2(\tau) = \gamma e^{-\alpha\tau} \qquad 0 \leqslant \gamma \leqslant 1$$

The negative-exponential form of $G(\tau)$ expresses the Markov property of the stochastic process. In defining the transformation probabilities, the variable τ represents the length of the interval of time between the

---

[1] Cf. Woods and Bharucha-Reid [91].

formation of an individual and his transformation. When transformation takes place, the following may occur: (1) The individual can die with probability $q_0(\tau)$, where $q_0(\tau)$ is an increasing function of τ that is equal to $1 - \beta$ when $\tau = 0$ and equal to unity when $\tau = \infty$. (2) The individual can be transformed into two new individuals with probability $q_2(\tau)$, where $q_2(\tau)$ is a decreasing function of τ that is equal to γ when $\tau = 0$ and equal to zero when $\tau = \infty$. (3) The individual can continue to live with probability $q_1(\tau)$, where $q_1(\tau)$ is also a decreasing function of τ that is equal to $(\beta - \gamma)$ when $\tau = 0$ and equal to zero when $\tau = \infty$. In order for $q_1(\tau)$ to be nonnegative, we must have $\beta \geqslant \gamma$. Hence, the equation for the generating function becomes

$$F(s,t) = \int_0^t \sum_{n=0}^{2} q_n(\tau) F^n(s, t - \tau) g(\tau)\, d\tau$$

$$= \int_0^t \{1 - \beta e^{-\alpha\tau} + (\beta - \gamma)e^{-\alpha\tau}F(s, t - \tau)$$

$$+ \gamma e^{-\alpha\tau}F^2(s, t - \tau)\}\lambda e^{-\lambda\tau}\, d\tau + se^{-\lambda t} \qquad (2.195)$$

This integral equation can be reduced to the differential equation

$$\frac{\partial F(s,t)}{\partial t} = \lambda\gamma F^2(s,t) - \{\alpha + \lambda[1 - (\beta - \gamma)]\}F(s,t)$$

$$+ \alpha + \lambda(1 - \beta) + \alpha(s - 1)e^{-\lambda t} \qquad (2.196)$$

A solution of this nonlinear and nonhomogeneous equation which can be expressed as a series expansion in s is difficult to obtain; hence, we restrict our attention to the integral equations which will yield the moments.

Differentiation of (2.195) gives the integral equation

$$m_1(t) = \lambda(\beta + \gamma)\int_0^t m_1(t - \tau)e^{-(\alpha + \lambda)\tau}\, d\tau + e^{-\lambda t} \qquad (2.197)$$

which can, in turn, be reduced to the differential equation

$$\frac{dm_1(t)}{dt} = -\{\alpha + \lambda[1 - (\beta + \gamma)]\}m_1(t) + \alpha e^{-\lambda t} \qquad (2.198)$$

The solution of Eq. (2.198), assuming $m_1(0) = 1$, gives

$$m_1(t) = \mathscr{E}\{X(t)\} = \frac{1}{\alpha - (\beta + \gamma)}\,[\alpha e^{-\lambda t} - \alpha(\beta + \gamma)$$

$$\times \exp\{-[\alpha + \lambda(1 - (\beta + \gamma))]t\}] \qquad (2.199)$$

To turn now to the variance, we see that the second derivative of $F(s,t)$ satisfies

$$m_2(t) = 2\lambda\gamma \int_0^t [m_1(t-\tau)]^2 e^{-(\alpha+\lambda)\tau} \, d\tau + \lambda(\beta+\gamma) \int_0^t m_2(t-\tau) e^{-(\alpha+\lambda)\tau} \, d\tau \tag{2.200}$$

This integral equation can be reduced to the differential equation

$$\frac{dm_2(t)}{dt} = -[\alpha + \lambda(1-(\beta+\gamma))]m_2(t) + 2\lambda\gamma[m_1(t)]^2 \tag{2.201}$$

Since $m_1(t)$ is known, the solution of Eq. (2.201), satisfying the initial condition $m_2(0) = 0$, is

$$m_2(t) = 2\lambda\gamma B^{-2} e^{At} \{\alpha^2(2B+A)^{-1} e^{(2B+A)t}$$
$$- 2\alpha(B-\alpha)\lambda^{-1} e^{-\lambda t} + (B-\alpha)^2 A^{-1} e^{At}\} + Ce^{At} \tag{2.202}$$

where $A = -\{\alpha + \lambda[1-(\beta+\gamma)]\}$

$B = \alpha - \lambda(\beta+\gamma)$

$C = -2\lambda\gamma B^{-2}\{\alpha^2(2B+A)^{-1} - 2\alpha(B-\alpha)\lambda^{-1} + (B-\alpha)^2 A^{-1}\}$

Since $\mathscr{D}^2\{X(t)\} = m_2(t) + m_1(t) - [m(t)]^2$, the rather involved expression for the variance of $X(t)$ can be obtained from (2.199) and (2.202).

For other properties of this particular process we refer to Ref. 91.

2.6   Limit Theorems

A.   Introduction. In this section we shall study some limit theorems for temporally homogeneous Markov processes with a denumerable number of states. In particular, we shall consider certain ergodic theorems, limit theorems associated with $\sup_{t \geqslant 0} \omega(t)$ and $\max_{t \geqslant 0} X(t)$, theorems for birth-and-death processes, and some theorems for branching processes depending on a continuous time parameter.

B.   Some Ergodic Theorems. The first theorem we consider, due to Lévy [62], is concerned with the existence of limiting transition probabilities, i.e., $\lim_{t \to 0} P_{ij}(t)$.

Theorem 2.9: If $\{X(t), t \geqslant 0\}$ is a Markov process with transition probabilities satisfying the continuity condition $\lim_{t \to 0} P_{ij}(t) = \delta_{ij}$, then

$$\lim_{t \to \infty} P_{ij}(t) = \pi_{ij} \tag{2.203}$$

exists for all i and j.[1]

----

[1] This result is simpler than the analogous result in the discrete parameter case, because here Cesàro limits are not needed.

Proof: Let τ be an arbitrary positive number; then the process $\{X(n\tau), n = 0, 1, \ldots\}$ is a stationary Markov *chain* with transition probabilities

$$p_{ij}^{(n)} = P_{ij}(n\tau)$$

and, for $n = 1$, the matrix of transition probabilities $P = (P_{ij}(\tau))$, $i, j = 0, 1, \ldots$ for any one $\tau > 0$. The chain defined by the above transition probabilities is aperiodic, since $P_{ii}(n\tau) > 0$, and, therefore, $\lim_{n\to\infty} P_{ij}(n\tau)$ exists for each fixed $\tau > 0$. Now, from the uniform continuity[1] of the transition probabilities for i and j fixed, it follows that this limit is independent of τ and that the limit (2.203) exists.

Let $\Pi = (\pi_{ij})$, $i, j = 0, 1, \ldots$, denote the matrix of limiting transition probabilities. A method of calculating Π from the infinitesimal transition probabilities a_{ij} has been given by Kendall and Reuter [55]. Their methods utilize the theory of semigroups of operators. The results they obtain, which are based on the properties of the Markov chain defined by the matrix $P = (P_{ij}(\tau))$, can also be obtained by utilizing the following result due to Doob:[2]

$$\sum_{k=0}^{\infty} \pi_{ik} P_{kj}(t) = \sum_{k=0}^{\infty} P_{ik}(t)\pi_{kj} = \sum_{k=0}^{\infty} \pi_{ik}\pi_{kj} = \pi_{ij} \qquad (2.204)$$

for each $t \geqslant 0$. In matrix form the above becomes

$$\Pi P = P\Pi = \Pi^2 = \Pi \qquad (2.205)$$

As in the discrete parameter case,[3] the states of the process can be classified on the basis of the elements of Π. The state i is called *positive* if $\pi_{ii} > 0$ or *dissipative* if $\pi_{ii} = 0$. Also, two positive states i and j are in the same *positive class* if $\pi_{ij} > 0$.

The next theorem we consider, due to Karlin and McGregor [46], is concerned with the limiting behavior of the ratio of transition probabilities $P_{ij}(t)/P_{rs}(t)$ in the case of birth-and-death processes. Their result, which is an *ergodic theorem of the ratio type*, is given by the following theorem.

Theorem 2.10: For a birth-and-death process with $\mu_0 \geqslant 0$,

$$\lim_{t\to\infty} \frac{P_{ij}(t)}{P_{rs}(t)}$$

exists and is finite and positive.

[1] Cf. Doob [21].
[2] *Ibid.*
[3] Cf. Sec. 1.6B.

We omit the proof of Theorem 2.10, because it utilizes the integral representation of transition probabilities given in Ref. 45 and unfortunately we have not given in this text an exposition of this approach. It has been shown that in the case of recurrent processes (with $\mu_0 = 0$)

$$\lim_{t\to\infty} \frac{P_{ij}(t)}{P_{rs}(t)} = \frac{\pi_j}{\pi_s}$$

$$= \frac{\lambda_0\lambda_1 \cdots \lambda_{j-1}}{\mu_1\mu_2 \cdots \mu_{j-1}} \bigg/ \frac{\lambda_0\lambda_1 \cdots \lambda_{s-1}}{\mu_1\mu_2 \cdots \mu_s} \qquad (2.206)$$

while in the case of transient processes the limit may not be independent of the initial states i and r.

C. Limit Theorems Associated with $\sup\limits_{t\geqslant 0} \omega(t)$ and $\max\limits_{t\geqslant 0} X(t)$.
Let $\{X(t), t \geqslant 0\}$ be a discontinuous Markov process with state space $\mathfrak{X}$. We do not require at the outset that $\mathfrak{X}$ be denumerable. We define

$$S(\omega) = \sup_{t\geqslant 0} \omega(t) \qquad (2.207)$$

that is, $S(\omega)$ is the supremum of the sample function $\omega(t)$ and represents the largest value assumed by $X(t)$ for all $t \geqslant 0$. In applications, $S(\omega)$ might, for example, represent the maximum number of organisms in a bacterial population, the maximum number of electrons in a cosmic-ray shower, or the maximum number of people having to wait in a queue. In this section we shall consider some limit theorems associated with $S(\omega)$. The theorems we consider are due to Urbanik [84–87]. The first theorem is concerned with the conditional distribution function of $S(\omega) = \sup\limits_{t\geqslant 0} \omega(t)$.

Theorem 2.11: If the state space $\mathfrak{X}$ is the space of nonnegative integers, the conditional distribution functions

$$F_n(s) = \mathscr{P}\{S(\omega) < s \mid X(0) = n\} \qquad (2.208)$$

satisfy the system of linear equations

$$F_n(s) + \sum_{r\,\in\,R_s^{(n)}} F_r(s)A_n^{(r)}(s) = B_n(s) \qquad n, s = 0, 1, \ldots \quad (2.209)$$

where

$$R_s^{(n)} = \{m\colon m \in M_s^{(n)},\, m < s\}$$

$$A_n^{(r)} = \int_0^\infty \left\{ \sum_{u\,\in\,R_s^{(n)}} a_{ur} P_{nu}(t) \right\} dt$$

$$B_n(s) = \lim_{t\to\infty} \sum_{u\,\in\,R_s^{(n)}} P_{nu}(t)$$

Before proving this theorem, we state certain definitions, lemmas, and theorems which will be used in the sequel. In addition to these results being required in the sequel, they are of independent interest. For details we refer to Ref. 87.

For a general-state space $\mathfrak{X}$, we first define the intensity functions[1]

$$q(x) = \lim_{t \to 0} \frac{1 - P(x, t; \{x\})}{t}$$

and
$$q(x,A) = \lim_{t \to 0} \frac{P(x,t;A)}{t} \qquad A \in \mathfrak{X} - \{x\}$$

where $P(x,t;A)$ is the probability of a transition from the set $\{x\}$, consisting of the single element x, to a set A in time t.

We next introduce the concepts of critical point and limit point of a Markov process. The point $x \in \mathfrak{X}$ is a *critical point* if for every $t \geqslant 0$

$$P(x, t; \{x\}) = 1$$

In terms of the intensity function $q(x)$, we can state that a point x is a critical point if and only if $q(x) = 0$. The point $x \in \mathfrak{X}$ is a *limit point* if for every neighborhood U of x

$$\lim_{t \to \infty} \mathscr{P}\left\{ \bigcup_{T \geqslant t} \{\omega: \ \omega(T) \in U\} \right\} > 0$$

If $\mathfrak{X}$ is discrete, the following theorem characterizes the noncritical limit points of the process in terms of the transition probabilities.

Theorem 2.12: If the state space $\mathfrak{X}$ is discrete, the point $x \in \mathfrak{X}$ is a limit point of the Markov process if and only if

$$\int_0^\infty P(x_0,t;x) = \infty \qquad \text{or} \qquad \int_0^\infty \mathscr{P}\{\omega(t) = x\} \, dt = \infty$$

The next concept we require is that of the limit set of the sample function $\omega(t)$. The set

$$L(\omega) = \bigcap_{t \geqslant 0} \overline{\bigcup_{T \geqslant t} \{x: \ \omega(T) = x\}}$$

is termed the *limit set of the sample function* $\omega(t)$.[2] If the state space $\mathfrak{X}$ is discrete, the closure is considered in the space of all nonnegative integers completed by the limit point ∞. If $\mathfrak{X}$ is compact, then $L(\omega) = \{l\}$, a one-point set, if and only if there exists a limit in the ordinary sense, i.e., $\lim_{t \to \infty} \omega(t) = l$.

[1] Cf. also Sec. 2.2.
[2] The idea of the limit set of a sample function is due to E. Marczewski.

The next two theorems describe the properties of limit sets of sample functions of Markov processes.

Theorem 2.13: For almost all sample functions $\omega \in \Omega$ the set $L(\omega)$ consists of the limit points of the process.

Theorem 2.14: There exists a sequence of disjoint open sets $\Gamma_1, \Gamma_2, \ldots$ such that for almost any

$$\text{Int } L(\omega) = \Gamma_1, \text{ or } \Gamma_2, \text{ or } \cdots$$

In Theorem 2.14 Int $L(\omega)$ denotes the interior of the limit set, and in Theorem 2.11 the symbol $M_s^{(n)}$, occurring in the definition of the set $R_s^{(n)}$, denotes the sum of all sets $\Gamma_1, \Gamma_2, \ldots$ of the process satisfying the initial condition $x_0 = x$ (cf. Theorem 2.12) and containing at least one number not less than s.

Finally, we need the following lemmas:

Lemma 2.6: If x is not a critical point and not a limit point of the process, then there exists a neighborhood U of x which is compact, after closure, for which

$$\lim_{t \to \infty} H(x,t,U) > 0 \qquad \text{or} \qquad P(x,t;U) \equiv 0$$

In the above, the function $H(x,t,U)$ is defined as

$$H(x,t,U) = \int_{\mathfrak{X}-U} \mathscr{P}\left\{ \bigcap_{0 \leqslant \tau \leqslant t} \{\omega(\tau) \notin U\} \mid \omega(0) = y \right\} q(x,dy)$$

and for the intensity function $q(x,A)$, $x \in U$, and $A \subset \mathfrak{X} - U$.

Lemma 2.7: If $m \in \mathfrak{X} - A(n)$, then

$$F_m(s) = \mathscr{P}\left\{ \left(\bigcap_{t \geqslant 0} \{\omega(t) < s\} \right) \bigcap \left(\bigcap_{T \geqslant 0} \{\omega(T) \notin M_s^{(n)}\} \right) \mid \omega(0) = n \right\}$$

In the above lemma $A(n) = \bigcup_{A \in \alpha(n)} A$, the class $\alpha(n)$ consisting of open sets A such that $P(x,t;A) = 0$ for each $t \geqslant 0$. Since the set A is open, the theorem of Lindelöf[1] yields the result $A(n) = \bigcup_{i=1} A_i$, $A_i \in \alpha(n)$. Hence we have $P(x,t;A(n)) = 0$ for each $t \geqslant 0$.

We can now proceed to the proof of Theorem 2.11.

Proof: Let

$$C_n(B,t) = \mathscr{P}\left\{ \bigcap_{0 \leqslant \tau \leqslant t} \{\omega(\tau) \in B\} \mid \omega(0) = n \right\} \qquad B \subset \mathfrak{X} \quad (2.210)$$

[1] Cf. C. Kuratowski, "Topologie," vol. 1, 4th ed., Pańtswowe Wydawnictwo Naukowe, Warsaw, 1958.

Since $\mathfrak{X}$ is the space of nonnegative integers, the above expression can be written in terms of the transition probabilities $P_{ij}(t)$ and the infinitesimal transition probabilities as

$$C_n(B,t) = \sum_{k \in B} P_{nk}(t) - \sum_{r \in B} \sum_{k \notin B} \int_0^t C_r(B, t - \tau) a_{kr} P_{nk}(\tau)\, d\tau \qquad (2.211)$$

From Lemma 2.7 we obtain, by using (2.210),

$$F_r(s) = \lim_{t \to \infty} C_r(R_s^{(n)}, t) \qquad (2.212)$$

for $r \in \mathfrak{X} - A(n)$, where $R_s^{(n)} = \{m : m \in M_s^{(n)},\ m < s\}$. Since the set $R_s^{(n)}$ is finite, we have, by using (2.211),

$$C_n(R_s^{(n)}, t) = \sum_{k \in R_s^{(n)}} P_{nk}(t) - \sum_{r \in R_s^{(n)}} \sum_{k \notin R_s^{(n)}} \int_0^t C_r(R_s^{(n)}, t - \tau) a_{kr} P_{nk}(\tau)\, d\tau$$

$$(2.213)$$

We shall now show that for each $r \in R_s^{(n)}$ the following inequality obtains for each $t \geqslant 0$:

$$A_n^{(r)} = A_n^{(r)}(s,t) = \int_0^t \left\{ \sum_{k \in R_s^{(n)}} a_{kr} P_{nk}(\tau) \right\} d\tau \leqslant s < \infty \qquad (2.214)$$

If r is not a limit point of the process [i.e., the process satisfying that initial condition $\omega(0) = n$], then from Theorem 2.12 we obtain

$$\int_0^\infty P_{nr}(\tau)\, d\tau < \infty \qquad (2.215)$$

From the Kolmogorov equations

$$\frac{dP_{ij}(t)}{dt} = a_{jj} P_{ij}(t) + \sum_{k \neq i} a_{kj} P_{ik}(t)$$

we obtain the inequality

$$\sum_{k \notin R_s^{(n)}} a_{kr} P_{nk}(\tau) \leqslant a_{rr} P_{nr}(\tau) + \frac{dP_{nr}(\tau)}{d\tau}$$

which in view of (2.215) proves (2.214) for points r which are not limit points.

We now assume that $r \in R_s^{(n)}$ and that r is a limit point [therefore, in particular, $r \in \mathfrak{X} - A(n)$]. Now, (2.212), (2.210), and Lemma 2.6 imply the identity $C_r(R_s^{(n)}, t) \equiv 1$ for r a limit point. From this identity, and from (2.213), we obtain the inequality (2.214) for limit points r. Hence we have proved (2.214) for all points r. From (2.214) we obtain, as $t \to \infty$,

$$\lim_{t \to \infty} C_n(R_s^{(n)}, t) = \lim_{t \to \infty} \sum_{k \in R_s^{(n)}} P_{nk}(t) - \sum_{r \in R_s^{(n)}} \lim_{t \to \infty} C_r(R_s^{(n)}, t) A_n^{(r)}(s,t)$$

Hence, in view of (2.212) and the fact that $A_n^{(r)}(s,t) = 0$ for $r \in A(n)$, Theorem 2.11 is proved.

The next theorem we consider is concerned with the random variable which represents the difference between the largest and smallest times at which the random variable $X(t)$ assumes its maximum value. The process $\{X(t), t \geqslant 0\}$ is assumed to be of the birth-and-death type. Let

$$M(X) = \max_{t \geqslant 0} X(t) \qquad (2.216)$$

and let
$$\tau_1(X) = \inf_{X(t) = M(X)} t \qquad \tau_2(X) = \sup_{X(t) = M(X)} t$$

We now define the random variable

$$Y(X) = \tau_2(X) - \tau_1(X) \qquad (2.217)$$

Theorem 2.15: The conditional expectation of $Y(X)$ under the assumption $M(X) = m$ is given by

$$\mathscr{E}\{Y(X) \mid M(X) = m\} = \frac{1}{a_{10}} \sum_{k=1}^{m} \frac{U_{k-1}}{k^n}\left(\frac{U_{m-k}}{U_m} - \frac{U_{m-k-1}}{U_{m-1}}\right) \qquad (2.218)$$

where
$$U_i = 0 \qquad \text{for } i < 0$$
$$U_0 = 1$$

$$U_i = (-a_{10})^{-1} \begin{vmatrix} a_{11} & a_{10} & 0 & 0 & \cdots & 0 \\ a_{12} & a_{11} & a_{10} & 0 & \cdots & 0 \\ \cdot & \cdot & & \cdot & \cdots & \\ \cdot & \cdot & & \cdot & \cdots & \\ \cdot & \cdot & & \cdot & \cdots & \\ a_{1i} & a_{1,i-1} & a_{1,i-2} & \cdots & & a_{11} \end{vmatrix}$$

for $i \geqslant 1$, and the a_{ij} are the infinitesimal transition probabilities.

In order to prove the above theorem we need the following lemma, which we state without proof.

Lemma 2.8: For k in the interval $1 \leqslant k \leqslant m$ the following formula holds:

$$\int_0^\infty \mathscr{P}\{X(t) = k, \max_{t \geqslant \tau} X(\tau) = m \mid X(0) = r\} = \frac{U_{k-1}}{ka_{10}}\left[\frac{U_{m-r}}{U_m} - \frac{U_{m-r-1}}{U_{m-1}}\right]$$

For the proof of this lemma we refer to Ref. 85.
We can now proceed to the proof of Theorem 2.15.

Proof: Since the integral

$$\int_0^\infty P_{rk}(t)\, dt$$

converges for $k \geqslant 1$, it is not difficult to obtain the inequality

$$\mathscr{P}\{\tau_i(X) \geqslant t \mid M(X) = m, X(0) = r\}$$
$$\leqslant [\mathscr{P}\{M(X) = m \mid X(0) = r\}]^{-1} \sum_{k=1}^m P_{rk}(t) \quad (2.219)$$

for $i = 1, 2$. From (2.219) it follows that

$$\lim_{t \to \infty} \inf_t \mathscr{P}\{\tau_i(X) \geqslant t \mid M(X) = m, X(0) = r\} = 0$$

Hence we obtain the equality

$$\mathscr{E}\{Y(X) \mid M(X) = m, X(0) = r\}$$
$$= \int_0^\infty [\mathscr{P}\{\tau_2(X) \geqslant t \mid M(X) = m, X(0) = r\}$$
$$- \mathscr{P}\{\tau_1(X) \geqslant t \mid M(X) = m, X(0) = r\}]\, di \quad (2.220)$$

From the formula

$$\mathscr{P}\{\tau_1(X) \geqslant t, M(X) = m \mid X(0) = r\}$$
$$= \sum_{k=1}^m \mathscr{P}\left\{\max_{\tau \geqslant t} X(\tau) = m, X(t) = k, \max_{t \geqslant \tau} X(\tau) < m \mid X(0) = r\right\}$$

it follows that

$$\mathscr{P}\{\tau_1(X) \geqslant t, M(X) = m \mid X(0) = r\}$$
$$= \sum_{k=1}^m \mathscr{P}\{M(X) = m \mid X(0) = k\}$$
$$\times \mathscr{P}\{X(t) = k, \max_{t \geqslant \tau} X(\tau) < m \mid X(0) = r\}$$

Similarly, we obtain for $\tau_2(X)$ the equality

$$\mathscr{P}\{\tau_2(X) \geqslant t, M(X) = m \mid X(0) = r\}$$
$$= \sum_{k=1}^m \mathscr{P}\{M(X) = m \mid X(0) = k\}$$
$$\times \mathscr{P}\{X(t) = k, \max_{t \geqslant \tau} X(\tau) \leqslant m \mid X(0) = r\}$$

Hence, from (2.220) it follows that

$$\mathscr{E}\{Y(X) \mid M(X) = m, X(0) = r\}$$

$$= \sum_{k=1}^{m} \frac{\mathscr{P}\{M(X) = m \mid X(0) = k\}}{\mathscr{P}\{M(X) = m \mid X(0) = r\}} \int_{0}^{\infty} \mathscr{P}\{X(t)$$

$$= k, \max_{t \geqslant \tau} X(\tau) = m \mid X(0) = r\} \, dt \qquad (2.221)$$

By using a result due to Zolotarev [92],[1] namely,

$$\mathscr{P}\{M(X) = m \mid X(0) = k\} = \frac{U_{m-k}}{U_m} - \frac{U_{m-k-1}}{U_{m-1}} \qquad m \geqslant 1$$

together with Lemma 2.8 and Eq. (2.221), we obtain the equality

$$\mathscr{E}\{Y(X) \mid M(X) = m, X(0) = r\}$$

$$= \frac{1}{a_{10}} \sum_{k=1}^{n} \frac{U_{k-1}}{k} \left[\frac{U_{m-k}}{U_m} - \frac{U_{m-k-1}}{U_{m-1}} \right] \qquad (2.222)$$

Since the right-hand side of (2.222) is independent of the initial condition $X(0) = r$, we have proved the statement of the theorem.

We close this subsection with the statement of several results for a birth-and-death process in which the death rate is proportional to the birth rate. That is, if $\lambda > 0$ is the birth rate, then the death rate $\mu = a\lambda$, $a > 0$. As before, let $M(X) = \max_{t \geqslant 0} X(t)$. We have the following theorem.

Theorem 2.16: If $\{X(t), t \geqslant 0\}$ is a birth-and-death process with birth rate λ and death rate $\mu = a\lambda$, then

$$\mathscr{P}\{M(X) < m \mid X(0) = n\} = \frac{a^n - a^m}{1 - a^m} \qquad \text{for } a \neq 1$$

$$= \frac{m - n}{m} \qquad \text{for } a = 1 \qquad (2.223)$$

Now let $$T = \min_{X(t) = M} t \qquad (2.224)$$

that is, T is the smallest time at which $X(t)$ assumes its maximum value.

The next result concerns the conditional expectation of T.

---

[1] Cf. also Urbanik [84].

Theorem 2.17: For a birth-and-death process with parameters as defined in Theorem 2.16

$$\mathscr{E}\{T \mid M(X) \leqslant m, X(0) = n, m > n\}$$

$$= \frac{(m+1)n}{2\mu(m+1-n)} \sum_{i=1}^{m} \frac{(i-n)(i-n+1)}{i^2(i+1)} \qquad \text{for } \lambda = \mu$$

$$= \frac{(1+a^{m+1})(1-a)}{(\mu-\lambda)(a^n-a^{m+1})} \sum_{i=n}^{m} \sum_{j=1}^{i-1} \frac{a^i(1-a^j)(a^{i-j}-1)(1-a^n)}{j(1-a^{i+1})(1-a^i)^2} \qquad \text{for } \lambda \neq \mu$$

$$(2.225)$$

It is of interest to note that on the basis of the above theorem we can conclude that the expected time for the random variable $X(t)$ to achieve its maximum value is not a decreasing function of the death rate. We also remark that in order for T to be finite, it is necessary and sufficient that $\lambda \neq \mu$ (i.e., $a = 1$).

D. Some Limit Theorems for Branching Stochastic Processes. In this section we shall consider several limit theorems for branching stochastic processes with a continuous time parameter. The first theorem, due to Sevastyanov [79], is concerned with a branching process of a special form; hence we describe this process before stating and proving the theorem. The second theorem we consider, due to Bellman and Harris [9], is concerned with age-dependent branching processes.

1. *A Branching Process of Special Form.* We consider a branching process the development of which can be described as follows: Each individual (or particle) in the population at time t has probability

$$\delta_{k1} + p_k \, \Delta t + o(\Delta t)$$

of producing k individuals in the interval $(t, t + \Delta t)$. This probability is assumed to be independent of the age and origin of the individual and the histories of the other individuals in the population. Also, k individuals are formed with probability

$$\delta_{k0} + q_k \, \Delta t + o(\Delta t)$$

in the interval $(t, t + \Delta t)$, this probability being independent of the number of other individuals in the population. The probabilities p_k and q_k satisfy the conditions

$$\sum_{k=1}^{\infty} p_k = 0 \qquad \sum_{k=0}^{\infty} q_k = 0$$

Let

$$f(s) = \sum_{k=0}^{\infty} p_k s^k \qquad g(s) = \sum_{k=0}^{\infty} q_k s^k$$

be the generating functions of the p_k and q_k, respectively, and let

$$a_1 = f'(1) \qquad b_1 = f''(1) \qquad c_1 = f'''(1)$$
$$a_2 = g'(1) \qquad b_2 = g''(1)$$

Now let the random variable $X(t)$ denote the number of individuals in the population at time t, if $X(0) = 0$; let $P_x(t) = \mathscr{P}\{X(t) = x\}$, $x = 0, 1, \ldots$; and let

$$F(s,t) = \sum_{x=0}^{\infty} P_x(t)s^x \qquad |s| \leqslant 1 \qquad (2.226)$$

denote the generating function of the $P_x(t)$.

As an example of the type of process here described, consider a population consisting of two types of individuals, say T_1 and T_2. Then, in the interval $(t, t + \Delta t)$ we have the following transitions and associated probabilities:

$$\mathscr{P}\{T_1 \to T_1\} = 1 + q_0\,\Delta t + o(\Delta t)$$
$$\mathscr{P}\{T_1 \to T_1 + kT_2\} = q_k\,\Delta t + o(\Delta t) \qquad\qquad k \neq 0$$
$$\mathscr{P}\{T_2 \to T_2\} = 1 + p_1\,\Delta t + o(\Delta t)$$
$$\mathscr{P}\{T_2 \to kT_2\} = p_k\,\Delta t + o(\Delta t) \qquad\qquad k \neq 1$$

These transitions can be expressed more conveniently as follows:

$$\mathscr{P}\{T_i \to \alpha_1 T_1 + \alpha_2 T_2\} = E_i^{(\alpha_1,\alpha_2)} + p_i^{(\alpha_1,\alpha_2)}\,\Delta t + o(\Delta t) \qquad \text{for } i = 1, 2$$
$$(2.227)$$

where $E^{(1,0)} = 1$, $E_2^{(0,1)} = 1$, and $E_i^{(1,2)} = 0$ otherwise. Let

$$f_i(s_1,s_2) = \sum_{(\alpha_1,\alpha_2)} p_i^{(\alpha_1,\alpha_2)} s_1^{\alpha_1} s_2^{\alpha_2} \qquad i = 1, 2$$

denote the generating function of the $p_i^{(\alpha_1,\alpha_2)}$. In terms of the p_k and q_k introduced earlier, we have

$$p_1^{(1,k)} = q_k \qquad p_2^{(0,k)} = p_k$$

and

$$f_1(s_1,s_2) = s_1 g(s_2)$$
$$f_2(s_1,s_2) = f(s_2)$$

Finally, let $P_i^{(\alpha_1,\alpha_2)}(t)$ denote the probability given by (2.227) and let

$$F_i(s_1,s_2,t) = \sum_{(\alpha_1,\alpha_2)} P_i^{(\alpha_1,\alpha_2)}(t)s_1^{\alpha_1} s_2^{\alpha_2}$$

It is easy to verify that the above generating function satisfies the functional equation

$$F_i(s_1, s_2, t + u) = F_i[F_1(s_1,s_2,t), F_2(s_1,s_2,t),u] \qquad (2.228)$$

for $t,u > 0$, $i = 1, 2$. It can also be shown that as $\Delta t \to 0$ we have the asymptotic relation

$$F_i(s_1,s_2,\Delta t) = s_i + f_i(s_1,s_2)\,\Delta t + o(\Delta t) \qquad (2.229)$$

It can be shown that $F_i(s_1,s_2,t)$ satisfies the differential equation

$$\frac{dF_i}{dt} = f_i(F_1,F_2) \qquad i = 1, 2 \tag{2.230}$$

with initial condition

$$F_i(s_1,s_2,0) = s_i \tag{2.231}$$

or the partial differential equation

$$\frac{\partial F_i}{\partial t} = f_1(s_1,s_2)\frac{\partial F_i}{\partial s_1} + f_2(s_1,s_2)\frac{\partial F_i}{\partial s_2} \tag{2.232}$$

From (2.230) and the definition of $f_i(s_1,s_2)$ we obtain the system of equations

$$\frac{dF_1}{dt} = F_1 g(F_2) \qquad \frac{dF_2}{dt} = f(F_2) \tag{2.233}$$

Similarly, Eq. (2.232) can be written as

$$\frac{\partial F_i}{\partial t} = s_1 g(s_2)\frac{\partial F_i}{\partial s_1} + f(s_2)\frac{\partial F_i}{\partial s_2} \tag{2.234}$$

In terms of the generating functions $F(s,t)$, defined by (2.226) and $F_1(s_1,s_2,t)$, we have the relation

$$F_1(s_1,s_2,t) = s_1 F(s,t) \tag{2.235}$$

Therefore, from (2.234) and (2.235) we obtain

$$\frac{\partial F}{\partial t} = g(s)F(s,t) + f(s)\frac{\partial F}{\partial s} \tag{2.236}$$

with initial condition $F(s,0) = 1$.

Now let $\Phi(s,t)$ denote the solution of the second equation in (2.233) satisfying the initial condition (2.231). Therefore, from (2.233), (2.231), and (2.235), we have

$$F(s,t) = \exp\left\{\int_0^t g[\Phi(s,u)]\,du\right\} \tag{2.237}$$

The above relation enables us to obtain the probabilities $P_x(t)$ from a knowledge of the generating function associated with individuals of type T_2.

We now state and prove the following theorem.

Theorem 2.18: If $a_1 < 0$ and $a_2 < \infty$, then the limiting probabilities

$$\lim_{t\to\infty} \mathscr{P}\{X(t) = x\} = \pi_x \qquad x = 0, 1, \ldots$$

exist, and the limiting generating function

$$F(s) = \sum_{x=0}^{\infty} \pi_x s^x$$

is given by

$$F(s) = \exp\left\{\int_s^1 \frac{g(\tau)}{f(\tau)}\,d\tau\right\} \tag{2.238}$$

Proof: We shall first prove that for $a_1 < 0, a_2 < \infty$, and $|s| \leqslant 1$

$$\lim_{t\to\infty} F(s,t) = \lim_{t\to\infty} \exp\left\{\int_0^t g[\Phi(s,u)]\,du\right\}$$

$$= \exp\left\{\int_0^\infty g[\Phi(s,u)]\,du\right\}$$

exist. In order to establish the above, it is necessary to prove that for $|s| \leqslant 1$ the improper integral

$$\int_0^\infty g[\Phi(s,u)]\,du \tag{2.239}$$

converges. Now, for $|s| \leqslant 1$ we have the inequalities

$$|g(s)| \leqslant s_2\,|s-1| \qquad |\Phi(s,u) - 1| \leqslant e^{a_1 u}\,|s-1| \tag{2.240}$$

since

$$\frac{\partial \Phi(1,u)}{\partial s} = e^{a_1 u}$$

therefore

$$|g[\Phi(s,u)]| \leqslant 2a_2 e^{a_1 u}$$

and, consequently, the integral (2.239) converges uniformly with respect to s in the unit circle $|s| \leqslant 1$. Thus the existence of the limiting generating function $F(s)$ is established.

We now show that

$$\int_0^\infty g[\Phi(s,u)]\,du = \int_s^1 \frac{g(\tau)}{f(\tau)}\,d\tau$$

Differentiation of (2.239) with respect to s yields

$$\frac{d}{ds}\int_0^\infty g[\Phi(s,u)]\,du = \int_0^\infty \frac{dg[\Phi(s,u)]}{d\Phi}\frac{\partial \Phi(s,u)}{\partial s}\,du$$

$$= \int_0^\infty \frac{dg}{d\Phi}\frac{\partial \Phi}{\partial u}\frac{1}{f(s)}\,du$$

$$= \int_0^\infty \frac{\partial g[\Phi(s,u)]}{\partial u}\frac{1}{f(s)}\,du$$

$$= \frac{g[\Phi(s,\infty)] - g[\Phi(s,0)]}{f(s)} = -\frac{g(s)}{f(s)}$$

In the above we have made use of the equality

$$\frac{\partial \Phi(s,u)}{\partial u} = f(s)\frac{\partial \Phi(s,u)}{\partial s}$$

which follows from (2.231), (2.232) (for $i = 2$), the initial condition $\Phi(s,0) = s$, and the fact that

$$\lim_{t\to\infty} \Phi(s,t) = 1 \qquad (2.241)$$

We remark that differentiation with respect to s in (2.239) is permitted, since the convergence in (2.241) is uniform with respect to s for $|s| \leqslant 1$. This completes the proof of the theorem.

Another theorem due to Sevastyanov, which we state without proof, concerns the limiting distribution of the normed random variable $X(t)/e^{a_1 t}$.

Theorem 2.19: If $a_1 > 0$ and a_2, b_1, and b_2 are finite, then the limiting distribution function

$$\lim_{t\to\infty} \mathscr{P}\left\{\frac{X(t)}{e^{a_1 t}} < x\right\} = \lim_{t\to\infty} S_t(x) = S(x) \qquad (2.242)$$

exists, the characteristic function of $S(x)$ being defined by

$$\psi(\tau) = \exp\left\{\frac{1}{a_1}\int_0^\tau \frac{g(u(z))}{z}\,dz\right\} \qquad (2.243)$$

where the function $u(\tau)$ is given by

$$(1 - u)\exp\left\{-\int_1^u \frac{f(v) - a_1(v-1)}{f(v)(v-1)}\,dv\right\} = -i\tau \qquad (2.244)$$

2. *Age-dependent Branching Processes.* In this subsection we consider a limit theorem associated with age-dependent branching processes. In particular, we consider the convergence of the normed random variable $X(t)/m(t)$, where $X(t)$ is the fundamental random variable, and $m(t) = \mathscr{E}\{X(t)\}$. We now state two lemmas which give certain properties of the process, and are required in the proof of the theorem.

Lemma 2.9: If $v(t)$ satisfies the integral equation

$$v(t) = \int_0^t v(t - \tau)\,dH(\tau) + K(t)$$

where $dH \geqslant 0$ $H(0+) = 0$ $H(\infty) = \alpha < 1$

$$|K(t)| \leqslant c_1 \qquad 0 \leqslant t \leqslant \infty$$

$$K(t) \to c_2 \qquad \text{as } t \to \infty$$

then

$$\lim_{t \to \infty} v(t) = \frac{c_2}{1-\alpha} \qquad (2.245)$$

If $|K(t) - c_2| = 0 \, (e^{-\varepsilon t})$ and $|\alpha - H(t)| = 0 \, (e^{-\alpha t})$ as $t \to \infty$, then there exists a $\delta > 0$ such that

$$\left| v(t) - \frac{c_2}{1-\alpha} \right| = 0(e^{-\delta t}) \qquad t \to \infty \qquad (2.246)$$

If for the set of functions $K(t,h)$ the quantities c_1 and $|K(t) - c_2|$ can be bounded independently of h, then relations (2.245) and (2.246) hold uniformly in h.

Lemma 2.10: If $dG \geqslant 0$, $G(0+) = 0$, $G(\infty) = 1$, and $G(t)$ is not a simple step function, then the solution of the integral equation for the mean $m(t)$ satisfies

$$m(t) \sim n_1 e^{\beta t} \qquad t \to \infty \qquad (2.247)$$

where

$$n_1 = \frac{1}{4\beta \displaystyle\int_0^\infty t e^{-\beta t} \, dG(t)}$$

and β is defined as the positive root of

$$1 = 2 \int_0^\infty e^{-\beta t} \, dG(t)$$

We now state and prove the following theorem.

Theorem 2.20: Under the hypotheses of Lemma 2.10 concerning the distribution function $G(t)$, the normed random variable $X(t)/m(t)$ converges in mean square to a random variable y as $t \to \infty$.

Proof: Let

$$Y(t) = \frac{X(t)}{n_1 e^{\beta t}}$$

By Lemma 2.10 it is sufficient to show that $Y(t)$ converges in mean square, which we do by showing that

$$\lim_{t_1, t_2 \to \infty} \mathscr{E}\{[Y(t_1) - Y(t_2)]^2\} = 0 \qquad (2.248)$$

We now introduce the joint generating function of $X(t_1)$ and $X(t_2)$:

$$F_2(s_1, s_2, t_1, t_2) = \sum_{x_1, x_2 = 0}^{\infty} \mathscr{P}\{X(t_1) = x_1, X(t_2) = x_2\} s_1^{x_1} s_2^{x_2} \qquad (2.249)$$

when $t_1 \leqslant t_2$.

By using the same reasoning employed in Sec. 2.5, it is easy to show that $F_2(s_1, s_2, t_1, t_2)$ satisfies the integral equation

$$F_2(s_1, s_2, t_1, t_2) = \int_0^t F_2^2(s_1, s_2, t_1 - \tau, t_2 - \tau) \, dG(\tau)$$

$$+ s_1 \int_{t_1}^{t_2} F^2(s_2, t_2 - \tau) \, dG(\tau) + s_1 s_2 [1 - G(t)] \qquad (2.250)$$

where $F(s, t)$ is the generating function of $X(t)$. If we now differentiate Eq. (2.250) with respect to s_1 and s_2 and put $s_1 = s_2 = 1$, we obtain the cross moment

$$\mathscr{E}\{X(t_1) X(t_2)\} = m^*(t_1, t_2) = 2 \int_0^{t_1} m^*(t_1 - \tau, t_2 - \tau) \, dG(\tau)$$

$$+ 2 \int_0^{t_1} m(t_1 - \tau) m(t_2 - \tau) \, dG(\tau)$$

$$+ 2 \int_{t_1}^{t_2} m(t_2 - \tau) \, dG(\tau) + 1 - G(t_2) \qquad t_1 \leqslant t_2$$

$$(2.251)$$

If we put $t_1 = t$, $t_2 = t + h$ ($h \geqslant 0$), and let

$$m^*(t, t + h) = e^{\beta h} e^{2\beta t} u(t, h)$$

then Eq. (2.251) becomes, as $t \to \infty$,

$$u(t, h) = 2 \int_0^t u(t - \tau, h) e^{-2\beta \tau} \, dG(\tau) + c_3 + o(1) \qquad (2.252)$$

where

$$c_3 = 2n_1^2 \int_0^{\infty} e^{-2\beta \tau} \, dG(\tau)$$

By estimation, it can be shown that the order term $o(1)$ in (2.252) is independent of h. If we now apply Lemma 2.9, we have

$$\lim_{t \to \infty} u(t, h) = c_3 \left[1 - 2 \int_0^{\infty} e^{-2\beta \tau} \, dG(\tau) \right]^{-1} = n_2 \qquad (2.253)$$

uniformly in h. Therefore

$$\mathscr{E}\{X(t) X(t + h)\} = n_2 e^{\beta h} e^{2\beta t} [1 + o(1)] \qquad h \geqslant 0 \qquad (2.254)$$

Equation (2.248) now follows from (2.254).

3. *Additional Studies.* For additional limit theorems associated with branching process (one- and N-dimensional, age-dependent, etc.)

we refer to the papers of Bellman and Harris [9], Čistyakov [19], Harris [33], Jiřina [39], Sevastyanov [77–80], Zolotarev [92,93], and the monograph of T. E. Harris.

2.7 N-dimensional Discontinuous Processes

In Sec. 1.8 we considered N-dimensional discrete processes; we now consider N-dimensional processes with continuous time parameter. As in the discrete case, we replace the scalar random variable $X(t)$ by the vector random variable $\mathbf{X}(t) = (X_1(t), X_2(t), \ldots, X_N(t))$, where the scalar components $X_i(t)$, $i = 1, 2, \ldots, N$, represent the number of individuals of type i in the population at time t. Similarly, the state space $\mathfrak{X}$ is N-dimensional, the states being represented by the points $\mathbf{x} = (x_1, x_2, \ldots, x_N)$.

In this section we derive the Feller integrodifferential equations for N-dimensional processes with general state space $\mathfrak{X}$. We then obtain as a special case the Kolmogorov differential equations for N-dimensional processes. For a detailed discussion of the general theory of N-dimensional processes we refer to Arley [1].

Let the events $E_1, E_2, \ldots, E_N$ be subsets of $\mathfrak{X}$ and let

$$P(E_1, \ldots, E_N, t; \xi_1, \ldots, \xi_N, \tau)$$
$$= \mathscr{P}\{X_1(t) \in E_1, \ldots, X_N(t) \in E_N \mid X_1(\tau) = \xi_1, \ldots, X_N(\tau) = \xi_N\} \quad (2.255)$$

That is, $P(E_1, \ldots, E_N, t; \xi_1 \ldots, \xi_N, \tau)$ is the conditional probability of the simultaneous occurrence of the events $X_1(t) \in E_1$, etc., given that $X_1(\tau) = \xi_1$, etc., where $t > \tau$. The above set function, which is a transition probability function, is assumed to be a continuous function of t and τ.[1] In addition, (2.255) satisfies the usual conditions

$$0 \leqslant P(E_1, \ldots, E_N, t; \xi_1, \ldots, \xi_N, \tau) \leqslant 1 \quad t \geqslant \tau \quad (2.256)$$
$$\int_{\mathfrak{x}_1, \ldots, \mathfrak{x}_N} d_E P(E_1, \ldots, E_N, t; \xi_1, \ldots, \xi_N, \tau) = 1 \quad t \geqslant \tau \quad (2.257)$$

The integral in (2.257) is an N-dimensional Lebesgue-Stieltjes integral. The Chapman-Kolmogorov equation in the N-dimensional case is given by

$$P(E_1, \ldots, E_N, t; \xi_1, \ldots, \xi_N, \tau)$$
$$= \int_{\mathfrak{x}_1, \ldots, \mathfrak{x}_N} P(E_1, \ldots, E_N, t; \omega_1, \ldots, \omega_N, s)$$
$$\times d_\Omega(\Omega_1, \ldots, \Omega_N, s; \xi_1, \ldots, \xi_N, \tau) \quad t \geqslant \tau, \tau \leqslant s \leqslant t \quad (2.258)$$

[1] The transition probability function is also assumed to be a Borel measurable function of $\xi_1, \xi_2, \ldots, \xi_N$.

For $t = \tau$, we have, since $P(E_1, \ldots, E_N, t; \xi, \ldots, \xi_N, \tau)$ is a continuous function of time,

$$\lim_{t \to \tau} P(E_1, \ldots, E_N, t; \xi_1, \ldots, \xi_N, \tau)$$

$$= \lim_{t \to \tau} P(E_1, \ldots, E_N, t; \xi_1, \ldots, \xi_N, \tau) = \chi(E_1) \cdots \chi(E_N) \quad (2.259)$$

where
$$\chi(E_i) = 1 \qquad \text{for } \xi_i \in E_i$$

$$= 0 \qquad \text{for } \xi_i \in \mathfrak{X} - E_i \quad (2.260)$$

In order to obtain the fundamental functional equations, we first introduce, as before, the *intensity function* $q(\xi_1, \ldots, \xi_N, t)$ and the *relative transition probability function* $Q(E_1, \ldots, E_N; \xi_1, \ldots, \xi_N, t)$. Analogously to the one-dimensional case, $q(\xi_1, \ldots, \xi_N, t) \, \Delta t + o \, (\Delta t)$ is the probability that $\mathbf{X}(t)$ will undergo a random change in the interval $(t, t + \Delta t)$ when $\mathbf{X}(t) = \boldsymbol{\xi} = (\xi_1, \ldots, \xi_N)$, and $Q(E_1, \ldots, E_N; \xi_1, \ldots, \xi_N, t)$ is the conditional probability of $X_i(t)$ assuming a value in E_i at time $t + \Delta t$, i.e., $X_i(t) \in E_i$, $i = 1, 2, \ldots, N$, when it is given that $\mathbf{X}(t) = \boldsymbol{\xi}$ and has undergone a change in the interval $(t, t + \Delta t)$. Both of the above functions are assumed to be continuous functions of t. We also require these functions to satisfy the conditions

$$q(\xi_1, \ldots, \xi_N, t) \geqslant 0 \qquad \text{for all } \xi_1, \ldots, \xi_N \text{ and } t \geqslant \tau \quad (2.261)$$

$$0 \leqslant Q(E_1, \ldots, E_N; \xi_1, \ldots, \xi_N, t) \leqslant 1$$

$$\text{for all } E_1, \ldots, E_n, \xi_1, \ldots, \xi_N, \text{ and } t \geqslant \tau \quad (2.262)$$

$$Q(E_1, \ldots, E_N; \xi_1, \ldots, \xi_N, t) = 0$$

$$\text{when } X_i(t) \in E_i, \text{ for } i = 1, 2, \ldots, N \quad (2.263)$$

$$\int_{\mathfrak{X}_1, \ldots, \mathfrak{X}_N} d_E Q(E_1, \ldots, E_N; \xi_1, \ldots, \xi_N, t) = 1$$

$$\text{for all } \xi_1, \ldots, \xi_N \text{ and } t \geqslant \tau \quad (2.264)$$

Finally, we have the regularity condition that for small increments Δt

$$P(E_1, \ldots, E_N, t + \Delta t; \xi_1, \ldots, \xi_N, t)$$

$$= [1 - q(\xi_1, \ldots, \xi_N, t) \, \Delta t] \chi(E_1) \cdots \chi(E_N)$$

$$+ Q(E_1, \ldots, E_N; \xi_1, \ldots, \xi_N, t) q(\xi_1, \ldots, \xi_N, t) \, \Delta t + o(\Delta t) \quad (2.265)$$

If we introduce (2.265) into the Chapman-Kolmogorov equation (2.258) and proceed as in the one-dimensional case, we obtain the pair of integrodifferential equations

$$\frac{\partial}{\partial t} P(E_1, \ldots, E_N, t; \xi_1, \ldots, \xi_N, \tau)$$

$$= -\int_{E_1, \ldots, E_N} q(\omega_1, \ldots, \omega_N, t)\, d_\Omega P(\Omega_1, \ldots, \Omega_N, t; \xi_1, \ldots, \xi_N, \tau)$$

$$+ \int_{\mathfrak{X}_1, \ldots, \mathfrak{X}_N} Q(E_1, \ldots, E_N; \omega_1, \ldots, \omega_N, t) q(\omega_1, \ldots, \omega_N, t)$$

$$\times d_\Omega P(\Omega_1, \ldots, \Omega_N, t; \xi_1, \ldots, \xi_N, \tau) \tag{2.266}$$

$$\frac{\partial}{\partial \tau} P(E_1, \ldots, E_N, t; \xi_1, \ldots, \xi_N, \tau)$$

$$= q(\xi_1, \ldots, \xi_N, \tau) P(E_1, \ldots, E_N, t; \xi_1, \ldots, \xi_N, \tau)$$

$$- q(\xi_1, \ldots, \xi_N, \tau) \int_{\mathfrak{X}_1, \ldots, \mathfrak{X}_N} P(E_1, \ldots, E_N, t; \omega_1, \ldots, \omega_N, \tau)$$

$$\times d_\Omega Q(\Omega_1, \ldots, \Omega_N; \xi_1, \ldots, \xi_N, \tau) \tag{2.267}$$

The above equations are, respectively, the *forward and backward functional equations for N-dimensional discontinuous Markov processes.*

We now consider the case when the scalar random variables $X_1(t)$, $\ldots, X_N(t)$ can assume only a denumerable number of values. In this case the set function $P(E_1, \ldots, E_N, t; \xi_1, \ldots, \xi_N, \tau)$ is reduced to the point function $P_{ij}(\tau, t)$, where $\mathbf{i} = (i_1, \ldots, i_N)$ and $\mathbf{j} = (j_1, \ldots, j_N)$ represent the initial and final states. Similarly, the function $Q(E_1, \ldots, E_N; \xi_1, \ldots, \xi_N, t)$ becomes $Q_{ij}(t)$. As in the one-dimensional case, the integrodifferential equations are replaced by the differential equations

$$\frac{\partial P_{ij}(\tau, t)}{\partial t} = -q_j(t) P_{ij}(\tau, t) + \sum_{k=0}^{\infty} P_{ik}(\tau, t) q_k(t) Q_{kj}(t) \tag{2.268}$$

$$\frac{\partial P_{ij}(\tau, t)}{\partial t} = q_i(\tau) \left\{ P_{ij}(\tau, t) - \sum_{k=0}^{\infty} Q_{ik}(\tau) P_{kj}(\tau, t) \right\} \tag{2.269}$$

In the above we have put $\mathbf{k} = (k_1, \ldots, k_N)$. Equations (2.268) and (2.269) are respectively the *forward and backward Kolmogorov equations for N-dimensional processes.*

In the above case the conditions (2.256) to (2.259) and the regularity condition (2.265) are replaced by

$$0 \leqslant P_{ij}(\tau,t) \leqslant 1 \tag{2.270}$$

$$\sum_{j_1=0}^{\infty} \cdots \sum_{j_N=0}^{\infty} P_{ij}(\tau,t) = 1 \tag{2.271}$$

$$\sum_{k_1=0}^{\infty} \cdots \sum_{k_N=0}^{\infty} P_{ik}(\tau,s)P_{kj}(s,t) \tag{2.272}$$

$$\lim_{t \to \tau} P_{ij}(\tau,t) = \lim_{\tau \to t} P_{ij}(\tau,t) = \delta_{i_1 j_1} \cdots \delta_{i_N j_N} \tag{2.273}$$

$$P_{ij}(t, t + \Delta t) = [1 - q_i(t)\, \Delta t]\, \delta_{i_1 j_1} \cdots \delta_{i_N j_N} + Q_{ij}(t)q_i(t)\, \Delta t + o(\Delta t) \tag{2.274}$$

Problems

2.1 By using the method of Reuter and Ledermann [74],[1] carry out the existence proof for the Kolmogorov differential equations with nonconstant coefficients, i.e., when the matrix of infinitesimal transition probabilities $A = A(t)$.

2.2 Let $\{X(t),\ t \geqslant 0\}$ be a time-homogeneous stochastic process with independent increments,[2] and let

$$Y_i(t) = \mathscr{P}\{[X(t) - X(0)] = i\}$$

satisfy the conditions

$$\lim_{t \to 0} \frac{Y_1(t)}{t} = \lambda > 0 \quad \text{and} \quad \lim_{t \to 0} \frac{1 - Y_0(t) - Y_1(t)}{t} = 0$$

Show that

$$Y_i(t) = \frac{(\lambda t)^i}{i!}\, e^{-\lambda t} \qquad i = 0, 1, \ldots$$

and, therefore, that

$$\mathscr{P}\{X(t) = i \mid X(0) = 0\} = \frac{(\lambda t)^i}{i!}\, e^{-\lambda t} \qquad i = 0, 1, \ldots$$

2.3 If $F(s,t)$ is the generating function associated with the random variable $X(t)$, show that $F(1/s, t)$ is the generating function associated with $-X(t)$.

2.4 Let $\{X(t),\ t \geqslant 0\}$ and $\{Y(t),\ t \geqslant 0\}$ be Poisson processes with parameters λ_1 and λ_2, respectively. Let $Z(t) = X(t) - Y(t)$, and let

$$F(s,t) = \sum_{x=0}^{\infty} P_x(t)s^x = e^{\lambda_1(1-s)t}$$

$$G(s,t) = \sum_{y=0}^{\infty} P_y(t)s^y = e^{\lambda_2(1-s)t}$$

(*a*) Show that

$$H(s,t) = \sum_{z=\infty}^{\infty} P_z(t)s^z = \exp\{(\lambda_1 + \lambda_2)t\} \exp\left\{-\left(\frac{\lambda_1 + s + \lambda_2}{s}\right)t\right\}$$

(*b*) Find $P_z(t)$. [*Hint:* Expand $H(s,t)$ in a Laurent series.]

---

[1] Cf. also Hille [36].

[2] A process $\{X(t),\ t \geqslant 0\}$ is said to have *independent increments* if for arbitrary times t_i $(i = 0, 1, \ldots, n)$, where $t_0 < t_1 < \cdots < t_n$, the random variables $X(t_1) - X(t_0)$, $X(t_2) - X(t_1)$, $\ldots$, $X(t_n) - X(t_{n-1})$ are independent.

2.5 Solve the equations for the birth process with

$$\lambda_x = \lambda + \gamma x \qquad x = 0, 1, \ldots \qquad (\lambda, \gamma > 0)$$

and with initial conditions $P_1(0) = 1$, $P_x(0) = 0$ for $x \neq 1$.

2.6 The *Ehrenfest process* is a finite birth-and-death process with $\lambda_x = (N - x)p$, $\mu_x = xq$ for $0 \leqslant x \leqslant N$, $0 < p < 1$, $q = 1 - p$. Solve the system of differential equations for this process, with initial conditions $P_1(0) = 1$, $P_x(0) = 0$ for $x \neq 1$, and discuss its properties.

2.7 Solve the system of differential equations

$$\frac{dP_x(t)}{dt} = \lambda P_{x-1}(t) - (\lambda + \mu x)P_x(t) + (x + 1)\mu P_{x+1}(t)$$

for $x > 0$ and

$$\frac{dP_0(t)}{dt} = -\lambda P_0(t) + \mu P_1(t)$$

with initial conditions $P_0(0) = 1$, $P_x(0) = 0$ for $x \geqslant 1$.

2.8 Solve the differential equations for the birth-and-death process with

$$\begin{aligned}
\lambda_x &= \lambda + \alpha x & x = 0, 1, \ldots \\
&= \lambda - \alpha' x & x = -1, -2, \ldots \\
\mu_x &= \mu + \beta x & x = 0, 1, \ldots \\
&= \mu - \beta' x & x = -1, -2, \ldots
\end{aligned}$$

and with initial conditions $P_1(0) = 1$, $P_x(0) = 0$ for $x \neq 1$. (*Hint:* Obtain the equation for the generating function and apply the bilateral Laplace transformation with respect to the new state variable.)

2.9 For the Prendiville model of the logistic birth-and-death process $\Big[$with

$$\lambda_x = \alpha\left(\frac{x_2}{x} - 1\right), \mu_x = \beta\left(1 - \frac{x_1}{x}\right), 0 < x_1 \leqslant X \leqslant x_2, x \leqslant x_0 \leqslant x_2, X(0) = x_0\Big],$$

the generating function $F(s,t)$ satisfies the partial differential equation

$$\frac{\partial F(s,t)}{\partial t} = (1 - s)(\alpha s + \beta)\frac{\partial F(s,t)}{\partial s} + (s - 1)\left(\alpha x_2 + \frac{\beta x_1}{s}\right)F(s,t)$$

Show that the solution of this equation is

$$F(s,t) = \frac{1}{(\alpha + \beta)^{x_2 - 1}} s^{x_1}\{(\alpha + \beta e^{-(\alpha+\beta)t})s$$
$$+ \beta(1 - e^{-(\alpha+\beta)t})\}^{x_0 - x_1}\{\alpha(1 - e^{-(\alpha+\beta)t})s + (\alpha e^{-(\alpha+\beta)t} + \beta)\}^{x_2 - x_0}$$

2.10 Consider a birth-and-death process with parameters λ_x and μ_x. Let T_n denote the time required for the random variable to increase from n to $n + 1$ and let $T_n^* = \mathscr{E}\{T_n\}$. If $\mu_1 > 0$, T_n^* is the conditional expected time, conditioned upon nonabsorption or extinction. Show that[1]

$$T_n = \frac{1}{\lambda_n} + \frac{\mu_n}{\lambda_n} T_{n-1}^*$$

[1] Cf. John [43].

2.11 Let $\{X(t), Y(t), t \geqslant 0\}$ be a bivariate birth-and-death process with $n + 1$ states and let $P_{x,y}(t) = \mathscr{P}\{X(t) = x, Y(t) = y\}$. In connection with a problem in epidemiology Foster[1] has utilized a two-dimensional random-walk representation of this process. In particular, if $X(t)$ and $Y(t)$ represent the numbers of susceptible and infected individuals in the population at time t, given that $X(0) = n$, $Y(0) = a$, then the development of the epidemic can be regarded as a random walk from the point (n,a) to the points $(n - u, 0)$, $u = 0, 1, \ldots, n$, with an absorbing barrier at $x = 0$, and where the possible transitions from the points (x,y) are

$$\mathscr{P}\{(x,y) \to (x - 1, y + 1)\} = \frac{x}{x + \rho} \qquad \mathscr{P}\{(x,y) \to (x,y - 1)\} = \frac{\rho}{x + \rho}$$

where ρ (a nonnegative constant) is the ratio of the removal and infection rates.

By considering the sum of the probabilities of all possible paths from (n,a) to $(n - u, 0)$, show that

$$Q(u) = \lim_{t \to \infty} P_{n-u,0}(t)$$

$$= \frac{\rho^{i+u}}{\rho + n - u} \frac{\binom{n}{u}}{\binom{n + \rho}{u}} \sum_{\alpha} (\rho + n)^{-\alpha_0}(\rho + n - 1)^{-\alpha_1} \ldots (\rho + n - u)^{-\alpha_u}$$

where the summation is extended over all compositions of $a + u - 1$ into $u + 1$ parts such that $0 \leqslant \alpha_1 \leqslant a + i - 1$ for $0 \leqslant i \leqslant u - 1$ and $1 \leqslant \alpha_u \leqslant a + u - 1$.

2.12 Find $\mathscr{P}\{X(t) > x\}$ for (a) the Poisson process, (b) the simple birth process, and (c) the simple birth-and-death process. (*Hint:* Expand the associated generating function as a power series in s and then use Maclaurin's theorem with remainder in integral form.)

2.13 The *coefficient of variation* of the random variable $X(t)$ is defined as

$$\mathscr{V}\{X(t)\} = \frac{\mathscr{D}\{X(t)\}}{\mathscr{E}\{X(t)\}}$$

i.e., the ratio of the standard deviation and mean of $X(t)$. Determine the coefficient of variation and its asymptotic behavior for the processes listed in Prob. 2.12.

2.14 It is well known that for the simple birth-and-death process $X(\infty) = 0$ with probability less than one when $\lambda > \mu$. Now consider a process $\{\tilde{X}(t), t > 0\}$ subject to the condition $\tilde{X}(\infty) = 0$ with probability one. Prove that, in order to obtain probabilities conditional on the hypothesis of extinction, it is only necessary to interchange the parameters λ and μ in any formulas derived for the unconditioned process $\{X(t), t \geqslant 0\}$.[2]

2.15 For certain Markov branching processes the generating function $F(s,t)$ satisfies a partial differential equation of the form

$$\frac{\partial F(s,t)}{\partial t} = \varphi(s) \frac{\partial F(s,t)}{\partial s}$$

where $\varphi(s)$ is a power series in s convergent in the circle $|s| < 1$. What is the relationship between the infinitesimal transition probabilities a_{ij} and the coefficients α_n of s^n in $\varphi(s)$?[3]

---

[1] Cf. Sec. 4.4C.

[2] Cf. Waugh [90].

[3] Cf. Bharucha-Reid and Rubin [14].

2.16   A Markov process with transition matrix $P(t)$ is said to be *periodic* with period ω ($\omega > 0$) if $P(t + \omega) = P(t)$ and ω is the smallest positive real number with this property. Determine the limit matrix $\Pi = \lim\limits_{t \to \infty} P(t)$ for periodic processes.[1]

2.17   Consider a birth-and-death process with infinitesimal transition probabilities $0 < a_{1j} \leqslant 1$ for $j = 0, 1, 2$ and $a_{1j} = 0$ for $j \geqslant 3$. With reference to Theorem 2.15, show that for $m \geqslant 0$

$$U_i = \frac{1 - a^{m+1}}{1 - a} \qquad \text{for } a_{10} \neq a_{12}$$

$$= m + 1 \qquad \text{for } a_{10} = a_{12}$$

and

$$\mathcal{E}\{Y(X) \mid M(X) = m\} = \frac{a^n}{a_{10}(1 - a^{m+1})(1 - a^m)} \sum_{i=1}^{m} \frac{(1 - a^i)^2}{ia^i} \qquad \text{for } a_{10} \neq a_{12}$$

$$= \frac{1}{2a_{10}} \qquad \text{for } a_{10} = a_{12}$$

In the above we have put $a = a_{12}/a_{10}$.

2.18   With reference to Theorem 2.17, find $\lim\limits_{m \to \infty} \mathcal{E}\{T \mid M(X) \leqslant m, X(0) = n\}$.

2.19   Under the hypothesis of Theorem 2.20, is $\{X(t), t \geqslant 0\}$ a continuous parameter martingale?[2] If so, utilize the martingale convergence theorem to prove $\lim\limits_{t \to \infty} Y(t) = y$ with probability one.

Bibliography

1   Arley, N.: "On the Theory of Stochastic Processes and Their Applications to the Theory of Cosmic Radiation," John Wiley & Sons, Inc., New York, 1948.

2   Arley, N., and V. Borchsenius: On the Theory of Infinite Systems of Differential Equations and Their Applications to the Theory of Stochastic Processes and the Perturbation Theory of Quantum Mechanics, *Acta Math.*, vol. 76, pp. 261–322, 1945.

3   Armitage, P.: A Note on the Time-homogeneous Birth Process, *J. Roy. Statist. Soc.*, ser. B, vol. 15, pp. 90–91, 1953.

4   Austin, D. G.: On the Existence of the Derivative of Markoff Transition Probability Functions, *Proc. Natl. Acad. Sci. U.S.*, vol. 41, pp. 224–226, 1955.

5   Austin, D. G.: Some Differentiation Properties of Markoff Transition Probability Functions, *Proc. Am. Math. Soc.*, vol. 7, pp. 751–761, 1956.

6   Bailey, N. T. J.: A Simple Stochastic Epidemic, *Biometrika*, vol. 40, pp. 177–185, 1953.

7   Bellman, R.: "Stability Theory of Differential Equations," McGraw-Hill Book Company, Inc., New York, 1953.

8   Bellman, R.: "A Survey of the Mathematical Theory of Time-lag, Retarded Control, and Hereditary Processes," RAND Monograph R-256, 1954.

9   Bellman, R., and T. E. Harris: On Age-dependent Binary Branching Processes, *Ann. Math.*, vol. 55, pp. 280–295, 1952.

---

[1] Cf. Bharucha-Reid [13].
[2] Cf. the treatise of Doob.

10 Bharucha-Reid, A. T.: An Age-dependent Stochastic Model of Population Growth, *Bull. Math. Biophys.*, vol. 15, pp. 361–365, 1953.

11 Bharucha-Reid, A. T.: Age-dependent Branching Stochastic Processes in Cascade Theory, *Phys. Rev.*, vol. 96, pp. 751–753, 1954.

12 Bharucha-Reid, A. T.: On the Stochastic Theory of Epidemics, *Proc. Third Berkeley Symposium on Math. Statistics and Probability*, vol. 4, pp. 111–119, 1956.

13 Bharucha-Reid, A. T.: Ergodic Projections for Semi-groups of Periodic Operators, *Studia Math.*, vol. 17, pp. 189–197, 1958.

14 Bharucha-Reid, A. T., and H. Rubin: Generating Functions and the Semi-group Theory of Branching Markov Processes, *Proc. Natl. Acad. Sci. U.S.*, vol. 44, pp. 1057–1060, 1958.

15 Bourbaki, N.: "Fonctions d'une variable réele," chap. 4, Hermann & Cie, Paris, 1951.

16 Breiman, L.: On Transient Markov Chains with Application to the Uniqueness Problem for Markov Processes, *Ann. Math. Statist.*, vol. 28, pp. 499–503, 1957.

17 Chung, K. L.: Foundations of the Theory of Continuous Parameter Markov Chains, *Proc. Third Berkeley Symposium on Math. Statistics and Probability*, vol. 2, pp. 29–40, 1956.

18 Chung, K. L.: New Developments in Markov Chains, *Trans. Am. Math. Soc.*, vol. 81, pp. 195–210, 1956.

19 Čistyakov, V. P.: Local Limit Theorems for Branching Processes (in Russian), *Teor. Veroyatnost. i Primenen.*, vol. 2, pp. 360–374, 1957.

20 Dobrushin, R. L.: On Conditions of Regularity of Stationary Markov Processes with a Denumerable Set of Possible States (in Russian), *Uspehi Mat. Nauk.*, n.s., vol. 7, pp. 185–191, 1952.

21 Doob, J. L.: Topics in the Theory of Markoff Chains, *Trans. Am. Math. Soc.*, vol. 52, pp. 37–64, 1942.

22 Doob, J. L.: Markoff Chains—Denumerable Case, *Trans. Am. Math. Soc.*, vol. 58, pp. 455–473, 1945.

23 Feller, W.: Zur Theorie der stochastischen Prozesse (Existenz und Eindeutigkeitssätze), *Math. Ann.*, vol. 113, pp. 113–160, 1936.

24 Feller, W.: Die Grundlagen der Volterraschen Theorie des Kampfes ums Dasein in wahrscheinlichkeitstheoretischer Behandlung, *Acta Biotheoretica*, vol. 5, pp. 11–40, 1939.

25 Feller, W.: On the Integrodifferential Equations of Completely Discontinuous Markov Processes, *Trans. Am. Math. Soc.*, vol. 48, pp. 488–515, 1940.

26 Feller, W.: On the Integral Equation of Renewal Theory, *Ann. Math. Statist.*, vol. 12, pp. 243–267, 1941.

27 Feller, W.: On Boundaries and Lateral Conditions for the Kolmogorov Differential Equations, *Ann. Math.*, vol. 65, pp. 527–570, 1957.

28 Feller, W.: "An Introduction to Probability Theory and Its Applications," 2d ed., vol. 1, John Wiley & Sons, Inc., New York, 1957.

29 Fisz, M., and K. Urbanik: Analytical Characterization of Some Composed Non-homogeneous Poisson Processes, *Studia Math.*, vol. 15, pp. 328–336, 1956.

30 Florek, K., E. Marczewski, and C. Ryll-Nardzewski: Remarks on the Poisson Stochastic Process: I, *Studia Math.*, vol. 13, pp. 122–129, 1953.

31 Hall, W. J.: Some Hypergeometric Series Distributions Occurring in Birth-and-Death Processes at Equilibrium (Abstract), *Ann. Math. Statist.*, vol. 27, p. 221, 1956.

32 Halmos, P. R.: "Measure Theory," D. Van Nostrand Company, Inc., Princeton, N.J., 1950.

33 Harris, T. E.: Some Mathematical Models for Branching Processes, *Proc. Second Berkeley Symposium on Math. Statistics and Probability*, pp. 305–327, 1951.

34 Hellinger, E., and H. S. Wall: Contributions to the Analytic Theory of Continued Fractions and Infinite Matrices, *Ann. Math.*, vol. 44, pp. 103–127, 1943.

35 Hille, E.: On the Integration of Kolmogoroff's Differential Equations, *Proc. Natl. Acad. Sci. U.S.*, vol. 40, pp. 20–25, 1954.

36 Hille, E.: Perturbation Methods in the Study of Kolmogoroff's Equations, *Proc. Intern. Congr. of Mathematicians* (Amsterdam, 1954), vol. 3, pp. 365–376, 1956.

37 Hille, E., and R. S. Phillips: "Functional Analysis and Semi-groups," American Mathematical Society, New York, 1957.

38 Jensen, A.: "A Distribution Model," Munksgaard, Copenhagen, 1954.

39 Jiřina, M.: The Asymptotic Behavior of Branching Stochastic Processes (in Russian), *Czechoslov. Math. J.*, vol. 7, pp. 130–153, 1957.

40 John, P. W. M.: On the Feller-Lundberg Phenomenon in the Birth-and-Death Processes (Abstract), *Bull. Am. Math. Soc.*, vol. 61, pp. 443–444, 1955.

41 John, P. W. M.: Quadratic Time Homogeneous Birth-and-Death Processes (Abstract), *Ann. Math. Statist.*, vol. 27, p. 550, 1956.

42 John, P. W. M.: The Quadratic Birth Process (Abstract), *Ann. Math. Statist.*, vol. 27, p. 865, 1956.

43 John, P. W. M.: Divergent Time Homogeneous Birth-and-Death Processes, *Ann. Math. Statist.*, vol. 28, pp. 514–517, 1957.

44 Karlin, S., and J. L. McGregor: Representation of a Class of Stochastic Processes, *Proc. Natl. Acad. Sci. U.S.*, vol. 41, pp. 387–391, 1955.

45 Karlin, S., and J. L. McGregor: The Differential Equations of Birth-and-Death Processes and the Stieltjes Moment Problem, *Trans. Am. Math. Soc.*, vol. 85, pp. 489–546, 1957.

46 Karlin, S., and J. L. McGregor: The Classification of Birth-and-Death Processes, *Trans. Am. Math. Soc.*, vol. 86, pp. 366–400, 1957.

47 Karlin, S., and J. L. McGregor: Linear Growth, Birth-and-Death Processes, *J. Math. Mech.*, vol. 1, pp. 643–662, 1958.

48 Kato, T.: On the Semi-groups Generated by Kolmogoroff's Differential Equations, *J. Math. Soc. Japan*, vol. 6, pp. 1–15, 1954.

49 Kendall, D. G.: On the Generalized Birth-and-Death Process, *Ann. Math. Statist.*, vol. 19, pp. 1–15, 1948.

50 Kendall, D. G.: On the Role of Variable Generation Time in the Development of a Stochastic Birth Process, *Biometrika*, vol. 35, pp. 316–330, 1948.

51 Kendall, D. G.: Les processus stochastiques de croissance en biologie, *Ann. inst. H. Poincaré*, vol. 13, pp. 43–108, 1952.

52 Kendall, D. G.: Some Analytical Próperties of Continuous Markov Transition Functions, *Trans. Am. Math. Soc.*, vol. 78, pp. 529–540, 1955.

53 Kendall, D. G.: Some Further Pathological Examples in the Theory of Denumerable Markov Processes, *Quart. J. Math. Oxford*, ser. 2, vol. 7, pp. 39–56, 1955.

54 Kendall, D. G., and G. E. H. Reuter: Some Pathological Markov Processes with a Denumerable Infinity of States and the Associated Semi-groups of Operators on *l*, *Proc. Intern. Congr. of Mathematicians* (Amsterdam, 1954), vol. 3, pp. 377–415, 1956.

55 Kendall, D. G., and G. E. H. Reuter: The Calculation of the Ergodic Projection for Markov Chains and Processes with a Countable Infinity of States, *Acta Math.*, vol. 97, pp. 103–144, 1957.

56 Kolmogorov, A. N.: Über die analytischen Methoden in der Wahrscheinlichkeitsrechnung, *Math. Ann.*, vol. 104, pp. 415–458, 1931.

57 Kolmogorov, A. N.: On Some Problems Concerning the Differentiability of the Transition Probabilities in a Temporally Homogeneous Markov Process Having a Denumerable Set of States (in Russian), *Ucenve Zapiski Mat. Moskov. Gos. Univ.*, vol. 148, pp. 53–59, 1951.

58 Koopman, B. O.: Uniqueness Conditions for a Class of Stochastic Differential Equations (Abstract), *Bull. Am. Math. Soc.*, vol. 63, p. 141, 1957.

59 Lamens, A.: Sur le processus non-homogène de naissance et de mort à deux variables aléatoires, *Acad. roy. Belg., Bull. classe sci.*, ser. 5 vol. 43, pp. 711–719, 1957.

60 Lamens, A., and R. Consael: Sur le processus non-homogène de naissance et de mort, *Acad. roy. Belg., Bull. classe sci.*, ser. 5, vol. 43, pp. 597–605, 1957.

61 Ledermann, W., and G. E. H. Reuter: Spectral Theory for the Differential Equations of Simple Birth-and-Death Processes, *Phil. Trans. Roy. Soc. London*, ser. A, vol. 246, pp. 321–369, 1954.

62 Lévy, P.: Systèmes Markoviens et stationnaires. Cas dénombrable, *Ann. sci. école norm. super.*, vol. 69, pp. 327–381, 1951.

63 Marczewski, E.: Remarks on the Poisson Stochastic Process: II, *Studia Math.*, vol. 13, pp. 130–136, 1953.

64 Morgenstern, D.: Über die Differentialgleichung des reinen Geburtsprocesses in der Wahrscheinlichkeitsrechnung, *Math. Nachr.*, vol. 13, pp. 57–58, 1955.

65 Moyal, J. E.: Discontinuous Markoff Processes, *Acta Math.*, vol. 98, pp. 221–264, 1957.

66 Opatowski, I.: An Inverse Problem Concerning a Chain Process, *Proc. Natl. Acad. Sci. U.S.*, vol. 28, pp. 83–88, 1942.

67 Opatowski, I.: Markoff Chains with Reverse Transitions, *Proc. Natl. Acad. Sci. U.S.*, vol. 31, pp. 411–414, 1945.

68 Pinney, E.: "Ordinary Difference-Differential Equations," University of California Press, Berkeley, 1958.

69 Prékopa, A.: On Composed Poisson Distributions: IV, *Acta Math. Acad. Sci. Hung.*, vol. 3, 317–325, 1952.

70 Ramakrishnan, A.: Some Simple Stochastic Processes, *J. Roy. Statist. Soc.*, ser. B, vol. 13, pp. 131–140, 1951.

71 Rényi, A.: On Composed Poisson Distributions: II, *Acta Math. Acad. Sci. Hung.*, vol. 2, pp. 83–98, 1951.

72 Rényi, A.: On Some Problems Concerning Poisson Processes, *Publ. Math.*, vol. 2, pp. 66–73, 1951.

73 Reuter, G. E. H.: Denumerable Markov Processes and the Associated Contraction Semigroups on *l*, *Acta Math.*, vol. 97, pp. 1–46, 1957.

74 Reuter, G. E. H., and W. Ledermann: On the Differential Equations for the Transition Probabilities of Markov Processes with Enumerably Many States, *Proc. Cambridge Phil. Soc.*, vol. 49, pp. 247–262, 1953.

75 Ryll-Nardzewski, C.: On the Non-homogeneous Poisson Process: I, *Studia Math.*, vol. 14, pp. 124–128, 1953.

76 Ryll-Nardzewski, C.: Remarks on the Poisson Stochastic Process: III, *Studia Math.*, vol. 14, pp. 314–318, 1953.

77 Sevastyanov, B. A.: Theory of Branching Stochastic Processes (in Russian), *Uspehi Mat. Nauk*, n.s., vol. 6, no. 6, pp. 47–99, 1951.

128 THEORY OF MARKOV PROCESSES

78 Sevastyanov, B. A.: Final Probabilities for Branching Stochastic Processes (abstract, in Russian), *Teor. Veroyatnost. i Primenen.*, vol. 2, pp. 140–141, 1957.

79 Sevastyanov, B. A.: Limit Theorems for Branching Stochastic Processes of Special Form (in Russian), *Teor. Veroyatnost. i Primenen.*, vol. 2, pp. 339–348, 1957.

80 Sevastyanov, B. A.: Transient Phenomena in Branching Stochastic Processes (in Russian), *Teor. Veroyatnost. i Primenen.*, vol. 4, pp. 121–135, 1959.

81 Sneddon, I. N.: "Elements of Partial Differential Equations," McGraw-Hill Book Company, Inc., New York, 1957.

82 Täcklind, S.: Fourieranalytische Behandlung von Erneuerungsproblem, *Skand. Akturietidskr.*, vol. 28, pp. 68–105, 1945.

83 Takashima, M.: Note on Evolutionary Processes, *Bull. Math. Statist.*, vol. 7, pp. 18–24, 1956.

84 Urbanik, K.: Limit Properties of Homogeneous Markoff Processes with a Denumerable Set of States, *Bull. acad. polon. sci.*, Classe (III), vol. 2, pp. 371–373, 1954.

85 Urbanik, K.: On a Problem from the Theory of Birth-and-Death Processes (in Russian), *Acta Math. Acad. Sci. Hung.*, vol. 7, pp. 99–106, 1956.

86 Urbanik, K.: Remarks on the Maximum Number of Bacteria in a Population (in Polish), *Zastos. Mat.*, vol. 2, pp. 341–348, 1956.

87 Urbanik, K.: "Limit Properties of Markov Processes" (in Polish), *Rozprawy Mat.*, vol. 13, Państwowe Wydawnictwo Naukowe, Warsaw, 1957.

88 Wang, S.: On the Differentiabilities of the Transition Probabilities in Temporally-homogeneous Markoff Process with Enumerable Number of States (in Chinese), *Acta Math. Sinica*, vol. 4, pp. 359–364, 1954.

89 Waugh, W. A. O'N.: An Age-dependent Birth-and-Death Process, *Biometrika*, vol. 42, pp. 291–306, 1955.

90 Waugh, W. A. O'N.: Conditioned Markov Processes, *Biometrika*, vol. 45, pp. 241–249, 1958.

91 Woods, W. M., and A. T. Bharucha-Reid: Age-dependent Branching Stochastic Processes in Cascade Theory: II, *Nuovo Cimento*, ser. 10, vol. 10, pp. 569–578, 1958.

92 Zolotarev, V. M.: On a Problem from the Theory of Branching Stochastic Processes (in Russian), *Uspehi Mat. Nauk*, n.s., vol. 9, no. 2, pp. 147–156, 1954.

93 Zolotarev, V. M.: More Exact Statements of Several Theorems in the Theory of Branching Processes (in Russian), *Teor. Veroyatnost. i Primenen.*, vol. 2, pp. 256–266, 1957.

3

Processes Continuous in Space and Time

3.1 Introduction

In the preceding chapter we considered discontinuous Markov processes defined on the nonnegative integers. The outstanding characteristic of these processes may be described as follows: for a small time interval Δt the probability of no change of state exceeds the probability of a change of state; however, should a change take place, it may be rather striking. This chapter will be devoted to the study of continuous Markov processes, called diffusion processes, defined on the real line. In contrast to the discontinuous processes described above, here it is certain that some change will take place in *any interval* Δt; however, if Δt is small, it is certain that the change in state will be small.

In Sec. 3.2 we derive the forward and backward Kolmogorov equations for diffusion processes on the real line. This derivation is based on the fundamental paper of Kolmogorov [21]. We also consider in this section methods of solving the Kolmogorov equations and obtain solutions for two well-known diffusion processes. Section 3.3 is devoted to an exposition of some of the results of Feller [9,10] on diffusion processes. In particular, we consider the general form of the forward Kolmogorov equation for one-dimensional processes.

The first-passage time problem for diffusion processes is studied in Sec. 3.4. We also consider the related problem of determining the absorption probabilities for a diffusion process on a finite interval. In Sec. 3.5 the diffusion-equation representation of discrete processes is considered. We here consider the diffusion analogues of the processes studied in the first two chapters. Finally, in Sec. 3.6, we consider the Kolmogorov equations for N-dimensional diffusion processes.

For additional studies on diffusion processes we refer to the books of Blanc-Lapierre and Fortet, Doob, Feller, and Lévy and the papers of Bochner [1], Feller [6], Ito [18], Khintchine [20], Maruyama [24], McKean [25], Neveu [26], and Ray [27]. We also refer to the paper of Litwiniszyn [23] for new and interesting applications of diffusion processes in geology.

3.2 Diffusion Processes on the Real Line: The Theory of Kolmogorov

A. Introduction. Let $\{X(t), t \geqslant 0\}$ be a continuous stochastic process of the Markov type defined on the real line; that is, $X(t)$ is a Markovian random variable, depending on a continuous parameter t, which assumes values in the state space $\mathfrak{X} = \{x: -\infty < x < \infty\}$. In this section we derive and study the Kolmogorov diffusion equations associated with continuous Markov processes on the real line.

B. The Kolmogorov Equations for Diffusion Processes. The backward equation, which we first derive, is satisfied by the transition probabilities of the process $\{X(t), t \geqslant 0\}$, while the forward equation is satisfied by the probability density of the process. Let

$$F(t,x;\tau,y) = \mathscr{P}\{X(\tau) < y \mid X(t) = x\} \qquad \tau > t \qquad (3.1)$$

denote the transition probabilities of the process $\{X(t), t \geqslant 0\}$. For t and x fixed, $F(t,x;\tau,y)$ is a continuous function of τ. In addition, $F(t,x;\tau,y)$ is a (conditional) distribution function in y satisfying the usual conditions

$$\lim_{y \to -\infty} F(t,x;\tau,y) = 0 \qquad\qquad (3.2)$$

$$\lim_{y \to \infty} F(t,x;\tau,y) = 1 \qquad\qquad (3.3)$$

An additional condition, which expresses the Markov property, is that for x, y, t, τ, and $s \in (t,\tau)$

$$F(t,x;\tau,y) = \int_{-\infty}^{\infty} F(s,z;\tau,y)\, d_z F(t,x;s,z) \qquad (3.4)$$

This is the Chapman-Kolmogorov equation for Markov processes of the diffusion type. In analogy with the processes considered in Chap. 2, we have the following continuity assumption:

$$\lim_{\tau \to t+0} F(t,x;\tau,y) = \lim_{t \to \tau-0} F(t,x;\tau,y) = \chi(x,y) = 0 \qquad y \leqslant x$$
$$= 1 \qquad y > x \qquad (3.5)$$

This condition states that the probability mass clusters about the point it started out from.

We also assume that $F(t,x;\tau,y)$ admits a density function $f(t,x;\tau,y)$, that is,

$$f(t,x;\tau,y) = \frac{\partial F(t,x;\tau,y)}{\partial y} \tag{3.6}$$

which satisfies the condition

$$F(t,x;\tau,y) = \int_{-\infty}^{y} f(t,x;\tau,z)\, dz \tag{3.7}$$

$$\int_{-\infty}^{\infty} f(t,x;\tau,z)\, dz = 1 \tag{3.8}$$

and

$$f(t,x;\tau,y) = \int_{-\infty}^{\infty} f(s,z;\tau,y) f(t,x;s,z)\, dz \tag{3.9}$$

In order to derive the backward equation, we assume the following:

(a) $$\lim_{\Delta t \to 0} \frac{1}{\Delta t} \int_{|y-x| \geqslant \delta} d_y F(t, x, t + \Delta t, y)$$

$$= \lim_{\Delta t \to 0} \frac{1}{\Delta t} \int_{|y-x| \geqslant \delta} d_y F(t - \Delta t, x; t, y) = 0 \tag{3.10}$$

where $\delta > 0$. The above condition states that the probability that $|X(\tau) - X(t)| \geqslant \delta$, given $X(t) = x$, during an infinitesimal time interval Δt is small compared to Δt.

(b) The first and second partial derivation of $F(t,x;\tau,y)$ with respect to the backward state variable x exist and are continuous functions in x:

$$\frac{\partial F(t,x;\tau,y)}{\partial x} \qquad \frac{\partial^2 F(t,x;\tau,y)}{\partial x^2} \tag{3.11}$$

(c) If at some time $t - \Delta t$, $X(t - \Delta t) = x$, then the mean and variance of the change in $X(t)$ during the following time interval of duration Δt are

$$\int_{-\infty}^{\infty} (y - x)\, d_y F(t - \Delta t, x; t, y) \tag{3.12}$$

and

$$\int_{-\infty}^{\infty} (y - x)^2\, d_y F(t - \Delta t, x; t, y) \tag{3.13}$$

respectively. Since the above integrals may diverge, we define the *truncated moments*

$$\lim_{\Delta t \to 0} \frac{1}{\Delta t} \int_{|y-x| < \delta} (y - x)\, d_y F(t - \Delta t, x; t, y)$$

$$= \lim_{\Delta t \to 0} \frac{1}{\Delta t} \int_{|y-x| < \delta} (y - x)\, d_y F(t, x; t + \Delta t, y) = b(t,x) \tag{3.14}$$

and $\quad\displaystyle\lim_{\Delta t\to 0}\frac{1}{\Delta t}\int_{|y-x|<\delta}(y-x)^2\,d_y F(t-\Delta t,\,x;\,t,\,y)$

$$=\lim_{\Delta t\to 0}\frac{1}{\Delta t}\int_{|y-x|<\delta}(y-x)^2\,d_y F(t,\,x;\,t+\Delta t,\,y)=a(t,x)\geqslant 0\quad(3.15)$$

The above are called the *infinitesimal mean and variance of the change in* $X(t)$, respectively. We remark that the existence of the above limits does not necessarily imply that $F(t,x;\tau,y)$ has first- and second-order moments.

Under the above assumptions we now derive the *backward Kolmogorov equation*

$$-\frac{\partial F(t,x;\tau,y)}{\partial t}=\frac{a(t,x)}{2}\frac{\partial^2 F(t,x;\tau,y)}{\partial x^2}+b(t,x)\frac{\partial F(t,x;\tau,y)}{\partial x}\qquad(3.16)$$

From (3.4) we have

$$F(t-\Delta t,\,x;\,\tau,\,y)=\int_{-\infty}^{\infty}F(t,z;\tau,y)\,d_z F(t-\Delta t,\,x;\,t,\,z)$$

Similarly,

$$F(t,x;\tau,y)=\int_{-\infty}^{\infty}F(t,x;\tau,y)\,d_z F(t-\Delta t,\,x;\,t,\,z)$$

Hence,

$$\frac{F(t-\Delta t,\,x;\,\tau,\,y)-F(t,x;\tau,y)}{\Delta t}$$

$$=\frac{1}{\Delta t}\int_{-\infty}^{\infty}[F(t,z;\tau,y)-F(t,x;\tau,y)]\,d_z F(t-\Delta t,\,x;\,t,\,z)$$

$$=\frac{1}{\Delta t}\int_{|z-x|\geqslant\delta}[F(t,z;\tau,y)-F(t,x;\tau,y)]\,d_z F(t-\Delta t,\,x;\,t,\,z)$$

$$+\frac{1}{\Delta t}\int_{|z-x|<\delta}[F(t,z;\tau,y)-F(t,x;\tau,y)]\,d_z F(t-\Delta t,\,x;\,t,\,z)\qquad(3.17)$$

Now, from (3.10) the first integral approaches zero as $\Delta t\to 0$. In the second integral, we can use (3.11) and Taylor's formula; hence,

$$F(t,z;\tau,y)-F(t,x;\tau,y)=(z-x)\frac{\partial F(t,x;\tau,y)}{\partial x}$$

$$+\frac{1}{2}(z-x)^2\frac{\partial^2 F(t,x;\tau,y)}{\partial x^2}+o((z-x)^2)\qquad(3.18)$$

By introducing (3.18) in (3.17), we have

$$\frac{F(t - \Delta t, x; \tau, y) - F(t,x;\tau,y)}{\Delta t}$$

$$= \frac{1}{\Delta t} \int_{|z-x|\geqslant\delta} [F(t,z;\tau,y) - F(t,x;\tau,y)] \, d_z F(t - \Delta t, x; t, z)$$

$$+ \frac{1}{\Delta t} \frac{\partial F(t,x;\tau,y)}{\partial x} \int_{|z-x|<\delta} (z - x) \, d_z F(t - \Delta t, x; t, z)$$

$$+ \frac{1}{\Delta t} \frac{\partial^2 F(t,x;\tau,y)}{\partial x^2} \int_{|z-x|<\delta} \{(z - x)^2 + o((z - x)^2)\} \, d_z F(t - \Delta t, x; t, z)$$

$$(3.19)$$

Hence, by passing to the limit and using (3.10), (3.14), and (3.15), we obtain the *backward Kolmogorov equation* (3.16). Similarly, the density function $f(t,x;\tau,y)$ satisfies

$$-\frac{\partial f(t,x;\tau,y)}{\partial t} = \frac{1}{2} a(t,x) \frac{\partial^2 f(t,x;\tau,y)}{\partial x^2} + b(t,x) \frac{\partial f(t,x;\tau,y)}{\partial x} \qquad (3.20)$$

We now consider the derivation of the forward Kolmogorov equation, which is the adjoint of the backward equation. The *forward Kolmogorov equation*, which is also called the *Fokker-Planck equation*, is given by

$$\frac{\partial f(t,x;\tau,y)}{\partial \tau} = \frac{1}{2} \frac{\partial^2 [a(\tau,y)f(t,x;\tau,y)]}{\partial y^2} - \frac{\partial [b(\tau,y)f(t,x;\tau,y)]}{\partial y} \qquad (3.21)$$

As we remarked earlier, the forward equation is satisfied by the probability density of the process $\{X(t), t \geqslant 0\}$, i.e., $f(t,x;\tau,y) = \mathscr{P}\{X(\tau) = y \mid X(t) = x\}$.

In deriving (3.21) we assume the existence of the following continuous partial derivatives:

$$\frac{\partial f(t,x;\tau,y)}{\partial \tau} \qquad \frac{\partial [b(\tau,y)f(t,x;\tau,y)]}{\partial y} \qquad \frac{\partial^2 [a(\tau,y)f(t,x;\tau,y)]}{\partial y^2} \qquad (3.22)$$

Now let $R(y)$ be a nonnegative continuous function such that

$$R(y) = 0 \quad \text{for } y < y_1 \text{ and } y > y_2 \qquad y_1 < y_2 \qquad (3.23)$$

and $\quad \cdot R(y_1) = R(y_2) = R'(y_1) = R'(y_2) = R''(y_1) = R''(y_2) = 0 \qquad (3.24)$

In terms of the function $R(y)$, it is clear that

$$\lim_{\Delta\tau\to0} \int_{-\infty}^{\infty} \frac{f(t, x; \tau + \Delta\tau, y) - f(t,x;\tau,y)}{\Delta\tau} R(y) \, dy$$

$$\doteq \int_{y_1}^{y_2} \frac{\partial f(t,x;\tau,y)}{\partial \tau} R(y) \, dy = \frac{\partial}{\partial \tau} \int_{y_1}^{y_2} f(t,x;\tau,y) R(y) \, dy \qquad (3.25)$$

From the Markov property (3.9) we have

$$f(t, x; \tau + \Delta\tau, y) = \int_{-\infty}^{\infty} f(t,x;\tau,z)f(\tau, z; \tau + \Delta\tau, y) \, dz \qquad (3.26)$$

By using (3.26) in (3.25), we have

$$\int_{y_1}^{y_2} \frac{\partial f(t,x;\tau,y)}{\partial \tau} R(y) \, dy$$

$$= \lim_{\Delta\tau \to 0} \frac{1}{\Delta\tau} \left\{ \int_{-\infty}^{\infty} \int_{-\infty}^{\infty} f(t,x,\tau,z)f(\tau, z; \tau + \Delta\tau, y)R(y) \, dz \, dy \right.$$

$$\left. - \int_{-\infty}^{\infty} f(t,x;\tau,y) R(y) \, dy \right\}$$

$$= \lim_{\Delta\tau \to 0} \frac{1}{\Delta\tau} \left\{ \int_{-\infty}^{\infty} f(t,x;\tau,z) \int_{-\infty}^{\infty} f(\tau, z; \tau + \Delta\tau, y)R(y) \, dy \, dz \right.$$

$$\left. - \int_{-\infty}^{\infty} f(t,x;\tau,y) R(y) \, dy \right\}$$

$$= \lim_{\Delta\tau \to 0} \frac{1}{\Delta\tau} \int_{-\infty}^{\infty} f(t,x;\tau,y) \left\{ \int_{-\infty}^{\infty} f(\tau, y; \tau + \Delta\tau, z)R(z) \, dz - R(y) \right\} dy \qquad (3.27)$$

Now consider the Taylor's expansion of $R(z)$ about y, i.e.,

$$R(z) = R(y) + (z - y)R'(y) + \tfrac{1}{2}(z - y)^2 R''(y) + o[(z - y)^2] \qquad (3.28)$$

Since $R(z)$ is bounded and because of (3.10),

$$\int_{|y-z| \geqslant \delta} f(\tau, y; \tau + \Delta\tau, z)R(z) \, dz = o(\Delta\tau) \qquad (3.29)$$

and

$$\int_{|y-z| \leqslant \delta} f(\tau, y; \tau + \Delta\tau, z) \, dz = 1 + o(\Delta\tau) \qquad (3.30)$$

By use of (3.28) to (3.30), the last expression in braces of (3.27) becomes

$$\int_{-\infty}^{\infty} f(\tau, y; \tau + \Delta\tau, z)R(z) \, dz - R(y)$$

$$= R'(y) \int_{|y-z| < \delta} (z - y)f(\tau, y; \tau + \Delta\tau, z) \, dz$$

$$+ \tfrac{1}{2}R''(y) \int_{|y-z| < \delta} [(z - y)^2 + o((z - y)^2)]f(\tau, y; \tau + \Delta\tau, z) \, dz + o(\Delta\tau) \qquad (3.31)$$

Hence, (3.27) can be rewritten as

$$\int_{y_1}^{y_2} \frac{\partial f(t,x;\tau,y)}{\partial \tau} R(y)\, dy$$

$$= \lim_{\Delta t \to 0} \int_{-\infty}^{\infty} f(t,x;\tau,y) \bigg\{ R'(y) \int_{|y-z|<\delta} (z-y)f(\tau,y;\tau+\Delta\tau,z)\, dz$$

$$+ \tfrac{1}{2}R''(y)\int_{|y-z|<\delta} [(z-y)^2 + o((z-y)^2)]f(\tau,y;\tau+\Delta\tau,z)\, dz + o(\Delta\tau) \bigg\}\, dy$$

If we now pass to the limit and use (3.14) and (3.15), we have

$$\int_{y_1}^{y_2} \frac{\partial f(t,x;\tau,y)}{\partial \tau} R(y)\, dy = \int_{-\infty}^{\infty} f(t,x;\tau,y)\{b(\tau,y)R'(y) + \tfrac{1}{2}a(\tau,y)R''(y)\}\, dy \quad (3.32)$$

Since $R'(y) = R''(y) = 0$ in the region $y \leqslant y_1$, $y \geqslant y_2$, (3.32) becomes

$$\int_{y_1}^{y_2} \frac{\partial f(t,x;\tau,y)}{\partial \tau} R(y)\, dy = \int_{y_1}^{y_2} f(t,x;\tau,y)\{b(\tau,y)R'(y) + \tfrac{1}{2}a(\tau,y)R''(y)\}\, dy \quad (3.33)$$

By partial integration we obtain the following relations:

$$\int_{y_1}^{y_2} f(t,x;\tau,y)b(\tau,y)R'(y)\, dy = \int_{y_1}^{y_2} R(y) \frac{\partial[b(\tau,y)f(t,x;\tau,y)]\, dy}{\partial y^2}$$

$$\int_{y_1}^{y_2} f(t,x;\tau,y)a(\tau,y)R''(y)\, dy = \int_{y_1}^{y_2} R(y) \frac{\partial^2[a(\tau,y)f(t,x;\tau,y)]\, dy}{\partial y^2}$$

From the above, (3.33) becomes

$$\int_{y_1}^{y_2} \frac{\partial f(t,x;\tau,y)}{\partial \tau} R(y)\, dy = \int_{y_1}^{y_2} \bigg\{ - \frac{\partial[b(\tau,y)f(t,x;\tau,y)]}{\partial y}$$

$$+ \frac{1}{2}\frac{\partial^2[a(\tau,y)f(t,x;\tau,y)]}{\partial y^2} \bigg\} R(y)\, dy \quad (3.34)$$

In view of the above equality, we can rewrite (3.34) as

$$\int_{y_1}^{y_2} \bigg\{ - \frac{\partial f(t,x;\tau,y)}{\partial \tau} - \frac{\partial[b(\tau,y)f(t,x;\tau,y)]}{\partial y}$$

$$+ \frac{1}{2}\frac{\partial^2[a(\tau,y)f(t,x;\tau,y)]}{\partial y^2} \bigg\} R(y)\, dy = 0 \quad (3.35)$$

Now, since the function $R(y)$ is arbitrary, (3.21) follows from (3.35). If it were not so, there would exist four numbers (t, x, τ, y) such that the expression in (3.35) which is within braces would be different from zero. Since we assumed all functions to be continuous, it follows that there exists an interval (a,b) such that for $y \in (a,b)$ the sign does not change. If $y_1 \leq a$ and $y_2 \geq b$, we put $R(y) = 0$ for $y \leq a$ and $y \geq b$, and $R(y) > 0$ for $y \in (a,b)$. Choosing $R(y)$ in this way, the left side of (3.35) must be different from zero. Hence we have a contradiction; and it follows that our assumption is false. Therefore, (3.21) follows from (3.35).

C.   Singular Diffusion Equations. In most physical applications of the Kolmogorov diffusion equations the coefficient $a(x)$ is essentially positive. However, in certain applications we encounter equations such that $a(x)$ vanishes at one (or possibly both) of the boundaries or one of the coefficients has no finite limit. Equations with coefficients such as described are called *singular diffusion equations*. As an example, consider the diffusion process on the interval $(0,1)$ with $a(x) = \alpha x(1 - x)$, $\alpha > 0$, $b(x) = 0$. Here $a(x)$ vanishes at both boundaries.

In the field of biology, for example, singular diffusion equations arise in connection with the study of population growth and the diffusion of gene frequencies. These problems will be considered in Chap. 4.

D.   Some Remarks on the Existence and Uniqueness of Solutions of the Kolmogorov Diffusion Equations. Unfortunately there does not exist for the Kolmogorov diffusion equations an elementary treatment of the existence and uniqueness theory as in the case of the Kolmogorov differential equations for discontinuous processes. The classical existence and uniqueness theorems for the solutions of the Kolmogorov diffusion equations were proved by Feller [5]. However, most of the recent results are based on the theory of semigroups of operators (cf. [17]), and therefore they utilize methods not developed in this text and not assumed to be familiar to the average reader. Hence, we shall simply refer to the relevant literature: Feller [6], Hille [14–16], and Yosida [35]. We also refer to Sec. 3.3B, where we present some results which relate to the problem of classifying the boundaries associated with a given diffusion process.

E.   Methods of Solving the Kolmogorov Diffusion Equations. Since the Kolmogorov diffusion equations are parabolic differential equations, the usual methods employed for solving equations of the type can be used.[1] In this section we give a sketch of two of the methods that are standard and most useful. These methods will be demonstrated in greater detail in the next section and in certain applications to be considered in Part II.

A standard method of obtaining a solution of a partial differential equation assumes that the variables are separable. That is, we attempt to find a solution which is the product of a function of y alone and a

---

[1] Cf. Sneddon [29].

function of t alone, say. To demonstrate this method we refer to the forward Kolmogorov equation

$$\frac{\partial f}{\partial \tau} = \frac{1}{2} \frac{\partial^2 [a(y)f]}{\partial y^2} - \frac{\partial [b(y)f]}{\partial y} \tag{3.36}$$

where $f = f(t,x;\tau,y)$. Throughout, we assume that the coefficients $a(y)$ and $b(y)$ are known. Let us now rewrite (3.36) as

$$\frac{\partial f}{\partial \tau} = \frac{1}{2} a(y) \frac{\partial^2 f}{\partial y^2} + [a'(y) - b(y)] \frac{\partial f}{\partial y} + \left[\frac{1}{2} a''(y) - b'(y) \right] f \tag{3.37}$$

where the primes denote differentiation with respect to y. In applying the method of separation of variables, we assume

$$f(\tau,y) = Y(y)T(\tau) \tag{3.38}$$

where Y and T are functions of y and τ, respectively. Substitution of (3.38) in (3.37) yields

$$Y \frac{dT}{d\tau} = \frac{1}{2} a(y)T \frac{d^2 Y}{dy^2} + [a'(y) - b(y)]T \frac{dY}{dy} + \left[\frac{1}{2} a''(y) - b'(y) \right] T Y$$

which may be rewritten

$$\frac{1}{T} \frac{dT}{d\tau} = \frac{1}{Y} \left\{ \frac{1}{2} a(y) \frac{d^2 Y}{dy^2} + [a'(y) - b(y)] \frac{dY}{dy} + \left[\frac{1}{2} a''(y) - b'(y) \right] Y \right\} \tag{3.39}$$

From the above, we see that the left-hand side is an expression depending on τ alone, while the right-hand side depends on y alone. Both sides must therefore be equal to the same constant, which we take as $-\lambda^2$. Hence from (3.39) we obtain the two ordinary differential equations

$$\frac{1}{T} \frac{dT}{d\tau} = -\lambda^2 \tag{3.40}$$

and

$$\frac{1}{2} a(y) \frac{d^2 Y}{dy^2} + [a'(y) - b(y)] \frac{dY}{dy} + \left[\frac{1}{2} a''(y) - b'(y) + \lambda^2 \right] Y = 0 \tag{3.41}$$

Equation (3.40) is a simple first-order equation with solution

$$T(\tau) = e^{-\lambda^2 \tau} \tag{3.42}$$

while Eq. (3.41) is a second-order equation, the solution of which can be found by the usual methods for ordinary differential equations with variable coefficients. Since (3.36) is a linear equation, the most general solution is given by

$$f(\tau,y) = \sum_{n=0}^{\infty} Y(y;A_n,B_n,\lambda_n)e^{-\lambda_n^2 \tau} \tag{3.43}$$

where A_n, B_n, and λ_n are determined by the initial and boundary conditions which must be specified for any particular diffusion problem.

We now consider the application of the Laplace transformation[1] to the Kolmogorov diffusion equations. Again we consider the forward Kolmogorov equation, the procedure being the same for the backward equation. Application of the Laplace transform to the Kolmogorov diffusion equations removes the time variable t (or τ), leaving an ordinary differential equation the solution of which yields the transform of the probability distribution (or density) as a function of the space variable x (or y). Let

$$f^*(t,x;s,y) = \mathscr{L}\{f(t,x;\tau,y)\} = \int_0^\infty f(t,x;\tau,y)e^{-s\tau}\,d\tau$$

be the Laplace transform of $f(t,x;\tau,y)$. Applying the Laplace transformation to the forward equation (3.37) yields

$$sf^* - f_0 = \frac{1}{2}\,a(y)\,\frac{d^2f^*}{dy^2} + [a'(y) - b(y)]\frac{df^*}{dy} + \left[\frac{1}{2}\,a''(y) - b'(y)\right]f^*$$

or $\quad \frac{1}{2}\,a(y)\,\frac{d^2f^*}{dy^2} + [a'(y) - b(y)]\frac{df}{dy} + \left[\frac{1}{2}\,a''(y) - b'(y) - s\right]f^* + f_0 = 0$

$$\tag{3.44}$$

In the above, f_0 is the initial value of f. Equation (3.44), the solution of which yields the Laplace transform of the density function, is a nonhomogeneous ordinary differential equation of second order. While (3.44) incorporates the initial condition, the *transformed boundary conditions* must be imposed on the solution of (3.44). Having obtained the solution of (3.44), application of the inversion theorem yields the solution $f(t,x;\tau,y)$. In some simple cases, tables of the Laplace transform may be useful, but in general, integration in the complex plane is required.

We have remarked that, in order to obtain a particular solution of the diffusion equation, the initial and boundary values associated with a given diffusion process must be known. As an example, consider a diffusion process on the finite interval (r_1,r_2), $r_1 < r_2$. The initial condition might be of the form

$$f(0,y) = \xi \qquad \xi \in (r_1,r_2) \tag{3.45}$$

or $$f(0,y) = g(y) \tag{3.46}$$

In addition to the initial condition, two boundary conditions of the form

$$\alpha_i f(\tau,r_i) + (-1)^i(1 - \alpha_i)\frac{\partial f(\tau,y)}{\partial y}\bigg]_{y=r_i} = 0 \tag{3.47}$$

[1] Cf. Appendix B.

for $i = 1,2$ are required. This condition is referred to as the elastic barrier condition. An elastic barrier may be described by saying that, whenever the diffusing particle reaches the boundary r_i, it has probability α_i of being absorbed and probability $1 - \alpha_i$ of being reflected. For $\alpha_i = 0$ boundary r_i is a *reflecting* boundary, and for $\alpha_i = 1$ it is an *absorbing* boundary. In Sec. 3.3 we consider the classification of boundaries and the formulation of appropriate boundary conditions for diffusion processes.

F. Two Diffusion Processes: The Wiener-Lévy Process and the Uhlenbeck-Ornstein Process. In this section we consider two well-known diffusion processes. The processes, which are of great interest in their own right, will be used to illustrate the methods of solving diffusion equations outlined in Sec. 3.2E.

1. *The Wiener-Lévy Process: The Heat Equation.* The differential equation

$$\frac{\partial u}{\partial t} = D \frac{\partial^2 u}{\partial x^2} \qquad u = u(t,x) \tag{3.48}$$

where D is the coefficient of diffusion, is well known in the theories of diffusion and heat condition. In the theory of diffusion the solution of (3.48) represents, say, the concentration of a diffusing substance as a function of the time variable t and the space variable x. In the theory of heat conduction, $u(t,x)$ represents the amount of heat in the system as a function of t and x.

In the theory of stochastic processes the solution of Eq. (3.48) gives the probability density $f(t,x;\tau,y)$ associated with the *Wiener-Lévy* (or *Brownian motion*) *process*, i.e.,

$$\frac{\partial f}{\partial \tau} = \frac{1}{2} \frac{\partial^2 f}{\partial y^2} \qquad f = f(t,x;\tau,y) \tag{3.49}$$

is the forward Kolmogorov equation with coefficients $a(x) = 1$ and $b(x) = 0$. It is of interest to note that the forward and backward equations of the Wiener-Lévy process are self-adjoint. We now obtain the solution of (3.49) for $f(\tau,y)$, which satisfies the initial condition

$$f(0,y) = 0 \tag{3.50}$$

and the boundary conditions

$$f(\tau,-\infty) = f(\tau,\infty) = 0 \tag{3.51}$$

Let $f^*(s,y)$ be the Laplace transform of $f(\tau,y)$. Then $f^*(s,y)$ satisfies

$$\frac{1}{2} \frac{d^2 f^*}{dy^2} - sf^* = 0 \tag{3.52}$$

140 THEORY OF MARKOV PROCESSES

since $f^*(0,y) = sf^*$. The two linearly independent solutions of (3.52) are

$$f_1^*(s,y) = e^{-\sqrt{2s}\,y}$$
$$f_2^*(s,y) = e^{\sqrt{2s}\,y}$$

Because of (3.51), we have

$$f^*(s,y) = f_1^*(s,y) = e^{-\sqrt{2s}\,y}$$

By inverting the above, we obtain

$$f(\tau,y) = \frac{1}{[2\pi\tau]^{\frac{1}{2}}} \exp\left\{-\frac{y^2}{2\tau}\right\} \qquad (3.53)$$

If we consider the density $f(t,x;\tau,y)$ instead of $f(\tau,y)$, we have

$$f(t,x;\tau,y) = \frac{1}{[2\pi(\tau-t)]^{\frac{1}{2}}} \exp\left\{-\frac{(y-x)^2}{2(\tau-t)}\right\} \qquad (3.54)$$

Let us now consider the case when an absorbing barrier is placed at $y = a$ $(a > 0)$. This requires that we put

$$\left.\frac{\partial f}{\partial y}\right]_{y=a} = 0 \qquad (3.55)$$

In this case the solution of (3.49) is

$$f(\tau,y) = \frac{1}{[2\pi\tau]^{\frac{1}{2}}} \left[\exp\left\{-\frac{y^2}{2\tau}\right\} - \exp\left\{\frac{2a-y}{2\tau}\right\}^2\right] \qquad (3.56)$$

In the case of absorbing barriers at $y = -a$ and $y = a$, i.e., we require

$$f(\tau,-a) = f(\tau,a) = 0 \qquad (3.57)$$

the solution of (3.49) is

$$f(\tau,y) = \frac{1}{[2\pi\tau]^{\frac{1}{2}}} \sum_{i=-\infty}^{\infty} (-1)^i \exp\left\{-\frac{(y-2ai)^2}{2\tau}\right\} \qquad (3.58)$$

In the above, we have assumed, by putting $a(x) = 1$, that the infinitesimal variance is unity. In general we have $a(x) = \sigma^2 > 0$; hence, (3.54) becomes

$$\frac{1}{\sigma[2\pi(\tau-t)]^{\frac{1}{2}}} \exp\left\{\frac{(y-x)^2}{2(\tau-t)\sigma^2}\right\} \qquad (3.59)$$

with similar changes in (3.56) and (3.58).

2. *The Uhlenbeck-Ornstein Process.*[1] The physical problem which leads to the Uhlenbeck-Ornstein process is the determination of the probability that a free particle in Brownian motion after time t has a

[1] Uhlenbeck and Ornstein [31].

velocity between x and $x + dx$, when at $t = 0$ the velocity was $x = x_0$. Let $X(t)$ denote the velocity of the particle at time t and let

$$f(x_0;t,x) = \mathscr{P}\{X(t) = x \mid X(0) = x_0\} \qquad -\infty < x < \infty \qquad (3.60)$$

It has been shown that $f(x_0;t,x)$ satisfies

$$\frac{\partial f}{\partial t} = \frac{\alpha}{2}\frac{\partial^2 f}{\partial x^2} + \beta\frac{\partial(xf)}{\partial x} \qquad f = f(x_0;t,x) \qquad (3.61)$$

where $\alpha = 2\beta k T m^{-1}$ and $\beta = \bar{f}m^{-1}$. Here, m is the mass of the particle, $\bar{f}$ the coefficient of friction, k is the coefficient of viscosity, and T is the absolute temperature. Equation (3.61) is the forward Kolmogorov equation associated with the Uhlenbeck-Ornstein process. Hence, Eq. (3.61) is the forward Kolmogorov equation with coefficients $a(x) = \alpha$ and $b(x) = -\beta x$. We now obtain the solution of Eq. (3.61) which satisfies the initial condition

$$f(0,x) = h(x) \qquad (3.62)$$

and the boundary conditions

$$f(t,-\infty) = f(t,\infty) = 0 \qquad (3.63)$$

Let $\tau = \beta t$ and $y = x\left(\frac{2\beta}{\alpha}\right)^{\frac{1}{2}}$. With these changes in variables, Eq. (3.61) becomes

$$\frac{\partial f}{\partial \tau} = \frac{\partial^2 f}{\partial y^2} + y\frac{\partial f}{\partial y} + f \qquad (3.64)$$

We now solve Eq. (3.64) with the initial condition $f(0,y) = {}'h(y)$ and boundary conditions $f(t,y) = 0$ when $y = \pm\infty$. By using the method of separation of variables, we obtain the solution

$$A_n e^{-n\tau} D_n(y) e^{-y^2/4} \qquad (3.65)$$

where $D_n(y)$ is the Weber function of nth order,[1] i.e.,

$$D_n(y) = (-1)^n e^{-y^2/4}\frac{d^n}{dy^n}e^{y^2/2}$$

Since $f(0,y) = h(y)$, we have

$$h(y) = \sum_{n=0}^{\infty} A_n D_n(y) e^{-y^2/4}$$

This gives

$$A_n = \frac{1}{(2\pi)^{\frac{1}{2}}n!}\int_{-\infty}^{\infty} D_n(\lambda)h(\lambda)e^{-\lambda^2/4}\,d\lambda \qquad (3.66)$$

Cf. Whittaker and Watson [33].

Therefore,

$$f(\tau,y) = \frac{1}{(2\pi)^{\frac{1}{2}}} \int_{-\infty}^{\infty} \left[\sum_{n=0}^{\infty} \frac{D_n(y)D_n(\lambda)}{n!} e^{-n\tau} e^{-(\lambda^2-y^2)/4} h(\lambda) \right] d\lambda \qquad (3.67)$$

It is known that

$$\sum_{n=0}^{\infty} \frac{D_n(y)D_n(\lambda)}{n!} e^{-n\tau} = \frac{e^{(y^2+\lambda^2)/4}}{(1-e^{-2\tau})^{\frac{1}{2}}} \exp\left\{ -\frac{(y^2+\lambda^2-2y\lambda e^{-\tau})}{\alpha(1-e^{-2\tau})} \right\}$$

hence (3.67) becomes

$$f(\tau,y) = \frac{1}{(2\pi)^{\frac{1}{2}}} \int_{-\infty}^{\infty} \exp\left\{ -\frac{(y-\lambda e^{-\tau})^2}{2(1-e^{-2\tau})^{\frac{1}{2}}} \right\} \frac{1}{(1-e^{-2\tau})} h(\lambda)\, d\lambda$$

The fundamental solution can now be written as

$$f(y_0;\tau,y) = \frac{1}{[2\pi(1-e^{-2\tau})]^{\frac{1}{2}}} \exp\left\{ -\frac{(y-y_0 e^{-\tau})^2}{2(1-e^{-2\tau})} \right\} \qquad (3.68)$$

On replacing τ and y by t and x, we obtain

$$f(x_0;t,x) = \left[\frac{m}{2kT(1-e^{-2\beta t})} \right]^{\frac{1}{2}} \exp\left\{ -\frac{m}{2kT} \frac{(x-x_0 e^{-\beta t})^2}{1-e^{-2\beta t}} \right\} \qquad (3.69)$$

as the solution of the forward equation.

Since $a(x) = \alpha$ and $b(x) = -\beta x$, the backward equation of the Uhlenbeck-Ornstein process is given by

$$\frac{\partial F}{\partial t} = \frac{\alpha}{2} \frac{\partial^2 F}{\partial x^2} - \beta x \frac{\partial F}{\partial x} \qquad (3.70)$$

3.3 Diffusion Processes on the Real Line: The Theory of Feller

A. Introduction. In this section we shall be concerned with diffusion processes of the Markov type on the interval $I = (r_1, r_2)$, where $-\infty \leqslant r_1 < r_2 \leqslant \infty$. When I is finite, or semi-infinite, it is necessary to specify boundary conditions in addition to an initial condition for the differential equations. In Sec. 3.3B we present some results of Feller [9] on the classification of boundaries associated with one-dimensional diffusion process. This classification is very important, for it enables us to formulate the correct forward Kolmogorov equation (Fokker-Planck equation) and the associated boundary conditions for a given diffusion process. In order to consider general diffusion processes on I, we define in Sec. 3.3C the elementary return process.

We remark that throughout this section we assume that for $x \in I$ the functions $a(x)$, $a'(x)$, and $b(x)$ are defined and continuous, with $a(x) > 0$.

B. Classification of Boundaries. We have seen that the coefficients $a(x)$ and $b(x)$ characterize a particular Markov process of the diffusion type. Given a pair $(a(x), b(x))$, it is possible for the process to exhibit various types of behavior at the boundaries r_1 and r_2. First, the coefficients may be such that the random variable $X(t)$ never takes on the value $x = r_1$ or $x = r_2$. In these cases no boundary conditions have to be imposed, and the boundaries are called *natural*. Second, the coefficients may be such that $X(t)$ has positive probability of taking the value r_1 or r_2. For these processes there are two possible cases: (1) The drift toward the boundaries can be such that the boundaries automatically act as absorbing barriers, and no boundary conditions can be imposed. In this case the boundaries are called *exit* boundaries. (2) The process may behave like the classical diffusion process on a finite interval, and various boundary conditions can be imposed. In this case the boundaries are termed *regular*. Boundaries which are regular or exit are called *accessible;* other boundaries are called *inaccessible*. As we shall see in Sec. 3.4, it is only for processes admitting accessible boundaries that we can consider first-passage time problems.

The classification of boundaries depends on the Lebesgue integrability of the function

$$f(x) = \exp\left\{-\int_{x'}^{x} \frac{b(z)}{a(z)}\, dz\right\} \tag{3.71}$$

where $x' \in (r_1, r_2)$ is fixed, and related functions on a prescribed open interval contained in (r_1, r_2). The above function was introduced by Feller [9] following Hille [16]. Let I_i ($i = 1, 2$) denote the interval (x', r_i). The function $f(x)$ is Lebesgue integrable on I_i (written $f(x) \in L(I_i)$) if

$$\int_{I_i} f(x)\, dx < \infty$$

i.e., the integral of $f(x)$ over the interval I_i is bounded. Before giving the classification criteria, we introduce the following functions:

$$g(x) = \frac{1}{a(x)f(x)} \qquad h(x) = f(x) \int_{x'}^{x} g(z)\, dz$$

Feller's Classification Criteria

1. The boundary r_i is *regular* if $f(x) \in L(I_i)$ and $g(x) \in L(I_i)$.
2. The boundary r_i is an *exit* boundary if $g(x) \notin L(I_i)$ and $h(x) \in L(I_i)$.

3. The boundary r_i is an *entrance* boundary if $g(x) \in L(I_i)$ and
$g(x) \int_{x'}^{x} f(z)\, dz \in L(I_i)$.

4. In all other cases the boundary is called *natural*.

To illustrate the application of the above criteria, we consider the classification of the boundaries associated with a diffusion process with coefficients[1] $a(x) = \alpha x(1 - x), \alpha > 0,$ and $b(x) = 0$ and with boundaries $r_1 = 0, r_2 = 1$. In this case $f(x) = 1$, since $b(x) = 0$. We have

$$\int_{I_1} f(x)\, dx = \int_0^{x'} dx = x' \qquad x' \in (0,1)$$

Therefore, $f(x) \in L(I_1)$. However,

$$\int_{I_1} g(x)\, dx = \int_0^{x'} g(x)\, dx = \frac{1}{\alpha} \int_0^{x'} \frac{dx}{x(1 - x)} = -\infty$$

Hence, $g(x) \notin L(I_1)$. We next consider the integrability of $h(x)$. We have

$$h(x) = \frac{1}{\alpha} \int_{x'}^{x} \frac{dz}{z(1 - z)} = \frac{1}{\alpha} \left(\log \frac{x}{1 - x} - \log \frac{x'}{1 - x'} \right)$$

If we now consider the integral of the first term on the right, we have

$$\int_0^{x'} \log \frac{x}{1 - x}\, dx = \int_0^{x'} \log x\, dx - \int_0^{x'} \log (1 - x)\, dx$$

Since both integrals on the right are bounded, $h(x) \in L(I_1)$. It follows, therefore, that $r_1 = 0$ is an exit boundary. Similar calculations show that $r_2 = 1$ is an exit boundary also.

By utilizing the theory of semigroups, Feller [9] has obtained the following results which relate the existence and uniqueness problem to that of classifying the boundaries:

1. When none of the boundaries is regular, there exists exactly one fundamental solution (or Green's function) common to the forward and backward diffusion equations, even though the initial value problem as such may have many solutions.

2. When one boundary is regular, or when both boundaries are regular, there exist infinitely many common fundamental solutions.

3. When both boundaries are natural, the initial value problem for both the forward and backward diffusion equations is uniquely determined, and the solutions are generated by a common fundamental solution.

---

[1] The diffusion equation with $a(x)$ and $b(x)$ so defined arises in genetics, and it will be considered in Chap. 4.

4. When r_1, say, is a natural boundary and r_2 is an exit boundary, the initial value problem for the backward diffusion equation has infinitely many solutions, but that for the forward diffusion equation is uniquely determined.

5. When r_1 is a natural boundary but r_2 is a regular boundary, there exist infinitely many solutions for the initial value problem for both the forward and backward diffusion equations.

6. When neither boundary is natural, there are two sources for non-uniqueness, and in these cases two lateral conditions[1] must be imposed.

C. The Elementary Return Process. Fokker-Planck Equations. Feller [9,10] has shown that for diffusion processes on an interval which admit accessible boundaries, the classical Fokker-Planck equation (3.21) must be replaced by a system of differential equations which gives a complete description of the underlying stochastic process. Because of the role of the Fokker-Planck equation in various applied fields, it is important that the correct form of this equation be utilized in order to obtain the proper representation of the process.

In order to develop the correct form of the Fokker-Planck equation, it is necessary to introduce the *elementary return process*. This process can be described as follows: Assume that at time t, $X(t) = r_i$ ($i = 1,2$); then $X(s) = r_i$ for $t \leqslant s \leqslant t + T$, where T is a random variable independent of the past, with the negative exponential distribution (i.e., the Markov property) e^{-t/σ_1}, $\sigma_i > 0$. At time $t + T$ a jump occurs to a position x_k, $r_1 < x_k < r_2$, or to the boundary r_j ($j = 1,2$) according to the rule

$$p_{kj} = \mathscr{P}\{X(t + T) = j\} \qquad \tau_k p_k(x) = \mathscr{P}\{0 < X(t + T) \leqslant x\} \qquad (3.72)$$

and the process starts again. Here $p_{k1} + p_{k2} + \tau_k \leqslant 1$, where $p_{kj} \geqslant 0$, $p_k(x)$ is a monotone function with $p_k(x) \to 0$ as $x \to r_1$ and $p_k(x) \to 1$ as $x \to r_2$. We define τ_k as the probability that the process, upon reaching the boundary r_k ($k = 1,2$), starts again from a point ρ_k in the neighborhood of r_k. Hence, with probability $1 - \tau_k$ the process terminates at the boundary r_k. Should $\sigma_i = 0$, we have the *instantaneous return process*. For this process, whenever the boundary r_k is reached, an instantaneous return to the interior of the interval takes place, and the process starts again from a point x_k. The point x_k is a random variable with distribution function $p_k(x)$. In case $\sigma_i = \infty$, we have the *absorbing barrier process*. In this case the process terminates as soon as r_i is reached.

---

[1] The requirement that the range of the resolvent of the differential operator coincide with a prescribed set is called a *lateral condition*.

Now let us assume that the boundaries r_1 and r_2 are accessible; then for every set $\Omega \subset (r_1,r_2)$ we have the system of equations

$$\frac{\partial}{\partial t}\int_\Omega f(t,x)\,dx = \int_\Omega \frac{\partial}{\partial x}\left\{\left(\frac{\partial[a(x)f(t,x)]}{\partial x} - b(x)f(t,x)\right)\right.$$
$$\left. + \left(\frac{\tau_1}{\sigma_1}\right)p_1(x)m_1(t) + \left(\frac{\tau_2}{\sigma_2}\right)p_2(x)m_2(t)\right\} \qquad \sigma_1, \sigma_2 > 0 \quad (3.73)$$

$$\frac{dm_1(t)}{dt} = -\left(\frac{1-p_{11}}{\sigma_1}\right)m_1(t) + \left(\frac{p_{21}}{\sigma_2}\right)m_2(t)$$
$$+ \lim_{x\to r_1}\left\{\frac{\partial[a(x)f(t,x)]}{\partial x} - b(x)f(t,x)\right\} \quad (3.74)$$

$$\frac{dm_2(t)}{dt} = \left(\frac{p_{12}}{\sigma_1}\right)m_1(t) - \left(\frac{1-p_{22}}{\sigma_2}\right)m_2(t)$$
$$- \lim_{x\to r}\left\{\frac{\partial[a(x)f(t,x)]}{\partial x} - b(x)f(t,x)\right\} \quad (3.75)$$

In the above, $f(t,x)$ is the probability density of $X(t)$ and $m_1(t)$ and $m_2(t)$ represent the masses at the boundaries r_1 and r_2 at time t, respectively. For the above system the initial values $f(0,x)$, $m_1(0)$, and $m_2(0)$ are assumed to be known. The boundary conditions to be imposed depend on the nature of the boundary. If the r_i are exit boundaries, no boundary conditions are needed. However, if r_i is a regular boundary, we impose the condition

$$\tau_i \lim_{x\to r} a(x)f(x)f(t,x)$$
$$+ (1-\tau_i)(-1)^i \lim_{x\to r_i}\left\{\frac{\partial[a(x)f(t,x)]}{\partial x} - b(x)f(t,x)\right\} = 0 \qquad i = 1,2$$
$$(3.76)$$

where, as in Sec. 3.3B, $f(x) = \exp\left\{-\int_{x'}^x [b(z)/a(z)]\,dz\right\}$. The above condition, which can be considered as the *generalized classical boundary condition*, is applicable to singular diffusion equations.

It is of interest to discuss the physical interpretation of the system Eqs. (3.73) to (3.75). Let $\Omega = (\omega_1,\omega_2)$ be an open interval, with $\Omega \in (r_1,r_2)$. Clearly the left-hand side of (3.73) represents the change of mass in Ω. On the right, the term

$$\lim_{x\to\omega_2}\left\{\frac{\partial[a(x)f(t,x)]}{\partial x} - b(x)f(t,x)\right\} - \lim_{x\to\omega_1}\left\{\frac{\partial[a(x)f(t,x)]}{\partial x} - b(x)f(t,x)\right\}$$

represents the net flux through the boundaries into Ω, while the term $(\tau_i m_i(t)/\sigma_i)[p(\omega_2) - p(\omega_1)]$ represents the direct flow into Ω from r_i. Now, the flow of mass at the boundaries can be described as follows: the rate at which r_1 loses mass is given by the product of $(1 - p_{11})/\sigma_1$ with the instantaneous mass. Of this mass the fraction $p_{12}/(1 - p_{11})$ flows to the boundary r_2 and the fraction $\tau_1/1 - p_{11}$ into the interior. In addition, r_1 gains mass at a rate $p_{21}m_2(t)/\sigma_2$ from the boundary r_2 and at a rate

$$\lim_{x \to r_1} \left\{ \frac{\partial[a(x)f(t,x)]}{\partial x} - b(x)f(t,x) \right\}$$

from Ω.

From the system of forward equations (3.73) to (3.75) it is possible to consider several special cases:

1. If $\sigma_1 = \sigma_2 = 0$ (*instantaneous return process*), we put $m_1(t) = m_2(t) = 0$, and Eq. (3.73) is replaced by

$$\frac{\partial}{\partial t} \int_\Omega f(t,x) \, dx = \int \frac{\partial}{\partial x} \left\{ \frac{\partial[a(x)f(t,x)]}{\partial x} - b(x)f(t,x) \right.$$

$$+ \tau_1 p_1(x) \lim_{x \to r_1} \left\{ \frac{\partial[a(x)f(t,x)]}{\partial x} - b(x)f(t,x) \right\}$$

$$\left. - \tau_2 p_2(x) \lim_{x \to r_2} \left\{ \frac{\partial[a(x)f(t,x)]}{\partial x} - b(x)f(t,x) \right\} \right\} \quad (3.77)$$

From the above equation we see that for $\tau_1 = \tau_2 = 0$ Eq. (3.73) reduces to the classical Fokker-Planck equation (3.21). It is also clear that, if the $p_i(x)$, $i = 1, 2$, are differentiable, the integrals in (3.73) and (3.77) can be omitted, and we obtain a partial differential equation.

2. If $\sigma_1 = \sigma_2 = \infty$ (*absorbing barrier process*), the system of forward equations becomes

$$\frac{\partial}{\partial t} \int_\Omega f(t,x) \, dx = \int_\Omega \frac{\partial}{\partial x} \left\{ \frac{\partial[a(x)f(t,x)]}{\partial x} - b(x)f(t,x) \right\} \quad (3.78)$$

$$\frac{dm_1(t)}{dt} = \lim_{x \to r} \left\{ \frac{\partial[a(x)f(t,x)]}{\partial x} - b(x)f(t,x) \right\} \quad (3.79)$$

$$\frac{dm_2 t}{dt} = -\lim_{x \to r_2} \left\{ \frac{\partial[a(x)f(t,x)]}{\partial x} - b(x)f(t,x) \right\} \quad (3.80)$$

Applications of these various systems will be given in Chap. 4 in connection with diffusion processes in genetics.

3.4   First-passage Time Problem for Diffusion Processes

A.   Introduction. In Chap. 1 we defined the first-passage time for a state $x = \xi$, say, as a random variable T which represents the time required for the system to enter the state ξ for the first time, when at time $t = 0$ the system was in some state $x = x_0 \neq \xi$, say. In the discrete case we say $T = n$ if n is an integer which represents the smallest number of steps for the system to pass from $x = x_0$ to $x = \xi$. In this section we wish to consider the first-passage time for diffusion processes. We first give some definitions and establish the notation which will be used in this section. In Sec. 3.4B we obtain the Laplace transform of the first-passage time distribution, and in Sec. 3.4C we consider the probabilities of ultimate absorption in the case where the states $x = r_1$ and $x = r_2$ are absorbing states. This section is based on the results of Darling and Siegert [4] and Feller [10].[1]

Let $\{X(t), t \geqslant 0\}$ be a diffusion process defined on the open interval (r_1,r_2), $r_2 > r_1$. Throughout this section we assume: (1) $\{X(t), t \geqslant 0\}$ has an associated transition probability function $F(t,x;\tau,y)$ which satisfies the Chapman-Kolmogorov equation and the backward Kolmogorov equation. (2) $F(t,x;\tau,y)$ admits a density function $f(t,x;\tau,y)$ which satisfies the forward Kolmogorov equation. (3) The random variable $X(t)$ is continuous with probability one.

We define the *first-passage time* $T(x_0;r_1,r_2)$ as the random variable

$$T = T(x_0;r_1,r_2) = \sup \{t \mid X(\tau) \in (r_1,r_2) \qquad 0 \leqslant \tau \leqslant t\} \qquad (3.81)$$

where $X(0) = x_0$. Now let $G(t,x_0;r_1,r_2)$ denote the distribution function of T, i.e.,

$$G(t,x_0;r_1,r_2) = \mathscr{P}\{T(x_0;r_1,r_2) < t\} \qquad (3.82)$$

We also assume that $G(t,x_0;r_1,r_0)$ admits a density function, i.e.,

$$g(t,x_0;r_1,r_2) = \frac{\partial G(t,x_0,r_1,r_2)}{\partial t} \qquad (3.83)$$

In case either $r_1 = -\infty$ or $r_2 = \infty$, we have only one boundary to consider. We therefore introduce a new random variable $T(x_0;p)$ such that

$$T(x_0;p) = T(x_0,p,\infty) \qquad x_0 > p$$

$$= T(x_0,-\infty,p) \qquad x_0 < p \qquad (3.84)$$

Also let $g_p(t,x_0)$ denote the density of $T(x_0;p)$.

[1] Cf. also Neveu [26].

In addition to the distribution of T we are interested in obtaining the conditional probability that the system, starting from $x = x_0$, will within time t reach r_2 without previously passing through r_1, say. Let

$$H_1(t,x_0;r_1,\bar{r}_2) = \mathscr{P}\{T < t, X(T) = r_1 \mid X(0) = x_0\} \qquad (3.85)$$

$$H_2(t,x_0;\bar{r}_1,r_2) = \mathscr{P}\{T < t, X(T) = r_2 \mid X(0) = x_0\} \qquad (3.86)$$

Writing $\bar{r}_i$ means that the boundary r_i is taboo. We denote by $h_1(t,x_0; r_1,\bar{r}_2)$ and $h_2(t,x_0;\bar{r}_1,r_2)$ the density functions associated with (3.85) and (3.86), respectively. It is clear that

$$G(t,x_0;r_1,r_2) = H_1(t,x_0;r_1,\bar{r}_2) + H_2(t,x_0;\bar{r}_1,r_2) \qquad (3.87)$$

In order that a first-passage time problem be defined for the boundary r_i $(i = 1,2)$, it is necessary that r_i be an accessible boundary; i.e., there must be a positive probability that $X(t)$ will equal r_i within a finite time from any initial position $x_0 \in (r_1,r_2)$. Hence, throughout this section we assume that the boundary being considered is accessible.

B. The Laplace Transform of the First-passage Time Distribution. In this section we obtain the Laplace transform of the first-passage time distribution. Thus, with the Laplace transform obtained, an application of the inversion theorem yields the first-passage time distribution. We first show that the Laplace transform of the probability density $f^*(x_0;s,y)$ can be expressed as the product of two independent functions. Next the expressions for the Laplace transforms of $G(t,x_0;r_1,r_2)$, $H_1(t,x_0;r_1,\bar{r}_2)$, and $H_2(t,x_0;\bar{r}_1,r_2)$ are obtained in terms of these functions. Finally, we show that the solutions are obtained by solving a second-order differential equation obtained from the backward Kolmogorov equation.

Theorem 3.1: If $X(t)$ is continuous with probability one and the transition probabilities associated with $\{X(t), t \geqslant 0\}$ satisfy the Chapman-Kolmogorov equation, then the Laplace transform of $f(x_0;t,y), f^*(x_0;s,y)$ can be expressed as the product

$$f^*(x_0;s,y) = \xi_1(x_0)\bar{\xi}_1(y) \qquad x_0 < y$$

$$= \xi_2(x_0)\bar{\xi}_2(y) \qquad x_0 > y \qquad (3.88)$$

and

$$g_p^*(x_0,s) = \frac{\xi_1(x_0)}{\xi_1(p)} \qquad x < p$$

$$= \frac{\xi_2(x_0)}{\xi_2(p)} \qquad x > p \qquad (3.89)$$

Proof: The proof is based on renewal theory. By the hypotheses of the theorem, we have for $p \in (x_0, y)$

$$f(x_0; t, y) = \int_0^t g_p(x_0; \tau) f(p; t - \tau, y) \, d\tau$$

By applying the Laplace transformation, we obtain

$$f^*(x_0; s, y) = g_p^*(x_0; s) f^*(p; s, y) \qquad p \in (x_0, y)$$

Hence $f^*(x_0; s, y)$ is expressed as the product of two functions, one depending on x_0 and the other on y. Call these functions $\xi_1(x_0)$ and $\bar{\xi}_1(y)$. Then $f^*(x_0; s, y) = \xi_1(x_0) \bar{\xi}_1(y)$, for $p \in (x_0, y)$. Similarly,

$$g_p^*(x_0, s) = \frac{\xi_1(x_0)}{\bar{\xi}_1(p)}$$

For $p \in (y, x_0)$, we have

$$g_p^*(x_0, s) = \frac{\xi_2(x_0)}{\bar{\xi}_2(p)}$$

Hence for any p and x_0 the conclusions of the theorem hold.

Theorem 3.2: Under the same hypotheses as Theorem 3.1, we have

$$h_1^*(s, x_0; r_1, \bar{r}_2) = \frac{\xi_1(r_2)\xi_2(x_0) - \xi_2(r_1)\xi_1(x_0)}{\xi_1(r_2)\xi_2(r_1) - \xi_1(r_1)\xi_2(r_2)} \tag{3.90}$$

$$h_2^*(s, x_0; \bar{r}_1, r_2) = \frac{\xi_2(r_1)\xi_1(x_0) - \xi_1(r_1)\xi_2(x_0)}{\xi_1(r_2)\xi_2(r_1) - \xi_1(r_1)\xi_2(r_2)} \tag{3.91}$$

$$g^*(s, x_0; r_1, r_2) = \frac{\xi_2(x_0)[\xi_1(r_2) - \xi_2(r_1)] - \xi_1(x_0)[\xi_2(r_2) - \xi_2(r_1)]}{\xi_1(r_2)\xi_2(r_1) - \xi_1(r_1)\xi_2(r_2)} \tag{3.92}$$

Proof: By using the renewal-theory approach employed in the proof of Theorem 3.1, we obtain the integral equations

$$g_{r_1}(t; x_0) = h_1(t, x_0; r_1, \bar{r}_2) + \int_0^t h_2(\tau, x_0; \bar{r}_1, r_2) g_{r_1}(t - \tau, r_2) \, d\tau$$

$$g_{r_2}(t; x_0) = h_2(t, x_0; \bar{r}_1, r_2) + \int_0^t h_1(\tau, x_0; r_1, \bar{r}_2) g_{r_2}(t - \tau, r_1) \, d\tau \tag{3.93}$$

The above equations can be considered as a simultaneous system of integral equations in two unknowns, $h_1(t, x_0; r_1, \bar{r}_2)$ and $h_2(t, x_0; \bar{r}_1, r_2)$. By applying the Laplace transformation to Eqs. (3.93), we obtain the linear system

$$g_{r_1}^*(s, x_0) = h_1^*(s, x_0; r_1, \bar{r}_2) + g_{r_1}^*(s, r_2) h_2^*(s, x_0; \bar{r}_1, r_2)$$

$$g_{r_2}^*(s, x_0) = g_{r_2}^*(s, x_0) h_1^*(s, x_0; r_1, \bar{r}_2) + h_2^*(s, x_0; \bar{r}_1, r_2) \tag{3.94}$$

If we now introduce (3.89) for $g_{r_1}^*(s;x_0)$ and $g_{r_2}^*(s,x_0)$ in the above system, we obtain (3.90) and (3.91). Finally, (3.92) is obtained from the above by noting that $g^*(s,x_0;r_1,r_2) = h_1^*(s,x_0;r_1,\bar{r}_2) + h_2^*(s,x_0; \bar{r}_1,r_2)$, since from (3.87) $G(t,x_0,r_1,r_2) = H_1(t,x_0;r_1,\bar{r}_2) + H_2(t,x_0;\bar{r}_1,r_2)$.

Having obtained the Laplace transforms of the various first-passage time distributions as functions of two functions $\xi_1(x)$ and $\xi_2(x)$, we now consider the differential equation which yields $\xi_1(x)$ and $\xi_2(x)$ as solutions. The result we need is given by the following theorem.

Theorem 3.3: If $f(x_0;t,y)$ uniquely satisfies the backward Kolmogorov equation with boundary conditions $f(\infty;t,y) = f(-\infty; t,y) = 0$ and $X(t)$ is continuous with probability one, the functions $\xi_1(x)$ and $\xi_2(x)$ can be chosen as any two linearly independent solutions of

$$\tfrac{1}{2}a(x)\frac{d\xi^2}{dx^2} + b(x)\frac{d\xi}{dx} - s\xi = 0 \tag{3.95}$$

Proof: If $f(x_0;t,y)$ satisfies the backward Kolmogorov equation, it is clear that its Laplace transform $f^* = f^*(x_0;s,y)$ satisfies

$$sf^* = \tfrac{1}{2}a(x_0)\frac{df^{*2}}{dx_0^2} + b(x_0)\frac{df^*}{dx_0} \tag{3.96}$$

and $-f^*$ is Green's solution[1] for this equation for $x_0 \in (-\infty,\infty)$. Hence, if $\xi_1(\infty) = \xi_2(-\infty) = 0$ and $\xi_1(x_0)$ and $\xi_2(x_0)$ satisfy (3.96), we obtain, to a constant factor,

$$
\begin{aligned}
f(x_0;s,y) &= \xi_1(y)\xi_2(x_0) \qquad x_0 \leqslant y \\
&= \xi_1(x_0)\xi_2(y) \qquad x_0 \geqslant y
\end{aligned}
\tag{3.97}
$$

Hence, we obtain (3.89) for $g_p^*(x_0,s)$, and we also obtain (3.90) to (3.92). The proof is completed by noting that h_1^*, h_2^*, and g^* are invariant under any nonsingular transformation of ξ_1 and ξ_2.

As a simple example, we consider the application of the above results to the Wiener-Lévy process. In this case $f(x_0;t,y)$ satisfies

$$\frac{\partial f}{\partial t} = \frac{1}{2}\frac{\partial^2 f}{\partial x^2}$$

From (3.95), what we seek are two linearly independent solutions of

$$\frac{1}{2}\frac{d^2\xi}{dx_0} - s\xi = 0 \tag{3.98}$$

[1] Cf. Sneddon [29].

Two linearly independent solutions of Eq. (3.98) are

$$\xi_1(x_0) = e^{-\sqrt{2s}x_0} \qquad \xi_2(x_0) = e^{\sqrt{2s}x_0} = \xi_1(-x_0) \qquad (3.99)$$

Now, for symmetric processes, i.e., processes for which $f(-x_0;t,-y) = f(x_0;t,y)$, we have from Theorem 3.2

$$g^*(s,x_0;\eta,-\eta) = \frac{\xi_1(x_0) + \xi_1(-x_0)}{\xi_1(\eta) + \xi_1(-\eta)} \qquad |x_0| < \eta \qquad (3.100)$$

Hence, from (3.99)

$$g^*(s,x_0;\eta,-\eta) = \frac{\cosh \sqrt{2s}x_0}{\cosh \sqrt{2s}\eta} \qquad |x_0| < \eta \qquad (3.101)$$

Application of the inversion theorem to (3.101) yields the density

$$g(t,x_0;\eta,-\eta) = \frac{\pi}{\eta^2} \sum_{i=0}^{\infty} (-1)^i (i + \tfrac{1}{2}) \cos \left[(i + \tfrac{1}{2}) \frac{\pi x_0}{\eta} \right] \exp \left\{ -(i + \tfrac{1}{2})^2 \frac{\pi^2 t}{2\eta^2} \right\}$$

$$(3.102)$$

By integrating, we obtain the distribution

$$G(t,x_0;\eta,-\eta) = \mathscr{P}\{T(x_0,-\eta,\eta) < t\}$$

$$= 1 - \frac{2}{\pi} \sum_{i=0}^{\infty} \frac{(-1)^i}{(i + \tfrac{1}{2})} \cos \left[(i + \tfrac{1}{2}) \frac{\pi x_0}{\eta} \right] \exp \left\{ -(i + \tfrac{1}{2})^2 \frac{\pi^2 t}{2\eta^2} \right\}$$

$$(3.103)$$

For the Wiener-Lévy process the general result can be obtained from (3.103) since

$$G(t,x_0;r_1,r_2) = G\left(t, x_0 - \frac{r_1 + r_2}{2} ; \frac{r_2 - r_1}{2} , \frac{-(r_2 - r_1)}{2} \right) \qquad (3.104)$$

C. Probabilities of Ultimate Absorption at Boundaries $x = r_1$ and $x = r_2$.

In addition to the problem considered in the preceding section, it is of interest to determine the probabilities of ultimate absorption at the boundaries $x = r_1$ and $x = r_2$. We have defined $H_1(t,x_0;r_1,\bar{r}_2)$ and $H_2(t,x_0;\bar{r}_1,r_2)$ as probabilities of absorption at r_1 and r_2 within time t, respectively, when $X(0) = x_0 \in (r_1,r_2)$. Now let

$$H_1(x_0;r_1,\bar{r}_2) = \lim_{t \to \infty} H_1(t,x_0;r_1,\bar{r}_2) \qquad (3.105)$$

and

$$H_2(x_0;\bar{r}_1,r_2) = \lim_{t \to \infty} H_2(t,x_0;\bar{r}_1,r_2) \qquad (3.106)$$

be the probabilities of ultimate absorption at $x = r_1$ and $x = r_2$, respectively. Now, since $h_1^*(x_0,0;r_1,r_2)$ and $h_2^*(x_0,0;r_1,r_2)$ are the probabilities of absorption at r_1 before r_2 and r_2 before r_1, respectively, it follows from the theory of the Laplace transform that these probabilities are given by the solution of the differential equation

$$\tfrac{1}{2}a(x_0)\frac{d^2\xi}{dx_0^2} + b(x_0)\frac{d\xi}{dx_0} = 0 \tag{3.107}$$

which satisfies the boundary condition

$$h_i^*(x_0;r_i,r_{i\pm1}) = 1 \qquad \text{for } x_0 = r_i$$
$$= 0 \qquad \text{for } x_0 \neq r_i, \, i = 1, 2 \tag{3.108}$$

We note that $h_1^*(x_0;r_1,\bar{r}_2) + h_2^*(x_0;\bar{r}_1,r_2) = 1$, that is, absorption in one of the absorbing states must ultimately take place.

As a simple example, we consider the case when[1] $a(x) = \alpha x(1 - x)$, $b(x) = 0$, $\alpha > 0$, $r_1 = 0$, and $r_2 = 1$. In this case (3.107) becomes

$$\alpha x_0(1 - x_0)\frac{d^2\xi}{dx_0^2} = 0 \qquad x_0 \in (0,1) \tag{3.109}$$

The solution of (3.109) is

$$\xi(x_0) = C_1 + C_2 x_0 \tag{3.110}$$

where C_1 and C_2 are constants. On applying the boundary conditions, we obtain

$$h_0^*(x_0;0,\bar{1}) = 1 - x_0 \tag{3.111}$$

and

$$h_1(x_0;\bar{0},1) = x_0 \tag{3.112}$$

D. The Laplace Transform of the Distribution of the Maximum of $X(t)$. A random variable of great interest in many applications is the maximum of $X(t)$. We define this random variable as follows:

$$M(t,x_0) = \sup_{0 \leqslant s \leqslant t} |X(s)| \tag{3.113}$$

where $X(0) = x_0$. In applied problems $M(t,x_0)$ might represent, for example, the maximum population size up to time t or the maximum gene frequency in a population up to time t, etc. The distribution function of $M(t,x_0)$ can be expressed in terms of distribution of $T(x_0; r_1,r_2)$; hence, the Laplace transform of the distribution of $M(t;x_0)$ can be obtained in terms of the Laplace transform of $G(t,x_0,r_1,r_2)$.

Let $K(t,x_0;\eta)$ denote the distribution function of $M(t,x_0)$; then clearly

$$K(t,x_0;\eta) = \mathscr{P}\{M(t,x_0) < \eta\}$$
$$= \mathscr{P}\{T(x_0;\eta,-\eta) > t\}$$
$$= 1 - G(t,x_0;\eta,-\eta) \qquad \eta > |x_0| \tag{3.114}$$

[1] This example occurs in the theory of gene frequencies; cf. Sec. 4.5.

so that the distribution of $M(t,x_0)$ can be obtained from the distribution of $T(x_0;\eta,-\eta)$. If we apply the Laplace transformation to both sides of (3.104), we obtain

$$K^*(s,x_0;\eta) = \frac{1}{s[1 - g^*(s,x_0,\eta,-\eta)]} \tag{3.115}$$

where g^* is given by (3.92). For a symmetric process we use (3.100).

For the Wiener-Lévy process we have from (3.103)

$$
\begin{aligned}
K(t,x_0;\eta) &= \mathscr{P}\{M(t,x_0) < \eta\} \\
&= \frac{2}{\pi} \sum_{i=0}^{\infty} \frac{(-1)^i}{i + \frac{1}{2}} \cos\left[(i + \tfrac{1}{2})\frac{\pi x_0}{\eta}\right] \exp\left\{-(i + \tfrac{1}{2})^2 \frac{\pi^2 t}{2\eta^2}\right\}
\end{aligned}
\tag{3.116}
$$

3.5 Diffusion-equation Representation of Discrete Processes

In Sec. 1.7 we considered the representation of discrete branching processes as random-walk processes and in Chap. 2 we saw that the discontinuous Markov processes could be considered as random-walk processes depending on a continuous time parameter. In this section we wish to consider the relationship between the generating-function approach of Chap. 1, the differential-difference equation approach of Chap. 2, and the Kolmogorov diffusion equations.[1] We first consider the generating-function approach.[2] Let $F(s) = \sum_{x=0}^{\infty} p(x)s^x$, $|s| \leqslant 1$, where $p(x)$ is the probability that in the first generation x individuals will be formed from a single individual present at time zero. Our purpose is to pass from the generating function associated with the simple discrete branching process to the diffusion equation for the probability density of the population size. Let

$$G(s) = 1 - F(s) \tag{3.117}$$

Hence as $s \to 1$, $G(s)$ admits the expansion

$$G(s) = G(1) + (s - 1)G'(1) + \frac{(s - 1)^2}{2} G''(1) + 0((s - 1)^3) \tag{3.118}$$

Now let $F'(1) = m_1 > 1$ and $F''(1) = m_2$. We then have from (3.117) and (3.118)

$$G(s) = 1 - F(s) = m_1(1 - s) - \frac{m_2}{2!}(1 - s)^2 - \cdots \tag{3.119}$$

[1] Cf. also A. N. Kolmogorov, Transition of Branching Processes into Diffusion Processes and Some Problems in Genetics (in Russian), *Teor. Veroyatnost. i Primenen.*, vol. 4, pp. 233–236, 1959; and W. Feller, The Birth and Death Processes as Diffusion Processes, *J. Math. Pure Appl.*, vol. 38, pp. 301–345, 1959.

[2] Due to Feller [7].

We now introduce new units for measuring time and population size such that during small intervals of time the changes in population size will be small. Hence the quantity $m_1 = 1 + \delta$ must be small. We also put $\Delta X = \Delta t = \delta$; that is, an individual in the old counting and the time of one generation are both equal to δ.

In view of the above assumptions, the random variable X representing the population size will no longer be integer-valued; hence, we must pass from the generating function $F(s)$ to the characteristic function $F(e^{iz})$. Because of our new units, the characteristic function becomes $F(e^{iz\delta})$.

From (3.119), we have for the characteristic function

$$1 - F(e^{iz\delta}) = m_1(1 - e^{iz\delta}) - \frac{m_2}{2}(1 - e^{iz\delta})^2 + 0(\delta^3)$$

$$= -i\delta m_1 z + \frac{\delta^2 z^2}{2}(m_1 + m_2) + 0(\delta^3) \tag{3.120}$$

Therefore

$$\log F(e^{iz\delta}) = \log\left\{1 + i\delta m_1 z - \frac{\delta^2 z^2}{2}(m_1 + m_2) + 0(\delta^3)\right\}$$

$$= \log\left\{1 + i\delta m_1 z + i\delta^2 z + \frac{\delta^2 z^2}{2}(\sigma^2 + 1) + 0(\delta^3)\right\}$$

$$= \log\left\{1 + i\delta z\left[1 + + \frac{i\delta z}{2}(\sigma^2 + 1)\right] + 0(\delta^3)\right\}$$

$$= i\delta z\left\{1 + + \frac{i\delta z}{2}(\sigma^2 + 1)\right\} + \frac{\delta^2 z^2}{2} + 0(\delta^3)$$

$$= i\delta z\left\{1 + \delta + \frac{i\delta z}{2}\sigma^2\right\} + 0(\delta^3) \tag{3.121}$$

In the above we have put $\sigma^2 = m_2 + m_1 - m_1^2$.

Now, let

$$\varphi(t,z) = F_n(e^{iz\delta}) \tag{3.122}$$

where $t = n\delta$. Then

$$\varphi(t + \delta, z) = F_{n+1}(e^{iz\delta})$$

$$= F_n[F(e^{iz\delta})]$$

$$= \varphi\left(t, \frac{1}{i\delta}\log F(e^{iz\delta})\right)$$

$$= \varphi\left(t, z + \delta z + \frac{i\delta z^2}{2}\sigma^2 + 0(\delta^3)\right) \tag{3.123}$$

From the above we have

$$\frac{\varphi(t + \delta, z) - \varphi(t,z)}{\delta} = \frac{\{\varphi(t, z + \delta z + (i\delta z^2/2)\sigma^2 + 0(\delta^3)) - \varphi(t,z)\}}{\delta} \quad (3.124)$$

Hence, as $\delta \to 0$, the left-hand side becomes

$$\frac{\partial\varphi(t,z)}{\partial t} \quad (3.125)$$

and the right-hand side becomes

$$\lim_{\delta \to 0} \frac{\varphi(t, z + \delta z + (i\,\delta z^2/2)\sigma^2 + 0(\delta^3)) - \varphi(t,z)}{\delta\left\{z + \frac{iz^2}{2}\sigma^2 + 0(\delta^3)\right\}} \left\{z + \frac{iz^2}{2}\sigma^2 + 0(\delta^3)\right\}$$

$$= \left\{z + \frac{iz^2}{2}\sigma^2\right\}\frac{\partial\varphi(t,z)}{\partial z} \quad (3.126)$$

Therefore, $\varphi(t,z)$ satisfies the partial differential equation

$$\frac{\partial\varphi(t,z)}{\partial t} = \left\{z + \frac{iz^2}{2}\sigma^2\right\}\frac{\partial\varphi(t,z)}{\partial z} \quad (3.127)$$

An application of the inversion theorem for characteristic functions[1] shows that $\varphi(t,z)$ is the characteristic function of a probability density $f(t,x)$ which satisfies the diffusion equation

$$\frac{\partial f(t,x)}{\partial t} = \frac{\sigma^2}{2}\frac{\partial^2\{xf(t,x)\}}{\partial x^2} - \frac{\partial\{xf(t,x)\}}{\partial x} \quad (3.128)$$

and the boundary condition $f(t,0) = 0$. Equation (3.128) is clearly the forward Kolmogorov diffusion equation with coefficients $a(x) = \sigma^2 x$ and $b(x) = x$. Hence, we see that the representation of the simple branching process as a continuous process leads to a singular diffusion equation, the coefficients (i.e., the infinitesimal mean and variance) being proportional to the instantaneous population size. The boundary condition is required in this case because the "probability mass" flowing out into the origin tends to zero.

We now consider the relationship between the Kolmogorov differential-difference equation and the Kolmogorov diffusion equations. In many applied problems it is not necessary to use the diffusion equation representation of a process, since, as an approximation, the exact position of a diffusing particle, say, might be neglected and it would suffice to know the probability that the particle is contained in a given interval. To fix ideas, consider the positive half of the real line. Let us

[1] Cf., for example, treatise of Loève.

now divide the real line into intervals of fixed length ϵ and denote by $\{E_j\}$ a sequence of states such that $j\epsilon < x < (j + 1)\epsilon$ denotes that $X(t) \in E_j$.

We now consider the transition probabilities of the process defined on the set of intervals $\{E_j\}$. As before, let

$$P_{ij}(\tau,t) = \mathscr{P}\{X(t) \in E_j \mid X(\tau) \in E_i\} \qquad t > \tau \tag{3.129}$$

Since our process is now discrete in space, the above transition probabilities will satisfy a system of Kolmogorov differential equations

$$\frac{\partial P_{ij}(\tau,t)}{\partial \tau} = q_i(\tau)\{P_{ij}(\tau,t) - \sum_k Q_{ik}(\tau)P_{kj}(\tau,t)\} \tag{3.130}$$

$$\frac{\partial P_{ij}(\tau,t)}{\partial t} = -q_j(t)P_{ij}(\tau,t) + \sum_k q_k(t)Q_{kj}(t)P_{ik}(\tau,t) \tag{3.131}$$

One would expect, therefore, that, as ϵ approaches zero, the solutions $P_{ij}(\tau,t)$ of the backward Kolmogorov differential equation (3.130), say, would approach a solution $F(\tau,x;t,y)$ of the backward diffusion equation

$$\frac{\partial F}{\partial \tau} = \frac{1}{2}a(\tau,x)\frac{\partial^2 F}{\partial x^2} + b(\tau,x)\frac{\partial F}{\partial x} \tag{3.132}$$

That is, if the state relations

$$\lim_{\epsilon \to 0} \epsilon i = x \qquad \lim_{\epsilon \to 0} \epsilon j = y \tag{3.133}$$

obtain, then it is reasonable to expect that

$$\lim_{\epsilon \to 0} \frac{P_{ij}(\tau,t)}{\epsilon} = F(\tau,x;t,y) \tag{3.134}$$

In order to investigate the relation between (3.130) and (3.132), it is necessary to make certain assumptions concerning the relations between the intensity functions $q_i(\tau)$, the relative transition probabilities $Q_{ij}(\tau)$, and the infinitesimal mean and variance $a(\tau,x)$ and $b(\tau,x)$. Now, as the length of the intervals ϵ decreases, the probability of a transition between the intervals or states E_j increases; hence, we assume that

$$\lim_{\epsilon \to 0} q_i(\tau)\epsilon = 1 \tag{3.135}$$

We also assume that

$$\lim_{\epsilon \to 0} Q_{ij}(j - i)\epsilon = b(\tau,x) \tag{3.136}$$

and

$$\lim_{\epsilon \to 0} Q_{ij}(j - i)^2\epsilon^2 = a(\tau,x) \tag{3.137}$$

Also, since for small intervals of time small changes in state are more probable, we assume that

$$\lim_{\epsilon \to 0} Q_{ij} |j - i|^3 \epsilon^3 = 0 \qquad (3.138)$$

If we now put $k = z\epsilon$, we have from (3.134) the following relations:

$$\frac{P_{kj}(\tau, t)}{\epsilon} \sim F(\tau, z; t, y)$$

$$= F(\tau, x; t, y) + (z - x) \frac{\partial F(\tau, x; t, y)}{\partial x} + \frac{(z - x)^2}{2} \frac{\partial^2 F(\tau, x; t, y)}{\partial x^2} + \cdots$$

$$\sim F(\tau, x; t, y) + (k - j)\epsilon \frac{\partial F(\tau, x; t, y)}{\partial x} + \frac{(k - j)^2 \epsilon^2}{2} \frac{\partial^2 F(\tau, x; t, y)}{\partial x^2}$$

$$(3.139)$$

If we now substitute (3.139) into (3.130) and use (3.136) and (3.137), we see that (3.130) passes formally into (3.132).

Hence, we can conclude that every continuous Markov process can be considered as a limiting case of a discontinuous Markov process and that the solutions of the Kolmogorov diffusion equations can be approximated by solutions of the Kolmogorov differential equations.

3.6   *N*-dimensional Diffusion Processes

In this section we consider the Kolmogorov equation for N-dimensional diffusion processes. Let $X_1(t), X_2(t), \ldots, X_N(t)$ be N (finite) random variables, and let $\mathbf{X}(t) = (X_1(t), X_2(t) \ldots, X_N(t))$. The state space $\mathfrak{X}$ associated with the process $\{\mathbf{X}(t), t \geqslant 0\}$ is N-dimensional Euclidean space. The states $\mathbf{x} \in \mathfrak{X}$ are determined by the N coordinates

$$(x_1, x_2, \ldots, x_N) \quad -\infty < x_i < \infty \qquad (3.140)$$

Now let

$$F(t, x_1, x_2, \ldots, x_N; \tau, y_1, y_2, \ldots, y_N)$$
$$= \mathscr{P}\{X_1(\tau) < y_1, X_2(\tau) < y_2, \ldots, X_N(\tau) < y_N \mid X_1(t) = x_1,$$
$$X_2(t) = x_2, \ldots, X_N(t) = x_N\} \qquad \tau > t$$

denote the transition probabilities of the N-dimensional process $\{\mathbf{X}(t), t \geqslant 0\}$. As before, $F(t, x_1, x_2, \ldots, x_N; \tau, y_1, y_2, \ldots, y_N)$ is a distribution function; hence

$$\lim_{y_i \to \infty} F(t, x_1, \ldots, x_N; \tau, y_1, \ldots, y_N)$$
$$= F(t, x_1, \ldots, x_N; \tau, y_1, \ldots, y_{i-1}, -\infty, y_{i+1}, \ldots, y_N) = 0 \quad (3.141)$$

and $\qquad \lim_{y_1 \to \infty, \ldots, y_N \to \infty} F(t, x_1, \ldots, x_N; \tau, y_1, \ldots, y_N) = 1 \qquad (3.142)$

We also assume that $F(t, x_1, \ldots, x_N; \tau, y_1, \ldots, y_N)$ admits a density function $f(t, x_1, \ldots, x_N; \tau, y_1, \ldots, y_N)$, that is,

$$f(t, x_1, \ldots, x_N; \tau, y_1, \ldots, y_N) = \frac{\partial^N F(t, x_1, \ldots, x_N; \tau, y_1, \ldots, y_N)}{\partial y_1 \cdots \partial y_N} \quad (3.143)$$

which satisfies the conditions

$$F(t, x, \ldots, x_N; \tau, y_1, \ldots, y_N)$$
$$= \int_{-\infty}^{y_1} \cdots \int_{-\infty}^{y_N} f(t, x_1, \ldots, x_N; \tau, z_1, \ldots, z_N) \, dz_1 \cdots dz_N \quad (3.144)$$

$$\int_{-\infty}^{\infty} \cdots \int_{-\infty}^{\infty} f(t, x_1, \ldots, x_N; \tau, z_1, \ldots, z_N) \, dz_1 \cdots dz_N = 1 \quad (3.145)$$

In order to obtain the backward Kolmogorov equation, it is necessary to replace conditions (3.10), (3.11), (3.14), and (3.15) by their N-dimensional analogues. We have:

1. $\lim\limits_{\Delta t \to 0} \dfrac{1}{\Delta t} \int_{E_N - S_\delta} d_{y_1, \ldots, y_N} F(t, x_1, \ldots, x_N; t + \Delta t, y_1, \ldots, y_N)$

$$= \lim_{\Delta t \to 0} \frac{1}{\Delta t} \int_{E_N - S_\delta} d_{y_1, \ldots, y_N} F(t - \Delta t, x_1, \ldots, x_N; t, y_1, \ldots, y_N) = 0$$

where $S_\delta \in E_N$ is a sphere of radius δ ($\delta > 0$) with center $(x_1, \ldots, x_N)$.

2. The partial derivatives

$$\frac{\partial F(t, x_1, \ldots, x_N; \tau, y_1, \ldots, y_N)}{\partial x_i}$$

and $\quad \dfrac{\partial^2 F(t, x_1, \ldots, x_N; \tau, y_1, \ldots, y_N)}{\partial x_i \, \partial x_j} \quad (3.146)$

exist and are continuous.

3. The limits

$$\lim_{\Delta t \to 0} \frac{1}{\Delta t} \int_{S_\delta} (y_i - x_i) \, d_{y_1 \ldots y_N} F(t, x_1, \ldots, x_N; t + \Delta t, y_1, \ldots, y_N)$$

$$= \lim_{\Delta t \to 0} \int_{S_\delta} (y_i - x_i) \, d_{y_1 \ldots y_N} F(t - \Delta t, x_1, \ldots, x_N; t, y_1, \ldots, y_N)$$

$$= b_i(t, x_1, \ldots, x_N) \qquad i = 1, \ldots, N \quad (3.147)$$

and

$$\lim_{\Delta t \to 0} \frac{1}{\Delta t} \int_{S_\delta} (y_i - x_i)(y_j - x_j) \, d_{y_1 \ldots y_N} F(t, x_1, \ldots, x_N; t + \Delta t, y_1, \ldots, y_N)$$

$$= \lim_{\Delta t \to 0} \frac{1}{\Delta t} \int_{S_\delta} (y_i - x_i)(y_j - x_j) \, d_{y_1 \ldots y_N} F(t - \Delta t, x_1, \ldots, x_N; t, y_1, \ldots, y_N)$$

$$= a_{ij}(t, x_1, \ldots, x_N) \qquad i, j = 1, \ldots, N \quad (3.148)$$

exist and are independent of δ.

The above relations define the *infinitesimal mean and variance of the changes in* $\mathbf{X}(t)$. We assume, as in the one-dimensional case, $a_{ij} \geqslant 0$.

Under the above conditions, and proceeding as in the one-dimensional case, we obtain the *backward Kolmogorov equation for N-dimensional diffusion processes*

$$-\frac{\partial F(t, x_1, \ldots, x_N; \tau, y_1, \ldots, y_N)}{\partial t}$$

$$= \frac{1}{2} \sum_{i=1}^{N} \sum_{j=1}^{N} a_{ij}(t, x_1, \ldots, x_N) \frac{\partial^2 F(t, x_1, \ldots, x_N; \tau, y_1, \ldots, y_N)}{\partial x_i \, \partial x_j}$$

$$+ \sum_{i=1}^{N} b_i(t, x_1, \ldots, x_N) \frac{\partial F(t, x_1, \ldots, x_N; \tau, y_1, \ldots, y_N)}{\partial x_i} \quad (3.149)$$

Similarly, we can obtain the *forward Kolmogorov equation*

$$\frac{\partial f}{\partial \tau} = \sum_{i=1}^{N} \sum_{j=1}^{N} \frac{\partial^2 [a_{ij}(\tau, y_1, \ldots, y_N) f(t, x_1, \ldots, y_N; \tau, y_1, \ldots, y_N)]}{\partial y_i \, \partial y_j}$$

$$- \sum_{i=1}^{N} \frac{\partial [b_i(\tau, y_1, \ldots, y_N) f(t, x_1, \ldots, x_N; \tau, y_1, \ldots, y_N)]}{\partial y_i}$$

In Chap. 4 we consider in detail a two-dimensional diffusion process which occurs in the theory of gene frequencies.

Problems

3.1 Classify the boundaries associated with the following diffusion processes:

(a) $a(x) = \alpha x^2 (1 - x)^2 \qquad \alpha > 0$

$b(x) = \beta x (1 - x) \qquad \beta > 0$

$r_1 = 0 \qquad r_2 = 1$

(b) $a(x) = \alpha x (1 - x) \qquad \alpha > 0$

$b(x) = \beta(\gamma - x) \qquad \beta > 0, 0 \leqslant \gamma \leqslant 1$

$r_1 = 0 \qquad r_2 = 1$

(c) $a(x) = \alpha x \qquad \alpha > 0$

$b(x) = \beta x + \gamma \qquad \gamma \leqslant 0, \text{ or } 0 < \gamma < \alpha, \text{ or } \alpha \geqslant \gamma$

$r_1 = 0 \qquad r_2 = +\infty$

(d) $a(x) = \gamma x \qquad \gamma > 0$

$b(x) = \alpha x - \beta x^2 \qquad \alpha, \beta > 0$

$r_1 = 0 \qquad r_2 = +\infty$

(e) $a(x) = 1$

$b(x) = 0$

(1) $r_1 = -\infty \qquad r_2 = +\infty$

(2) r_1, r_2 finite

3.2 Čerkasov [3] has shown that a continuous Markov process of the diffusion type can be transformed into a Wiener-Lévy process, with density given by (3.54), if $D = 0$, where

$$D = \begin{vmatrix} \alpha(t,x) & \beta(t,x) & \gamma(t,x) \\ \alpha_x(t,x) & \beta_x(t,x) & \gamma_x(t,x) \\ \alpha_{xx}(t,x) & \beta_{xx}(t,x) & \gamma_{xx}(t,x) \end{vmatrix}$$

and

$$\alpha(t,x) = \sqrt{a(t,x)}$$

$$\beta(t,x) = \sqrt{a(t,x)} \int_0^x [a(t,y)]^{-\frac{1}{2}} \, dy$$

$$\gamma(t,x) = 2b(t,x) - a_x(t,x) - \sqrt{a(t,x)} \int_0^x a_t(t,y)[a(t,y)]^{-\frac{3}{2}} \, dy$$

In the above, the subscripts denote differentiation with respect to the indicated variable. Determine if the following processes can be transformed:

(a)
$$a(t,x) = \frac{2t}{(\cos x + 2t)^2}$$

$$b(t,x) = \frac{2(\sin x)t}{(\cos x + 2t)^3} - \frac{2x}{\cos x + 2t}$$

(b)
$$a(t,x) = a(x) = kx^2(1 - x)^2 \qquad k > 0$$
$$b(t,x) = 0$$

3.3 Tanaka [30] has proved the following theorem: Let $\{X(t), t \geq 0\}$ be a diffusion process on the real line satisfying the backward Kolmogorov equation. If the following condition is satisfied,

(1)
$$\int_{r_1}^{x'} \frac{dz}{f(z)} = \int_{x'}^{r_2} \frac{dz}{f(z)} = \infty$$

(2)
$$\int_{r_1}^{r_2} \frac{f(z)}{a(z)} \, dz < \infty$$

where $f(z)$ is defined by (3.71), $x' \in (r_1, r_2)$, and r_1 and r_2 are inaccessible boundaries, then the transition probability defining the process admits a limiting distribution as $t \to \infty$, independent of x, with density

$$P(y) = \left[\int_{r_1}^{r_2} \frac{f(z)}{a(z)} \, dz \right]^{-1} \frac{f(y)}{a(y)}$$

The density $P(y)$ is a solution of the differential equation

$$\frac{dP(y)}{dy} = \left[\frac{b(y) - a'(y)}{a(y)} \right] P(y)$$

satisfying the condition $\int_{r_1}^{r_2} P(y) \, dy = 1$.

(a) For r_1, r_2 finite show that

$$P(y) = A(x - r_1)^\alpha (r_2 - x)^\beta$$

where A is a normalizing constant, when $a(x) = 1$ and $b(x) = \alpha(x - r_1)^{-1} - \beta(r_2 - x)^{-1}$, $\alpha \geq 1$, $\beta \geq 1$.

(b) Determine $P(y)$ for $r_1 = -\infty$, $r_2 = +\infty$, and $a(x) = 1$, $b(x) = 0$.

162   THEORY OF MARKOV PROCESSES

Bibliography

1   Bochner, S.: Diffusion Equation and Stochastic Processes, *Proc. Natl. Acad. Sci. U.S.*, vol. 35, pp. 368–370, 1949.
2   Čerkasov, I. D.: On Kolmogorov's Equations (in Russian), *Uspehi Mat. Nauk*, n.s., vol. 12, no. 5, pp. 237–244, 1957.
3   Čerkasov, I. D.: On Transforming the Diffusion Process to a Wiener Process (in Russian), *Teor. Veroyatnost. i Primenen.*, vol. 2, pp. 384–388, 1957.
4   Darling, D. A., and A. J. F. Siegert: The First Passage Problem for a Continuous Markov Process, *Ann. Math. Statist.*, vol. 24, pp. 624–639, 1953.
5   Feller, W.: Zur Theorie der stochastischen Prozesse (Existenz- und Eindeutigkeitssätze), *Math. Ann.*, vol. 113, pp. 113–160, 1936.
6   Feller, W.: Some Recent Trends in the Mathematical Theory of Diffusion, *Proc. Intern. Congr. of Mathematicians* (Cambridge, Mass.), vol. 2, pp. 322–339, 1950.
7   Feller, W.: Diffusion Processes in Genetics, *Proc. Second Berkeley Symposium on Math. Statistics and Probability*, pp. 227–246, 1951.
8   Feller, W.: Two Singular Diffusion Problems, *Ann. Math.*, vol. 54, pp. 173–182, 1951.
9   Feller, W.: The Parabolic Differential Equations and the Associated Semigroups of Transformations, *Ann. Math.*, vol. 55, pp. 468–519, 1952.
10 Feller, W.: Diffusion Processes in One Dimension, *Trans. Am. Math. Soc.*, vol. 77, pp. 1–31, 1954.
11 Feller, W.: On Differential Operators and Boundary Conditions, *Commun. Pure Appl. Math.*, vol. 8, pp. 203–216, 1955.
12 Feller, W., and H. P. McKean: A Diffusion Equivalent of a Countable Markov Chain, *Proc. Natl. Acad. Sci. U.S.*, vol. 42, pp. 351–354, 1956.
13 Fortet, R.: Les fonctions aléatoires du type de Markoff associées à certaines équations linéaires aux dérivées partielles du type parabolique, *J. Math. Pure Appl.*, 9th ser., vol. 22, pp. 177–244, 1942.
14 Hille, E.: On the Integration Problem for Fokker-Planck's Equation in the Theory of Stochastic Processes, *Onzième congr. des math. scand.*, pp. 183–194, 1949.
15 Hille, E.: "Explosive" Solution of Fokker-Planck's Equation, *Proc. Intern. Congr. of Mathematicians* (Cambridge, Mass.), vol. 1, p. 435, 1950.
16 Hille, E.: Les probabilités continues en chaîne, *Compt. rend. acad. sci. Paris*, vol. 230, pp. 34–35, 1950.
17 Hille, E., and R. S. Phillips: "Functional Analysis and Semi-groups," American Mathematical Society, New York, 1957.
18 Ito, K.: On Stochastic Differential Equations, *Mem. Am. Math. Soc.*, vol. 4, 1951.
19 Kac, M.: Random Walk and the Theory of Brownian Motion, *Am. Math. Monthly*, vol. 54, pp. 369–391, 1947.
20 Khintchine, A. Ya.: "Asymptotische Gesetze der Wahrscheinlichkeitsrechnung," Springer-Verlag, Berlin, Vienna, 1933.
21 Kolmogorov, A. N.: Über die analytischen Methoden der Wahrscheinlichkeitsrechnung, *Math. Ann.*, vol. 104, pp. 415–458, 1931.
22 Kolmogorov, A. N.: Zur Umkehrbarkeit der statistischen Naturgesetze, *Math. Ann.*, vol. 113, pp. 766–722, 1937.
23 Litwiniszyn, T.: Application of the Equation of Stochastic Processes to Mechanics of Loose Bodies, *Arch. Mech. Stos.*, vol. 8, pp. 393–411, 1956.

24 Maruyama, G.: Continuous Markov Processes and Stochastic Equations, *Rend. circ. mat. Palermo*, ser. 2, vol. 4, pp. 48–90, 1955.

25 McKean, H. P.: Elementary Solutions for Certain Parabolic Partial Differential Equations, *Trans. Am. Math. Soc.*, vol. 82, pp. 519–548, 1956.

26 Neveu, J.: Théorie des semi-groups de Markov, *Univ. Calif. (Berkeley) Publs., Statistics*, vol. 2, no. 14, pp. 319–394, 1958.

27 Ray, D.: Stationary Markov Processes with Continuous Paths, *Trans. Am. Math. Soc.*, vol. 82, pp. 452–493, 1956.

28 Siegert, A. J. F.: On the First Passage Time Probability Problem, *Phys. Rev.*, vol. 81, pp. 617–623, 1951.

29 Sneddon, I. N.: "Elements of Partial Differential Equations," McGraw-Hill Book Company, Inc., New York, 1957.

30 Tanaka, H.: On Limiting Distributions for One-dimensional Diffusion Processes, *Bull. Math. Statist.*, vol. 7, pp. 84–91, 1957.

31 Uhlenbeck, G. E., and L. S. Ornstein: On the Theory of the Brownian Motion, *Phys. Rev.*, vol. 36, pp. 823–841, 1930.

32 Wang, M. C., and G. E. Uhlenbeck: On the Theory of the Brownian Motion: II, *Revs. Modern Phys.*, vol. 17, pp. 323–342, 1945.

33 Whittaker, E. T., and G. N. Watson: "Modern Analysis," Cambridge University Press, New York, 1927.

34 Yosida, K.: An Operator-theoretical Treatment of Temporally Homogeneous Markoff Process, *J. Math. Soc. Japan*, vol. 1, pp. 244–253, 1949.

35 Yosida, K.: Semi-group Theory and the Integration Problem of Diffusion Equations, *Proc. Intern. Congr. of Mathematicians* (Amsterdam, 1954), vol. 1, pp. 405–420, 1957.

PART II

Applications

Representation is a compromise with chaos.
Bernard Berenson

4

Applications in Biology

4.1 Introduction

It is only recently that a somewhat systematic attack on biological problems by using mathematical methods has taken place. That biology should be somewhat late, as compared with physics, for example, in reaching a higher stage in its development is due in the main to the complexity of the phenomena with which it deals and the fact that the mathematical methods that worked so well in providing a description of much of the physical world are not adequate for the treatment of the problems presented by the biological world.

In this chapter we consider some applications of the theory of stochastic processes in biology. The methods made available by this theory are of great importance in the development of a mathematical biology, and in recent years many workers have utilized the methods to construct models of biological phenomena. A large body of the present theory of stochastic processes is immediately applicable to the study of biological problems, but as we shall see, there are many problems that require the development of new stochastic processes and methods. The applications we consider in this chapter are representative of the types of biological problems that can be treated by using stochastic methods. It will be clear to the reader that there are a large number of problems that remain unsolved, since, in spite of the great advances made, we do not as yet have a satisfactory or complete stochastic theory of any biological phenomena.

This chapter has five sections, some with several subsections, devoted to a stochastic treatment of the following: the growth of populations, mutation processes, the spread of epidemics, the theory of gene frequencies, and the effect of radiation on biological systems.

4.2   Growth of Populations

A.   Introduction. A property common to biological systems is that of growth. Hence, whether we are concerned with animal or plant populations, bacterial or human populations, a useful theory of growth processes is extremely desirable. Owing to the complex nature of the growth process, it is necessary to formulate and integrate models that attempt to describe the growth process at different levels. First, we should concern ourselves with models which generate the number of organisms in a population at time t. Models of this type might be called *macro*, or *external, models*. Should an adequate model of this type be formulated, it is then necessary to formulate models which deal with the physical, chemical, or biological processes within the organism that might be responsible for generating the population we observe. Models of this type might be called *micro*, or *internal, models*.

Early studies in the mathematical theory of population growth were primarily concerned with the development of deterministic models of the macro type. In this approach a functional equation (differential or integral equation, etc.) for the number of organisms in the population at time t, say $x(t)$, is derived on the basis of a postulated mechanism by which growth is assumed to take place. (The postulated mechanism may or may not reflect the internal processes responsible for changes in the population size.) This equation, together with some initial condition (the number of organisms in the population at time $t = 0$), is then solved to obtain $x(t)$. There are at least two objections to this approach: (1) It assumes that $x(t)$ is a real-valued continuous function of time, rather than an integer-valued function of time. (2) By assuming that population growth is deterministic, one obtains the result that the population size at a given time t will always be the same if the initial conditions are not changed. Of these two objections the second is by far the more serious one, since it does not take into consideration the large number of random or chance factors that can influence the growth of a population.

In 1939, Feller [24] considered the problems of population growth within the framework of the theory of stochastic processes. This study was the first to treat systematically stochastic models of population growth processes. In this approach a random variable $X(t)$, integer- (or real-) valued, was used to denote the population size at time t and the postulated mechanism for the growth of the population was expressed in terms of the probabilities of certain elementary events occurring in a small interval of time. The development or evolution of the population can then be described by the stochastic process $\{X(t), t \geqslant 0\}$, and it is assumed that the process is Markovian.

In this section, which has six subsections, we consider some of the stochastic models of population growth that have been developed by using the theory of Markov processes. These subsections are concerned with birth-and-death type models, models for the growth of sexes, models for competitive and predatory populations, and models for population migration. We also consider age-dependent models of population growth and the diffusion equation representation of growth processes.

B. Simple Stochastic Models for Population Growth. In this section we discuss some properties of certain *simple* stochastic models for population growth. The adjective "simple" is employed because the models we consider are simplified representations of the complex population growth process; it is not intended to convey the impression that from the mathematical point of view the models are necessarily simple or trivial. We restrict our attention to birth-and-death type processes, the basic properties of which are discussed in Chap. 2.

1. *The Nonhomogeneous Birth-and-Death Process.* The birth-and-death process with time-dependent birth and death rates is characterized by the following system of differential-difference equations (cf. [45]):

$$\frac{dP_x(t)}{dt} = \lambda(t)(x-1)P_{x-1}(t) - (\lambda(t) + \mu(t))xP_x(t)$$

$$+ \mu(t)(x+1)P_{x+1}(t) \qquad x = 1, 2, \ldots \quad (4.1)$$

$$\frac{dP_0(t)}{dt} = \mu(t)P_1(t)$$

where $P_x(t) = \mathscr{P}\{X(t) = x\}$, $x = 0, 1, \ldots$, and the random variable $X(t)$ denotes the population size at time t. In Chap. 2 we showed that the solution of this system satisfying the initial conditions

$$P_x(0) = 1 \qquad \text{for } x = 1$$
$$= 0 \qquad \text{for } x \neq 1 \qquad (4.2)$$

is
$$P_x(t) = [1 - \alpha(t)][1 - \beta(t)][\beta(t)]^{x-1} \qquad x = 1, 2, \ldots$$
$$P_0(t) = \alpha(t) \qquad (4.3)$$

where
$$\alpha(t) = 1 - \frac{e^{-\gamma(t)}}{w(t)} \qquad \beta(t) = 1 - \frac{1}{w(t)}$$

and
$$\gamma(t) = \int_0^t [\mu(\tau) - \lambda(\tau)]\, d\tau \qquad w(t) = e^{-\gamma(t)}\left[1 + \int_0^t \mu(\tau)e^{\gamma(\tau)}\right] d\tau$$

In applications of birth-and-death processes to the growth of populations it is not the probabilities $P_x(t)$ that are usually of primary

importance but the moments of $X(t)$ and the probability of extinction.
From Chap. 2 we have

$$\mathscr{E}\{X(t)\} = e^{-\gamma(t)} \tag{4.4}$$

$$\mathscr{D}^2\{X(t)\} = e^{-2\gamma(t)} \int_0^t [\lambda(\tau) + \mu(\tau)]e^{\gamma(\tau)}\,d\tau \tag{4.5}$$

Hence, for given $\lambda(t)$ and $\mu(t)$ the above yield explicit expressions for
the mean and variance of $X(t)$. In the special case when $\lambda(t)$ and $\mu(t)$
are independent of time we obtain

$$\mathscr{E}\{X(t)\} = e^{(\lambda-\mu)t} \tag{4.6}$$

$$\mathscr{D}^2\{X(t)\} = \frac{\lambda + \mu}{\lambda - \mu} e^{(\lambda-\mu)t}(e^{(\lambda-\mu)t} - 1) \tag{4.7}$$

The asymptotic properties of these expressions are discussed in Chap. 2.

From (4.6) we can obtain the expected population size for a pure birth
(or death) process by putting μ (or λ) equal to zero. In the birth process
we obtain

$$\mathscr{E}\{X(t)\} = e^{\lambda t} \tag{4.8}$$

It is of interest to note that the above expression is identical with the
solution of the differential equation

$$\frac{dx(t)}{dt} = \lambda x(t) \qquad \lambda > 0, x(0) = 1$$

which describes the simple *deterministic* pure birth process. Here $x(t)$ is
not an integer-valued random variable, but a real-valued continuous
function of time. In spite of this correspondence, we cannot conclude
that the solution of the functional equation for a deterministic growth
process will in general be the same as the expression for the mean of the
random variable associated with the analogous stochastic process
(cf. [24]). The same relation holds for the pure death process. In this
case

$$\mathscr{E}\{X(t)\} = x_0 e^{-\mu t} \qquad X(0) = x_0 \tag{4.9}$$

and the differential equation for the deterministic death process is

$$\frac{dx(t)}{dt} = -\mu x(t) \qquad \mu > 0, x(0) = x_0$$

From (4.3), and the definition of $\alpha(t)$, the probability that the population will die out by time t, i.e., $\mathscr{P}\{X(t) = 0\}$ for some $t > 0$, is

$$P_0(t) = \frac{\displaystyle\int_0^t \mu(\tau)e^{\gamma(\tau)}\,d\tau}{1 + \displaystyle\int_0^t \mu(\tau)e^{\gamma(\tau)}\,d\tau} \tag{4.10}$$

If $X(0) = x_0$, then, because of the independence of the x_0 populations, the probability of extinction is obtained by raising (4.10) to the (x_0)th power. For the case when $\lambda(t) = \lambda$ and $\mu(t) = \mu$, (4.10) yields

$$P_0(t) = \frac{\mu e^{(\lambda-\mu)t} - \mu}{\lambda e^{(\lambda-\mu)t} - \mu} \tag{4.11}$$

Hence, as $t \to \infty$, the probability that the population will eventually die out is

$$\lim_{t\to\infty} P_0(t) = 1 \qquad \text{for } \lambda < \mu$$

$$= \frac{\mu}{\lambda} \qquad \text{for } \lambda > \mu \tag{4.12}$$

We now consider the *cumulative process* associated with a birth-and-death process. In addition to the random variable $X(t)$ representing the number of individuals in the population at time t, we introduce the random variable $Y(t)$, which represents the *total number of births* in the population up to time t. In view of the above, $\{Y(t), t \geqslant 0\}$ is a pure birth process, since a change in $Y(t)$ is induced by a unit increase in $X(t)$ and $Y(t)$ can assume only increasing integer values. The study of cumulative processes, introduced by Kendall [45], is very important in many applications. For example, in the study of bacterial growth $X(t)$ represents the number of viable organisms at time t, while $Y(t)$ represents the total count up to time t in which living and dead organisms are not distinguished.

Let $\{X(t), Y(t), t \geqslant 0\}$ denote the cumulative process; and let

$$P_{x,y}(t) = \mathscr{P}\{X(t) = x, Y(t) = y\} \qquad x, y = 0, 1, 2, \ldots$$

In order to discuss this bivariate process, we introduce the generating function

$$G(r,s,t) = \sum_{x=0}^{\infty} \sum_{y=0}^{\infty} P_{x,y}(t) r^x s^y \qquad |r| \leqslant 1, |s| \leqslant 1$$

If we denote by $F(r,t)$ the generating function associated with the univariate process $\{X(t), t \geqslant 0\}$, it has been shown that $F(r,t)$ satisfies the partial differential equation

$$\frac{\partial F(r,t)}{\partial t} = [\lambda r^2 - (\lambda + \mu)r + \mu] \frac{\partial F(r,t)}{\partial r} \tag{4.13}$$

Since $Y(t)$ shares only the positive jumps of $X(t)$, it follows that $G(r,s,t)$ satisfies

$$\frac{\partial G(r,s,t)}{\partial t} = [\lambda s r^2 - (\lambda + \mu)r + \mu] \frac{\partial G(r,s,t)}{\partial r} \tag{4.14}$$

This equation is to be solved with the initial condition $G(r,s,0) = rs$ when $X(0) = Y(0) = 1$. The solution of Eq. (4.14) for general $\lambda(t)$ and $\mu(t)$ is not known; however, the solution for the simple case of constant birth and death rates has been given by Kendall. Here, we restrict our attention to the moments of $X(t)$ and $Y(t)$. In order to obtain these moments we utilize the cumulant generating function.

Let $K(u,v,t) = \log G(e^u, e^v, t)$ denote the cumulant generating function. From Eq. (4.14) we see that $K(u,v,t)$ satisfies

$$\frac{\partial K(u,v,t)}{\partial t} = [\lambda(e^{u+v} - 1) - \mu(1 - e^{-u})]\frac{\partial K(u,v,t)}{\partial u} \tag{4.15}$$

By definition

$$K(u,v,t) = u\mathscr{E}\{X(t)\} + v\mathscr{E}\{Y(t)\} + \tfrac{1}{2}u^2\mathscr{D}^2\{X(t)\}$$
$$+ \tfrac{1}{2}v^2\mathscr{D}^2\{Y(t)\} + uv\,\text{Cov}\,\{X(t),\,Y(t)\} + \cdots$$

If we now expand both sides of Eq. (4.15) in powers of u and v and equate coefficients, we obtain the following differential equations for the moments:

$$\frac{d\mathscr{E}\{X(t)\}}{dt} = (\lambda - \mu)\mathscr{E}\{X(t)\} \tag{4.16}$$

$$\frac{d\mathscr{D}^2\{X(t)\}}{dt} = (\lambda + \mu)\mathscr{E}\{X(t)\} + 2(\lambda - \mu)\mathscr{D}^2\{X(t)\} \tag{4.17}$$

$$\frac{d\mathscr{E}\{Y(t)\}}{dt} = \lambda\mathscr{E}\{X(t)\} \tag{4.18}$$

$$\frac{d\mathscr{D}^2\{Y(t)\}}{dt} = \lambda\mathscr{E}\{X(t)\} + 2\lambda\,\text{Cov}\,\{X(t),\,Y(t)\} \tag{4.19}$$

$$\frac{d\,\text{Cov}\,\{X(t),\,Y(t)\}}{dt} = \lambda\mathscr{E}\{X(t)\} + \lambda\mathscr{D}^2\{X(t)\} + (\lambda - \mu)\,\text{Cov}\,\{X(t),\,Y(t)\} \tag{4.20}$$

We give the solutions of these equations for time-dependent birth and death rates. Equations (4.16) and (4.17) are the well-known differential equations for the mean and variance of the birth-and-death process. From Chap. 2, or referring to (4.4) and (4.5), we have

$$\mathscr{E}\{X(t)\} = e^{-\gamma(t)} = \exp\left\{-\int_0^t [\mu(\tau) - \lambda(\tau)]\,d\tau\right\} \tag{4.21}$$

and
$$\mathscr{D}^2\{X(t)\} = e^{-2\gamma(t)}\int_0^t e^{\gamma(\tau)}[\lambda(\tau) + \mu(\tau)]\,d\tau \tag{4.22}$$

By using (4.21), we see that the solution of (4.18) is

$$\mathscr{E}\{Y(t)\} = 1 + \int_0^t e^{-\gamma(\tau)}\lambda(\tau)\,d\tau \tag{4.23}$$

By using (4.21) and (4.22), we see that the solution of (4.20) is

$$\text{Cov}\,\{X(t),\,Y(t)\} = e^{-\gamma(t)}\int_0^t \left[1 + \frac{\mathscr{D}^2\{X(\tau)\}}{\mathscr{E}\{X(\tau)\}}\right]\lambda(\tau)\,d\tau \tag{4.24}$$

Hence, the solution of (4.19) is

$$\mathscr{D}^2\{Y(t)\} = \int_0^t [\mathscr{E}\{X(\tau)\} + 2\,\text{Cov}\,\{X(\tau),\,Y(\tau)\}]\lambda(\tau)\,d\tau \tag{4.25}$$

In the special case when $\lambda(t) = \lambda$ and $\mu(t) = \mu$ the above solutions become

$$\mathscr{E}\{X(t)\} = e^{(\lambda-\mu)t} \tag{4.26}$$

$$\mathscr{D}^2\{X(t)\} = \frac{\lambda + \mu}{\lambda - \mu}\,e^{(\lambda-\mu)t}(e^{(\lambda-\mu)t} - 1) \tag{4.27}$$

$$\mathscr{E}\{Y(t)\} = \frac{1}{\mu - \lambda}\,(\mu - \lambda e^{(\lambda-\mu)t}) \tag{4.28}$$

$$\mathscr{D}^2\{Y(t)\} = \frac{\lambda(\mu + \lambda)}{(\mu - \lambda)^2}\,(1 - e^{(\lambda-\mu)t})\frac{4\lambda^2\mu t e^{(\lambda-\mu)t}}{(\mu - \lambda)^2}$$

$$+ \frac{\lambda^2(\mu + \lambda)}{(\mu - \lambda)^3}\,(1 - e^{2(\lambda-\mu)t}) \tag{4.29}$$

and

$$\text{Cov}\,\{X(t),\,Y(t)\} = \frac{\lambda e^{(\lambda-\mu)t}}{\mu - \lambda}\left[2\mu t - \frac{\mu + \lambda}{\mu - \lambda}\,(1 - e^{(\lambda-\mu)t})\right] \tag{4.30}$$

The limiting behavior of $\mathscr{E}\{X(t)\}$ and $\mathscr{D}^2\{X(t)\}$ has been considered earlier. For the other moments we have

$$\lim_{t\to\infty} \mathscr{E}\{Y(t)\} = \frac{\mu}{\lambda - \mu} \qquad \lim_{t\to\infty} \mathscr{D}^2\{Y(t)\} = \frac{\lambda\mu(\lambda + \mu)}{(\mu - \lambda)^3}$$

and

$$\lim_{t\to\infty} \text{Cov}\,\{X(t),\,Y(t)\} = 0$$

If the initial population size is $x_0 > 1$ and if we assume $y_0 = x_0$, then all of the solutions obtained above must be multiplied by x_0.

2. *The Simple Birth, Death, and Immigration Process.* A birth-and-death type process that is of interest in application is the *birth, death, and immigration process* (cf. [45]). The assumptions underlying the development of this process are the same as in the simple birth-and-death process, with an additional assumption which states that within

the time interval $(t, t + \Delta t)$ the probability of a unit increase in the population size due to immigration from outside the population is $\nu \Delta t + o(\Delta t)$, where $\nu > 0$ is the immigration rate. While immigration increases the population size, the effect of the immigration rate ν differs from that of the birth rate λ in that it is independent of the population size.

Let $F_I(s,t)$ denote the generating function associated with this process. It is easily shown that $F_I(s,t)$ satisfies the partial differential equation

$$\frac{\partial F_I(s,t)}{\partial t} = [\lambda s^2 - (\lambda + \mu)s + \mu]\frac{\partial F_I(s,t)}{\partial s} + \nu(s-1)F_I(s,t) \quad (4.31)$$

It has been shown (cf. [46]) that Eq. (4.31) can be written in the form

$$\frac{\partial F_I(s,t)}{\partial t} = \nu F_I(s,t)[F(s,t) - 1] \quad (4.32)$$

where $F(s,t)$ is the generating function associated with the birth-and-death process. If $X(0) = 0$, then Eq. (4.31), or Eq. (4.32), is to be solved with the initial condition

$$F_I(s,0) = 1 \quad (4.33)$$

The solution of Eq. (4.31) is

$$F_I(s,t) = \left(\frac{\lambda - \mu}{\lambda e^{(\lambda-\mu)t} - \mu}\right)^{\nu/\lambda}\left[1 - s\frac{\lambda e^{(\lambda-\mu)t} - 1}{\lambda e^{(\lambda-\mu)t} - \mu}\right]^{-\nu/\lambda} \quad (4.34)$$

when $\lambda \neq \mu$, and

$$F_I(s,t) = (1 + \lambda t - \lambda ts)^{-\nu/\lambda} \quad (4.35)$$

when $\lambda = \mu$. Differentiation of the above expressions yields the expected population size at time t, that is,

$$\frac{\partial F_I(s,t)}{\partial s}\Bigg]_{s=1} = \mathscr{E}\{X(t)\} = \frac{\nu}{\lambda - \mu}(e^{(\lambda-\mu)t} - 1) \quad (4.36)$$

when $\lambda \neq \mu$, and

$$\mathscr{E}\{X(t)\} = \nu t \quad (4.37)$$

when $\lambda = \mu$. As $t \to \infty$, we have

$$\lim_{t\to\infty} \mathscr{E}\{X(t)\} = \frac{\nu}{\lambda - \mu} \qquad \text{for } \lambda < \mu$$

$$= \infty \qquad \text{for } \lambda \geqslant \mu$$

3. *The Stochastic Analogue of the Logistic Law of Growth.* The logistic law, which is well known in the deterministic theory of population growth, is expressed by the *Pearl-Verhulst equation*

$$\frac{dx(t)}{dt} = [\lambda - \mu x(t)]x(t) \tag{4.38}$$

where λ and μ are positive constants. The stochastic analogue of the logistic process was first studied by Feller [24]. Kendall [45,46] has studied the logistic process within the framework of the theory of birth-and-death processes by assuming that the former is characterized by birth and death rates which are linear functions of the instantaneous population size; i.e., he assumes

$$\lambda_x = \alpha(x_2 - x) \qquad \mu_x = \beta(x - x_1)$$

where α, β, x_1, and x_2 are constants with $0 \leqslant x_1 \leqslant x \leqslant x_2$.

With the above birth and death rates the generating function $F(s,t)$ associated with the logistic process satisfies

$$\frac{\partial F(s,t)}{\partial t} = \beta(1 - s) \frac{\partial}{\partial s} \left[s \frac{\partial F(s,t)}{\partial s} - x_1 F(s,t) \right]$$

$$- \alpha(1 - s)s \frac{\partial}{\partial s} \left[x_2 F(s,t) - s \frac{\partial F(s,t)}{\partial s} \right] \tag{4.39}$$

with the initial condition

$$F(s,0) = s^{x_0} \tag{4.40}$$

if $X(0) = x_0$.

We see that Eq. (4.39) is a linear equation of second order; hence, no simple general solution can be obtained. By assuming the birth and death rates have the form

$$\lambda_x = \alpha\left(\frac{x_2}{x} - 1\right) \qquad \mu_x = \beta\left(1 - \frac{x_1}{x}\right) \qquad 0 < x_1 \leqslant x \leqslant x_2$$

Prendiville (cf. [45]) has obtained an explicit expression for the generating function $F(s,t)$. Takashima [88] has also obtained the solution of the differential equation for $F(s,t)$ and discussed its behavior in various limiting cases (cf. Prob. 2.9). For a discussion of the asymptotic properties of the logistic process we refer to the paper of Kendall [45], and for a-discussion of an approximation method applicable to this process we refer to Whittle [95].

C. Stochastic Models for the Population Growth of Sexes. In the preceding section we restricted our attention to the growth of populations made up of one type of individual. In many problems that are of interest in the study of human population growth it is not the total number of individuals in the population that is of primary concern, but the growth of each sex in the general population. The problem of the growth of sexes, however, is of interest in the study not only of human populations, but of other animal populations as well. In this section we present some stochastic models for the population growth of sexes. Since we are dealing with populations made up of two types of individuals, the stochastic processes will be two-dimensional, or bivariate. The results of this section are due to Goodman [35] and Joshi [41].[1] For a discussion of deterministic models we refer to Goodman and also Kendall [45].

Let the random variables $X(t)$ and $Y(t)$ denote the numbers of females and males in the population at time t and let $P_{x,y}(t) = \mathscr{P}\{X(t) = x, Y(t) = y\}$, $x \geqslant 0$, $y \geqslant 0$, denote the joint probability distribution of $X(t)$ and $Y(t)$. In order to write down the differential-difference equations for the probabilities $P_{x,y}(t)$, we assume that the population develops according to the following mechanism:

1. The subpopulations generated by two coexisting individuals are statistically independent of one another.

2. An individual alive at time t has probability $\lambda p \, \Delta t + o(\Delta t)$ of reproducing a female and probability $\lambda q \, \Delta t + o(\Delta t)$ of reproducing a male in the interval $(t, t + \Delta t)$, where $\lambda > 0$, $p + q = 1$.

3. A female alive at time t has probability $\mu \, \Delta t + o(\Delta t)$ of dying in the interval $(t, t + \Delta t)$, and a male alive at time t has probability $\mu' \, \Delta t + o(\Delta t)$ of dying in the interval $(t, t + \Delta t)$, where $\mu \geqslant 0$, $\mu' \geqslant 0$.

Because of the above assumptions, we see that the stochastic process for the growth of sexes here considered is a bivariate process of the birth-and-death type.

It is clear from the above assumptions that the probabilities $P_{x,y}(t)$ satisfy the following system of differential-difference equations:

$$\frac{\partial P_{x,y}(t)}{\partial t} = -[(\lambda + \mu)x + \mu'y]P_{x,y}(t) + [\lambda p(x - 1)]P_{x-1,y}(t)$$
$$+ \lambda q x P_{x,y-1}(t) + \mu(x + 1)P_{x+1,y}(t) + \mu'(y + 1)P_{x,y+1}(t)$$
$$x, y = 1, 2, \ldots$$

$$\frac{\partial P_{x,0}(t)}{\partial t} = -(\lambda + \mu)x P_{x,0}(t) + \lambda p(x - 1)P_{x-1,0}(t)$$
$$+ \mu(x + 1)P_{x+1,0}(t) + \mu' P_{x,1}(t) \qquad x = 1, 2, \ldots \qquad (4.41)$$

$$\frac{\partial P_{0,y}(t)}{\partial t} = -\mu'y P_{0,y}(t) + \mu P_{1,y}(t) + \mu'(y + 1)P_{0,y+1}(t) \qquad y = 1, 2, \ldots$$

$$\frac{\partial P_{0,0}(t)}{\partial t} = \mu P_{1,0}(t) + \mu' P_{0,1}(t)$$

[1] Cf. also Lamens [61].

If we assume that at time $t = 0$ the population is made up of only one female, then the initial conditions to be imposed are

$$P_{x,y}(0) = 1 \qquad \text{for } x = 1$$

$$= 0 \qquad \text{for } x \neq 1 \tag{4.42}$$

We do not attempt to solve the above system in order to obtain explicit expressions for the probabilities $P_{x,y}(t)$, but utilize the method of generating functions to obtain the moments of $X(t)$ and $Y(t)$. Let

$$F(r,s,t) = \sum_{x=0}^{\infty} \sum_{y=0}^{\infty} P_{x,y}(t) r^x s^y \qquad |r| \leqslant 1, |s| \leqslant 1$$

be the generating function of the probabilities $P_{x,y}(t)$. From Eq. (4.41) we see that $F(r,s,t)$ satisfies the partial differential equation

$$\frac{\partial F}{\partial t} = (\lambda p r^2 + \lambda q r s - \lambda r - \mu r + \mu)\frac{\partial F}{\partial r} + \mu'(1 - s)\frac{\partial F}{\partial s} \tag{4.43}$$

If we now write $r = e^u$ and $s = e^v$ and let $K(u,v,t)$ denote the cumulant generating function, we obtain from Eq. (4.43)

$$\frac{\partial K}{\partial t} = (\lambda p e^u + \lambda q e^v - \lambda - \mu + \mu e^{-u})\frac{\partial K}{\partial u} + \mu'(e^{-v} - 1)\frac{\partial K}{\partial v} \tag{4.44}$$

If we now let $m_{i,j}(t)$ denote the moments of order (i,j) of $X(t)$ and $Y(t)$, we obtain from Eq. (4.44) the following differential equations:

$$\frac{dm_{1,0}(t)}{dt} = (\lambda p - \mu)m_{1,0}(t)$$

$$\frac{dm_{0,1}(t)}{dt} = \lambda q m_{1,0}(t) - \mu' m_{0,1}(t)$$

$$\frac{dm_{2,0}(t)}{dt} = (\lambda p + \mu)m_{1,0}(t) + 2(\lambda p - \mu)m_{2,0}(t) \tag{4.45}$$

$$\frac{dm_{1,1}(t)}{dt} = (p - \mu - \mu')m_{1,1}(t) + \lambda q m_{2,0}(t)$$

$$\frac{dm_{0,2}(t)}{dt} = 2\lambda q m_{1,1}(t) + \lambda q m_{1,0}(t) + \mu' m_{0,1}(t) - 2\mu' m_{0,2}(t)$$

On solving these equations, we obtain the following expressions for the moments:

$$m_{1,0}(t) = \mathscr{E}\{X(t)\} = e^{(\lambda p - \mu)t} \tag{4.46}$$

$$m_{0,1}(t) = \mathscr{E}\{Y(t)\} = \frac{q}{\lambda p - \mu + \mu'}(e^{(\lambda p - \mu)t} - e^{-\mu't}) \tag{4.47}$$

$$m_{2,0}(t) = \mathscr{D}^2\{X(t)\}$$
$$= \frac{\lambda p + \mu}{\lambda p - \mu} e^{(\lambda p - \mu)t}(e^{(\lambda p - \mu)t} - 1) \tag{4.48}$$

$$m_{0,2}(t) = \mathscr{D}^2\{Y(t)\}$$
$$= \frac{(\lambda p + \mu)\lambda^2 q^2}{(\lambda p - \mu + \mu')^2}\left\{\frac{e^{2(\lambda p - \mu)t}}{(\lambda p - \mu)} + \frac{2e^{(\lambda p - \mu - \mu')t}}{\mu'} - \frac{e^{-2\mu't}}{\lambda p - \mu + 2\mu'}\right\}$$
$$- \frac{2\lambda^2 q^2(\lambda p + \mu)e^{(\lambda p - \mu)t}}{\mu'(\lambda p - \mu)(\lambda p - \mu + 2\mu')} + \mathscr{E}\{Y(t)\} \tag{4.49}$$

and

$$m_{1,1}(t) = \text{Cov}\,\{X(t),\, Y(t)\}$$
$$= \frac{\lambda q(\lambda p + \mu)}{\mu'(\lambda p - \mu)(\lambda p - \mu + \mu')}\{(\lambda p - \mu)e^{-\mu't} + \mu'e^{(\lambda p - \mu)t}$$
$$- (\lambda p - \mu + \mu')\}e^{(\lambda p - \mu)t} \tag{4.50}$$

The asymptotic behavior of the moments of $t \to \infty$ is easily obtained. Of particular interest is the asymptotic behavior of the ratio of the expected numbers of males and females in the population. Let

$$R(t) = \frac{\mathscr{E}\{X(t)\}}{\mathscr{E}\{Y(t)\}} = \frac{(\lambda p - \mu + \mu')e^{(\lambda p - \mu)t}}{\lambda q(e^{(\lambda p - \mu)t} - e^{-\mu't})}$$

We see that in the case where $(\lambda p - \mu + \mu') > 0$

$$\lim_{t \to \infty} R(t) = \frac{\lambda p - \mu + \mu'}{\lambda q}$$

Similarly, we observe that in the case where the male and female death rates are equal, i.e., $\mu = \mu'$,

$$R(t) = \frac{pe^{\lambda pt}}{q(e^{\lambda pt} - 1)}$$

Hence in this case

$$\lim_{t \to \infty} R(t) = \frac{p}{q}$$

It is also of interest to consider the probability distribution of the females in the population. From the assumptions on which the above model is based, and from the expression for the expected number of females in the population, it is clear that the growth of the female subpopulation follows a simple birth-and-death process. To demonstrate this, we let

$$P_x(t) = \mathscr{P}\{X(t) = x\} = \sum_{y=0}^{\infty} P_{x,y}(t) \qquad x = 0, 1, \ldots$$

denote the marginal distribution of $X(t)$. From the differential-difference equations for $P_{x,y}(t)$ we obtain

$$\frac{\partial P_x(t)}{\partial t} = -(\lambda p + \mu)x P_x(t) + \lambda p(x - 1) P_{x-1}(t)$$
$$+ \mu(x + 1) P_{x+1}(t) \qquad x = 1, 2, \ldots \qquad (4.51)$$
$$\frac{\partial P_0(t)}{\partial t} = \mu P_1(t)$$

The initial conditions are

$$P_x(0) = 1 \qquad \text{for } x = 1$$
$$= 0 \qquad \text{otherwise} \qquad (4.52)$$

From the theory developed in Chap. 2, we have

$$P_x(t) = [1 - \alpha(t)][1 - \beta(t)][\beta(t)]^{x-1} \qquad x = 1, 2, \ldots$$
$$P_0(t) = \alpha(t) \qquad\qquad (4.53)$$

where in this case

$$\alpha(t) = \frac{\mu(e^{(\lambda p - \mu)t} - 1)}{\lambda p e^{(\lambda p - \mu)t} - \mu} \qquad \beta(t) = \frac{\lambda p(e^{(\lambda p - \mu)t} - 1)}{\lambda p e^{(\lambda p - \mu)t} - \mu}$$

If we now let $Q = \lim_{t \to \infty} P_0(t)$, we have

$$Q = 1 \qquad \text{for } \lambda p \leqslant \mu$$
$$= \frac{\mu}{\lambda p} \qquad \text{for } \lambda p > \mu \qquad (4.54)$$

For a discussion of additional properties of the process considered in this section we refer to the papers of Goodman and Joshi.

D. Stochastic Models for the Growth of Competitive and Predatory Populations. In addition to the study of mathematical models of population growth involving only one type of individual, it is of great importance in population ecology to consider models which describe the growth of populations made up of two or more types of individuals which associate or interact with one another. The phe- nomena of biological association or interaction are usually referred to as "struggle for existence" phenomena. The classical studies of Lotka [67] and Volterra [91] on the deterministic theory of struggle for existence phenomena are well known. For a complete discussion of the Lotka-Volterra theory we refer to the works of D'Ancona [22] and Kostitzin [60].

In this section we formulate a general stochastic model of interaction between two species and then consider several special cases of this model that are of interest. Before considering the stochastic model, we give a brief sketch of two deterministic models the stochastic analogues of which will be special cases of the general stochastic model.

The first model we consider is for the *growth of two species which compete for the same food*. We first consider the growth of a single species S. Let $x(t)$ denote the number of individuals of species S in the population at time t. If the growth of the species can be described by the logistic curve, then $x(t)$ satisfies the logistic equation

$$\frac{dx(t)}{dt} = [\lambda - \mu f(x)]x(t) \tag{4.55}$$

where λ and μ, the birth and death rates, are assumed to be positive constants. The function $f(x)$ is a strictly increasing function of x which assumes the value zero when x is equal to zero. This function is interpreted as representing the "environmental resistance," in the sense that it can be regarded as the amount of food consumed, or the space occupied, by the x individuals in existence at time t.

We now consider the growth of the species S_1 and S_2, which consume the same food, making the assumption that the growth of each species can be described by an equation of the logistic form. Let $x_1(t)$ and $x_2(t)$ denote the numbers of individuals in S_1 and S_2, respectively, at time t. By generalizing the relation expressed by Eq. (4.55) to two species, we obtain the system of differential equations:

$$\frac{dx_1(t)}{dt} = [\lambda_1 - \mu_1 f(x_1,x_2)]x_1(t) \qquad \frac{dx_2(t)}{dt} = [\lambda_2 - \mu_2 f(x_1,x_2)]x_2(t) \tag{4.56}$$

The positive constants λ_1 and λ_2 are the birth rates of S_1 and S_2, respectively, and the positive constants μ_1 and μ_2 are the offense (or voracity) coefficients of the two species. The function $f(x_1,x_2)$ has the same interpretation as in the case of a single species. A simple model is obtained by taking the function $f(x_1,x_2)$ to be a linear function of x_1 and x_2; for example

$$f(x_1,x_2) = \alpha_1 x_1 + \alpha_2 x_2$$

The second model we consider is for the *growth of two species in which one species feeds upon the other*. We call S_1 the *prey* (or hosts) and S_2 the *predator* (or parasites). The simplest Lotka-Volterra model for a predator-prey system is based on the assumption that the growth of the mixed population depends on the frequency of encounters between the two species, the frequency of encounters in turn being a function of the population sizes of both species.

To obtain the system of equations describing the growth of the two species, we proceed as follows: We first consider the factors which influence the growth of S_1. In the interval $(t, t + \Delta t)$ the number of individuals in S_1 will change owing to the following factors: (1) the birth of individuals of the first species in $(t, t + \Delta t)$, (2) the death of individuals of the first species in $(t, t + \Delta t)$ not caused by members of S_2, and (3) the death of individuals of the first species killed by members of S_2 in $(t, t + \Delta t)$. If we now assume that both the numbers of births and deaths are proportional to the population size of S_1, then the net increase will be proportional to the population size of S_1. Since the death of individuals of S_1 due to the predators depends on the frequency of encounters between members of S_1 and S_2, the decrease in the population size of S_1 is proportional to the product of the population sizes of the two species. In view of the above, we obtain the following relation:

$$\Delta x_1(t) = [\lambda_1 x_1(t) - \mu_1 x_1(t) x_2(t)] \, \Delta t$$

where λ_1 and μ_1 are constants. Similar reasoning applied to S_2 yields

$$\Delta x_2(t) = [\lambda_2 x_1(t) x_2(t) - \mu_2 x_2(t)] \, \Delta t$$

From the above we obtain the system of differential equations

$$\frac{dx_1(t)}{dt} = \lambda_1 x_1(t) - \mu_1 x_1(t) x_2(t)$$

$$\frac{dx_2(t)}{dt} = \lambda_2 x_1(t) x_2(t) - \mu_2 x_2 t$$

$$(4.57)$$

We refrain from discussing the solution of this system, except to remark that it is periodic. This, however, is obvious from the nature of the predator-prey relationship assumed.

We now consider the formulation of a general stochastic model of interactions between two species, a formulation that is due to Chiang [19]. Let the random variables $X(t)$ and $Y(t)$ denote the population sizes of S_1 and S_2, respectively. What we seek is a differential equation for the joint probability distribution function $P_{x,y}(t) = \mathscr{P}\{X(t) = x, Y(t) = y\}$, $x, y = 0, 1, \ldots$. In order to derive the differential equation for $P_{x,y}(t)$, we first consider the events, and the associated conditional probabilities, that can cause a change in the population sizes of two species. We assume the following:

1. The probability of a unit increase in the population size of S_1 in the interval $(t, t + \Delta t)$, given that there are exactly x individuals in S_1 at time t, is $\lambda_x \Delta t + o(\Delta t)$.

2. The probability of a unit decrease in the population size of S_1 in the interval $(t, t + \Delta t)$, given that there are x individuals in S_1 at time t, is $\mu_x \Delta t + o(\Delta t)$.

3. The probability of a unit increase in the population size of S_2 in the interval $(t, t + \Delta t)$, given that there are y individuals in S_2 at time t, is $\lambda_y \Delta t + o(\Delta t)$.

4. The probability of a unit decrease in the population size of S_2 in the interval $(t, t + \Delta t)$, given that there are y individuals in S_2 at time t, is $\mu_y \Delta t + o(\Delta t)$.

5. The probability of a change in either S_1 or S_2 of absolute value greater than unity in the interval $(t, t + \Delta t)$ is $o(\Delta t)$.

6. The probability of no change in either S_1 or S_2 in the interval $(t, t + \Delta t)$, given that there are x individuals in S_1 and y individuals in S_2 at time t, is $1 - [\lambda_x + \mu_x + \lambda_y + \mu_y] \Delta t + o(\Delta t)$.

The above assumptions imply a bivariate birth-and-death process; hence, we can immediately write the system of differential equations

$$\frac{dP_{x,y}(t)}{dt} = -(\lambda_x + \mu_x + \lambda_y + \mu_y)P_{x,y}(t) + \lambda_{x-1}P_{x-1,y}(t)$$
$$+ \mu_{x+1}P_{x+1,y}(t) + \lambda_{y-1}P_{x,y-1}(t) + \mu_{y+1}P_{x,y+1}(t) \quad (4.58)$$

for $x, y = 0, 1, \ldots$. The joint distribution function $P_{x,y}(t)$ is defined as zero for either $x < 0$ or $y < 0$, or for both.

The system of differential equations (4.58) describes the stochastic development of a population of two species. In order to investigate the development of the two species for a given interaction mechanism, it is necessary to specify the set of functions $(\lambda_x, \mu_x, \lambda_y, \mu_y)$.

We now consider the application of the above system of differential equations to two special cases. The first case we consider is that of two species which compete for the same food. The set of functions in this case is given by

$$\lambda_x = \lambda_1 x \qquad \mu_x = \mu_{11}x^2 + \mu_{12}xy$$
$$\lambda_y = \lambda_2 y \qquad \mu_y = \mu_{21}xy + \mu_{22}y^2$$

where λ_1, λ_2, μ_{11}, μ_{12}, μ_{21}, μ_{22} are constants. The constants λ_1 and λ_2 have the same interpretation as in the deterministic case, while the linear functions $\mu_{11}x + \mu_{12}y$ and $\mu_{21}x + \mu_{22}y$ correspond to the function $f(x_1, x_2)$. If we substitute the above functions in Eq. (4.58), we obtain the stochastic analogue of the deterministic system (4.57).

It is difficult to obtain an explicit solution of Eq. (4.58); however, an examination of the equations for the moments of $X(t)$ and $Y(t)$ enables us to obtain a limited amount of information about the stochastic model. Let $m_{11}(t) = \mathscr{E}\{X(t)\}$, $m_{12}(t) = \mathscr{E}\{X^2(t)\}$, $m_{21}(t) = \mathscr{E}\{Y(t)\}$, $m_{22}(t) = \mathscr{E}\{Y^2(t)\}$, and $\bar{m}_{11}(t) = \mathscr{E}\{X(t)Y(t)\}$. If we now use the method of generating functions to compute the moments from Eq. (4.58), we obtain for the first two moments

$$\frac{dm_{11}(t)}{dt} = \lambda_1 m_{11}(t) - \mu_{11} m_{12}(t) - \mu_{12}\bar{m}_{11}(t)$$

$$\frac{dm_{21}(t)}{dt} = \lambda_2 m_{21}(t) - \mu_{21}\bar{m}_{11}(t) - \mu_{22} m_{22}(t)$$

It is of interest to compare this system of equations for the expected population sizes with the deterministic system

$$\frac{dx_1(t)}{dt} = \lambda_1 x_1(t) - \mu_1 \alpha_1 x_1^2(t) - \mu_1 \alpha_2 x_1(t) x_2(t)$$

$$\frac{dx_2(t)}{dt} = \lambda_2 x_2(t) - \mu_2 \alpha_1 x_1(t) x_2(t) - \mu_2 \alpha_2 x_2^2(t)$$

We first observe that for the type of interaction being considered the population size of one species varies inversely with the population size of the other. This means that there is a negative correlation between $X(t)$ and $Y(t)$, which implies that $\mathscr{E}\{X(t)Y(t)\} < \mathscr{E}\{X(t)\}\mathscr{E}\{Y(t)\}$. Hence we can conclude that the deterministic system does not describe the growth of two species competing for the same food as one would expect from the stochastic system.

We now consider the stochastic analogue of the deterministic predator-prey system. The differential equations for $P_{x,y}(t)$ in this case are obtained by substituting the following set of functions in Eq. (4.58):

$$\lambda_x = \lambda_1 x \qquad \mu_x = \mu_1 xy$$
$$\lambda_y = \lambda_2 xy \qquad \mu_y = \mu_2 y$$

In the above, x is the population size of the prey, y is the population size of the predator, and $\lambda_1, \lambda_2, \mu_1, \mu_2$ are constants.

For this case it can be shown that the differential equations for the expected values of $X(t)$ and $Y(t)$ are

$$\frac{dm_{11}(t)}{dt} = \lambda_1 m_{11}(t) - \mu_1 \bar{m}_{11}(t)$$

$$\frac{dm_{21}(t)}{dt} = \lambda_2 \bar{m}_{11}(t) - \mu_2 m_{21}(t)$$

This system can be compared with the system of equations in the deterministic case, that is,

$$\frac{dx_1(t)}{dt} = \lambda_1 x_1(t) - \mu_2 x_1(t) x_2(t)$$

$$\frac{dx_2(t)}{dt} = \lambda_2 x_1(t) x_2(t) - \mu_2 x_2(t)$$

In view of the above equations one might expect $\mathscr{E}\{X(t)Y(t)\} = \mathscr{E}\{X(t)\}\mathscr{E}\{Y(t)\}$ at any given time t. However, owing to the fact the one species feeds on the other, the population sizes of S_1 and S_2 are mutually dependent. Therefore, $\mathscr{E}\{X(t)Y(t)\} \neq \mathscr{E}\{X(t)\}\mathscr{E}\{Y(t)\}$. It is also important to note that, owing to the periodic nature of $x_1(t)$ and $x_2(t)$, the correlation coefficient between $X(t)$ and $Y(t)$ will fluctuate in sign; hence, the relative magnitude of the two members on both sides of the above inequality cannot be uniquely determined.

From this brief discussion of stochastic models of interaction between two species, it is clear that the main difficulty in obtaining a complete description of the stochastic development of interacting species is the intractability of the nonlinear differential equations for the joint probability distribution. This is also true of the equation for the generating function. In order to circumvent these difficulties, it is necessary that there be found new stochastic representations or approximation systems which incorporate the same probability mechanism for the development of the two species, but which can be handled analytically.[1] Another approach involves the use of Monte Carlo techniques.[2]

For a discussion of experimental struggle for existence studies we refer to the paper of Neyman, Park, and Scott [77]. This paper reviews the experimental results on the interaction of flour beetles (*Tribolium*) obtained by Park and his research group, and it also discusses the statistical problems arising in studies of this type. Another paper of interest is that of Chiang [20], which considers the "egg-laying–egg-eating" stochastic process of flour beetles. The model used in this paper is a univariate birth-and-death process with time-dependent birth and death rates.

E. Some Stochastic Models for Population Migration. We now consider several models arising in the theory of population migration. The situation we wish to study can be described as follows: Consider two disjoint regions, say R_1 and R_2, and assume that only one-way migration from R_1 to R_2 can occur. We also assume that at

[1] Cf. Whittle [95].
[2] Appendix C.

time $t = 0$ there are n_1 individuals in R_1 and n_2 individuals in R_2. The problem is to determine the probability that x individuals will migrate from R_1 to R_2 in the time interval $[0,t)$. The models considered in this section are due to Pyke.[1]

Let the random variable $X(t)$ denote the number of individuals in region R_2 at time t that have migrated from region R_1. Since at $t = 0$ there are n_2 individuals in R_2, at any time $t > 0$ there will be $n_2 + x$ individuals in R_2, where $x = 0, 1, \ldots, n_1$. Hence, the growth of the population in R_2 due to migration can be described by a finite birth process, the birth of an individual in R_2 being due to the migration of an individual from R_1.

Now, let $p(x,t) \, \Delta t$ denote the probability that a single individual will migrate from R_1 to R_2 in the interval $(t, t + \Delta t)$, given that x individuals have migrated in the interval $[0,t)$. The function $p(x,t)$ is called the *migration probability function*, and it is this function which will characterize a particular stochastic model for population migration. Let $q_x(y,t) \, \Delta t$ denote the probability that y individuals will migrate from R_1 to R_2 in the interval $(t, t + \Delta t)$ given that x individuals have migrated in the interval $[0,t)$. If we now assume that the migrating individuals do not interact, i.e., are statistically independent, we obtain

$$q_x(1,t) = p(x,t)(n_1 - x) \, \Delta t + o(\Delta t)$$

$$q_x(0,t) = 1 - q_x(1,t) = 1 - p(x,t)(n_1 - x) \, \Delta t + o(\Delta t)$$

Let $P_x(t) = \mathscr{P}\{X(t) = x\}$, $x = 0, 1, \ldots, n_1$. Given the migration probability function $p(x,t)$, differential equations can be derived for $P_x(t)$. The solution of these equations will then yield the probability that x individuals will migrate from R_1 to R_2 in the interval $[0,t)$. We now consider several models based on different migration probability functions.

1. Let $p(x,t) = f(x)$ be an almost arbitrary function of time, subject to the condition that

$$(n_1 - j)f(j) \neq (n_1 - k)f(k) \qquad j, k = 0, 1, \ldots, n_1$$

In this case $P_x(t)$ satisfies the differential-difference equations

$$\frac{dP_x(t)}{dt} = \lambda_{x-1}P_{x-1}(t) - \lambda_x P_x(t) \qquad x = 1, 2, \ldots, n_1$$

$$\frac{dP_0(t)}{dt} = -\lambda_0 P_0(t)$$

(4.59)

---

[1] Unpublished M.S. thesis, University of Washington, 1955.

where $\lambda_x = (n_1 - x)f(x)$. The solution[1] of the above system is

$$P_x(t) = \beta_x \sum_{k=0}^{\infty} \left(\frac{1}{\alpha_{xk}}\right) e^{-\lambda_k t} \qquad x = 1, 2, \ldots, n$$

(4.60)

$$P_0(t) = e^{-\lambda_0 t}$$

where
$$\alpha_{xk} = \prod_{\substack{i=0 \\ i \neq k}}^{x} \gamma_{ik} \qquad \text{for } k \leqslant x$$

$$\gamma_{ik} = \lambda_i - \lambda_k$$

$$\beta_x = \prod_{i=0}^{x-1} \lambda_i$$

2. Let

$$p(x,t) = \frac{f(t)}{n_1 - x} \qquad \text{for } x < n_1$$

$$= 0 \qquad \text{for } x = n_1$$

where $f(t)$ is an almost arbitrary function of time. In this case $P_x(t)$ satisfies Eq. (4.59) with $\lambda_x(t) = f(t)(n_1 - x)^{-1}$, for $x = 0, 1, \ldots, n_1 - 1$. The solution of Eq. (4.50) in this case is

$$P_x(t) = \frac{[A(t)]^x}{x!} e^{-A(t)} \qquad x = 0, 1, \ldots, n_1 - 1$$

(4.61)

where
$$A(t) = \int_0^t f(\tau) \, d\tau$$

For $x = n_1$ we have

$$P_{n_1}(t) = 1 - \sum_{x=0}^{n_1-1} P_x(t)$$

(4.62)

Hence, in this case $P_x(t)$ is the Poisson distribution. We remark that the probability distribution is Poisson only approximately for large values of n_1.

While we have restricted our attention to one-way migration, it is clear that models for two-way migration could also be considered. In these cases the migration models would be formulated as birth-and-death processes.

F. Age-dependent Stochastic Models for Population Growth. The stochastic models for population growth considered so far have all been of the Markov type; i.e., if the random variable T denotes the generation time for the individuals in the population, then distribution function $G(\tau)$ is of the form $G(\tau) = \mathscr{P}\{T \leqslant \tau\} = 1 - e^{-\lambda \tau}$, where $\lambda > 0$ is the birth rate. Hence the density function $g(\tau)$ is

$$g(\tau) = \lambda e^{-\lambda \tau}$$

---

[1] Cf. Sec. 2.4C for methods of solving the equations of a pure birth process.

While the assumption that the stochastic growth process is Markovian may be a reasonable one for the study of some populations, there is considerable experimental evidence that many populations, especially bacterial ones, do not exhibit generation times which reflect the Markov property.[1] Hence it is necessary to consider more general models which are non-Markovian.

The most general approach to the problem of bacterial growth is due to Bellman and Harris. In Sec. 2.5 the *Bellman-Harris process*, termed *age-dependent*, was introduced and certain of its properties were studied. It was shown that the generating function $F(s,t)$ for the probabilities $P_x(t)$ satisfies the Stieltjes integral equation

$$F(s,t) = \int_0^t h[F(s, t - \tau)] \, dG(\tau) + s[1 - G(t)] \tag{4.63}$$

In (4.63) $G(\tau)$, $0 < \tau < \infty$, is the distribution function of the generation times, and

$$h(s) = \sum_{n=0}^{\infty} q_n s^n \tag{4.64}$$

is the generating function of the transformation probabilities q_n. Since $G(\tau)$ is arbitrary, the Bellman-Harris process is, in general, non-Markovian. In order to characterize a given model for population growth, it is necessary to specify the functions $G(\tau)$ and $h(s)$, since these functions reflect the probability mechanism by which the population is assumed to develop.

In Sec. 2.5 we considered several models that could be obtained as special cases of the Bellman-Harris process. Of these models the one that is of particular interest in the study of bacterial growth is the non-Markovian birth process developed by Kendall [46]. Several other distribution functions have been examined by Powell, who considered their genetic interpretations.

For most of the distribution functions that are of biological importance it is difficult, if not impossible, to obtain a solution of Eq. (4.63) which, in turn, will yield explicit expressions for the probabilities $P_x(t)$. However, in many experimental situations it is not the probabilities that are of interest, but the moments, and we have shown that these can be obtained by the solution of integral equations of the renewal type. In cases when it is also difficult to obtain the moments, one can usually discuss the asymptotic properties of the moments by utilizing the theory of Feller and Täcklind.

[1] We refer to the paper of Powell [81] for a discussion of bacterial generation times and for references to earlier literature.

Bellman and Harris have also considered the age distribution of a population. In this problem it is necessary to introduce the random variable $Y(t,x)$, which represents the number of individuals in the population at time t of age less than or equal to x. The *age distribution of the population at time t* is defined by the ratio $\dfrac{Y(t,x)}{Y(t)}$, where $Y(t)$ is the number of individuals in the population of time t. Define

$$A(x) = \frac{\displaystyle\int_0^x e^{-\beta t}[1 - G(t)]\, dt}{\displaystyle\int_0^\infty e^{-\beta t}[1 - G(t)]\, dt}$$

$$= 2\beta \int_0^x e^{-\beta t}[1 - G(t)]\, dt \tag{4.65}$$

and

$$D(t) = \sup_{0 < x < \infty} \frac{A(x) - Y(t,x)}{Y(t)}$$

where β in (4.65) is the positive number satisfying

$$\tfrac{1}{2} = \int_0^\infty e^{-\beta t} g(t)\, dt$$

The function $A(x)$ is the stochastic analogue of the stable age distribution considered in the deterministic theory of population growth (cf. [65]). We state the following theorem for pure birth processes:

Theorem 4.1: If

(a) $h(s) = s^2$

(b) $G(t) = \displaystyle\int_0^t g(\tau)\, d\tau$

(c) $\displaystyle\int_0^\infty |dg(\tau)| < \infty$

then $\lim\limits_{t \to \infty} D(t) = 0$ with probability one.

Hence, the limiting distribution of the age structure of population is given by $A(x)$. We refer to Ref. 36 for a proof of Theorem 4.1. For additional studies on the age structure of population we refer to Refs. 41 and 47.

Waugh (cf. Sec. 2.5) has also considered age-dependent stochastic processes of interest in population growth. In the Waugh theory the transformation probabilities q_n are no longer assumed to be constants, as in the Bellman-Harris theory, but are assumed to be explicit functions of time. Waugh has considered an age-dependent birth-and-death process in which

$$q_2(t) = \frac{\lambda(t)}{\lambda(t) + \mu(t)}$$

$$q_0(t) = 1 - q_2(t) = \frac{\mu(t)}{\lambda(t) + \mu(t)}$$

and where the birth and death rates are given by

$$\lambda(t) = \frac{k\lambda}{(k-1)!} \frac{(k\lambda t)^{k-1}}{1 + k\lambda t + \cdots + (k\lambda t)^{k-1}/(k-1)!} \qquad \lambda > 0, k \geqslant 1$$

and

$$\mu(t) = \frac{m\mu}{(m-1)!} \frac{(m\mu t)^{m-1}}{1 + m\mu t + \cdots + (m\mu t)^{m-1}/(m-1)!} \qquad \mu > 0, m \geqslant 1$$

The distribution function of the generation times is given by

$$G(t) = 1 - e^{-(\lambda+\mu)t}\left[1 + k\lambda t + \cdots + \frac{(k\lambda t)^{k-1}}{(k-1)!}\right]$$

$$\times \left[1 + m\mu t + \cdots + \frac{(m\mu t)^{m-1}}{(m-1)!}\right]$$

This model has the following interpretation: Following the formation of a new individual, two internal processes start within it; one process consists of k phases and the other of m phases. If the k-phase process is completed first, the individual undergoes binary division and the population size is increased by one; if the m-phase process is completed first, the individual dies and the population size is decreased by one. For $k \geqslant 1$, $m = 1$, and $\mu = 0$ this process is identical with the multiphase birth process considered by Kendall, and for $k = m = 1$ this process is identical with the simple birth-and-death process.

For a discussion of a simple birth-and-death process of the Waugh type we refer to Sec. 2.5.

G. Diffusion-equation Representation of Population Growth Processes. In Sec. 3.5 we considered the diffusion-equation representation of discrete processes. In particular, we showed that every continuous Markov process can be considered as a limiting distribution of a discontinuous Markov process. Population growth is a discontinuous process, but if we restrict our attention to large populations, the assumption that the process is continuous is a reasonable one. In these cases the probability density of the population size at time t is given by the solution of the forward Kolmogorov equation

$$\frac{\partial f(t,x)}{\partial t} = \frac{\partial^2}{\partial x^2}\left[a(x)f(t,x)\right] - \frac{\partial[b(x)f(t,x)]}{\partial x} \qquad 0 < x < \infty \qquad (4.66)$$

where $f(t,x) = f(t,x;x_0) = \mathscr{P}\{X(t) = x \mid X(0) = x_0\}$. Hence, in order to characterize a particular growth process, it is necessary to specify the coefficients $a(x)$ and $b(x)$.

In this section we consider the diffusion-equation representation of two population growth processes. The processes we consider are the simple linear birth process and the logistic process. The formulation of these growth processes as diffusion processes is due to Feller [24,26].

In Sec. 4.2B we saw that the simple deterministic model for a birth process is expressed by the differential equation

$$\frac{dx(t)}{dt} = \lambda x(t) \qquad x(0) = x_0$$

where $x(t)$ is the population size at time t and $\lambda > 0$ is the birth or growth rate. The stochastic analogue of the above equation is the system of differential-difference equations

$$\frac{dP_x(t)}{dt} = -\lambda x P_x(t) + \lambda(x-1)P_{x-1}(t) \qquad x = 1, 2, \ldots$$

where $P_x(t) = \mathscr{P}\{X(t) = x\}$.

The diffusion-equation analogue of the above equations is obtained by assuming that the individuals in the population are independent of each other and have a fertility function which is independent of the total population size. Under these assumptions the infinitesimal variance and mean, i.e., the coefficients $a(x)$ and $b(x)$, are proportional to the instantaneous population size. Hence, we can write

$$a(x) = \alpha x \qquad b(x) = \beta x$$

where α and β are constants, $\alpha > 0$. The numerical value of these constants will depend on the choice of units in which t and x are expressed. The coefficient β can be interpreted as representing the "drift" in the population, and it will be positive or negative depending on whether the expected population size increases or decreases.

With the coefficients as defined above, Eq. (4.66) becomes

$$\frac{\partial f(t,x)}{\partial t} = \alpha \frac{\partial^2 [xf(t,x)]}{\partial x^2} - \beta \frac{\partial [xf(t,x)]}{\partial x} \tag{4.67}$$

$0 < x < \infty$. We remark that an application of Feller's classification criteria for the boundaries shows that $x = 0$ is an exit boundary and $x = \infty$ is a natural boundary. Hence no boundary conditions at $x = 0$ can be imposed, and the initial value $f(t,0)$ can be prescribed arbitrarily.

Let $X(0) = x_0$ and let $\beta \neq 0$. Under these conditions it has been shown (cf. [24,26]) that the solution of (4.67) is

$$f(t,x;x_0) = \frac{\beta}{\alpha(e^{\beta t} - 1)} \left[\frac{x_0 e^{\beta t}}{x}\right]^{\frac{1}{2}} I_1\left(\frac{2\beta(x_0 e^{\beta t})^{\frac{1}{2}}}{\alpha(e^{\beta t} - 1)}\right) \exp\left\{\frac{-\beta(x_0 e^{\beta t} + x)}{\alpha(e^{\beta t} - 1)}\right\} \quad (4.68)$$

where $I(\cdot)$ is the Bessel function of first order. This solution also admits the series representation

$$f(t,x;x_0) = \frac{\beta^2 x_0 e^{\beta t}}{\alpha^2(e^{\beta t} - 1)^2} \exp\left\{\frac{-\beta(x_0 e^{\beta t} + x)}{\alpha(e^{\beta t} - 1)}\right\}$$

$$\times \sum_{i=0}^{\infty} \frac{1}{i!(i+1)!} \left[\frac{\beta(x_0 x e^{\beta t})^{\frac{1}{2}}}{\alpha(e^{\beta t} - 1)}\right]^{2i} \quad (4.69)$$

Now let $m(t)$ denote the expected population size at time t. Hence,

$$m(t) = \mathscr{E}\{X(t)\} = \int_0^\infty x f(t,x;x_0)\, dx \quad (4.70)$$

From (4.70), and assuming that $xf(t,x;x_0)$ and $x \dfrac{\partial f(t,x;x_0)}{\partial x}$ approach zero as $x \to 0$, (4.67) yields the differential equation

$$\frac{dm(t)}{dt} = \beta m(t) \quad (4.71)$$

Therefore

$$m(t) = x_0 e^{\beta t} \quad (4.72)$$

Similarly, it can be shown that

$$\mathscr{D}^2\{X(t)\} = 2\left(\frac{\alpha}{\beta}\right) e^{\beta t}(e^{\beta t} - 1) \quad (4.73)$$

It is of interest to note that, except for a multiplicative factor in (4.73), (4.72) and (4.73) are, when $\alpha = \beta = \lambda$, the mean and variance of the Yule-Furry process. Let

$$P(t;x_0) = \int_0^\infty f(t,x;x_0)\, dx \quad (4.74)$$

Then $Q(t,x_0) = 1 - P(t;x_0)$ is the probability that a population of size x_0 at $t = 0$ will die out before time t. Hence $Q(t;x_0)$ is the absorption or extinction probability. From (4.64) and (4.74), we obtain

$$Q(t;x_0) = \exp\left\{\frac{-\beta x_0 e^{\beta t}}{\alpha(e^{\beta t} - 1)}\right\} \quad (4.75)$$

If we now let $t \to \infty$, we obtain the probability of ultimate extinction

$$Q(x_0) = \lim_{t \to \infty} Q(t;x_0) = 1 \qquad \text{for } \beta \leqslant 0$$
$$= e^{-x_0 \beta / \alpha} \qquad \text{for } \beta > 0 \qquad (4.76)$$

For a detailed discussion of a singular diffusion process which includes the one just discussed as a special case, we refer to Ref. 27. For this process $a(x) = \alpha x$ and $b(x) = \beta x + \gamma$.

As a final example we give the diffusion-equation representation of the logistic growth process.[1] The logistic, or Pearl-Verhulst, equation is as follows:

$$\frac{dx(t)}{dt} = \alpha x(t) - \beta x^2(t) \qquad (4.77)$$

To obtain the associated diffusion equation, it is assumed that the infinitesimal variance of $X(t)$ is proportional to instantaneous population size and that the infinitesimal mean is given by the instantaneous change in the population size. Hence, we have

$$a(x) = \gamma x \qquad b(x) = \alpha x - \beta x^2$$

where α, β, and γ are constants. By inserting these coefficients in (4.66), we obtain

$$\frac{\partial f(t,x)}{\partial t} = \gamma \frac{\partial^2 [x f(t,x)]}{\partial x^2} - \frac{\partial [(\alpha x - \beta x^2) f(t,x)]}{\partial x} \qquad (4.78)$$

$0 < x < \infty$. Equation (4.78) is the diffusion analogue of the deterministic equation for the logistic process. As far as we know, an explicit expression for the probability density $f(t,x)$ for the logistic process has not been obtained.

4.3 Growth of Populations Subject to Mutation

A. Introduction. In Sec. 4.2 we considered the growth of populations consisting of either one type of organism (homogeneous populations) or two or more types of organism (heterogeneous populations). We did not, however, consider the important problem of the growth of heterogeneous populations when it is assumed that one type of organism can be transformed into an organism of a different type. A stochastic treatment of this problem is of great importance in the study of the growth of bacterial populations subject to *mutation*. In this section we consider several models for mutation processes by utilizing the theory of Markov processes.

As in the case of the growth of populations not subject to mutation,

[1] Cf. also Fortet [30].

the first mathematical studies of bacterial mutation were deterministic. We refer to the papers of Armitage [1] and Shapiro [84] for a discussion of the deterministic theory. Deterministic models, while adequate for describing the general trend of the mutation process, are clearly inadequate for a more refined analysis, since they do not take into consideration the randomness of mutations. The first stochastic models were introduced by Lea and Coulson [64] and Luria and Delbrück [66]. For a discussion of these models we refer to the original papers, and the paper of Armitage.

Before considering several stochastic models of mutation processes, we introduce some terminology which will be used throughout this section. Consider a population made up of two types of bacteria, say type A and type B. If mutation can occur from A to B only, we say that a *single mutation* takes place, and if mutation can also occur from B to A, we say that a *double mutation* takes place. Single mutations are also referred to as *forward mutations*, while double mutations are called *reverse mutations*.

Let $X(t)$ and $Y(t)$ denote the numbers of type A organisms (normal) and type B organisms (mutant) in the population at time t, respectively. If forward and reverse mutations can occur, we assume that in an interval of length dt each organism of type A gives rise to exactly $\alpha \, dt$ organisms of type A and $\beta \, dt$ organisms of type B and that each organism of type B gives rise to exactly $\gamma \, dt$ organisms of type B and $\delta \, dt$ organisms of type A. Hence, $(\alpha + \beta)$ and $(\gamma + \delta)$ represent the rates of division of type A and type B organisms, and β and δ represent the rates of *forward and reverse mutation*. The *relative mutation rate* from type A to type B, which is defined as the rate of mutation from type A to type B relative to the total growth rate of type A, is given by $\beta/(\alpha + \beta)$.

In Sec. 4.3B we consider a stochastic model for a single mutation process with equal growth rates for the normal and mutant populations, and in Sec. 4.3C we consider a general model for mutation processes developed within the framework of the Bellman-Harris theory of age-dependent stochastic processes.

B. Stochastic Model for a Single Mutation Process. In this section we consider a stochastic model for a single mutation process; the results of this section are due to Armitage. Let the random variable $X(t)$ denote the number of organisms of type A (normal form) in the population at time t and let the random variable $Y(t)$ denote the number of organisms of type B (mutant form) in the population at time t. We now assume that:

1. The number of organisms in *each* population increases exponentially (i.e., deterministically) as a function of time, the rate of growth being $\alpha + \beta$ for each population; i.e., the growth rates are equal.

2. In any interval $(t, t + \Delta t)$ the probability of a mutation from A to B is $\beta x\,\Delta t + o(\Delta t)$, where $X(t) = x$. Hence, when a mutation takes place, the number of organisms of type A is decreased by one, and the number of organisms of type B is increased by one.

From the above assumptions it is clear that the total population size at time t, $X(t) + Y(t)$, is not subject to random fluctuations and that

$$X(t) + Y(t) = (x_0 + y_0)e^{(\alpha+\beta)t} \tag{4.79}$$

where $X(0) = x_0$ and $Y(0) = y_0$. Hence the number of normal organisms in the population at time t is given by

$$X(t) = (x_0 + y_0)e^{(\alpha+\beta)t} - Y(t) \tag{4.80}$$

Because of the dependence of $X(t)$ on $Y(t)$, expressed by the above relations, it is not necessary to treat this mutation problem as a two-dimensional or bivariate stochastic process; hence, we consider the one-dimensional process $\{Y(t), t \geqslant 0\}$ only. It is of interest to point out that this process is called a *mixed process*, because the random variable $Y(t)$ experiences both *continuous changes* (due to the deterministic growth of the mutant population) and *discrete changes* (by mutation of an organism of type A).

Let $f(y,t) = \mathscr{P}\{Y(t) = y\}$ denote the probability density function of $Y(t)$. We assume that $f(y,t)$ exists and is differentiable with respect to both y and t. From assumption (2) we obtain the following relation:

$$\mathscr{P}\{y < Y(t + dt) < y + dy\}$$
$$= f(y, t + dt)\,dy$$
$$= (1 - \beta x\,dt)[1 - (\alpha + \beta)\,dt]f(y - (\alpha + \beta)y\,dt, t)\,dy$$
$$\quad + \beta(x + 1)f(y - 1, t)\,dy\,dt + (o(dt))\,dy + o(dy)$$

By passing to the limit, we obtain the differential-difference equation

$$\frac{\partial f(y,t)}{\partial t} + (\alpha + \beta)y\,\frac{\partial f(y,t)}{\partial y} = \beta(x + 1)f(y - 1, t) - (\alpha + \beta + \beta x)f(y,t) \tag{4.81}$$

By use of (4.80), this equation becomes

$$\frac{\partial f(y,t)}{\partial t} + (\alpha + \beta)y\,\frac{\partial f(y,t)}{\partial y} = \beta[(x_0 + y_0)e^{(\alpha+\beta)t} - y + 1]f(y - 1, t)$$
$$\quad - [\beta(x_0 + y_0)e^{(\alpha+\beta)t} - \beta y + \alpha + \beta]f(y,t) \tag{4.82}$$

The boundary conditions for Eq. (4.82) are

$$f(y,t) = 0 \qquad \text{for } y < 0, t \geqslant 0$$
$$\lim_{y \to \infty} y^i f(y,t) = 0 \qquad \text{for all } i \geqslant 0, t \geqslant 0 \tag{4.83}$$

Rather than attempt to solve (4.82), we turn our attention to the problem of determining the mean and variance of $Y(t)$; hence, we first introduce the generating function

$$F(s,t) = \int_{-\infty}^{\infty} f(y,t)s^y \, dy \qquad 0 \leqslant s \leqslant 1$$

The partial derivative of $F(s,t)$ with respect to s exists, and from the boundary conditions (4.83) we have

$$\int_{-\infty}^{\infty} ys^y \frac{\partial f(y,t)}{\partial y} \, dy = -\left[F(s,t) + s \log s \, \frac{\partial F(s,t)}{\partial s} \right]$$

If we now multiply both sides of Eq. (4.82) by s^y and integrate, we obtain the following differential equation for the generating function:

$$\frac{\partial F(s,t)}{\partial t} - [\beta s + (\alpha + \beta)s \log s - \beta s^2] \frac{\partial F(s,t)}{\partial s}$$
$$= \beta(s - 1)(x_0 + y_0)e^{(\alpha+\beta)t}F(s,t) \quad (4.84)$$

We now use the method discussed in Sec. 4.2 to obtain the cumulants of $Y(t)$. Let $K(u,t) = \log F(s,t)$, where $s = e^u$; then Eq. (4.84) yields

$$\frac{\partial K(u,t)}{\partial t} - [\beta + (\alpha + \beta)u - \beta e^u] \frac{\partial K(u,t)}{\partial u} = \beta(x_0 + y_0)e^{(\alpha+\beta)t}(e^u - 1) \quad (4.85)$$

By proceeding as before, i.e., by expanding $K(u,t)$ as a power series in u and equating the coefficients of u to obtain the differential equations for the cumulants, and solving these differential equations, we obtain

$$i_1(t) = \mathscr{E}\{Y(t)\} = (x_0 + y_0)e^{(\alpha+\beta)t} - x_0 e^{\beta t} \quad (4.86)$$

and

$$i_2(t) = \mathscr{D}^2\{Y(t)\} = \frac{\beta}{\alpha} x_0 e^{\alpha t}(e^{\alpha t} - 1) \quad (4.87)$$

We remark that these cumulants are obtained from the solutions of the differential equations for $i_1(t)$ and $i_2(t)$ by letting the relative mutation rate $\beta/(\alpha + \beta)$ approach zero. It is of interest to note that (4.86) is the expression for the number of mutant organisms in the population at time t when a deterministic model is assumed, and also that (4.87), except for the multiplicative constant β/α, is the variance of the Yule-Furry birth process.

In experimental studies it is of interest to determine the *correlation between the numbers of mutant organisms* in the population at different times. Let $Y(t_1) = y_1$ and $Y(t_2) = y_2$, where $t_2 > t_1$, and let $\mathscr{E}\{Y(t_2) \mid Y(t_1)\}$ denote the conditional expectation of $Y(t_2)$, given $Y(t_1)$. From (4.86) we have

$$\mathscr{E}\{Y(t_2) \mid Y(t_1)\} = (x_1 + y_1)e^{(\alpha+\beta)(t_2-t_1)} - x_1 e^{\alpha(t_2-t_1)} \quad (4.88)$$

By use of (4.79), (4.88) becomes

$$\mathscr{E}\{Y(t_2) \mid Y(t_1)\} = (x_0 + y_0)(e^{(\alpha+\beta)t_2} - e^{(\alpha t_2 + \beta t_1)}) + y_1 e^{\alpha(t_2 - t_1)} \quad (4.89)$$

The above expression is the equation of the regression line of $Y(t_2)$ on $Y(t_1)$.

Now let $\mathscr{D}^2\{Y(t_2) \mid Y(t_1)\}$ denote the conditional variance of $Y(t_2)$, given $Y(t_1)$. From (4.87) we have

$$\mathscr{D}^2\{Y(t_2) \mid Y(t_1)\} = \frac{\beta}{\alpha} x_1 e^{\alpha(t_2 - t_1)}(e^{\alpha(t_2 - t_1)} - 1)$$

$$= \frac{\beta}{\alpha}[(x_0 + y_0)e^{(\alpha+\beta)t_1} - y_1][e^{2\alpha(t_2 - t_1)} - e^{\alpha(t_2 - t_1)}] \quad (4.90)$$

From (4.89) we observe that the regression of $Y(t_2)$ on $Y(t_1)$ is linear; hence,

$$\mathscr{E}\{Y(t_2)[Y(t_1) - \mathscr{E}\{Y(t_1)\}]\} = b_{2.1}\mathscr{D}^2\{Y(t_1)\} \quad (4.91)$$

where $b_{2.1}$ is the regression coefficient of $Y(t_2)$ on $Y(t_1)$. From (4.89) we have

$$b_{2.1} = e^{\alpha(t_2 - t_1)} \quad (4.92)$$

hence the correlation coefficient ρ between $Y(t_1)$ and $Y(t_2)$ is given by

$$\rho^2 = \frac{[\mathscr{E}\{Y(t_2)[Y(t_1) - \mathscr{E}\{Y(t_1)\}]\}]^2}{\mathscr{D}^2\{Y(t_1)\}\mathscr{D}^2\{Y(t_2)\}}$$

$$= \frac{1 - e^{-\alpha t_1}}{1 - e^{-\alpha t_2}} \quad (4.93)$$

From the above we see that the following limiting values of ρ^2 obtain:

$$\rho^2 = 1 \qquad \text{when } t_1 \to t_2$$

$$= 1 - e^{-\alpha t_1} \qquad \text{when } t_2 \to \infty \quad (4.94)$$

Hence, except when t_1 is small, the correlation coefficient is approximately one for all $t_2 > t_1$. In view of this result we can state that the mutation process is almost uniquely determined by its early history.

For a discussion of the above model when the growth rates are unequal, and for a stochastic treatment of double mutation processes, we refer to the paper of Armitage.

C. Application of the Bellman-Harris Theory to Mutation Processes. We now consider the application of the Bellman-Harris theory of age-dependent branching processes[1] to the study of mutation

---

[1] Cf. Sec. 2.5.

processes (cf. [36]). The age-dependent model is based on the following assumptions:

1. The population of normal organisms grows deterministically, the number at time t being $X(t) = x_0 e^{\lambda t}$, where λ is the growth or birth rate and $X(0) = x_0$.

2. The probability that a mutation occurs in the interval $(t, t + \Delta t)$ is $\mu x_0 \lambda e^{\lambda t} \Delta t + o(\Delta t)$, where $\mu > 0$. Hence, the probability that a mutation does not take place in the interval $(0, t)$ is $\exp\{-\mu x_0(e^{\lambda t} - 1)\}$.

Let $Y(t)$ denote the number of mutant organisms in the population at time t and let $H(s,t;x_0)$ denote the generating function of the number of mutant organisms at time t, if at time $t = 0$ there were x_0 normal organisms and no mutants present. Assuming that the time at which the first mutant is formed is a regeneration point for the process, the usual reasoning shows that $H(s,t;x_0)$ satisfies the functional equation

$$H(s,t;x_0) = \int_0^t H(s, t - \tau; x_0 e^{\lambda \tau}) F(s, t - \tau) \mu x_0 \lambda e^{\lambda \tau} e^{-\mu x_0(e^{\lambda \tau}-1)}\, d\tau + e^{-\mu x_0(e^{\lambda t}-1)}$$

(4.95)

In the above equation $F(s,t)$ is the generating function of the number of descendants of a parent mutant organism in the population at time t after the formation of the parent. $F(s,t)$, for example, might be taken as the generating function associated with a birth process of the Yule-Furry type.

Consider the mutation stochastic process as depending on two parameters, x_0 and t. As a function of x_0 the process is infinitely divisible; hence, for each $x_{01} \geqslant 0$, $x_{02} \geqslant 0$,

$$H(s, t; x_{01} + x_{02}) = H(s,t;x_{01})H(s,t;x_{02})$$

(4.96)

From (4.96) we have

$$H(s,t;x_0) = e^{x_0 G(s,t)}$$

(4.97)

where $G(s,t)$ is the cumulant generating function of the number of mutations at time t. If we now substitute (4.97) into Eq. (4.95) and evaluate

$$\lim_{x_0 \to 0} \left\{ \frac{H(s,t;x_0) - e^{-\mu x_0(e^{\lambda \tau}-1)}}{x_0} \right\}$$

we have

$$G(s,t) = -\mu(e^{\lambda t} - 1) + \mu x_0 \lambda \int_0^t F(s, t - \tau)e^{\lambda \tau}\, d\tau$$

(4.98)

From (4.97) and (4.98) the expected number of mutants in the population is easily determined. We have

$$m(t) = \mathscr{E}\{Y(t)\} = \mu x_0 \lambda \int_0^t m(t - \tau)e^{\lambda \tau}\, d\tau$$

(4.99)

where $m(t)$ is the expected number of descendants of a parent mutant at a time t after the formation of the parent. Hence,

$$m(t) = \frac{\partial F(s,t)}{\partial s}\Bigg]_{s=1}$$

If we now assume that $m(t) \sim \alpha e^{\lambda t}$, it follows from (4.99) that

$$\mathscr{E}\{Y(t)\} \sim \mu x_0 \alpha \lambda t e^{\lambda t} \qquad (4.100)$$

In (4.100)

$$\alpha = \frac{1}{4\beta \displaystyle\int_0^\infty e^{-\beta\tau} \tau g(\tau)\, d\tau}$$

where β is defined by

$$\tfrac{1}{2} = \int_0^\infty e^{-\beta\tau} g(\tau)\, d\tau$$

and $g(\tau)$ is the density function of the generation times for the descendants of a parent mutant.

For additional applications of the Bellman-Harris theory to mutation problems we refer to the paper of Kendall [49].

D.   The Effect of Phenotypic Delay on Mutation Processes. The models discussed in the last two sections did not take into consideration the occurrence of *phenotypic delay*. This effect is very important in the study of mutation processes, since the appearance of the characteristic which differentiates normal and mutant forms, termed phenotypic expression, may actually be delayed for a certain length of time after a mutation occurs. Hence, it is necessary to distinguish between *true mutants*, those organisms which are genotypically of the mutant form, and *effective mutants*, those mutants which have achieved phenotypic expression.

Stochastic models for mutation processes which take into consideration the occurrence of phenotypic delay have been studied by Bartlett [8] and Kendall [49,50]. We refer the reader to these references for a discussion of the topic.

4.4   Stochastic Theory of Epidemics

A.   Introduction. Early work in the mathematical theory of epidemics, as in the theory of population growth, was mainly concerned with the development of deterministic models for the spread of disease through a population. In this approach a functional equation for the number of infected (or susceptible) individuals in the population at time t, say $x(t)$, is derived on the basis of certain assumptions concerning the mechanism by which the disease is to be transmitted among members of the population. This equation, together with some initial condition (the number of infected, or susceptible, individuals at the start of the epidemic), is then solved to obtain $x(t)$. In assuming a deterministic causal mechanism for the spread or development of the epidemic, the number of infected (or susceptible) individuals at some time $t > 0$ will always be the same if the initial conditions remain the same. Because of the large number of chance or random factors which determine the manner in which an epidemic develops, it became clear to workers in epidemic theory that probabilistic or stochastic models would have to be used to supplement or replace the existing deterministic models.

The development of the theory of stochastic processes has given the mathematical epidemiologist the proper theoretical framework within which his models for the spread of epidemics can be constructed. Of particular interest in the stochastic theory of epidemics are Markov processes with a finite or denumerable number of states. It is of interest to note that the first paper on stochastic epidemics seems to have been published by McKendrick [72], in 1926. Unfortunately, this paper did not attract much attention at the time of its publication and for many years thereafter, for it is only recently that workers in this area have learned of McKendrick's results.

In 1946, M. S. Bartlett studied the epidemic process introduced earlier by McKendrick. In particular, the equation for the generating function associated with the stochastic process was derived. Since that time many workers have contributed to the stochastic theory of epidemics. For a detailed discussion of the mathematical theory of epidemics we refer to the book of Bailey [5]. Other discussions of epidemic theory are given in Refs. 8, 15, and 17.

This section on the stochastic theory of epidemics has three subsections. In the first two we consider models for a simple stochastic epidemic and for a general stochastic epidemic. In the last subsection we give a brief discussion of some additional results in epidemic theory.

B. Model for a Simple Stochastic Epidemic: A Nonlinear Pure Death Process. The model we discuss in this section (cf. [2,5]) was developed in order to treat the case of the spread of a relatively mild infection through a *finite* population in which none of the infected individuals is removed from the population by isolation, recovery, or death. This model may be used to describe the spread of mild infections of the upper respiratory tract and it may be used (approximately) to represent the development of epidemics for which the time of removal from circulation is long compared with the time usually required for the epidemic to be completed.

Let the random variable $X(t)$ represent the number of *susceptibles* in a population of size $n + 1$ at time t and let $X(0) = n$; i.e., at the start of the epidemic there is only one infected individual in the population. Let $P_x(t) = \mathscr{P}\{X(t) = x\}$, $x = 0, 1, \ldots, n$. We now make the assumption of homogeneous mixing of the population and let

$$\mu x(n - x + 1) \, \Delta t + o(\Delta t)$$

denote the probability that one new infection will take place in the interval $(t, t + \Delta t)$. Since the random variable $X(t)$ can only decrease with time, the model we here consider is a pure death process with a finite number of states and death rate

$$\mu_x = \mu x(n - x + 1) \qquad \mu > 0$$

From the theory presented in Chap. 2, we see that the probabilities $P_x(t)$ satisfy the differential-difference equations[1]

$$\frac{dP_x(t)}{dt} = (x + 1)(n - x)P_{x+1}(t) - x(n - x + 1)P_x(t) \qquad x = 0, \ldots, n - 1$$

$$\frac{dP_n(t)}{dt} = -nP_n(t) \tag{4.101}$$

This system is to be solved with the initial condition

$$P_x(0) = 1 \qquad \text{for } x = n$$

$$= 0 \qquad \text{otherwise} \tag{4.102}$$

The above model is the stochastic analogue of the following deterministic model: Let $x(t)$ denote the number of susceptibles in a population of size $n + 1$ at time t and let $x(0) = n$. If μ is the infection rate, then the number of new infections in the interval $(t, t + \Delta t)$ is $\mu x(n - x + 1) \, \Delta t$. Hence, the differential equation for $x(t)$ (again assuming a change in the time variable) is

$$\frac{dx(t)}{dt} = -x(n - x + 1)$$

the solution of which, when $x(0) = n$, is

$$x(t) = \frac{n(n + 1)}{n + e^{(n+1)t}}$$

[1] In writing Eqs. (4.101) we have assumed that the time variable is changed in order to make the equations dimensionless with respect to the infection rate μ.

We now consider the solution of Eq. (4.101). Let $p_x(z)$ denote the Laplace transform of $P_x(t)$. By transforming Eq. (4.101) we obtain the subsidiary equations

$$zp_x(z) = (x + 1)(n - x)p_{x+1}(z) - x(n - x + 1)p_x(z)$$
$$x = 0, 1, \ldots, n - 1 \quad (4.103)$$
$$zp_n(z) = 1 - np_n(z)$$

From (4.103) we obtain the recurrence relations

$$p_x(z) = \frac{(x + 1)(n - x)}{z + (n - x + 1)} p_{x+1}(z) \qquad x = 0, 1, \ldots, n - 1$$
$$\qquad (4.104)$$
$$p_n(z) = \frac{1}{z + n}$$

These relations give

$$p_x(z) = \frac{n!(n - x)!}{x!} \prod_{i=1}^{n-x+1} [z + i(n - i + 1)]^{-1} \qquad x = 0, 1, \ldots, n \quad (4.105)$$

Hence, the formal solution of Eqs. (4.101) is given by the inverse Laplace transform

$$P_x(t) = \lim_{\beta \to \infty} \frac{1}{2\pi i} \int_{\alpha - i\beta}^{\alpha + i\beta} p_x(z)e^{zt}\, dz$$

Bailey [2] has shown that for $x > n/2$, n an even integer,

$$p_x(z) = \sum_{i=1}^{n-x+1} \frac{\alpha_{xi}}{z + i(n - i + 1)} \qquad (4.106)$$

where the coefficients α_{xi} are given by

$$\alpha_{xi} = \lim_{z \to -i(n-i+1)} [z + i(n - i + 1)]p_x(z)$$
$$= \frac{(-1)^{i-1}(n - 2i + 1)n!(n - x)!(x - i - 1)!}{x!(i - 1)!(n - i)!(n - x - i + 1)!} \qquad (4.107)$$

Inversion of (4.106) yields

$$P_x(t) = \sum_{i=1}^{n-x+1} \alpha_{xi}e^{-i(n-i+1)t} \qquad x > \frac{n}{2} \qquad (4.108)$$

Similarly, for $0 < x \leqslant n/2$, $p_x(z)$ has the form

$$p_x(z) = \sum_{i=1}^{n/2} \alpha_{xi}[z + i(n - i + 1)]^{-1} + \sum_{i=1}^{n/2} \beta_{xi}[z + i(n - i + 1)]^{-2} \qquad (4.109)$$

$$p_0(z) = z^{-1} + \sum_{i=1}^{n/2} \alpha_{0i}[z + i(n - i + 1)]^{-1} + \sum_{i=1}^{n/2} \beta_{0i}[z + i(n - i + 1)]^{-2} \quad (4.110)$$

In the above, the coefficients α_{xi} are given by (4.107) when $i < x$, but for $i \geqslant x$

$$\alpha_{xi} = \lim_{z \to -i(n-i+1)} \frac{\partial [z + i(n - i + 1)]^2}{\partial z} p_x(z)$$

$$= \frac{(-1)^x (n - 2i + 1) n! (n - x)!}{x! (i - 1)! (n - i)! (n - x - i + 1)! (i - x)!}$$

$$\times \left(\sum_{\lambda = i}^{n-i} \lambda^{-1} + \sum_{\lambda = i+x-1}^{n-i-x+1} \lambda^{-1} - \frac{2}{n - 2i + 1} \right) \qquad (4.111)$$

In (4.111) the first summation does not occur when $x = 1$ and the second summation does not occur when $x = n/2$. The coefficients β_{xi} are given by

$$\beta_{xi} = \lim_{z \to -i(n-i+1)} [z + i(n - i + 1)]^2 p_x(z)$$

$$= \frac{(-1)^{x+1} (n - 2i + 1)^2 n! (n - x)!}{x! (i - 1)! (n - i)! (n - i - x + 1)! (i - x)!} \qquad (4.112)$$

Inversion of (4.109) and (4.110) yields

$$P_x(t) = \sum_{i=1}^{n/2} \alpha_{xi} e^{-i(n-i+1)t} + \sum_{i=x}^{n/2} \beta_{xi} t e^{-i(n-i+1)t} \qquad 0 < x \leqslant \frac{n}{2} \qquad (4.113)$$

and $$P_0(t) = 1 + \sum_{i=1}^{n/2} \alpha_{0i} e^{-i(n-i+1)t} + \sum_{i=1}^{n/2} \beta_{0i} t e^{-i(n-i+1)t} \qquad (4.114)$$

The above expressions, (4.108), (4.113), and (4.114), are exact expressions for the probabilities $P_x(t)$ when n is an even integer. Similar expressions can be obtained when n is an odd integer.

We now consider the problem of determining the mean or expected number of susceptibles in the population at time t, given that there are n susceptibles in the population at the start of the epidemic. We first consider this problem, following Bailey, by utilizing the method of generating functions.

Let $F(s,t)$ denote the generating function of the $P_x(t)$. From (4.101) we see that $F(s,t)$ satisfies the partial differential equation

$$\frac{\partial F}{\partial t} = s(s - 1) \frac{\partial^2 F}{\partial s^2} + n(1 - s) \frac{\partial F}{\partial s} \qquad (4.115)$$

together with the initial condition

$$F(s,0) = s^n \qquad (4.116)$$

Now, let $f(s,z)$ denote the Laplace transform of $F(s,t)$. From our previous results we see that $f(s,z)$ is of the form

$$f(s,z) = z^{-1} + \sum_{x=1}^{n/2} \left\{ \frac{g(x,s)}{[z + x(n - x + 1)]^2} + \frac{h(x,s)}{[z + x(n - x + 1)]} \right\} \qquad (4.117)$$

BIOLOGY

where the functions $g(x,s)$ and $h(x,s)$ are polynomials in s. More precisely, they are the sums of the coefficients of $[z + x(n - x + 1)]^{-2}$ and $[z + x(n - x + 1)]^{-1}$, respectively, in the partial-fraction expansion of $f(s,z)$. Hence, $F(s,t)$ is of the form

$$F(s,t) = 1 + \sum_{x=1}^{n/2} \{tg(x,s) + h(x,s)\}e^{-x(n-x+1)t} \qquad (4.118)$$

From (4.118) a formal expression for the mean can be obtained by differentiation; that is,

$$m(t) = \mathscr{E}\{X(t)\} = \frac{\partial F}{\partial s}\bigg]_{s=1}$$

$$= \sum_{x=1}^{n/2} \left\{ t\frac{\partial g}{\partial s}\bigg]_{s=1} + \frac{\partial h}{\partial s}\bigg]_{s=1} \right\} e^{-x(n-x+1)t} \qquad (4.119)$$

From (4.119) the *epidemic curve*, given by $-\dfrac{dm(t)}{dt}$, can be obtained. Hence

$$-\frac{dm(t)}{dt} = \sum_{x=1}^{n/2} \left\{ [x(n - x + 1)t - 1]\frac{\partial g}{\partial s}\bigg]_{s=1} \right.$$

$$\left. + x(n - x + 1)\frac{\partial h}{\partial s}\bigg]_{s=1} \right\} e^{-x(n-x+1)t} \quad (4.120)$$

Bailey has shown that $g(x,s)$ and $h(x,s)$ satisfy the equations

$$s(1 - x)g''(x,s) - n(1 - s)g'(x,s) - x(n - x + 1)g(x,s) = 0 \qquad (4.121)$$

$$s(1 - s)h''(x,s) - n(1 - s)h'(x,s) - x(n - x + 1)h(x,s) = -g(x,s) \qquad (4.122)$$

where the primes denote differentiation with respect to s. Explicit expressions have been obtained for $\partial g(x,s)/\partial s]_{s=1}$, since (4.121) is satisfied by a hypergeometric function, and also for $\partial h(1,s)/\partial s]_{s=1}$, but not for $\partial h(x,s)/\partial s]_{s=1}$. For n fixed, an explicit series solution of $g(x,s)$ can be obtained.

Haskey [38], continuing the work of Bailey, has given a relatively simple expression for the mean number of susceptibles in the population at time t. After rather lengthy algebraic manipulations, Haskey obtained

$$m(t) = \sum_x \frac{n!}{(n - x)!(x - 1)!}$$

$$\times \left[t(n - 2x + 1)^2 + 2 - (n - 2x + 1)\sum_{j=x}^{n-x} j^{-1} \right] e^{-x(n-x+1)t} \quad (4.123)$$

where the summation is for $x = 1, 2, \ldots, n/2$, for n even, and $x = 1,2, \ldots, (n + 1)/2$, for n odd. In the latter case the term

$$\frac{n!}{\{[\frac{1}{2}(n - 1)]!\}^2}$$

must be added to the expression given by (4.123).

C. Model for a General Stochastic Epidemic: The Bartlett-McKendrick Process.

In this section we consider a more realistic model for the spread of an epidemic by taking into consideration the possibility of removal of infected individuals from the population by death or isolation. If we do not distinguish between death and isolation, both of these events can be considered as a generalized death or removal.

Let us now consider finite population with n susceptible individuals and assume that, at time $t = 0$, i $(i > 0)$ infected individuals are introduced into this population. Let the random variables $X(t)$ and $Y(t)$ denote the number of susceptible and infected individuals, respectively, in the population at time t. The mechanism for the spread of the epidemic is as follows:

1. The individuals in the population mix homogeneously.
2. In the interval $(t, t + \Delta t)$ the probability of one new infection is $\lambda xy\, \Delta t + o(\Delta t)$, where λ $(\lambda > 0)$ is the infection rate.
3. In the interval $(t, t + \Delta t)$, the probability of one removal is $\mu y\, \Delta t + o(\Delta t)$, where μ $(\mu > 0)$ is the removal rate.
4. In the interval $(t, t + \Delta t)$ a number of multiple transitions (multiple infections and removals) occur with probability $o(\Delta t)$.
5. In the interval $(t, t + \Delta t)$, the probability of no change is $1 - (\lambda x + \mu)y\, \Delta t + o(\Delta t)$.

Let $P_{x,y}(t) = \mathscr{P}\{X(t) = x, Y(t) = y\}, x \geqslant 0, y \geqslant 0, t \geqslant 0$. From (2) and (3) we have that the possible transitions are $(x \to x - 1, y \to y + 1)$ and $(y \to y - 1)$. Hence, the differential-difference equations characterizing the stochastic epidemic are given by

$$\frac{dP_{x,y}(t)}{dt} = (x + 1)(y - 1)P_{x+1,y-1}(t) - y(x + \rho)P_{x,y}(t)$$
$$+ \rho(y + 1)P_{x,y+1}(t) \quad (4.124)$$

$$\frac{dP_{n,i}(t)}{dt} = -i(n + \rho)P_{n,i}(t)$$

where $0 \leqslant x + y \leqslant n + i, 0 \leqslant x \leqslant n$, and $0 \leqslant y \leqslant n + i$. In writing (4.124) we have put $\rho = \mu/\lambda$, defined as the *relative removal rate*, and changed the time variable in order to make the equations dimensionless with respect to the infection rate λ. The above equations are to be solved with the initial condition

$$P_{x,y}(0) = 1 \quad \text{for } x = n, y = i$$
$$= 0 \quad \text{otherwise} \quad (4.125)$$

Hence we see that this general stochastic epidemic leads to the study of a bivariate, nonlinear birth-and-death process with a finite number of states.

The stochastic epidemic model characterized by Eq. (4.123) is the stochastic analogue of the following deterministic model: Consider a homogeneous population of n individuals, and let $x(t)$ represent the number of susceptibles at time t, $y(t)$ the number of infected individuals, and $z(t)$ the number of individuals who are dead, isolated, or recovered and immune. It is clear that $x(t) + y(t) + z(t) = n$ for all $t \geqslant 0$. Since the mechanism for the spread of the epidemic in both the stochastic and deterministic representations is the same, the system of differential equations characterizing the deterministic epidemic is

$$\frac{dx(t)}{dt} = -\lambda xy \qquad \frac{dy(t)}{dt} = \lambda xy - \mu y \qquad \frac{dz(t)}{dt} = \mu y \qquad (4.126)$$

The equations are called the *Kermack-McKendrick equations*, after W. O. Kermack and A. G. McKendrick, who formulated the deterministic model and studied certain of its properties. In particular, Kermack and McKendrick obtained an approximate solution of (4.126) and also obtained a relation between the total size of the epidemic (i.e., the total number of removals after an infinite period of time) and the ratio of the removal rate to the infection rate. This relation is expressed by the following theorem.

Theorem 4.2 (*Kermack-McKendrick Threshold Theorem*): Let $x(0) = x_0$ and $y(0) = y_0$ represent the numbers of susceptible and infected individuals in the population at the start of the epidemic; then

$$\lim_{y_0 \to 0} z(\infty) = \frac{2\rho}{x_0}(x_0 - \rho) \qquad \text{for } x_0 > \rho$$

$$= 0 \qquad \text{for } x_0 \leqslant \rho$$

In the case when $x_0 > \rho$, say $x_0 = \rho + \epsilon$, the total size of the epidemic is approximately 2ϵ.

The limiting relation given above asserts the following: (1) If $x_0 \leqslant \rho$, an epidemic will not develop; (2) if $x_0 = \rho + \epsilon$, the number of susceptibles is eventually reduced to $\rho - \epsilon$, that is, to a value as far below the threshold value ρ as it initially was above.

Kendall [51] has given an exact treatment of the epidemic situation studied by Kermack and McKendrick. In particular, he shows that in the case when $x_0 \gg \rho$ the limiting relation becomes

$$\lim_{y_0 \to 0} z(\infty) = 2(x_0 - \rho) \qquad \text{for } x_0 > \rho$$

$$= 0 \qquad \text{for } x_0 \leqslant \rho$$

Let us now return to the stochastic model and consider the solution of Eqs. (4.124). Let $F(r,s,t)$ denote the generating function of $P_{x,y}(t)$, that is,

$$F(r,s,t) = \sum_{x=0}^{\infty} \sum_{y=0}^{\infty} P_{x,y}(t) r^x s^y \qquad |r| \leqslant 1, |s| \leqslant 1$$

Then from (4.124) we see that $F(r,s,t)$ satisfies the partial differential equation

$$\frac{\partial F}{\partial t} = (s^2 - rs) \frac{\partial^2 F}{\partial r \, \partial s} + \rho(1 - s) \frac{\partial F}{\partial s} \qquad (4.127)$$

Let $f(r,s,z)$ denote the Laplace transform of $F(r,s,t)$. Then, from Eq. (4.127), $f(r,s,z)$ satisfies

$$(s^2 - rs) \frac{\partial^2 f}{\partial r \, \partial s} + \rho(1 - s) \frac{\partial f}{\partial s} - zf + r^n s^i = 0 \qquad (4.128)$$

Since
$$f(r,s,z) = \sum_{x=0}^{\infty} \sum_{y=0}^{\infty} p_{x,y}(z) r^x s^y \qquad (4.129)$$

where $p_{x,y}(z)$ is the Laplace transform of $P_{x,y}(t)$, we obtain, after introducing (4.129) into (4.128) and equating coefficients of $r^x s^y$, the recurrence relations

$$(x + 1)(y - 1)p_{x+1,y-1}(z) - [y(x + \rho) + z]p_{x,y}(z) + \rho(y + 1)p_{x,y+1}(z) = 0 \qquad (4.130)$$

$$-[i(n + \rho) + z]p_{n,i}(z) + 1 = 0$$

where $0 \leqslant x + y \leqslant n + i$, $0 \leqslant x \leqslant n$, $0 \leqslant y \leqslant n + i$. Any $p_{x,y}(z)$ whose suffixes fall outside the stipulated ranges is taken to be zero. The probabilities $P_{x,y}(t)$ can be obtained from (4.130) by the method used in Sec. 4.4B, i.e., by starting with $p_{n,i}(z)$ and calculating all the $p_{x,y}(z)$ in succession, obtaining the partial-fraction expansion of the $p_{x,y}(z)$, and finally applying the inverse Laplace transformation. While this procedure is straightforward, the calculations are rather involved, and the results are obtained in a form which is not very useful in applications.

However, the recurrence relations given by (4.130) can be used to obtain the *probability*, say $Q(u)$, *that the total size of the epidemic is u*. Here u is the value of $n - X(t)$ for $t = \infty$, not counting the i initial cases. By definition,

$$Q(u) = \lim_{t \to \infty} P_{n-u,0}(t) \qquad \text{for } 0 \leqslant u \leqslant n$$

By use of Theorem B.5, the above becomes

$$Q(u) = \lim_{z \to 0} z p_{n-u,0}(z)$$
$$= \lim_{z \to 0} \rho p_{n-u,1}(z)$$

Hence
$$Q(u) = \rho q_{n-u,1} = \rho[\lim_{z \to 0} p_{n-u,1}(z)] \tag{4.131}$$

where $1 \leqslant x + y \leqslant n + i$, $0 \leqslant x \leqslant n$, $1 \leqslant y \leqslant n + i$. By putting $z = 0$ in (4.130), we obtain the recurrence relations

$$(x + 1)(y - 1)q_{x+1,y-1} - y(x + \rho)q_{x,y} + \rho(y + 1)q_{x,y+1} = 0$$
$$-i(n + \rho)q_{ni} + 1 = 0 \tag{4.132}$$

with the same limits for x and y as given above. These relations may be simplified by putting

$$q_{x,y} = \frac{n!(x + \rho - 1)!\rho^{n+i-(x+y)}}{yx!(n + \rho)!} w_{x,y} \tag{4.133}$$

By introducing (4.133) into (4.132), we obtain

$$w_{x+1,y-1} - w_{x,y} + (x + \rho)^{-1}w_{x,y+1} = 0$$
$$w_{ni} = 1 \tag{4.134}$$

Bailey has used (4.131), (4.132), and (4.134) to calculate $Q(u)$ for $n = 10$, 20, and 40, when $i = 1$, by using various values of ρ.

Other workers have studied the distribution of $Q(u)$. In particular, Whittle [94] has obtained a set of singly recurrent relations to replace the set (4.132) given by Bailey. Foster [31] has utilized the relationship between random-walk processes on two-dimensional lattices and birth-and-death processes to study $Q(u)$. We refer to Prob. 2.11 for a statement of the result of this random-walk approach.

We close this section by giving the stochastic analogue of the Kermack-McKendrick threshold theorem. This result is due to P. Whittle. Before stating the theorem, we define π_j as the probability that a proportion j of the n susceptibles are eventually attacked. Then we clearly have that

$$\pi_j = \sum_{u=0}^{nj} Q(u)$$

where the probabilities $Q(u)$ are given by the set of equations

$$\sum_{u=0}^{k} \binom{n-u}{n-k} \left(\frac{n-k+\rho}{\rho}\right)^u Q(u) = \binom{n}{k} \left(\frac{n-k+\rho}{\rho}\right)^i \qquad 0 \leqslant k \leqslant n$$

Hence π_j is the probability that an epidemic will not develop.

Theorem 4.3: Let ρ denote the relative removal rate for a population of size n into which i infected individuals are introduced at time $t = 0$. The following relations obtain for the probability of no epidemic:

(a) $\left(\dfrac{\rho}{n}\right)^i \leqslant \pi_j \leqslant \left[\dfrac{\rho}{n(1-j)}\right]^i$ for $\rho < n(1-j)$

(b) $\left(\dfrac{\rho}{n}\right)^i \leqslant \pi_j \leqslant 1$ for $n(1-j) \leqslant \rho < n$

(c) $\pi_j = 1$ for $\rho \geqslant n$

These results have the following interpretations: If $\rho \geqslant n$, then the probability of an epidemic exceeding any preassigned intensity i is zero, while if $\rho < n$, the probability of an epidemic, for i small, is approximately $1 - (\rho/n)^i$.

D.   Some Additional Studies on Stochastic Epidemics

1.   *Generalization of the Model for a Simple Stochastic Epidemic to Two Populations.*   Haskey [39] has considered the following model for a simple stochastic epidemic, which is a generalization to two interacting populations of the model considered in Sec. 4.4B. Let P_1 and P_2 represent two populations of size n_1 and n_2, respectively. At time $t = 0$, i infected individuals are introduced into P_1. Let the random variables $X_1(t)$ and $X_2(t)$ denote the number of susceptible individuals in P_1 and P_2, respectively, at time t. The probability mechanism for the spread of the epidemic is as follows:

1. In the interval $(t, t + \Delta t)$, the probability that one of the susceptibles in P_1 will be infected by an infective in P_1 is

$$\mu_1 x_1 (n_1 + i - x_1)\,\Delta t + o(\Delta t) \qquad \mu_1 > 0$$

and the probability that one of the susceptibles in P_1 will be infected by an infective in P_2 is

$$\mu_2' x_1 (n_2 - x_2)\,\Delta t + o(\Delta t) \qquad \mu_2' > 0$$

2. In the interval $(t, t + \Delta t)$, the probability that one of the susceptibles in P_2 will be infected by an infective in P_2 is

$$\mu_2 x_2 (n_2 - x_2)\,\Delta t + o(\Delta t) \qquad \mu_2 > 0$$

and the probability that one of the susceptibles in P_2 will be infected by an infective in P_1 is

$$\mu_1' x_2 (n_1 + i - x_1)\,\Delta t + o(\Delta t) \qquad \mu_1' > 0$$

Hence, in P_1 the total probability of a new infection in $(t, t + \Delta t)$ is

$$x_1[\mu_1(n_1 + i - x_1) + \mu_2'(n_2 - x_2)]\,\Delta t + o(\Delta t)$$

and in P_2 the total probability of a new infection in $(t, t + \Delta t)$ is

$$x_2[\mu_2(n_2 - x_2) + \mu_1'(n_1 + i - x_1)] \Delta t + o(\Delta t)$$

3. In the interval $(t, t + \Delta t)$ the probability of no new infections in P_1 and P_2 is one minus the sum of the total probabilities for a new infection in P_1 and P_2.

Let

$$P_{x_1,x_2}(t) = \mathscr{P}\{X_1(t) = x_1, X_2(t) = x_2\} \qquad x_1 = 0, \ldots, n_1, x_2 = 0, \ldots, n_2$$

By considering the transitions $(x_1 + 1, x_2) \to (x_1, x_2 + 1) \to (x_1, x_2)$ and $(x_1, x_2) \to (x_1, x_2)$ and assuming that events in the two populations are independent, we have that $P_{x_1,x_2}(t)$ satisfies the differential-difference equations

$$\begin{aligned}
\frac{dP_{x_1,x_2}(t)}{dt} = &-\{((\mu_1 x_1 + \mu_1' x_2)(n_1 + i - x_1) \\
&+ (\mu_2 x_2 + \mu_2' x_1)(n_2 - x_2)\}P_{x_1,x_2}(t) \\
&+ \{\mu_2(x_2 + 1)(n_2 - x_2 - 1) + \mu_1'(x_2 + 1)(n_1 + i - x_1)\}P_{x_1,x_2+1}(t) \\
&+ \{\mu_1(x_1 + 1)(n_1 + i - x_1 - 1) + \mu_2'(x_1 + 1)(n_2 - x_2)\}P_{x_1+1,x_2}(t)
\end{aligned}$$

$$(4.135)$$

By using the method of the Laplace transform, Haskey has obtained expressions for the mean numbers of susceptibles at time t in the two populations P_1 and P_2.

2. *Approximating Systems for Stochastic Epidemics.* In order to obtain more information about an epidemic developing according to the general model discussed in Sec. 4.4C, several authors (Kendall [51,52] and Whittle [94]) have proposed various "approximating stochastic systems" which are simpler stochastic representations of the same process, but exhibit the same characteristic features as the general process.

The approximating stochastic system proposed by Kendall can be described as follows: Let the random variable $\tilde{Y}(t)$ denote the number of infected individuals in the population. Two cases are considered.

CASE 1: If $n \leqslant \rho$, then $\{\tilde{Y}(t), t \geqslant 0\}$ is a simple birth-and-death process, with birth rate $n\lambda$ and death rate μ, satisfying the initial condition $\tilde{Y}(0) = i$.

CASE 2: If $n > \rho$, then the system has two modes of behavior, say A and B, with

$$\mathscr{P}\{A\} = \left(\frac{\rho}{n}\right)^i \qquad \text{and} \qquad \mathscr{P}\{B\} = 1 - P\{A\} = 1 - \left(\frac{\rho}{n}\right)^i$$

In mode A, $\{\tilde{Y}(t),\, t \geqslant 0\}$ is a simple birth-and-death process, with birth rate $n\lambda$ and death rate μ, that satisfies the initial condition $\tilde{Y}(0) = i$ and is subject to the condition that $\tilde{Y}(\infty) = 0$, i.e., that ultimate extinction will occur. In mode B, $\{\tilde{Y}(t),\, t \geqslant 0\}$ can be replaced by the *associated deterministic process* that describes the change in the number of infected individuals in the population.

This approximating system has been utilized to calculate the mean or expected total size of the epidemic. Let u denote the total size of the epidemic based on the approximating system. Kendall has obtained the following results:

CASE 1: $\mathscr{E}\{u\} = \dfrac{i\rho}{\rho - n} - i = \dfrac{in}{\rho - n}$ for $\rho \geqslant n$

CASE 2: $\mathscr{E}\{u\} = \left(\dfrac{\rho}{n}\right)^{i}\left(\dfrac{i\rho}{n - \rho}\right) + \left[1 - \dfrac{\rho}{n}\right]^{i}(\xi - i)$ for $\rho < n$

where ξ, the total number of cases in the deterministic process, is the unique positive root of

$$i - \xi + n(1 - e^{-\xi/p}) = 0$$

Comparison of the results with the calculations obtained by Bailey shows that the agreement between the approximate and exact treatment is reasonably good, ignoring, of course, the case when $\rho = n$ (the threshold value).

Monte Carlo methods were also used by Kendall to check the usefulness of his approximating system. In particular, artificial realizations of the exact process $\{Y(t),\, t \geqslant 0\}$ were obtained, and the associated epidemic curve compared with $\mathscr{E}\{\tilde{Y}(t)\}$, $t \geqslant 0$. Again, there was good agreement between certain characteristics of the two processes.

The results obtained by Kendall indicate that the method of approximating stochastic systems should play an important role in the study of complex epidemic situations which seem to defy a complete stochastic treatment.

3. *Recurrent Epidemics.* An epidemic is said to be *recurrent* if the susceptible population is in one way or another replenished. One way to replenish the population of susceptibles is to permit the immigration of uninfected individuals into the population. In this case, the equation for the generating function (4.127) must include an additional term to account for the transition $(x \rightarrow x + 1)$. If we assume that in the interval $(t,\, t + \Delta t)$ the probability of a new susceptible being added to the population is $\nu\,\Delta t + o(\Delta t)$, then Eq. (4.127) becomes

$$\frac{\partial F}{\partial t} = \lambda(s^2 - rs)\frac{\partial^2 F}{\partial r\,\partial s} + \mu(1 - s)\frac{\partial F}{\partial s} + \nu(s - 1)F \qquad (4.136)$$

The treatment of this stochastic model presents considerable difficulties. In order to circumvent them, Bartlett [9] (cf. also [5]) has used
deterministic models as approximations to the full stochastic model.
Monte Carlo methods have also been employed to generate artificial
epidemics of the recurrent type.

4. *Age-dependent Branching Processes and Epidemic Models.*[1]
Three important factors associated with any infectious disease are
(1) *the latent period,* (2) *the infectious period,* and (3) the *incubation
period.* The latent period is defined as the interval of time, following
infection, during which the organisms are multiplying but the infected
individual is unable to infect other individuals. The infectious period is
defined as the interval of time during which organisms are discharged.
And the incubation period is defined as the period of time between the
receipt of infection and the appearance of symptoms. It is clear that
any realistic mathematical treatment of epidemics must take these
factors into account and that they should be introduced in a form that is
epidemiologically meaningful. The models considered in Secs. 4.4B and
C have made no allowance for the effect of a well-defined incubation
period. Also, the models were based on the assumption that the latent
period was zero; since the models were Markov processes, the infectious
period had a negative exponential distribution.

Stochastic models incorporating some of the above factors can be
formulated within the framework of the Bellman-Harris theory of age-
dependent branching processes.[2] In the Bellman-Harris approach the
infectious period is a random variable, say τ, with general distribution
function $G(\tau)$, $0 < \tau < \infty$. At the end of this period the infected
individual can infect n ($n = 0, 1, \dots$) other individuals with probabilities q_n, where the q_n sum to unity, and each newly infected individual
has the same distribution $G(\tau)$ for the time that will elapse before he
infects someone else.

In view of the above, we see that the Bellman-Harris theory permits
the formulation of more realistic models in two ways: First, through
the introduction of a general distribution function for the infectious
period and, second, by allowing death (if $q_0 \neq 0$) and infection of more
than one individual at a time (if $q_n \neq 0$, $n > 2$). Of course, the same
mathematical difficulties enter here as in the less general cases; i.e.,
it is difficult to solve the Bellman-Harris functional equation for the
generating function for arbitrary $G(\tau)$, and so far only the Markov case,
i.e., $G(\tau) = 1 - e^{-\lambda t}$, has been considered. This suggests the development of approximation methods for treating age-dependent epidemics,
or the use of Monte Carlo methods to generate artificial epidemics when

[1] Cf. Bharucha-Reid [14,17].
[2] See Sec. 2.5 for a discussion of the Bellman-Harris theory.

different distribution functions for the infectious period are used. However, the moments of the random variable (or random variables) associated with the stochastic epidemic can be obtained, since the moments are given by the solutions of integral equations of the renewal type.

The latent period can be taken into consideration by an appropriate definition of $G(\tau)$. For example, a latent period of length α and an infectious period with negative exponential distribution can be obtained by letting

$$G(\tau) = 0 \qquad \text{for } 0 < \tau < \alpha$$
$$= 1 - e^{-\lambda(\tau-\alpha)} \qquad \text{for } \tau > \alpha$$

It is also possible to utilize Waugh processes in the study of age-dependent epidemics. In these cases the infection probabilities q_n are no longer assumed to be constants, as in the Bellman-Harris theory, but are functions of τ. Hence, the probability of an infected individual infecting one or more susceptibles is a function of the time he is in an infectious state.

5. *Epidemics in Small Groups.* In our previous discussions we have been primarily concerned with the development of epidemics in large populations. It is also of great interest to develop models that are useful for the study of the spread of disease in small groups, such as a single household or family, servicemen in a barracks, etc.

In 1931, M. Greenwood (cf. [5]) investigated the spread of a within-group (or family) epidemic and showed that it could be represented by a *chain of binomial distributions.* That is, at each stage in the development of the epidemic the members of the population can be divided into two groups, namely, infectives and susceptibles; hence, it is reasonable to assume that at the next stage the susceptible group will yield a number of new infectives, this number being a random variable with a binomial distribution. Therefore, as the epidemic develops, we have a chain of binomial distributions, with the probability of a new infection at any stage depending on the numbers of infectives and susceptibles at the previous stage. The frequencies of the final number of cases can then be found in terms of a parameter γ which is a measure of infectiousness. A model of this type is quite adequate for measles.

Bailey [3] has considered the case in which the parameter is not the same for each family, but varies from family to family. He assumes that γ is distributed among the population of families as a beta distribution. An excellent agreement was found between theoretical and observed frequencies. For a detailed discussion of the chain-binomial approach we refer to Ref. 5.

As an application of the results given in Sec. 4.4C, Bailey [3] has considered the problem of determining the total size of an epidemic in a small population when there is an extended period of infectiousness. This condition is satisfied, for example, in the case of diphtheria. Equations (4.131), (4.133), and (4.134) were used to calculate $Q(u)$ for $n = 1, 2, \ldots, 5$, when $i = 1$, by using various values of ρ.

Gaffey [32] has developed a stochastic model of within-family contagion which is a modification of the logistic model in population growth. In this model a susceptible individual has risk λ of becoming infected from outside his family and risk $n\mu$ of infection from the n infectious individuals within his family. Agarwal[1] has generalized Gaffey's model by introducing an incubation period which is greater than zero. Two cases are considered: (1) the length of the incubation period is a fixed constant, say a, and (2) the length is uniformly distributed over the interval $(a, a + b)$, where $0 < b \leqslant a/2$.

4.5 Diffusion Processes in the Theory of Gene Frequencies

A. Introduction. In this section we consider some applications of diffusion theory in population genetics. The model considered can be described as follows: To fix ideas, consider a finite population of size N whose members have alleles A_1 and A_2 and let $x = A_1/(A_1 + A_2)$ denote the frequency of A_1 in the population. Let the random variable $X(t)$ represent the frequency of A_1 in the population at time t (i.e., the tth generation). What we wish to study is the stochastic process $\{X(t), t \geqslant 0\}$ which describes the change in the frequency of A_1 in the population as a function of time.[2] In the diffusion problems associated with gene frequencies, the state space $\mathfrak{X}$ is no longer the real line, but we have instead the process defined on the interval $(0,1)$, that is, $\mathfrak{X} = \{x: 0 < x < 1\}$. The positions (or states) $x = 0$ and $x = 1$ represent the fixed classes A_1 and A_2, respectively. The change in gene frequencies is best described by a stochastic process whose state space is made up of the discrete values $0, \dfrac{1}{2N}, \dfrac{2}{2N}, \ldots, 1 - \dfrac{1}{2N}, 1$. However, for N large the assumption that the gene frequency x is a continuous variable does not result in serious error.

Let $f(t,x;x_0)$ denote the probability density of $X(t)$, where $X(0) = x_0$, $0 < x_0 < 1$. In 1945, Wright [97] introduced the forward Kolmogorov (Fokker-Planck) diffusion equation in the study of population genetics

---

[1] Unpublished manuscript, 1955.

[2] We observe that $1 - x$ is the frequency of A_2; hence, in the case of two alleles it is only necessary to consider a one-dimensional diffusion process. In general, the study of gene frequencies in the N-allele case leads to an $(N - 1)$-dimensional process.

in order to obtain the probability density $f(t,x;x_0)$. Earlier, in 1935, Kolmogorov [58] had introduced the steady-state form of the forward equation in genetics and obtained a result that Wright had previously obtained by another method.[1] The forward Kolmogorov equation plays a fundamental role in the theory of gene frequencies, and it has been used by many workers in their theoretical and experimental studies of genetic situations.

In Chap. 3 we observed that the general description of the forward aspects of a diffusion process is given by the system of equations

$$\frac{\partial f(t,x)}{\partial t} = \frac{1}{2}\frac{\partial^2[a(x)f(t,x)]}{\partial x^2} - \frac{\partial[b(x)f(t,x)]}{\partial x} + \left(\frac{\tau_0}{\sigma_0}\right)p_0(x)m_0(t) + \left(\frac{\tau_1}{\sigma_1}\right)p_1(x)m_1(t)$$

(4.137)

$$\frac{dm_0(t)}{dt} = -\left(\frac{1-p_{00}}{\sigma_0}\right)m_0(t) + \left(\frac{p_{10}}{\sigma_1}\right)m_1(t) + \lim_{x\to 0}\left\{\frac{\partial[a(x)f(t,x)]}{\partial x} - b(x)f(t,x)\right\}$$

$$\frac{dm_1(t)}{dt} = \left(\frac{p_{01}}{\sigma_0}\right)m_0(t) - \left(\frac{1-p_{11}}{\sigma_1}\right)m_1(t) - \lim_{x\to 1}\left\{\frac{\partial[a(x)f(t,x)]}{\partial x} - b(x)f(t,x)\right\}$$

In the above, $f(t,x)$, with the initial frequency x_0 suppressed, is the probability density of the gene frequency, and $m_i(t)$, $i = 0,1$, represents the fraction of the population in the ith fixed class at time t. We refer to Chap. 3 for the meaning of the other symbols.

In order to characterize the diffusion process associated with a given genetic situation, it is necessary to specify the coefficients $a(x)$ and $b(x)$, as well as the other quantities, which occur in the forward system (4.137). As pointed out in Chap. 3, the coefficients $a(x)$ and $b(x)$ determine the nature of the boundaries $(x = 0,1)$ and the boundary conditions to be imposed on the general solution of the diffusion equation. In the theory of gene frequencies the coefficients $a(x)$ and $b(x)$ represent the mean and variance, respectively, of an infinitesimal change in the gene frequency x.

In illustrating the application of diffusion processes in the theory of gene frequencies we will restrict our attention to two diffusion processes, one of which is one-dimensional and the other two-dimensional. However, it is of interest to list some of the different genetic situations that have been studied by using the forward Kolmogorov equation. The genetic situations, and the associated coefficients, are the following:[2]

---

[1] For another early study in genetics utilizing the diffusion equation we refer to the paper of Kolmogorov, Petrovskiĭ, and Piscounov [59].

[2] We observe that in all the diffusion problems arising in the theory of gene frequencies the diffusion equations are singular; hence, the Feller theory of singular diffusion equations must be utilized in the treatment of these problems.

1. Random drift only: $a(x) = \dfrac{1}{2N}\,x(1-x)$, $b(x) = 0$, where N is the effective population size.

2. Random fluctuations in selection intensities: $a(x) = V_s x^2 (1-x)^2$, $b(x) = \bar{s}x(1-x)$, where $\bar{s}$ and V_s are the mean and variance, respectively, of the selection coefficient s (assumed to be a random variable).

3. Selection without dominance complicated by random sampling: $a(x) = \dfrac{1}{2N}\,x(1-x)$, $bx = sx(1-x)$.

4. Transformation of the population under linear pressures (mutation, migration): $a(x) = \dfrac{1}{2N}\,x(1-x)$, $b(x) = m(\xi - x)$, where m is the rate per generation at which the population exchanges individuals in a random sample taken from the total species and ξ $(0 \leqslant \xi \leqslant 1)$ is the frequency of A_1 in the immigrants.

These cases have been studied by several authors. In particular, Kimura [53–55,57] has studied cases 1, 2, and 3, and Feller [26], Goldberg [33], and Malécot [68] have studied case 4.[1] For a discussion of other genetic-situation processes we refer to the papers of Crow and Kimura [21] and Kimura [55,57].

For a discussion of additional applications of stochastic processes in genetics we refer to the following: Bartlett [8], Feller [26], Kempthorne [42], Kimura [57], Malécot [69,70], and Moran [73].

B. Random Drift in the Case of Two Alleles. We now consider the diffusion process associated with the genetic situation of random drift with two alleles. Consider a random mating population of N diploid parents and let A_1 and A_2 denote a pair of alleles with frequencies x and $1-x$, respectively. Let the random variable $X(t)$ denote the frequency of A_1 and let $f(t,x;x_0) = \mathscr{P}\{X(t) = x \mid X(0) = x_0\}$. In this case the coefficients $a(x)$ and $b(x)$ are given by

$$a(x) = \frac{1}{2N}\,x(1-x) \qquad b(x) = 0$$

Therefore, the system of equations (4.137) in this case is given by

$$\frac{\partial f(t,x)}{\partial t} = \frac{1}{4N}\frac{\partial^2 [x(1-x)f(t,x)]}{\partial x^2} + \frac{\tau_0}{\sigma_0}\,p_0(x)m_0(t) + \frac{\tau_1}{\sigma_1}\,p_1(x)m_1(t)$$

$$\frac{dm_0(t)}{dt} = -\left(\frac{1-p_{00}}{\sigma_0}\right)m_0(t) + \left(\frac{p_{10}}{\sigma_1}\right)m_1(t) + f(t,0) \qquad (4.138)$$

$$\frac{dm_1(t)}{dt} = \left(\frac{p_{01}}{\sigma_0}\right)m_0(t) - \left(\frac{1-p_{11}}{\sigma_1}\right)m_1(t) + f(t,1)$$

---

[1] A treatment of this case by utilizing the moments of the distribution about the origin is given by Crow and Kimura [21].

This general system, however, is not of great interest in the study of gene frequencies in the case of random drift, since the general solution does not admit an interpretation which is biologically meaningful. We are primarily interested in the solution of Eqs. (4.138) when $x = 0$ and $x = 1$ are *absorbing barriers*. For this case (cf. Chap. 3) $\sigma_0 = \sigma_1 = \infty$; hence, (4.138) becomes

$$\frac{\partial f(t,x)}{\partial t} = \frac{1}{4N} \frac{\partial^2 [x(1-x)f(t,x)]}{\partial x^2}$$

$$\frac{dm_0(t)}{dt} = f(t,0) \tag{4.139}$$

$$\frac{dm_1(t)}{dt} = f(t,1)$$

The solution of the first equation has been obtained by Kimura [54], and it is a special case of the solution obtained by Goldberg [33] in the case of transformation under linear pressures. We now consider the solution of the partial differential equation in (4.139).

Application of the method of separation of variables[1] leads to a solution of the form

$$f(t,x) = \sum_{i=1}^{\infty} C_i X_i(x) e^{-\lambda_i t/4N} \tag{4.140}$$

where the $X_i(x)$ are functions of x alone, the λ_i are the eigenvalues of the system, and the C_i are constants. If we now assume $f(t,x) = X_i(x)e^{-\lambda_i t/4N}$, and insert this expression in (4.139), we obtain the ordinary differential equation

$$x(1-x)\frac{d^2 X_i}{dx^2} + 2(1-2x)\frac{dX_i}{dx} - (2-\lambda_i)X_i = 0 \tag{4.141}$$

Equation (4.141) can be identified with the standard form of the *hypergeometric equation* (cf. Morse and Feshbach [74], Valiron [90])

$$x(1-x)\frac{d^2 X}{dx^2} + [c - (a+b+1)x]\frac{dX}{dx} - abX = 0 \tag{4.142}$$

by putting $c = 2$, $a + b = 3$, $ab = 2 - \lambda_i$; hence, $a = \frac{1}{2}(3 + \sqrt{1+4\lambda_i})$, $b = 3 - a$, and $c = 2$.

If we now introduce the variable $y = 1 - 2x$ $(-1 < y < 1)$, (4.141) is transformed into

$$(y^2 - 1)\frac{d^2 X_i}{dy^2} + 4y\frac{dX_i}{dy} - (\lambda_i - 2)X_i = 0 \tag{4.143}$$

---

[1] Cf. Sec. 3.2E.

which can be identified with the standard form of the *Gegenbauer equation* (cf. [74])

$$(y^2 - 1)\frac{d^2X_i}{dy^2} + 2(b+1)y\frac{dX_i}{dy} - a(a + 2b + 1)X_i = 0 \qquad (4.144)$$

by putting $2(b+1) = 4$, $a(a + 2b + 1) = 2 - \lambda_i$. The solution of (4.144) with parameters a and b is

$$X_i(y) = T_a^b(y) \qquad (4.145)$$

where $T_a^b(y)$ is the Gegenbauer function (cf. Szegö [87]). The Gegenbauer function can be expressed in terms of the hypergeometric function by the relation

$$T_a^b(y) = \frac{\Gamma(a + 2b + 1)}{2^b a!\,\Gamma(b+1)}\, F\left(-a, a + 2b + 1, 1 + b; \frac{1-y}{2}\right) \qquad (4.146)$$

where the hypergeometric function is defined by the series

$$F(a,b,c;y) = 1 + \frac{ab}{c}y + \frac{a(a+1)b(b+1)}{2!c(c+1)}y^2 + \cdots \qquad (4.147)$$

$F(a,b,c;y)$ is the analytic solution of (4.142) at $y = 0$, and it converges for $|y| < 1$, since the next singularity is at $y = 1$.

To determine the eigenvalues λ_i, we must utilize the boundary conditions for singular diffusion equations [cf. Eq. (3.76)]. At this stage it is most convenient to work with the hypergeometric function. The general solution of Eq. (4.142) is

$$X_i(x) = A\,F(a,b,c;x) + Bx^{1-c}F(a - c + 1, b + 1 - c, 2 - c; x) \qquad (4.148)$$

where A and B are arbitrary constants. The following transformation formula is useful:

$$F(a,b,c;x) = \frac{\Gamma(c)\Gamma(c-a-b)}{\Gamma(c-a)\Gamma(c-b)}\, F(a, b, 1 - c + a + b; 1 - x)$$

$$+ \frac{\Gamma(c)\Gamma(a+b-c)}{\Gamma(a)\Gamma(b)}(1-x)^{c-a-b}F(c - a, c - b, 1 + c - a - b; 1 - x) \qquad (4.149)$$

This formula expresses the solution relative to the singularity at $x = 0$ in terms of the solutions relative to the singularity at $x = 1$.

The boundary conditions require that

$$\lim_{x\to 0}\left\{\frac{\partial[a(x)f(t,x)]}{\partial x} - b(x)f(t,x)\right\} = \lim_{x\to 1}\left\{\frac{\partial[a(x)f(t,x)]}{\partial x} - b(x)f(t,x)\right\} = 0 \qquad (4.150)$$

that is, the flux is zero at $x = 0$ and $x = 1$. From (4.148), (4.149), and (4.150) and by using the relation

$$\frac{dF(a,b,c;x)}{dx} = \frac{ab}{c}\, F(a+1,\, b+1,\, c+1;\, x)$$

we obtain

$$A = 0 \qquad A\, \frac{\Gamma(c)\Gamma(c-a-b)}{\Gamma(c-a)\Gamma(c-b)} + B\, \frac{\Gamma(2-c)\Gamma(c-a-b)}{\Gamma(1-a)\Gamma(1-b)} = 0$$

Hence, (4.150) requires that A be zero and that the λ_i satisfy the equation

$$\frac{1}{\Gamma(1-a)\Gamma(1-b)} = 0 \qquad (4.151)$$

Equation (4.151) will be satisfied at those points where $\Gamma(1-a)$ $\Gamma(1-b)$ has a pole, i.e., when $1-a$ or $1-b$ is equal to zero or a negative integer. Since $a = \frac{1}{2}(3 + \sqrt{1+4\lambda_i})$, we find that the λ_i are given by

$$\lambda_i = i(i+1) \qquad i = 1, 2, \ldots$$

Returning to Eq. (4.143), we see that the Gegenbauer parameters are $a = i - 1$ and $b = 1$; hence, (4.145) becomes

$$X_i(y) = T_{i-1}^1(y) \qquad (4.152)$$

Therefore, (4.140) becomes

$$f(t,x) = \sum_{i=1}^{\infty} C_i T_{i-1}^1(y) e^{-i(i+1)t/4N} \qquad (4.153)$$

The constants C_i are determined by the initial condition $X(0) = x_0$, $0 < x_0 < 1$; i.e.,

$$\delta(x - x_0) = \sum_{i=1}^{\infty} C_i T_{i-1}^1(y)$$

where $\delta(x)$ is the Dirac delta function. If we now multiply both sides of (4.153) by $(1 - y^2)T_{i-1}^1(y)$, integrate over the interval $(-1,1)$, and use the orthogonality property

$$\int_{-1}^{1}(1 - y^2)T_j^1(y)T_{i-1}^1(y)\, dy = \delta_{j,i-1}\, \frac{2i(i+1)}{2i+1}$$

where δ_{mn} is the Kronecker delta, we obtain

$$C_i = 4x_0(1 - x_0)\, \frac{i+1}{i(i+1)}\, T_{i-1}^1(z) \qquad (4.154)$$

where $z = 1 - 2x_0$. On inserting (4.154) in (4.153), we see that the solution of the partial differential equation in (4.139) is

$$f(t,x;x_0) = \sum_{i=1}^{\infty} \frac{(2i+1)(1-z^2)}{i(i+1)}\, T_{i-1}^1(z)T_{i-1}^1(y)e^{-i(i+1)t/4N} \qquad (4.155)$$

By use of (4.146), the solution in terms of the hypergeometric function is given by

$$f(t,x;x_0) = x_0(1 - x_0) \sum_{i=1}^{\infty} i(i + 1)(2i + 1)F(i + 2, 1 - i, 2; x_0)$$
$$\times \, F(i + 2, c - 1, 2; x)e^{-i(i+1)t/4N} \quad (4.156)$$

Therefore, either (4.155) or (4.156) gives the probability density of $X(t)$, the frequency of A_1 in the population at time t.

Let us now return to the two remaining equations in (4.139). These equations give the rate of fixation of A_2 and A_1, respectively. By using the initial conditions $m_0(0) = m_1(0) = 0$ and the expression for $f(t,x;x_0)$, the solutions of these equations can be very easily obtained.

C. Random Drift in the Case of Three Alleles. In this section we consider the generalization of the results of the last section to the case in which there are three alleles at a single locus. The solution in this case has been obtained by Kimura [56].

Consider a random mating population of N diploid parents and let A_1, A_2, and A_3 denote the three alleles, with frequencies $x = A_1/(A_1 + A_2 + A_3)$, $y = A_2/(A_1 + A_2 + A_3)$, and $z = 1 - (x + y)$, respectively. Since $x + y + z = 1$, we need consider only the probability density for a pair of alleles. Let the random variables $X(t)$ and $Y(t)$ denote the frequencies of A_1 and A_2, respectively, and let $f(t,x,y;x_0,y_0) = \mathscr{P}\{X(t) = x, \ Y(t) = y \mid X(0) = x_0, \ Y(0) = y_0\}$. In treating the case of three alleles we restrict our attention to the partial differential equation for the probability density $f(t,x,y)$. In the case of three alleles the Fokker-Planck equation is

$$\frac{\partial f(t,x,y)}{\partial t} = \frac{1}{2} \frac{\partial^2[a(x)f(t,x,y)]}{\partial x^2} + \frac{\partial^2[c(x,y)f(t,x,y)]}{\partial x \, \partial y}$$
$$+ \frac{1}{2} \frac{\partial^2[a(y)f(t,x,y)]}{\partial y^2} - \frac{\partial[b(x)f(t,x,y)]}{\partial x}$$
$$- \frac{\partial[b(y)f(t,x,y)]}{\partial y} \quad (4.157)$$

where $a(\cdot)$, $c(\cdot,\cdot)$, and $b(\cdot)$ denote the variance, covariance, and mean of the quantities occurring in the arguments of these functions. In the case of random drift

$$a(x) = \frac{1}{2N}\, x(1 - x) \qquad a(y) = \frac{1}{2N}\, y(1 - y)$$
$$c(x,y) = -\frac{1}{2N}\,(xy) \qquad b(x) = b(y) = 0$$

Hence, Eq. (4.157) becomes

$$\frac{\partial f(t,x,y)}{\partial t} = \frac{1}{4N} \frac{\partial^2[x(1-x)f(t,x,y)]}{\partial x^2} - \frac{1}{2N} \frac{\partial^2[(xy)f(t,x,y)]}{\partial x\,\partial y}$$

$$+ \frac{1}{4N} \frac{\partial^2[y(1-y)f(t,x,y)]}{\partial y^2} \quad (4.158)$$

where $0 < x < 1$, $0 < y < 1$.

To obtain the solution of (4.158), we use the method of separation of variables. We first put

$$f(t,x,y) = g(x,y)e^{-\lambda_i t} \quad (4.159)$$

where $\qquad \lambda_i = \dfrac{(i+1)(i+2)}{4N} \qquad i = 1, 2, \ldots$

When we substitute (4.159) into (4.158) and introduce the new variables

$$x = r(1-s) \qquad y = rs$$

where $r \geqslant 0$, $s \leqslant 1$, Eq. (4.158) becomes

$$r(1-r)\frac{\partial^2 g(r,s)}{\partial r^2} + \frac{s(1-s)}{r}\frac{\partial^2 g(r,s)}{\partial s^2} + \frac{2(1-2s)}{r}\frac{\partial g(r,s)}{\partial s}$$

$$+ 2(2-3r)\frac{\partial g(r,s)}{\partial r} + (i-1)(i+4)g(r,s) = 0 \quad (4.160)$$

If we now assume that $g(r,s)$ is of the form

$$g(r,s) = R(r)S(s) \quad (4.161)$$

and apply the method of separation of variables, Eq. (4.160) yields the ordinary differential equations

$$r^2(1-r)\frac{d^2 R}{dr^2} + 2r(2-3r)\frac{dR}{dr} + [(i-1)(i+4)r - \mu]R = 0 \quad (4.162)$$

$$s(1-s)\frac{d^2 S}{ds^2} + 2(1-2s)\frac{dS}{ds} + \mu S = 0 \quad (4.163)$$

We see that (4.163) is the hypergeometric equation with parameters $a + b = 3$, $ab = -\mu$, and $c = 2$. Therefore $a = \frac{1}{2}(3 + \sqrt{a - 4\mu})$, $b = \frac{1}{2}(3 - \sqrt{9 + 4\mu})$, and $c = 2$. On using the transformation formula (4.149), we see that in order for $\lim_{s \to 1} F(a,b,2;s)$ to be finite, $2 - a$ must be a negative integer. These conditions are satisfied if and only if the eigenvalues μ_j are given by

$$\mu_j = (j-1)(j+2) \qquad j = 1, 2, \ldots$$

Corresponding to these eigenvalues we have the parameters $a_j = j + 1$, $b_j = 1 = j$. If we now put $s = (1 - z)/2$, the solution of (4.163) is

$$S(z) = F\left(j + 1, 1 - j, 2; \frac{1 - z}{2}\right) \qquad (4.164)$$

up to a multiplicative constant.

To solve Eq. (4.162), we first put

$$R(r) = r^{j-1}U(r) \qquad (4.165)$$

Then (4.162) is transformed into the hypergeometric equation

$$r(1 - r)\frac{d^2U}{dr^2} + [2(j + 1) - (2j + 4)r]\frac{dU}{dr} - (j - i)(j + i + 3)U = 0$$
$$(4.166)$$

Hence, the solution of (4.166) can be expressed in terms of the Jacobi polynomial

$$U(r) = J_{i-j}(2j + 3, 2m + 2; r) \qquad (4.167)$$

The Jacobi polynomial $J_n(a,c;r)$ is related to the hypergeometric function by relation

$$J_n(a,c;r) = F(a + n, -n, c; r) \qquad (4.168)$$

From (4.159), (4.161), (4.164), (4.165), and (4.167), after expressing (4.164) in terms of the Gegenbauer function, we obtain

$$f(t,x,y) = \sum_{n=0}^{\infty} \sum_{k=0}^{\infty} C_{n,k} u^n J_k(2n + 5, n + 4; u)$$
$$\times\ T_n^1(w) \exp\left\{-\frac{(k + n + 2)(j + n + 3)}{4N} t\right\} \qquad (4.169)$$

where $n = j - 1$ $(i - j \neq k)$, $w = (x - y)/(x + y)$, $u = x + y$, and the $C_{n,k}$ are constants to be determined by the initial conditions. Since $X(0) = x_0$ and $Y(0) = y_0$, the $C_{n,k}$ are determined by the condition

$$f(0,x,y) = \delta(x - x_0)\delta(y - y_0)$$

By using the orthogonality properties of the Gegenbauer functions and Jacobi polynomials (cf. Szegö [87]), we find

$$C_{n,k} = \frac{4(k + 2n + 3)!(k + 2n + 4)!(2k + 2n + 5)}{k!(k + 1)!(n + 1)(n + 2)(2n + 2)!(2n + 3)!}$$
$$\times\ x_0 y_0 z_0 (1 - z_0)^n T_n^1\left(\frac{x_0 - y_0}{1 - z_0}\right) J_k(2n + 5, 2n + 4; 1 - z)$$
$$\times\ \exp\left\{-\frac{(k + n + 2)(k + n + 3)}{4N} t\right\} \qquad (4.170)$$

Therefore, by inserting these $C_{n,k}$ in (4.169), we obtain the solution of (4.158).

D.   Some Additional Problems.   In this section we discuss briefly several additional problems in the theory of gene frequencies that can be treated by using the theory presented in Chap. 3.

1. *First-passage Time Problems for the Fixed Classes.*   A problem of importance in the theory of gene frequencies is that of determining the time, or number of generations, required for the population to enter one of the fixed classes.   In probabilistic terms, what we seek is the distribution of the time required for the system originally in the state x_0, $0 < x_0 < 1$, to enter either state $x = 0$ or $x = 1$.   The theory for handling these problems is presented in Sec. 3.4.   In that section we showed that the probabilities of fixation can be obtained from the solutions of the *backward Kolmogorov equation.*

Several first-passage time problems have been considered in the literature.   In particular, Kimura [54,55] has shown, by using both the method of moments and the backward Kolmogorov equation, that in the case of random drift with two alleles, the probability of A_1 being fixed by the tth generation is

$$q(t,x_0) = x_0 + \sum_{i=1}^{\infty} (2i + 1)x_0(1 - x_0)(-1)^i F(i + 2, 1 - i, 2; x_0)e^{-i(i+1)t/4N}$$

(4.171)

where, as before, $X(0) = x_0$.

2. *Probabilities of Ultimate Fixation.*   Another problem, which is clearly related to the one just discussed, is that of determining the probabilities of ultimate fixation.   That is, we wish to determine

$$q(x_0) = \lim_{t \to \infty} q(t,x_0)$$

It is clear that $q(x_0)$ is given by the steady-state solution of the backward Kolmogorov equation, i.e., when $\partial q/\partial t = 0$.   This problem for the case of random drift was given as an example in Sec. 3.4C, where we showed that the probability of ultimate absorption at $x = 1$ (A_1) was x_0, while the probability of ultimate absorption at $x = 0$ (A_2) was $1 - x_0$.   This first result follows from (4.171) by letting $t \to \infty$.

3. *Distribution of the Maximum Gene Frequency.*   As a final problem we consider the distribution of a random variable which is related to the first-passage time T and which may be of interest in the theory of gene frequencies.   The random variable we consider is the maximum gene frequency, which we denote by $M(t)$, defined as

$$M(t,x_0) = \sup_{0 \leqslant \tau \leqslant t} \{X(\tau), X(0) = x_0\} \qquad 0 < x_0 < 1$$

That is, $M(t,x_0)$ is the largest value assumed by $X(\tau)$ in the interval $[0,t]$, given that $X(0) = x_0$.

Following the notation of Sec. 3.4D, let

$$K(t,x_0;\eta) = \mathscr{P}\{M(t,x_0) < \eta\} \tag{4.172}$$

be the distribution function of $M(t,x_0)$. In Sec. 3.4D we showed that

$$K(t,x_0;\eta) = \mathscr{P}\{T(x_0;\eta,-\eta) > t\} \tag{4.173}$$

where $T(x_0;\eta,-\eta)$ is the random variable representing the time required for passage into either η or $-\eta$ when $X(0) = x_0$. Since $0 < x_0 < 1$ and $\eta > 0$, the condition $\eta > |x_0|$ in the case of gene frequencies is replaced by the conditions $x_0 - \eta > 0$ and $x_0 + \eta < 1$. Hence, (4.173) becomes

$$K(t,x_0;\eta) = \mathscr{P}\{T(x_0 - \eta, x_0 - \eta) > t\} \tag{4.174}$$

4.6 Radiobiology

A. Introduction. In developing a mathematical theory of radiobiological phenomena it is necessary to consider two separate, yet related, problems. The first problem is concerned with determining the probability of occurrence of effective ionizing radiation within the so-called sensitive volume (or mass) of a living organism. This probability depends on the geometry of the target and the absorption of radiation quanta. The quantum hit theory which has been advanced to explain this initial effect is discussed in detail by Lea [63]. The second problem is concerned with determining the probability that the initial damage to the system will cause a certain effect, e.g., the death of the organism. In this section we consider two stochastic models which have developed in connection with the second problem. Both models are of the birth-and-death type. The first model, developed by Opatowski [78,79], depends on a continuous time parameter, while the second model depends on a discrete time parameter and is of the random-walk type [12,14,18].

B. Opatowski's Model for Radiation Damage. Consider a microorganism which is capable of being in $n + 1$ states, say $0, 1, \ldots, n$. The state 0 represents the living organisms in normal conditions, and the state n represents dead organisms. We assume that the underlying stochastic process is of the birth-and-death type, with transitions from a state i to $i + 1$ and from i to $i - 1$. The forward transitions can be identified with the process of injury and damage, while the backward transitions represent a process of recovery.

Now consider a homogeneous group of N microorganisms and let $NX_i(t)$ denote the number of microorganisms in the ith state at time t.

Let $Nd_i(t)$ denote the rate with which the microorganisms pass from the state $i - 1$ to i and let $Nr_i(t)$ denote the rate with which the microorganisms pass from the state $i + 1$ to i. The usual reasoning leads to the differential-difference equation

$$\frac{dX_i(t)}{dt} = -d_{i+1} + d_i + r_i - r_{i-1} \tag{4.175}$$

We now assume that $d_i = \lambda_i X_i$, where λ_i, the intensity of damage, depends on the intensity of radiation and on the sensibility of the organism in the state $i - 1$ to a given radiation. Hence the λ_i depend on the probability of collision between a photon and the sensitive volume. We also assume that $r_i = \mu_i X_{i+1}$, where μ_i represents the intensity of recovery of the microorganisms from the state $i + 1$ to i. We can now rewrite (4.175) as

$$\frac{dX_i(t)}{dt} = -\lambda_{i+1}X_i + \lambda_i X_{i-1} + \mu_i X_{i+1} - \mu_i X_i$$

$$= \lambda_i X_{i-1} - (\lambda_{i+1} - \mu_{i-1})X_i + \mu_i X_{i+1} \tag{4.176}$$

This system of equations is to be solved with the initial condition

$$X_i(0) = 1 \qquad i = 0$$

$$= 0 \qquad i = 1, 2, \ldots, n \tag{4.177}$$

We assume throughout that λ_i and μ_i are nonnegative constants. In the general case we could put $\lambda_i = f(i,t)$ and $\mu_i = g(i,t)$, where f and g are arbitrary functions of the state i and time. We now have the relations

$$d_0 = \lambda_0 = 0$$

$$d_i > 0 \qquad \lambda_i > 0 \qquad i = 1, 2, \ldots, n$$

and $\qquad r_i \geqslant 0 \qquad \mu_i \geqslant 0 \qquad i = 0, 1, \ldots, n-1$

$$r_n = \mu_n = 0$$

The last relation states that the nth state is an absorbing state and that recovery of the organism is impossible once this state has been reached. Hence,

$$d_{n+1} = \lambda_{n+1} = r_n = \mu_n = 0$$

To solve the system (4.176), we introduce the Laplace transform $\mathscr{L}\{X_i(t)\} = x_i(s)$ and obtain the linear system

$$(s + 2I_0)x_0 - \mu_0 x_1 = 1$$

$$-\lambda_i x_{i-1} + (s + 2I_i)x_i - \mu_i x_{i+1} = 0 \qquad i = 1, 2, \ldots, n \tag{4.178}$$

In (4.178) we have put $I_i = (\lambda_{i+1} + \mu_{i-1})/2$, so that I_i is the arithmetic mean of the forward and reverse transition intensities from the state i.

The solution of the system (4.178) is

$$x_i(s) = A_i \frac{D(i+1, n)}{(0,n)} \qquad (4.179)$$

where
$$A_i = 1 \qquad i = 0$$

$$= \prod_{h=1}^{i} \lambda_h \qquad i = 1, 2, \ldots, n$$

and $D(i,n)$ is a determinant of order $n - i + 1$ given by

$$D(i,n) = \begin{vmatrix} s + 2I_i & \mu_i & 0 & 0 & \cdot & \cdot & & \cdot \\ \lambda_{i+1} & s + 2I_{i+1} & \mu_{i+1} & 0 & \cdot & \cdot & & \cdot \\ 0 & \lambda_{i+2} & s + 2I_{i+2} & \mu_{i+2} & \cdot & \cdot & & \cdot \\ \cdot & \cdot & \cdot & \cdot & \cdot & \cdot & \cdot & \cdot \\ \cdot & \cdot & \cdot & \cdot & \cdot & \cdot & \lambda_n & s + 2I_n \end{vmatrix}$$

satisfying the condition $D(n + 1, n) = 1$.

In order to invert (4.179), we require an expansion of $D(i,n)$ as a polynomial in s. Let $D_0(i,n)$ denote the value of $D(i,n)$ for $s = 0$ and let $M_j(i,n)$ denote a principal minor of order j of $D_0(i,n)$. We then have (cf. [75], p. 72)

$$D(i + 1, n) = s^{n-1}\left[1 + \sum_{j=1}^{n-1} s^{-j} \sum_{M} M_j(i + 1, n)\right] \qquad (4.180)$$

where $\sum_{M}$ stands for the sum of all the principal minors M_j of $D_0(i + 1, n)$. In this way each coefficient of s^{-j} in (4.180) is a sum of C_j^{n-1} determinants of order j satisfying the relation

$$M_{n-1}(i + 1, n) = D_0(i + 1, n)$$

Now, from (4.179) and (4.180) the expansion of $x_i(s)$ as a power series in s^{-1} can be obtained. In particular, for the nth state we obtain

$$x_n(s) = \frac{A_n}{D(0,n)} = A_n\left\{ s^{n+1}\left[1 + \sum_{M} M_1(0,n)s^{-1} \right.\right.$$

$$\left.\left. + \sum_{M} M_2(0,n)s^{-2} + \cdots + D_0(0,n)s^{-n-1}\right]\right\}^{-1}$$

$$= A_n s^{n-1}\left[1 + \sum_{j=1}^{\infty} \alpha_{nj}s^j\right] \qquad (4.181)$$

where the coefficients α_{nj} are obtained in terms of the $\sum_M$'s by the usual rules of division of power series, so that

$$\alpha_{n1} = -\sum_1 \qquad \alpha_{n2} = (\sum_1)^2 - \sum_2$$
$$\alpha_{n3} = -(\sum_1)^3 + 2\sum_1\sum_2 - \sum_3 \qquad \text{etc.}$$

where $\sum_j = \sum_M M_j(0,n)$. Now, from (4.181) we obtain by inversion[1]

$$X_n(t) = A_n t^n \left[\frac{1}{n!} + \sum_{j=1}^{\infty} \frac{\alpha_{nj} t^j}{(n+j)!} \right] \qquad (4.182)$$

The series in (4.182) is clearly convergent for finite t because of Eq. (4.175) and the form of d_i and r_i.

In most experimental situations the states 0 and n are the only observable ones, since it is impossible to distinguish organisms in the various intermediate states. In this case (4.182) can be used to obtain some information about the number of states and the transition intensities. First, we note that from (4.182) we obtain

$$\log X_n(t) = \log A_n - \log n! + n \log t + \log \left[1 + \sum_{j=1}^{\infty} \frac{n! \alpha_{nj} t^j}{(n+j)!} \right] \quad (4.183)$$

For sufficiently small values of t, the sum in braces is small compared with 1; hence (4.183) becomes

$$\log X_n(t) = n \log t + \log A_n - \log n! \qquad (4.184)$$

From (4.184) we see that, for small values of time, $\log X_n(t)$ plotted against $\log t$ is a straight line whose slope is equal to the *number of states* n and the intercept on the ordinate is the *geometric average of the intensities of damage* $A_n^{1/n}$. If larger values of time are taken into consideration, the approximation procedure requires taking second-order terms into account. However, the sum

$$\alpha_{n1} = \sum_{i=0}^{n} (\lambda_{i+1} + \mu_{i-1}) \qquad (4.185)$$

can also be obtained from the observed value of $X_n(t)$. If state n is the final state (i.e., $\lambda_{n+1} = 0$), we have from (4.185)

$$\alpha_{n1} \sim nA_n^{1/n} + nB \qquad (4.186)$$

where B is the arithmetic mean of the intensities of recovery $\mu_0, \ldots, \mu_{n-1}$. In (4.186) the arithmetic mean of the intensities of damage has been taken as approximately equal to its geometric mean. If the intensities of damage are all equal, i.e., $\lambda_1 = \cdots = \lambda_n$, then (4.186) is exact. In addition, if no recovery is possible from state n (i.e., $\mu_{n-1} = 0$), it is necessary to replace (4.186) by

$$\alpha_{n1} \sim nA_n^{1/n} + B(n-1) \qquad (4.187)$$

[1] Cf. Refs. 3 and 9 of Appendix B.

where B now denotes the arithmetic mean of the intensities of recovery $\mu_0, \ldots, \mu_{n-2}$.

In addition to obtaining the number of states n from the slope of the straight line given by (4.184), it is possible to obtain n by using the formula

$$n = \lim_{t \to 0} \left\{ t \frac{d}{dt} [\log X_n(t)] \right\} \tag{4.188}$$

This formula, which is obtained from (4.183), requires an accurate knowledge of $X_n(t)$ for small values of time. In the case of low radiation intensities this knowledge is not easy to obtain. Since n can be interpreted as the number of hits with photons (or radiation quanta) and is therefore independent of the radiation intensity, a sufficient degree of accuracy may be obtained if $X_n(t)$ is known for a number of different radiation intensities.

C. A Random-walk Model. We now consider a simple random-walk model for the transmission of primary radiation damage through a biological system. The physical mechanism postulated for transmission can be described as follows: Present in the organism is a control molecule (or group of molecules) to which chain macromolecules are connected. The control molecule can be identified with the "sensitive volume," and following a "hit" in this sensitive volume, the initial damage is transmitted through the system by the chain depolymerization of the macromolecules connected to the control molecule. The complete depolymerization of the macromolecules is assumed to be responsible for the observed damage to the organism.

The mathematical model considered is a finite Markov chain with $n + 1$ states: $0, 1, \ldots, n$. The system can be considered to be in state 0 initially (i.e., the macromolecules are intact), and following a "hit," it passes to state 1. The transitions $1 \to 2 \to \cdots \to n - 1 \to n$ represent the transmission of the primary radiation damage, and state n represents the state when the radiation damage becomes observable.

The transition probabilities that describe the stochastic process are assumed to be functions of the number of states n, and they are given by

$$\gamma_{i,i+1} = \mathscr{P}\{i \to i + 1\} = \frac{i}{n} \qquad i = 1, 2, \ldots, n - 1 \tag{4.189}$$

$$\gamma_{i,i-1} = \mathscr{P}\{i \to i - 1\} = 1 - \frac{i}{n} \qquad i = 1, 2, \ldots, n - 1 \tag{4.190}$$

$$\gamma_{i,i} = 1 \qquad \text{for } i = 0, n$$

$$= 0 \qquad \text{otherwise} \tag{4.191}$$

The forward transition probabilities (4.189) represent transmission, and the reverse transition probabilities (4.190) represent recovery. Conditions (4.191) mean that no change will occur after states 0 and n are reached, i.e., either the damage is complete or the organism recovers. Hence, 0 and n are absorbing states. From (4.189) we see that the probability of observable damage increases as a linear function of the extent of depolymerization. Similar unsymmetrical random-walk models have been considered in connection with the theory of Brownian motion.

Let Q_0 and Q_n denote the probabilities of recovery and observable damage, respectively. These probabilities can be obtained as follows: Let P_i denote the probability of absorption in state n given that the system was initially in state $i, i = 1, 2, \ldots, n - 1$. It is easy to verify that the P_i satisfy the difference equation

$$P_i = \gamma_{i,i+1}P_{i+1} + \gamma_{i,i-1}P_{i-1}$$

$$= \frac{i}{n} P_{i+1} + \left(1 - \frac{i}{n}\right)P_{i-1} \qquad 1 < i < n - 1 \qquad (4.192)$$

Equation (4.192) is also valid for all $i \geqslant 1$ if we define $P_{n+1} = 1$. This equation is to be solved with the boundary conditions

$$P_0 = 0 \qquad P_n = 1 \qquad (4.193)$$

If we rewrite (4.192) as

$$P_i = \left(\frac{i+1}{n}\right)P_{i+1} - \frac{1}{n} P_{i+1} + \left(1 - \frac{i-1}{n}\right)P_{i-1} - \frac{1}{n} P_{i-1}$$

and let $F(s)$ denote the generating function of the P_i, we obtain

$$\sum_{i=1}^{\infty} P_i s^i = \sum_{i=0}^{\infty} \frac{i}{n} P_i s^{i-1} - \frac{1}{n}\sum_{i=0}^{\infty} P_i s^{i-1} + \left(1 - \frac{1}{n}\right)\sum_{i=0}^{\infty} P_i s^{i+1} - \sum_{i=0}^{\infty} \frac{i}{n} P_i s^{i+1} \quad (4.194)$$

Since $F'(s) = \sum_{i=0}^{\infty} i P_i s^{i-1}$, (4.194) yields the differential equation

$$\frac{dF(s)}{ds} + \frac{(n-1)s - (1/s) - n}{1 - s^2} F(s) = 0$$

or

$$\frac{1}{F(s)} \frac{dF(s)}{ds} = \frac{1}{s} + \frac{1}{1 - s} + \frac{n - 1}{1 + s} \qquad (4.195)$$

The solution of (4.195) is

$$F(s) = C \frac{s(1 + s)^{n-1}}{1 - s} = Cs\left[\sum_{i=0}^{n-1} C\binom{n-1}{i} s^i\right]\sum_{j=0}^{\infty} s^j \qquad (4.196)$$

where C is the constant of integration. From the above we see that the coefficients P_i, $0 \leqslant i \leqslant n$, are given by

$$P_i = C \sum_{k=0}^{i-1} \binom{n-1}{k} \tag{4.197}$$

By utilizing the boundary conditions, we have

$$1 = C \sum_{k=0}^{i-1} \binom{n-1}{k} = C2^{n-1}$$

Hence

$$P_i = \sum_{k=0}^{i-1} \binom{n-1}{k} 2^{1-n} \qquad 0 \leqslant i \leqslant n \tag{4.198}$$

Since we have assumed that the process starts in state 1, we have

$$Q_n = P_1 = \binom{n-1}{0} 2^{1-n} = 2^{1-n} \tag{4.199}$$

since $\binom{n-1}{0} = 1$. Also, since $Q_0 + Q_n = 1$, we have

$$Q_0 = 1 - Q_n = 1 - 2^{1-n} \tag{4.200}$$

In order for the above results to be useful in experimental radiobiology, we now give several relationships that result from the application of the model to many organisms. In particular, let $S(n_0)$ denote the probability that, out of a group of N irradiated organisms, there will eventually be n_0 survivors. If the N organisms are statistically independent, $S(n_0)$ is given by the binomial distribution

$$S(n_0) = b(n_0;N,Q_0) = \binom{N}{n_0} Q_0^{n_0}(1 - Q_0)^{N-n_0}$$

If we wish to obtain an expression for the probability that there will be n_0 survivors after a finite period of time (i.e., a finite number of transitions), we must rewrite the above as

$$b(n_0;N,p_{10}^{(k)}) = \binom{N}{n_0} (p_{10}^{(k)})^{n_0} (p_{1n}^{(k)})^{N-n_0}$$

that is, we must replace Q_0 and Q_n by the k-step transition probabilities $p_{10}^{(k)}$ and $p_{1n}^{(k)}$, respectively. By using well-known relationships, we can now obtain the probability that there will be at most n_0 survivors at time k in terms of the incomplete beta functions:

$$B(n_0;N,p_{10}^{(k)}) = \sum_{i=0}^{n_0} b(i;N,p_{10}^{(k)})$$
$$= 1 - N\binom{N-1}{n_0} \int_0^{p_{10}^{(k)}} \xi^{n_0}(1 - \xi)^{N-n_0-1} \, d\xi$$

In all of the expressions given above the number of states n has been unspecified; hence, in order to compute the above probabilities it is necessary that n be known. From our remarks concerning the mechanism for the transmission of radiation damage, n should denote (or be a function of) the number of monomer units in the chain macromolecule. In Ref. 16 a method is suggested for obtaining the maximum-likelihood estimator of n from the observed times required for a given number of organisms to recover or die.

Bibliography

1   Armitage, P.: The Statistical Theory of Bacterial Populations Subject to Mutation, *J. Roy. Statist. Soc.*, ser. B, vol. 14, pp. 1–40, 1952.
2   Bailey, N. T. J.: A Simple Stochastic Epidemic, *Biometrika*, vol. 37, pp. 194–202, 1950.
3   Bailey, N. T. J.: The Total Size of a General Stochastic Epidemic, *Biometrika*, vol. 40, pp. 177–185, 1953.
4   Bailey, N. T. J.: Some Problems in the Statistical Analysis of Epidemic Data, *J. Roy. Statist. Soc.*, vol. ser. B, 17, pp. 35–58, 1955.
5   Bailey, N. T. J.: "The Mathematical Theory of Epidemics," Hafner Publishing Company, New York, 1957.
6   Bartholomay, A. F.: On the Linear Birth-and-Death Processes of Biology as Markoff Chains, *Bull. Math. Biophys.*, vol. 20, pp. 97–118, 1958.
7   Bartlett, M. S.: Some Evolutionary Stochastic Processes, *J. Roy. Statist. Soc.*, ser. B, vol. 11, pp. 211–229, 1949.
8   Bartlett, M. S.: "An Introduction to Stochastic Processes," Cambridge University Press, New York, 1955.
9   Bartlett, M. S.: Deterministic and Stochastic Models for Recurrent Epidemics, *Proc. Third Berkeley Symposium on Math. Statistics and Probability*, vol. 4, pp. 81–109, 1956.
10 Bartlett, M. S.: On Theoretical Models for Competitive and Predatory Biological Systems, *Biometrika*, vol. 44, pp. 27–42, 1957.
11 Bartlett, M. S.: Measles Periodicity and Community Size, *J. Roy. Statist. Soc.*, ser. A, vol. 120, pp. 48–70, 1957.
12 Bharucha-Reid, A. T.: A Probability Model of Radiation Damage, *Nature*, vol. 169, pp. 369–370, 1952.
13 Bharucha-Reid, A. T.: An Age-dependent Stochastic Model of Population Growth, *Bull. Math. Biophys.*, vol. 15, pp. 361–365, 1953.
14 Bharucha-Reid, A. T.: On Stochastic Processes in Biology, *Biometrics*, vol. 9, pp. 275–289, 1953.
15 Bharucha-Reid, A. T.: On the Stochastic Theory of Epidemics, *Proc. Third Berkeley Symposium on Math. Statistics and Probability*, vol. 4, pp. 111–119, 1956.
16 Bharucha-Reid, A. T.: Note on the Estimation of the Number of States in a Discrete Markov Chain, *Experientia*, vol. 12, p. 176, 1956.
17 Bharucha-Reid, A. T.: "An Introduction to the Stochastic Theory of Epidemics and Some Related Statistical Problems," USAF School of Aviation Medicine, Randolph Field, Texas, 1957.
18 Bharucha-Reid, A. T., and H. G. Landau: A Suggested Chain Process for Radiation Damage, *Bull. Math. Biophys.*, vol. 13, pp. 153–163, 1951.

19 Chiang, C. L.: Competition and Other Interactions between Species, in "Statistics and Mathematics in Biology," pp. 197–215, Iowa State College Press, Ames, Iowa, 1954.

20 Chiang, C. L.: An Application of Stochastic Processes to Experimental Studies on Flour Beetles, *Biometrics*, vol. 13, pp. 79–97, 1957.

21 Crow, J. F., and M. Kimura: Some Genetic Problems in Natural Populations, *Proc. Third Berkeley Symposium on Math. Statistics and Probability*, vol. 4, pp. 1–22, 1956.

22 D'Ancona, U.: "The Struggle for Existence," E. J. Brill, Leiden, 1954.

23 Darwin, J. H.: Population Differences between Species Growing According to Simple Birth-and-Death Processes, *Biometrika*, vol. 40, pp. 370–382, 1953.

24 Feller, W.: Die Grundlagen der Volterraschen Theorie des Kampfes ums Dasein in wahrscheinlichkeitstheoretischen Behandlung, *Acta Biotheoretica*, vol. 5, pp. 1–40, 1939.

25 Feller, W.: On the Theory of Stochastic Processes with Particular Reference to Applications, *Proc. First Berkeley Symposium on Math. Statistics and Probability*, pp. 403–432, 1949.

26 Feller, W.: Diffusion Processes in Genetics, *Proc. Second Berkeley Symposium on Math. Statistics and Probability*, pp. 227–246, 1951.

27 Feller, W.: Two Singular Diffusion Problems, *Ann. Math.*, vol. 54, pp. 173–182, 1951.

28 Feller, W.: "An Introduction to Probability Theory and Its Applications," vol. 1, 2d ed., John Wiley & Sons, Inc., New York, 1957.

29 Fisher, R. A.: "The Genetical Theory of Natural Selection," Oxford University Press, New York, 1930.

30 Fortet, R.: "Calcul des probabilités," Centre National de la Recherche Scientifique, Paris, 1950.

31 Foster, F. G.: A Note on Bailey's and Whittle's Treatment of a General Stochastic Epidemic, *Biometrika*, vol. 42, pp. 123–125, 1955.

32 Gaffey, W. R.: The Problem of Within-family Contagion, unpublished doctoral dissertation, University of California, Berkeley, 1954.

33 Goldberg, S.: A Singular Diffusion Equation (Abstract), *Bull. Am. Math. Soc.*, vol. 58, p. 181, 1952.

34 Goldberg, S.: Probability Models in Biology and Engineering, *J. Soc. Ind. Appl. Math.*, vol. 2, pp. 10–19, 1954.

35 Goodman, L. A.: Population Growth of the Sexes, *Biometrics*, vol. 9, pp. 212–225, 1953.

36 Harris, T. E.: Some Mathematical Models for Branching Processes, *Proc. Berkeley Symposium on Math. Statistics and Probability*, pp. 305–327, 1951.

37 Harris, T. E.: "Branching Processes," Springer-Verlag, Berlin, Vienna, in press.

38 Haskey, H. W.: A General Expression for the Mean in a Simple Stochastic Epidemic, *Biometrika*, vol. 41, pp. 272–275, 1954.

39 Haskey, H. W.: Stochastic Cross-infection between Two Otherwise Isolated Groups, *Biometrika*, vol. 44, pp. 193–204, 1957.

40 Iversen, S., and N. Arley: On the Mechanism of Experimental Carcinogenesis, *Acta Pathol. Microbiol. Scand.*, vol. 27, pp. 1–31, 1950.

41 Joshi, D. D.: Les processus stochastiques en démographie, *Publ. inst. statist. univ. Paris*, vol. 3, pp. 153–177, 1954.

42 Kempthorne, O.: "An Introduction to Genetic Statistics," John Wiley & Sons, Inc., New York, 1957.

43   Kendall, D. G.: A Review of Some Recent Work on Discontinuous Markoff
      Processes with Applications to Biology, Physics, and Actuarial Science,
      J. Roy. Statist. Soc., ser. A, vol. 110, pp. 130–137, 1947.
44   Kendall, D. G.: On Some Modes of Population Growth Leading to R. A.
      Fisher's Logarithmic Series Distribution, *Biometrika*, vol. 35, pp. 6–15, 1948.
45   Kendall, D. G.: On the Generalized "Birth-and-Death" Process, *Ann.
      Math. Statist.*, vol. 19, pp. 1–15, 1948.
46   Kendall, D. G.: Stochastic Processes and Population Growth, *J. Roy.
      Statist. Soc.*, ser. B, vol. 11, pp. 230–264, 1949.
47   Kendall, D. G.: Random Fluctuations in the Age-distribution of a Population
      Whose Development Is Controlled by a Simple Birth-and-Death Process,
      J. Roy. Statist. Soc., ser. B, vol. 12, pp. 278–285, 1950.
48   Kendall, D. G.: On the Choice of a Mathematical Model to Represent
      Normal Bacterial Growth, *J. Roy. Statist. Soc.*, ser. B, vol. 14, pp. 41–44, 1952.
49   Kendall, D. G.: Les Processus stochastiques de croissance en biologie,
      Ann. inst. H. Poincaré, vol. 13, pp. 43–108, 1952.
50   Kendall, D. G.: Stochastic Processes and the Growth of Bacterial Colonies,
      Symposium Soc. Exptl. Biology, no. 7, *Evolution*, pp. 55–65, 1953.
51   Kendall, D. G.: Deterministic and Stochastic Epidemics in Closed Popula-
      tions, *Proc. Third Berkeley Symposium on Math. Statistics and Probability*,
      vol. 4, pp. 149–165, 1956.
52   Kendall, D. G.: Contributions to the Theory of Stochastic Epidemics, in
      preparation.
53   Kimura, M.: Processes Leading to Quasi-fixation of Genes in Natural
      Populations Due to Random Fluctuations of Selection Intensities, *Genetics*,
      vol. 39, pp. 280–295, 1954.
54   Kimura, M.: Solution of a Process of Random Drift with a Continuous Model,
      Proc. Natl. Acad. Sci. U.S., vol. 41, pp. 144–150, 1955.
55   Kimura, M.: Stochastic Processes and Distribution of Gene Frequencies
      under Natural Selection, *Cold Spring Harbor Symposium on Quantitative
      Biology*, vol. 20, pp. 33–55, 1955.
56   Kimura, M.: Random Genetic Drift in a Triallelic Locus; Exact Solution
      with a Continuous Model, *Biometrics*, vol. 12, pp. 57–66, 1956.
57   Kimura, M.: Some Problems of Stochastic Processes in Genetics, *Ann.
      Math. Statist.*, vol. 28, pp. 882–901, 1957.
58   Kolmogorov, A. N.: Deviations from Hardy's Formula in Partial Isolation,
      Compt. rend. acad. sci. U.R.S.S., vol. 3, pp. 129–132, 1935.
59   Kolmogorov, A. N., I. Petrovskiĭ, and N. Piscounov: Etude de l'équation de
      la diffusion avec croissance de la quantité de matière et son application à un
      problème biologique, *Bull. State Univ. Moscow*, ser. A, vol. 1, pp. 1–25, 1937.
60   Kostitzin, V. A.: "Biologie mathématique," Librairie Armand Colin,
      Paris, 1937.
61   Lamens, A.: Sur le processus non-homogène de naissance et de mort à deux
      variables aléatoires, *Acad. roy. Belg., Bull. classe. sci.*, ser. 5, vol. 43, pp.
      711–719, 1957.
62   Lamens, A., and R. Consael: Sur le processus non-homogène de naissance et
      de mort, *Acad. roy. Belg., Bull. classe sci.*, ser. 5, vol. 43, pp. 597–605, 1957.
63   Lea, D. E.: "Actions of Radiations on Living Cells," 2d ed., Cambridge
      University Press, New York, 1955.
64   Lea, D. E., and C. A. Coulson: The Distribution of the Number of Mutants
      in Bacterial Populations, *J. Genetics*, vol. 49, pp. 264–285, 1949.

65 Leslie, P. H.: Some Further Notes on the Use of Matrices in Population Mathematics, *Biometrika*, vol. 35, pp. 213–245, 1948.
66 Luria, S. E., and M. Delbrück: Mutation of Bacteria from Virus Sensitivity to Virus Resistance, *Genetics*, vol. 28, pp. 491–511, 1943.
67 Lotka, A. J.: "Elements of Physical Biology," The Williams & Wilkins Company, Baltimore, 1925.
68 Malécot, G.: "Les Mathématiques de l'hérédité," Masson et Cie, Paris, 1948.
69 Malécot, G.: Les Processus stochastiques de la génétique, in "Le Calcul des probabilités et ses applications," pp. 121–126, Centre National de Recherche Scientifique, Paris, 1949.
70 Malécot, G.: Un Traitement stochastique des problèmes linéaires (mutation, linkage, migration) en génétique de population, *Ann. univ. Lyon*, vol. 14, pp. 79–117, 1951.
71 Marchand, H.: Essai d'étude mathématique d'une forme d'épidemic, *Ann. univ. Lyon*, vol. 19, pp. 13–46, 1956.
72 McKendrick, A. G.: Applications of Mathematics to Medical Problems, *Proc. Edinburgh Math. Soc.*, vol. 44, pp. 98–130, 1926.
73 Moran, P. A. P.: Random Processes in Genetics, *Proc. Cambridge Phil. Soc.*, vol. 54, pp. 60–71, 1958.
74 Morse, P. M., and H. Feshbach: "Methods of Theoretical Physics," McGraw-Hill Book Company, Inc., New York, 1953.
75 Muir, T., and W. H. Metzler: "A Treatise on the Theory of Determinants," Syracuse University Press, Syracuse, 1930.
76 Neyman, J., and E. L. Scott: On a Mathematical Theory of Populations Conceived as Conglomeration of Clusters, *Cold Spring Harbor Symposium on Quantitative Biology*, vol. 22, pp. 109–120, 1957.
77 Neyman, J., T. Park, and E. L. Scott: Struggle for Existence. The *Tribolium* Model: Biological and Statistical Aspects, *Proc. Third Berkeley Symposium of Math. Statistics and Probability*, vol. 4, pp. 41–79, 1956.
78 Opatowski, I.: Chain Processes and Their Biophysical Applications: Part I. General Theory; Part II. The Effect of Recovery, *Bull. Math. Biophysics*, vol. 7, pp. 161–180, 1945; vol. 8, pp. 7–13, 1946.
79 Opatowski, I.: The Probabilistic Approach to the Effects of Radiations and Variability of Sensitivity, *Bull. Math. Biophys.*, vol. 8, pp. 101–109, 1946.
80 Patil, V. T.: The Consistency and Adequacy of the Poisson-Markoff Model for Density Fluctuations, *Biometrika*, vol. 44, pp. 43–56, 1957.
81 Powell, E. O.: Some Features of Generation Times of Individual Bacteria, *Biometrika*, vol. 42, pp. 16–44, 1955.
82 Seiden, E.: On a Mathematical Model for a Problem in Epidemiology (Abstract), *Bull. Am. Math. Soc.*, vol. 63, pp. 142–143, 1957.
83 Serfling, R. E.: Historical Review of Epidemic Theory, *Human Biology*, vol. 24, pp. 145–166, 1952.
84 Shapiro, A.: The Kinetics of Growth and Mutation in Bacteria, *Cold Spring Harbor Symposium on Quantitative Biology*, vol. 11, pp. 228–235, 1946.
85 Skellam, J. G.: Random Dispersal in Theoretical Populations, *Biometrika*, vol. 38, pp. 196–218, 1951.
86 Skellam, J. G.: Studies in Statistical Ecology: I. Spatial Patterns, *Biometrika*, vol. 39, pp. 346–362, 1952.
87 Szegö, G.: "Orthogonal Polynomials," American Mathematical Society, New York, 1939.

88 Takashima, M.: Note on Evolutionary Processes, *Bull. Math. Statist.* vol. 7, pp. 18–24, 1957.

89 Urbanik, K.: Remarks on the Maximum Number of Bacteria in a Population (in Polish), *Zastos. Mat.*, vol. 2, pp. 341–348, 1956.

90 Valiron, G.: "Cours d'analyse mathématique: II. Equations fonctionnelles. Applications," Masson et Cie, Paris, 1950.

91 Volterra, V.: "Leçons sur la théorie mathématique de la lutte pour la vie," Gauthier-Villars, Paris, 1931.

92 Waugh, W. A. O'N.: Conditioned Markov Processes, *Biometrika*, vol. 45, pp. 241–249, 1958.

93 Whittle, P.: Certain Nonlinear Models of Population and Epidemic Theory, *Skand. Aktuar.*, vol. 14, pp. 211–222, 1952.

94 Whittle, P.: The Outcome of a Stochastic Epidemic—A Note on Bailey's Paper, *Biometrika*, vol. 42, pp. 116–122, 1955.

95 Whittle, P.: On the Use of the Normal Approximation in the Treatment of Stochastic Processes, *J. Roy. Statist. Soc.*, ser. B, vol. 19, pp. 268–281, 1957.

96 Wright, S.: "Statistical Genetics in Relation to Evolution," Hermann & Cie, Paris, 1939.

97 Wright, S.: The Differential Equations of the Distribution of Gene Frequencies, *Proc. Natl. Acad. Sci. U.S.*, vol. 31, pp. 382–389, 1945.

98 Wright, S.: On the Roles of Directed and Random Changes in Gene Frequency in the Genetics of Populations, *Evolution*, vol. 2, pp. 279–294, 1948.

5

Applications in Physics:
Theory of Cascade Processes

5.1 Introduction

It is well known that charged particles, as they pass through matter, lose energy by collisions and radiation. At low energies the main effects that enter into the physical process are the change in direction of motion of the particle (termed *scattering*) and loss of energy of the radiation particles due to their collision with the atoms of the absorber. At high energies the situation is more complex. To fix ideas, we consider a high-energy electron (primary particle) incident on some absorber. When an electron collides with an atom of the absorber, the deflection caused by the collision results in the radiation of high-energy quanta (photons). When its energy is low, the electron loses an amount of energy by emission of photons that is negligible compared to the energy it loses by ionization, i.e., by knocking out orbital electrons from the absorber atom. The above sources of energy loss become comparable only when the kinetic energy of the electron is of the same order as the energy equivalent, its rest mass. Another situation arises when the emitted photon has an energy greater than twice the rest energy of an electron, for in this case the photon can create a pair of electrons, one positive and one negative. This process is termed pair formation. The pair of electrons, if its energy is high enough, may radiate photons, which in turn will produce further electron pairs. Hence, a high-energy electron passing through an absorber will create an *electron-photon cascade*, with the cascade or multiplication process described above continuing until the energies of the secondary particles are too low for further multiplication. At this stage of the cascade the remaining

235

energy is rapidly absorbed. In the case of electrons the energy is lost through ionization; in the case of photons energy loss is due to Compton scattering. In Fig. 5.1 a schematic representation of an electron-photon cascade is given. In this case the cascade is initiated by an electron; however, it could equally well have been initiated by a photon.

The idea that the absorption of energetic electrons or photons leads to cascade or multiplicative processes of the type described above was put forward simultaneously by Bhabha and Heitler[13] and Carlson and Oppenheimer [19]. These authors also pointed out that the soft component of cosmic-ray showers, i.e., the electrons and photons, could be interpreted in terms of such cascades.

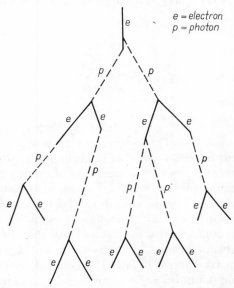

Figure 5.1 Schematic representation of an electron-photon cascade.

In Sec. 5.2 we develop the theory of electron-photon cascades, with particular emphasis on the fluctuation problem for electron-photon cascades.

In addition to electron-photon cascades, it is necessary to consider cascade processes due to nucleons, i.e., neutrons and protons. High-energy nucleons tend to penetrate the nuclei they collide with instead of being scattered. The nucleons that result from the cascade within the nucleus may then collide with other nuclei of the absorbing medium.

As the energy of the nucleons increases, the threshold for production of mesons in the individual nucleon-nucleon interactions is reached, and jets of fast nucleons and mesons are emitted. The mesons are unstable particles which decay spontaneously into other particles. Those produced in nucleon-nucleon interactions are primarily π mesons. which may be charged or neutral. The charged π mesons decay into μ mesons, which in turn decay into electrons. The neutral π mesons decay into photons, which in turn cause pair production. Therefore, the electrons and photons produced in this way form an electron-photon cascade.

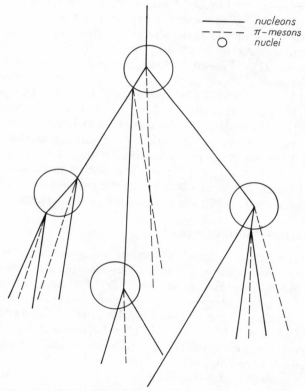

Figure 5.2 Schematic representation of a nucleon cascade in a finite absorber.

In view of the above, we are led to consider a "cascade within a cascade," one cascade taking place within the nuclei of the absorber, and the other taking place among the nuclei. This type of cascade process is termed a *nucleon cascade*, a schematic representation of which is given in Fig. 5.2. The theory of nucleon cascades is considered in Sec. 5.3.

The fact that the nucleons can cascade within a nucleus and in turn cascade among the nuclei themselves gives rise to an essential difference between the nucleon cascade and the electron-photon cascade, even though the multiplicative processes involved in the two cases are similar. In an electron-photon cascade the interior of the nucleus plays no role, and the cascade develops in a single medium.

It is perhaps clear from the very brief description given above that cascade processes are essentially statistical in character. In this chapter we consider the problem of formulating the development of a cascade process within the framework of the theory of stochastic processes. That is, we consider the problem of determining the probability distribution of the state variables which characterize a given cascade process. As we shall see, the stochastic theory of cascade processes is a rather difficult and complex subject, in spite of the fact that the physical processes involved in the production of cascades are rather well known. The difficulties involved are due to the fact that, in addition to considering the numbers of one or more types of particles at a given thickness or depth of the absorbing medium, we must consider the energy distribution of the particles. Hence, in characterizing the "state" of the cascade at some depth or thickness t, we must consider a discrete variable (number of particles) and a continuous variable (energy). In Secs. 5.2C and 5.3C we consider the mathematical methods that have been developed to handle problems of this type.

5.2   Electron-Photon Cascades

A. Introduction. In this section we consider the stochastic theory of electron-photon cascades. The first problem we consider is that of determining the mean number of electrons and photons produced in a given absorber by a primary electron or photon. The treatment of this problem is based on the work of Landau and Rumer [50] and Bhabha and Chakrabarty [11]. Before deriving the Landau-Rumer equations, we discuss the approximations under which electron-photon cascades are treated.

In the treatment that follows it is assumed that the cascade process is one-dimensional, i.e., we assume that all particles (electrons and photons) move in the same direction as the primary particle which initiates the cascade. This assumption is justified (especially for very high energies), since, as we remarked earlier, the change in path length due to scattering is, in general, negligible.

Since the phenomena mainly responsible for the production of cascades are radiation processes and pair production, it is useful to introduce as the unit of thickness in an absorber

$$\tau = \left(\frac{4NZ^2}{137} \, r_0^2 \log 183 Z^{-\frac{1}{3}} \right)^{-1}$$

In the above, N is the number of atoms per unit mass, Z is the atomic number of the absorber, and r_0 is the radius of the electrons. In the subsequent treatment the absorber depth or thickness, denoted by t, will be assumed to be expressed in terms of the *cascade unit* τ, rather than in terms of centimeters. The introduction of the cascade unit enables us to consider the cascade process as developing in the same manner regardless of the absorber.

Cascade theory has been developed under two approximations that depend on the relation of the energy range of the cascade process to the *critical energy*, the critical energy being defined as the energy equal to the energy lost by a fast particle due to ionization along one cascade unit. In case we consider energies which are large compared with the critical energy, we have:

APPROXIMATION A:

1. Radiation phenomena (Bremsstrahlung) and pair production can be described by the asymptotic formulas for very high energies and complete screening.[1]

2. No other processes are taken into account; in particular, energy loss due to collisions and Compton effect are neglected.

3. The development of the cascade process is one-dimensional.

In case energies in the neighborhood of the critical energy are considered, we have:

APPROXIMATION B:

1. Radiation phenomena and pair production can be described by the asymptotic formulas for very high energies and complete screening.

2. Energy loss due to the Compton effect is neglected.

3. Energy loss due to collisions is described as a constant energy dissipation,[2] i.e., $-dE/dt = \beta$.

B.   The Landau-Rumer Equations. Let $\pi(E,t)\, dE$ denote the mean number of electrons (positive and negative) at absorber depth t with energy in the interval $(E, E + dE)$ and let $\gamma(W,t)\, dW$ denote the near number of photons at absorber depth t with energy in the interval $(W, W + dW)$. $\pi(E,t)\, dE$ and $\gamma(W,t)\, dW$ are termed the *differential electron* and *differential photon spectra*, respectively.

The functional equations satisfied by $\pi(E,t)$ and $\gamma(W,t)$ can be obtained as follows: In a given thickness dt the number of electrons in the interval $(E, E + dE)$ can change owing to the following mutually exclusive processes:

1. Photons with energy $W(W > E)$ produce a number of electrons

---

[1] Bethe and Heitler [8]. Cf. also Rossi and Greisen [94].

[2] Collision loss is not a constant; here it is assumed to be so in order to simplify matters.

(positive and negative) in the interval $(E, E + dE)$ by pair production. This number is

$$2 \, dE \, dt \int_E^\infty \gamma(W,t) \psi_0 \left(\frac{E}{W}\right) \frac{dW}{W} = 2 \, dE \, dt \int_0^1 \gamma\left(\frac{E}{u}, t\right) \psi_0(u) \frac{du}{u} \qquad (5.1)$$

In (5.1) $\psi_0(u)$, which is the differential probability for pair production per radiation length in the case of complete screening, is given by

$$\psi_0(u) = [u^2 + (1 - u)^2] + [(\tfrac{2}{3} - 2b)u(1 - u)]$$

where $u = E/W$ and $b = (18 \log 183 Z^{-1/3})^{-1}$. The constant b is usually taken to be 0.0135 for all elements.

2. Some electrons with energy E' $(E' > E)$ enter the interval $(E, E + dE)$ by radiating part of their energy. This number is

$$dE \, dt \int_E^\infty \pi(E',t) \varphi_0 \left(\frac{E' - E}{E'}\right) \frac{dE'}{E'} = dE \, dt \int_0^1 \pi\left(\frac{E}{1 - v}, t\right) \varphi_0(v) \frac{dv}{1 - v} \qquad (5.2)$$

where $\varphi_0(v)$, the differential radiation probability per radiation length in the case of complete screening, is given by

$$\varphi_0(v) = \frac{1}{v}[1 + (1 - v)^2 - (1 - v)(\tfrac{2}{3} - 2b)]$$

where $v = (E' - E)/E'$.

3. Some electrons initially in the interval $(E, E + dE)$ leave this interval by radiation loss. This number is

$$\pi(E,t) \, dE \, dt \int_0^E \varphi_0 \left(\frac{W}{E}\right) \frac{dW}{E} = \pi(E,t) \, dE \, dt \int_0^1 \varphi_0(v) \, dv \qquad (5.3)$$

where $\varphi_0(v)$ is the same as before, but now $v = W/E$. It is important to note that the Bremsstrahlung cross sections given by (5.2) and (5.3) diverge, since $\varphi_0(v)$ behaves as $1/v$; however, their difference remains finite.

4. If every electron traversing the thickness dt loses by collision an amount β of energy (β a constant), then the change in the number of electrons in the interval $(E, E + dE)$ is given by[1]

$$\beta \left(\frac{\partial \pi}{\partial E}\right) dE \, dt \qquad (5.4)$$

---

[1] If $\beta = \beta(E)$, (5.4) should read $\dfrac{\partial(\beta\pi) \, dE \, dt}{\partial E}$.

In the thickness dt the number of photons with energy in the interval $(W, W + dW)$ can change owing to the following mutually exclusive processes:

1. Electrons with energy $E(E > W)$ radiate a certain number of photons in the interval $(W, W + dW)$. This number is

$$dW\, dt \int_W^\infty \pi(E,t)\varphi_0\left(\frac{W}{E}\right)\frac{dE}{E} = dW\, dt \int_0^1 \pi\left(\frac{W}{v}, t\right)\varphi_0(v)\frac{dv}{v} \quad (5.5)$$

where $v = W/E$.

2. Some photons initially in the interval $(W, W + dW)$ are absorbed by the process of pair production. This number is

$$dW\, dt\, \gamma(W,t)\sigma_0$$

where σ_0, the total probability for pair production per unit length in the case of complete screening, is given by

$$\sigma_0 = \frac{7}{9} - \frac{b}{3}$$

From the above considerations, we obtain the *Landau-Rumer equations:*[1]

$$\frac{\partial \pi(E,t)}{\partial t} = 2\int_0^1 \gamma\left(\frac{E}{u}, t\right)\psi_0(u)\frac{du}{u}$$

$$- \int_0^1\left[\pi(E,t) - \frac{1}{1-v}\pi\left(\frac{E}{1-v}, t\right)\right]\varphi_0(v)\, dv + \beta\,\frac{\partial \pi(E,t)}{\partial E} \quad (5.6)$$

$$\frac{\partial \gamma(W,t)}{\partial t} = \int_0^1 \pi\left(\frac{W}{v}, t\right)\varphi_0(v)\frac{dv}{v} - \sigma_0\gamma(W,t) \quad (5.7)$$

We now consider the solutions of the Landau-Rumer equations under Approximations A and B.

1. *Solution under Approximation A.* In this case energy loss due to collision loss is neglected; hence, we put $\beta = 0$ in Eq. (5.6). Since collision loss does not appear directly in the equation for the photon, its neglect leaves the photon equation unchanged.

Let

$$\bar{\pi}(s,t) = \int_0^\infty \pi(E,t)E^s\, dE \quad (5.8)$$

and

$$\bar{\gamma}(s,t) = \int_0^\infty \gamma(W,t)W^s\, dW \quad (5.9)$$

[1] These equations are also called the *diffusion equations* for the electron-photon cascade process.

represent the Mellin transforms[1] of $\pi(E,t)$ and $\gamma(W,t)$ with respect to the energy variables E and W, respectively. Application of the Mellin transformation to Eqs. (5.6) and (5.7) yields the system of partial differential equations

$$\frac{\partial \bar{\pi}(s,t)}{\partial t} = -A(s)\bar{\pi}(s,t) + B(s)\bar{\gamma}(s,t) \tag{5.10}$$

$$\frac{\partial \bar{\gamma}(s,t)}{\partial t} = C(s)\bar{\pi}(s,t) - \sigma_0\bar{\gamma}(s,t) \tag{5.11}$$

In the above we have put

$$A(s) = \int_0^1 [1 - (1 - v)^s]\varphi_0(v)\, dv \tag{5.12}$$

$$B(s) = 2\int_0^1 u^s\psi_0(u)\, du \tag{5.13}$$

$$C(s) = \int_0^1 v^s\varphi_0(v)\, dv \tag{5.14}$$

In order to solve the above system of partial differential equations, we utilize the Laplace transformation. Let $\Pi(s,p)$ and $\Gamma(s,p)$ denote the respective Laplace transforms of $\bar{\pi}(s,t)$ and $\bar{\gamma}(s,t)$ with respect to the thickness t. Application of the Laplace transformation to Eqs. (5.10) and (5.11) yields the system of algebraic equations

$$[p + A(s)]\Pi(s,p) - B(s)\Gamma(s,p) = \bar{\pi}_0(s) \tag{5.15}$$

$$-C(s)\Pi(s,p) + (p + \sigma_0)\Gamma(s,p) = \bar{\gamma}_0(s) \tag{5.16}$$

where $\bar{\pi}_0(s) = \bar{\pi}(s,0)$ and $\bar{\gamma}_0(s) = \bar{\gamma}(s,0)$. The solutions of the above equations are

$$\Pi(s,p) = \frac{B(s)\bar{\gamma}_0(s) + (p + \sigma_0)\bar{\pi}_0(s)}{[p + A(s)](p + \sigma_0) - B(s)C(s)} \tag{5.17}$$

$$\Gamma(s,p) = \frac{C(s)\bar{\pi}_0(s) + [p + A(s)]\bar{\gamma}_0(s)}{[p + A(s)](p + \sigma_0) - B(s)C(s)} \tag{5.18}$$

If we now put

$$p_1(s) = \tfrac{1}{2}\{[(A(s) - \sigma_0)^2 - B(s)C(s)]^{\frac{1}{2}} - [A(s) + \sigma_0]\}$$

$$p_2(s) = \tfrac{1}{2}\{[(A(s) - \sigma_0)^2 - B(s)C(s)]^{\frac{1}{2}} + [A(s) + \sigma_0]\}$$

---

[1] See Appendix B for some properties of the Mellin transformation.

Equations (5.17) and (5.18) can be rewritten in the form

$$\Pi(s,p) = \frac{1}{p_1(s) - p_2(s)} \left\{ \frac{B(s)\bar{\gamma}_0(s) + [\sigma_0 + p_1(s)]\bar{\pi}_0(s)}{p - p_1(s)} \right.$$

$$\left. - \frac{B(s)\bar{\gamma}_0(s) + [\sigma_0 + p_2(s)]\bar{\pi}_0(s)}{p - p_2(s)} \right\} \quad (5.19)$$

$$\Gamma(s,p) = \frac{1}{p_1(s) - p_2(s)} \left\{ \frac{C(s)\bar{\pi}_0(s) + [A(s) + p_1(s)]\bar{\gamma}(s)}{p - p_1(s)} \right.$$

$$\left. - \frac{C(s)\bar{\pi}_0(s) + [A(s) + p_2(s)]\bar{\gamma}_0(s)}{p - p_2(s)} \right\} \quad (5.20)$$

If we now apply the inversion theorem for Laplace transforms and use the result

$$e^{p_j t} = \frac{1}{2\pi i} \int_{c-i\infty}^{c+i\infty} \frac{e^{pt}}{p - p_j} dp \qquad j = 1, 2$$

we have

$$\bar{\pi}(s,t) = \frac{1}{p_1(s) - p_2(s)} \left\{ \{B(s)\bar{\gamma}_0(s) + [\sigma_0 + p_1(s)]\bar{\pi}_0(s)\} e^{p_1(s)t} \right.$$

$$\left. - \{B(s)\bar{\gamma}_0(s) + [\sigma_0 + p_2(s)]\bar{\pi}_0(s)\} e^{p_2(s)t} \right\} \quad (5.21)$$

$$\bar{\gamma}(s,t) = \frac{1}{p_1(s) - p_2(s)} \left\{ \{C(s)\bar{\pi}_0(s) + [A(s) + p_1(s)]\bar{\gamma}_0(s)\} e^{p_1(s)t} \right.$$

$$\left. - \{C(s)\bar{\pi}_0(s) + [A(s) + p_2(s)]\bar{\gamma}_0(s)\} e^{p_2(s)t} \right\} \quad (5.22)$$

If we now apply the inversion theorem for Mellin transform, we obtain the formal solutions of Eqs. (5.6) and (5.7), which yield the differential electron and photon spectra, respectively.

In obtaining the above solutions we have made no assumption concerning the primary particle which initiates the cascade process. Two cases are of special interest: (1) the cascade is initiated by a single primary electron of energy E_0 or (2) the cascade is initiated by a single primary photon of energy W_0. We now consider the solutions of the Landau-Rumer equations for the case in which the cascade is initiated by a single primary electron.

For an electron-initiated cascade the initial conditions for (5.6) and (5.7) are

$$\pi(E,0) = \delta(E - E_0) \qquad \gamma(W,0) = 0 \qquad (5.23)$$

where $\delta(E)$ is the Dirac delta function. From (5.23) it follows that

$$\bar{\pi}_0(s) = (E_0)^s \qquad \bar{\gamma}_0(s) = 0 \qquad (5.24)$$

On inserting (5.24) in (5.21), we see that the Mellin transform of the differential electron spectrum is given by

$$\bar{\pi}(s,t) = \frac{(E_0)^s}{p_1(s) - p_2(s)} \{[\sigma_0 + p_1(s)]e^{p_1(s)t} - [\sigma_0 + p_2(s)]e^{p_2(s)t}\} \quad (5.25)$$

Application of inversion theorem for Mellin transforms to (5.25) yields

$$\pi(E,t;E_0)\, dE = -\frac{dy}{2\pi i}\left\{\int_{c-i\infty}^{c+i\infty} [H_1(s)e^{ys+p_1(s)t} + H_2(s)e^{ys+p_2(s)t}]\right\} ds \quad (5.26)$$

In the above we have put $y = \log(E_0/E)$, $H_1(s) = [\sigma_0 + p_1(s)]/[p_1(s) - p_2(s)]$, and $H_2(s) = [\sigma_0 + p_2(s)]/[p_1(s) - p_2(s)]$.

By proceeding in the same way, we find that the differential photon spectrum is given by

$$\gamma(W,t;E_0)\, dW = -\frac{dy}{2\pi i}\left\{\int_{c-i\infty}^{c+i\infty} L(s)[\exp\{ys + p_1(s)t - \tfrac{1}{2}\log s\}\right.$$

$$\left. \times \exp\{ys + p_2(s)t - \tfrac{1}{2}\log s\}]\right\} ds \quad (5.27)$$

where $y = \log(E_0/W)$ and $L(s) = s^{\frac{1}{2}}C(s)/[p_1(s) - p_2(s)]$.

From the above we see that in order to obtain explicit expressions for the differential spectra, it is necessary first to evaluate the functions $A(s)$, $B(s)$, and $C(s)$. From (5.12) to (5.14) and the definition of the functions $\varphi_0(v)$ and $\psi(u)$, we obtain

$$A(s) = 1.36\frac{d[\log(s+1)!]}{ds}\frac{1}{(s+1)(s+2)} - 0.0750$$

$$B(s) = 2\left[\frac{1}{s+1} - \frac{1.36}{(s+2)(s+3)}\right]$$

$$C(s) = \frac{1}{s+2} + \frac{1.36}{s(s+1)}$$

From the above, and by using $\sigma_0 = \tfrac{7}{9} - b/3$, the functions $p_1(s)$ and $p_2(s)$ can be calculated; in turn, the functions $H_1(s)$, $H_2(s)$, $L(s)$, and $M(s)$ can be obtained. For the details of these calculations we refer to Refs. 39 and 94.

To sum up,[1] the expressions for the *differential electron and photon spectra* for the case in which the cascade is initiated *by a primary electron of energy E_0* are given by

$$\pi(E,t;E_0)\, dE = \frac{H_1(s)e^{p_1(s)t}}{[2\pi t p_1''(s)]^{\frac{1}{2}}}\left(\frac{E_0}{E}\right)^s \frac{dE}{E} \quad (5.28)$$

---

[1] The solutions given by Eqs. (5.28) to (5.31) are approximate solutions obtained by application of the saddle-point method.

where $t = -\log (E_0/E)/p_1'(s)$ and

$$\gamma(W,t;E_0) \, dW = \frac{L(s)e^{p_1(s)t}}{[2\pi t s p_1''(s) + \pi/s]^{\frac{1}{2}}} \left(\frac{E_0}{E}\right)^s \frac{dW}{W} \qquad (5.29)$$

where $t = -[\log(E_0/E) - \frac{1}{2}s]/p_1'(s)$. In the above expressions $p_1'(s)$ and $p_1''(s)$ denote the derivatives of $p_1(s)$ with respect to s.

Should the cascade be initiated by a *primary photon of energy* W_0, the differential electron and photon spectra are given by

$$\pi(E,t;W_0) \, dE = \frac{s^{\frac{1}{2}}M(s)e^{p_1(s)t}}{[2\pi p_1''(s)t - \pi/s^2]^{\frac{1}{2}}} \left(\frac{W_0}{E}\right)^s \frac{dE}{E} \qquad (5.30)$$

where

$$t = -\frac{1}{p_1'(s)}\left[\log\left(\frac{W_0}{E}\right) + \frac{1}{2}s\right]$$

$$\gamma(W,t;W_0) \, dW = \frac{H_2(s)e^{p_1(s)t}}{[2\pi p_1''(s)t]^{\frac{1}{2}}} \left(\frac{W_0}{W}\right)^s \frac{dW}{W} \qquad (5.31)$$

where $t = -\log (W_0/W)/p_1'(s)$.

In order to carry out the numerical calculations for a given electron-photon cascade, it is necessary to specify the energy of the primary particle (E_0 or W_0) and the energy (E or W) at which the functions $\pi(E,t)$ and $\gamma(W,t)$ are to be determined. In Ref. 94 tables of the functions occurring in the above solutions are given, and we refer to that paper for a discussion of numerical calculations.

In addition to the differential spectra, it is of interest to obtain the integral electron and photon spectra. The *integral electron spectrum* is defined as the number of electrons at thickness t which possess an energy greater than E. Hence

$$\pi_I(E,t) = \int_E^\infty \pi(E',t) \, dE' \qquad (5.32)$$

Similarly, the *integral photon spectrum* is given by

$$\gamma_I(W,t) = \int_W^\infty \gamma(W',t) \, dW' \qquad (5.33)$$

In the above expressions the integrand will depend on the type, and energy, of the particle initiating the cascade.

2. *Solutions under Approximation B.* We now consider the solutions of the Landau-Rumer equations when the energy loss due to collision is not neglected. Hence we seek the solutions of Eqs. (5.6) and (5.7) when $\beta \neq 0$ in (5.6).

We shall not consider Approximation B in detail, but shall describe the method of solution due to Bhabha and Chakrabarty [11,12]. They assume that the exact solution of (5.6) has the form

$$\pi(E,t) = \frac{1}{2\pi i E_0} \int_{c-i\infty}^{c+i\infty} \left[\frac{E_0}{E + \beta g(s,t)} \right]^s f(s,t,\beta)\, ds \tag{5.34}$$

where $g(s,t)$ and $f(s,t,\beta)$ are functions of s and t, but not of E. Under the assumption that $f(s,t,\beta)$ satisfies the initial conditions

$$f(s,0,\beta) = 1$$

$$\frac{\partial f(s,t,\beta)}{\partial t}\Bigg]_{t=0} = -A(s)$$

it has been shown that $g(s,t)$ must satisfy the boundary conditions

$$g(s,0) = 0 \qquad \frac{\partial g(s,t)}{\partial t}\Bigg]_{t=0} = 1$$

in order that (5.34) satisfy the initial conditions

$$\pi(E,0) = \delta(E - E_0) \qquad \gamma(W,0) = 0 \tag{5.23}$$

Introducing (5.34) into (5.6) and imposing suitable initial conditions enables the functions $f(s,t,\beta)$ and $g(s,t)$ to be determined. Both $f(s,t,\beta)$ and $g(s,t)$ can be expressed in terms of the solution of the Mellin transform of Eq. (5.6) with $\beta = 0$ and with $f(s,t,\beta)$ being expressed as a power series in β. Hence, the solution of (5.34) has the form

$$\pi(s,t) = \sum_{i=0}^{\infty} \pi_i(s,t) \tag{5.35}$$

For other studies on the cascade theory under Approximation B we refer to Refs. 39, 98, and 102.

C. The Fluctuation Problem for Electron-Photon Cascades. In experimental studies on *electron-photon cascades* it is of greatest interest to obtain the probability distribution of the number of particles with a given energy at a given depth of the absorber, since it is the probability distribution that is required for comparison with experiments. Because it is very difficult to obtain the probability distribution, the main problem in cascade theory has been that of determining the fluctuation of the actual numbers of particles around the average or mean number of particles.[1] Hence the problem that has attracted the attention of most workers in this field is that of obtaining

---

[1] This situation is now changed. By using Monte Carlo methods, the research group in the University of Sydney, under the direction of H. Messel, is preparing numerical results for the probability distribution. See Ref. 17 for the first results in this direction.

the mean and variance of the number of particles.[1] In this section we consider the methods developed by various investigators for the treatment of the fluctuation problem.

The first model in the study of the fluctuation problem was introduced by Bhabha and Heitler [13] in a fundamental paper on cascade theory. It was assumed that in the cascade shower the probability for a secondary electron to be produced with energy in the interval $(E_1, E_1 + dE_1)$ is independent of the probability for another secondary electron to be produced with energy in the interval $(E_2, E_2 + dE_2)$. Hence, it was assumed that the probability of finding n electrons above a specified energy E at thickness t, denoted by $P_n(E,t)$, is given by the Poisson distribution

$$P_n(E,t) = \frac{(\lambda t)^n}{n!} e^{-\lambda t} \qquad \lambda = \lambda(E) \tag{5.36}$$

where $\lambda \, \Delta t + o(\Delta t)$, $\lambda > 0$, is the probability that a secondary electron will be produced in the interval $(t, t + \Delta t)$. It is well known that for the Poisson process

$$\mathscr{E}\{X(t)\} = \mathscr{D}^2\{X(t)\} = \lambda t \tag{5.37}$$

where $X(t)$ represents the number of particles in the cascade above a specified energy E at thickness t. From this result alone it is clear that the Poisson process is not a realistic model, since one would not expect the average number of electrons to be proportional to the thickness of the absorber, and hence increase indefinitely. One would expect the average number to decrease with thickness, since the secondary electrons will lose energy with increasing thickness, and therefore will eventually be absorbed.

The Bhabha-Heitler model was criticized by Furry [27], who developed a multiplicative model for the cascade process. The model developed by Furry, which neglects the photons in the cascade, is based on the assumption that an electron traversing an absorber has probability $\lambda \, \Delta t + o(\Delta t)$, $\lambda > 0$, of being converted into two electrons in the interval $(t, t + \Delta t)$. It is also assumed that there is no collision loss. These assumptions lead to the distribution function

$$P_n(t) = e^{-\lambda t}(1 - e^{-\lambda t})^{n-1} \qquad n = 1, 2, \ldots \qquad t \geqslant 0 \tag{5.38}$$

This distribution function was considered in Chap. 2 and at that time referred to as the distribution function of the *Yule-Furry process* or the *simple birth process*. For this process

$$\mathscr{E}\{X(t)\} = e^{\lambda t} \tag{5.39}$$

and
$$\mathscr{D}^2\{X(t)\} = e^{\lambda t}(e^{\lambda t} - 1) \tag{5.40}$$

[1] The variance is usually referred to in the physical literature as the mean-square deviation.

In order to compare the two models considered thus far, we consider the *relative fluctuation* or *coefficient of variation*, the relative fluctuation being defined as the ratio

$$\mathscr{V}(t) = \frac{\mathscr{D}\{X(t)\}}{\mathscr{E}\{X(t)\}} \tag{5.41}$$

For the Poisson process (Bhabha-Heitler model) we obtain

$$\mathscr{V}(t) = [\lambda t]^{-\frac{1}{2}} \tag{5.42}$$

and for the Furry model we obtain

$$\mathscr{V}(t) = [1 - e^{-\lambda t}]^{\frac{1}{2}} \sim 1 \tag{5.43}$$

where λt is not too small. On comparing (5.42) and (5.43) we see that the relative fluctuation for the Furry distribution is much larger than the relative fluctuation for the Poisson distribution, except for small values of λt, in which case the two distributions are essentially the same.

It is clear that the Furry process is a better idealized model for the cascade process than the Poisson process; however, we see from (5.39) that the average number of particles is an exponential function of the thickness of absorber traversed. In this case the neglect of collision loss or absorption in the Furry process becomes increasingly noticeable for large thicknesses.

The Furry process was generalized by Nordsieck, Lamb, and Uhlenbeck [74] by taking into consideration the fact that the energy of the primary electron initiating the cascade is divided among all of the secondaries it produces. Their results were in agreement with earlier results, namely, that the fluctuation for the Furry distribution is much larger than the Poisson fluctuation.

In addition to the Poisson and Furry processes, several other simple stochastic processes have been introduced in cascade theory as approximations to the real cascade process. Of particular interest is the Pólya process, which was studied in Chap. 2. For this process

$$P_n(t) = \frac{(\lambda t)^n}{n!} (1 + \alpha \lambda t)^{-n-1/\alpha} \prod_{i=1}^{n-1}(1 + \alpha i) \qquad n = 1, 2, \ldots \tag{5.44}$$

This process, being a two-parameter one, has greater flexibility in its applications than the Poisson and Furry processes, and as we have shown in Chap. 2, the Poisson and Furry distributions can be obtained from (5.44) by putting $\alpha = 0$ and $\alpha = 1$, respectively. For applications of the Pólya process[1] to electron-photon cascades we refer to Arley [1] and Messel [58].

---

[1] We refer also to the paper of Mitra [69], in which the Pólya process is used to study the size-frequency distribution of μ-meson bursts.

The first realistic model of a cascade process was introduced by Scott and Uhlenbeck [97]. This model was the first to incorporate quantum-mechanical cross sections for radiation loss and pair production, and because of this represented the first attempt to solve the fluctuation problem for the real cascade process.

In order to obtain more realistic mathematical representations of cascade processes, several rather powerful techniques have been developed. The need for new techniques is pointed up by the fact that in cascade theory one studies the distribution of a random variable representing the number of particles in the cascade at thickness t, these particles being distributed in a continuous infinity of states which are characterized by the energy variable E. If the energy space were discrete, one could then define a probability density, say $f(X_1, X_2, \ldots ; E_1, E_2, \ldots ; t)$, representing the probability that at thickness t there are X_i particles in energy state E_i, $i = 1, 2, \ldots$. In this case the development or evolution of the cascade could be studied within the framework of the theory of Markov processes, since we would be able to define the "state" of the cascade at thickness t adequately and thus predict the behavior of the cascade in the interval $(t, t + \Delta t)$. However, the fact that the energy state is a continuum prevents us from defining a true probability density.

This difficulty was, of course, well known to the physicists who first formulated the theory of cascade processes, and this in part accounts for the relatively simple and unrealistic stochastic models which were first advanced in order to simulate some of the features of the real cascade process.

We now consider three of the methods that have been developed in order to permit a more realistic treatment of the fluctuation problem for cascade processes.

1. *The Method of Product Density Functions.* The method of product density functions, developed independently by Bhabha [9] and Ramakrishnan [79], can be described as follows: Let the random variable $X(E;t)$ denote the number of particles[1] with energy values less than E for arbitrary thickness t. Then the random variable $dX(E;t)$ denotes the number of particles in the elementary interval dE. Now let $f_1(E;t)$ be a function such that

$$f_1(E;t)\, dE = \mathscr{E}\{dX(E;t)\} \tag{5.45}$$

where $\mathscr{E}\{dX(E;t)\}$ denotes the expected or mean number of particles

---

[1] The term "particle" is used here in a generic sense, i.e., it is used to designate electrons, photons, nucleons, etc.

in the interval dE. If we now denote by P_n the probability that n particles are in the interval dE, then

$$P_1 = f_1(E;t)\,dE + 0(dE)^2 = \mathscr{E}\{dX(E;t)\} + 0(dE)^2$$

$$P_0 = 1 - P_1 = 1 - [f_1(E;t)\,dE + 0(dE)^2]$$

$$P_n = 0(dE)^n \qquad n > 1$$

Hence, we assume that the probability of having one particle in the interval dE is proportional to dE, and the probability of n ($n > 1$) particles being in dE is $0(dE)^n$.

Let

$$\mathscr{E}\{n^k\} = \sum_{n=0}^{\infty} P_n n^k \tag{5.46}$$

then $\qquad \mathscr{E}\{n^k\} = \mathscr{E}\{[dX(E;t)]^k\} = \mathscr{E}\{n\} = \mathscr{E}\{dX(E;t)\}$ (5.47)

Hence all of the moments of the random variable $dX(E;t)$ are equal to the probability that the random variable assumes the value 1. It must be pointed out that the function $f_1(E;t)$ is not a probability density. However, $f_1(E;t)\,dE$ is a probability magnitude; hence, for t fixed the integral of $f_1(E;t)\,dE$ over the energy range yields only the mean number of particles in the range of integration. This result obtains because the operation of integration in this case does not correspond to the summation of infinitesimal probabilities associated with mutually exclusive events. Hence we have

$$\mathscr{E}\{X(E_i;t) - X(E_j;t)\} = \int_{E_j}^{E_i} \mathscr{E}\{dX(E;t)\}$$

$$= \int_{E_j}^{E_i} f_1(E;t)\,dE \qquad E_i > E_j \tag{5.48}$$

With the above introductory remarks, we can now introduce the concept of a product density. Let $dX(E_1;t)$ and $dX(E_2;t)$ be two random variables. We can form the product of these random variables and define a function $f_2(E_1E_2;t)$ such that

$$f_2(E_1,E_2;t)\,dE_1\,dE_2 = \mathscr{E}\{dX(E_1;t)\,dX(E_2;t)\} \tag{5.49}$$

That is, $f_2(E_1,E_2;t)\,dE_1\,dE_2$ is the joint probability that there is one particle in the interval dE_1 and one particle in the interval dE_2, when dE_1 and dE_2 are disjoint intervals. If the two intervals are not disjoint, a degeneracy obviously occurs and we have

$$\mathscr{E}\{[dX(E_1;t)]^2\} = \mathscr{E}\{dX(E_1;t)\} = f_1(E;t)\,dE \tag{5.50}$$

The function $f_2(E_1,E_2;t)$ is called the *product density of degree* 2. Because of the degeneracy, and from (5.48) and (5.49), we have

$$\mathscr{E}\{[X(E_i;t) - X(E_j;t)]^2\} = \int_{E_j}^{E_i} \int_{E_j}^{E_i} \mathscr{E}\{dX(E_1;t)\,dX(E_2;t)\}$$

$$= \int_{E_j}^{E_i} f_1(E;t)\,dE + \int_{E_j}^{E_i} \int_{E_j}^{E_i} f_2(E_1,E_2;t)\,dE_1\,dE_2 \quad (5.51)$$

Similarly, we can define a *product density of degree n* as the function $f_n(E_1, E_2, \ldots, E_n; t)$ such that

$$f_n(E_1, E_2, \ldots, E_n; t)\,dE_1\,dE_2 \cdots dE_n = \mathscr{E}\{dX(E_1;t)\,dX(E_2;t) \cdots dX(E_n;t)\}$$
$$(5.52)$$

Hence $f_n(E_1, E_2, \ldots, E_n; t)\,dE_1\,dE_2 \cdots dE_n$ is the joint probability that there is one particle in dE_1, one in dE_2, ..., etc., when the intervals $dE_1, dE_2, \ldots, dE_n$ are pair-wise disjoint. By taking into consideration the degeneracies that can occur, it can be shown that the kth moment of the number of particles in any finite interval $\Delta E = E_i - E_j$ is

$$\mathscr{E}\{[X(E_i;t) - X(E_j;t)]^k\} = \mathscr{E}\{[X(\Delta E;t)]^k\}$$

$$= \sum_{n=1}^{k} C_n^k \int_{E_j}^{E_i} \cdots \int_{E_j}^{E_i} f_n(E_1, E_2, \ldots, E_n; t)\,dE_1\,dE_2 \cdots dE_n \quad (5.53)$$

where the coefficients C_n^k are defined by the identity[1]

$$m^k = \sum_{n=1}^{k} C_n^k \prod_{i=0}^{n-1} (m - i) \quad (5.54)$$

It follows from (5.53) that, in order to calculate the kth moment of $[X(E_i;t) - X(E_j;t)]$, it is necessary to know the product densities of degree less than or equal to k.

We now consider the application of the method of product density functions in the theory of cascade processes. We shall see that the fluctuation of the total number of particles in the entire energy range can be calculated through the determination of the density of particles in any particular energy interval and the correlation between particles in two different energy intervals. It is this feature which distinguishes the Bhabha-Ramakrishnan method from the earlier methods based on oversimplified models. The exception is the work of Scott and Uhlenbeck, which, as remarked earlier, was the first attempt to give a

---

[1] The coefficients C_n^k are the Stirling numbers of the second kind. Cf. Jordan. "Calculus of Finite Differences," Chelsea, New York, 1947.

statistical treatment of electron-photon cascades by taking into consideration the energy distribution of the particles.

In order to derive the Bhabha-Ramakrishnan equations for the product density functions, it is necessary to state the nature of the elementary events, and their associated probabilities, which can cause a change in the product density functions. The elementary events here considered are the same as those used in deriving the Landau-Rumer equations; however, we restate them because the notation and interpretation are different. The elementary events and their probabilities are as follows:

1. The probability of an electron of energy E radiating a photon of energy $E - E'$ in traversing a unit length of the absorber[1] is denoted by $R(E,E')\,dE'$.

2. The probability of a photon of energy E creating an electron pair, one of which has energy in the interval $(E', E' + dE')$, is $R'(E,E')\,dE'$.

3. The probability of an electron of energy E losing energy by collision (or ionization), after which its energy is in the interval $(E', E' + dE')$, is $\rho(E,E')\,dE'$.

The probabilities (cross sections) of the elementary events considered are, as before, given by the Bethe-Heitler theory; however, it is not necessary to specify the form of these functions for the derivation of the Bhabha-Ramakrishnan equations.

Since we are concerned with the electron-photon cascade, it is necessary to consider product density functions for the number of electrons and the number of photons in the cascade. Let $f_1(E,t)$ and $g_1(E,t)$ denote the product density functions of degree 1 for the number of electrons and the number of photons, respectively. The elementary events considered above are mutually exclusive; hence, we obtain the following equations for $f_1(E;t)$ and $g_1(E;t)$:

$$\frac{\partial f_1(E;t)}{\partial t} = -f_1(E;t)\left[\int_0^E R(E,E')\,dE' + \int_0^E \rho(E,E')\,dE'\right]$$

$$+ \int_E^\infty f_1(E';t)\{R(E',E) + \rho(E',E)\}\,dE'$$

$$+ 2\int_E^\infty g_1(E';t)R'(E',E)\,dE' \tag{5.55}$$

$$\frac{\partial g_1(E;t)}{\partial t} = -g_1(E;t)\int_0^E R'(E',E)\,dE$$

$$+ \int_E^\infty f_1(E',t)R(E', E' - E)\,dE' \tag{5.56}$$

[1] As before, the absorber depth is measured in cascade units.

As we would expect, the integrodifferential equations for the product density functions of degree 1 are the same as the equations obtained by Landau and Rumer for Approximation A and by Bhabha and Chakrabarty for Approximation B. In Approximation B the term $\beta\dfrac{\partial f_1(E;t)}{\partial E}$ was introduced for collision loss. This approximation of constant (or deterministic) energy loss due to collision can be obtained by assuming that the function $\rho(E,E')$ is such that it becomes vanishingly small for large differences between E and E'. Hence, as a first approximation we can put

$$\int_0^E (E - E')\rho(E,E')\,dE' = \beta$$

and replace all terms in Eq. (5.55) involving $\rho(E,E')$ by $\beta\dfrac{\partial f_1(E;t)}{\partial E}$.

In order to study the fluctuation of the total number of particles, it is necessary to obtain the product density functions of degree 2 for electrons and photons as well as the mixed-product density of electrons and photons. We will denote these functions by f_2, g_2, and h_1, respectively. The mixed-product density is defined as follows: $h_1(E_1,E_2;t)\,dE_1\,dE_2$ is the joint probability that there is one electron in the interval dE_1 and one photon in the interval dE_2. Now let $f_2(E_1,E_2;t)\,dE_1\,dE_2$ denote the joint probability of one electron in the interval dE_1 and one electron in the interval dE_2, and let $g_2(E_1,E_2;t)\,dE_1\,dE_2$ denote the joint probability of one photon in the interval dE_1 and one photon in the interval dE_2. From the definitions of the product densities it follows that the product densities of degree 2 for electrons and photons are symmetric in the energy variable, but the mixed-product density is not symmetric, i.e.,

$$f_2(E_1,E_2;t) = f_2(E_2,E_1;t)$$

$$g_2(E_1,E_2;t) = g_2(E_2,E_1;t)$$

$$h_1(E_1,E_2;t) \neq h_1(E_2,E_1;t)$$

In order to derive the equations for the product densities of degree 2, it is necessary, as in the case of densities of degree 1, to consider the elementary events which can take place and cause a change in the product densities of degree 2. We only consider the derivation of the equation for $f_2(E_1,E_2;t)$, since similar reasoning is employed to obtain the other functional equations. We have the following mutually exclusive events which contribute to a change in $f_2(E_1,E_2\,t)$:

1. Either the electron in state E_1 or the electron in state E_2 may move

out of the state by radiation or collision loss. This transition results in a decrease in the probability magnitude, the measure of which is

$$f_2(E_1,E_2;t)\left[\int_0^{E_1}\{R(E_1,E) + \rho(E_1,E)\}\,dE\right.$$
$$\left.+ \int_0^{E_2}\{R(E_2,E) + \rho(E_2,E)\}\,dE\right]dE_1\,dE_2\,dt$$

2. The electron in state E may radiate or lose energy by collision loss and change $f_2(E,E_1;t)$ to $f_2(E_2,E_1;t)$. The measure of this contribution to the probability magnitude is given by

$$\left[\int_{E_2}^\infty f_2(E,E_1;t)\{R(E,E_2) + \rho(E,E_2)\}\,dE\right]dE_1\,dE_2\,dt$$

3. The electron in state E may radiate or lose energy by collision loss and change $f_2(E,E_2;t)$ to $f_2(E_1,E_2;t)$. The measure of this contribution to the probability magnitude is given by

$$\left[\int_{E_1}^\infty f_2(E,E_2;t)\{R(E,E_1) + \rho(E,E_1)\}\,dE\right]dE_1\,dE_2\,dt$$

4. The photon of energy E may produce a pair of electrons, one of which has an energy E_2, or a photon of energy E associated with the mixed-product density $h_1(E_2,E_1;t)$ may produce a pair of electrons, one of which has an energy E_1. The measure of this contribution to $f_2(E_1,E_2;t)$ is given by

$$\left[2\int_{E_2}^\infty h_1(E_1,E;t)R'(E,E_2)\,dE + 2\int_{E_1}^\infty h_1(E_2,E)R'(E,E_1)\,dE\right]dE_1\,dE_2\,dt$$

In the above expression the factor 2 occurs because the function $R'(E,E')$ is symmetric as regards the production of positive and negative electrons.

5. A photon of energy $E_1 + E_2$ may produce a pair of electrons one of which has energy E_1 and the other energy E_2. The measure of this contribution is given by

$$2g_1(E_1 + E_2;\,t)R'(E_1 + E_2,\,E_1)\,dE_1\,dE_2\,dt$$

This contribution to the change in $f_2(E_1,E_2;t)$ reflects the fact that a photon of energy $E_1 + E_2$ producing an electron of energy E_1 will also produce an electron of energy E_2.

From the above we obtain the following integrodifferential equation for $f_2(E_1,E_2;t)$:

$$\frac{\partial f_2(E_1,E_2;t)}{\partial t} = -f_2(E_1,E_2;t)\left[\int_0^{E_1}\{R(E_1,E) + \rho(E_1,E)\}\,dE\right.$$

$$\left. + \int_0^{E_2}\{R(E_2,E) + \rho(E_2,E)\}\,dE\right]$$

$$+ \int_{E_1}^{\infty} f_2(E,E_2;t)\{R(E,E_1) + \rho(E,E_1)\}\,dE$$

$$+ \int_{E_2}^{\infty} f_2(E,E_2;t)\{R(E,E_2) + \rho(E,E_2)\}\,dE$$

$$+ 2\int_{E_2}^{\infty} h_1(E_1,E;t)R'(E,E_2)\,dE$$

$$+ 2\int_{E_1}^{\infty} h_1(E_2,E)R'(E,E_1)\,dE$$

$$+ 2g_1(E_1 + E_2; t)R'(E_1 + E_2; E_1) \quad (5.57)$$

Similar reasoning leads to the following integrodifferential equations for $h_1(E_1,E_2;t)$, $h_1(E_2,E_1;t)$, and $g_2(E_1,E_2;t)$:

$$\frac{\partial h_1(E_1,E_2;t)}{\partial t} = -h_1(E_1,E_2;t)\left[\int_0^{E_1}\{R(E_1,E) + \rho(E_1,E)\}\,dE + \int_0^{E_2} R'(E_2,E)\,dE\right]$$

$$+ \int_{E_1}^{\infty} h_1(E,E_2;t)\{R(E,E_1) + \rho(E,E_1)\}\,dE$$

$$+ 2\int_{E_1}^{\infty} g_2(E,E_2;t)R'(E,E_1)\,dE$$

$$+ \int_{E_2}^{\infty} f_2(E_1,E;t)R(E, E - E_2)\,dE$$

$$+ f_1(E_1 + E_2; t)R(E_1 + E_2, E_1) \quad (5.58)$$

$$\frac{\partial h_1(E_2,E_1;t)}{\partial t} = -h_1(E_2,E_1;t)\left[\int_0^{E_1} R'(E_1,E)\,dE + \int_0^{E_2}\{R(E_2,E) + \rho(E_2,E)\}\,dE\right]$$

$$+ \int_E^{\infty} h_1(E,E_1;t)\{R(E,E_2)\,dE + \rho(E,E_2)\}\,dE$$

$$+ 2\int_{E_2}^{\infty} g_2(E,E_1;t)R'(E,E_2)\,dE$$

$$+ \int_{E_1}^{\infty} f_2(E_2,E;t)R(E, E - E_1)\,dE$$

$$+ f_1(E_1 + E_2; t)R(E_1 + E_2, E_2) \quad (5.59)$$

$$\frac{\partial g_2(E_1,E_2;t)}{\partial t} = -g_2(E_1,E_2;t)\left[\int_0^{E_1} R'(E_1,E)\,dE + \int_0^{E_2} R'(E_2,E)\,dE\right]$$

$$+ \int_{E_2}^{\infty} h_1(E,E_1;t)R(E, E - E_2)\,dE$$

$$+ \int_{E_1}^{\infty} h_1(E,E_2;t)R(E, E - E_1)\,dE \quad (5.60)$$

Because of symmetry, the product densities $f_2(E_2,E_1;t)$ and $g_2(E_2,E_1;t)$ satisfy Eqs. (5.57) and (5.60), respectively.

If the cascade is initiated by a primary electron of energy E_0, the initial conditions to be imposed are

$$f_1(E;0) = \delta(E - E_0) \qquad g_1(E;0) = 0$$
$$f_2(E_1,E_2;0) = g_2(E_1,E_2;0) = h_1(E_1,E_2;0) = 0$$

(5.61)

Since Eqs. (5.55) and (5.56) are the equations of cascade theory considered in Sec. 5.2B, the solutions obtained will satisfy the equations for $f_1(E;t)$ and $g_1(E;t)$, respectively. The complete solutions of Eqs. (5.57) to (5.60) have been obtained by Bhabha and Ramakrishnan [14] by using the method of the Mellin transformation. We refer to their paper for details.

From the theory of product density functions [cf. Eq. (5.51)] the mean-square number of electrons in the cascade with energy greater than E is given by

$$\mathscr{E}\{X^2(E;t)\} = \mathscr{E}\{X(E;t)\} + \int_E^{E_0}\int_E^{E_0} f_2(E_1,E_2;t)\,dE_1\,dE_2 \qquad (5.62)$$

In the above, E_0 is the energy of the primary electron initiating the shower. Since the variance of the number of electrons with energy above E is given by

$$\mathscr{D}^2\{X(E;t)\} = \mathscr{E}\{X^2(E;t)\} - [\mathscr{E}\{X(E;t)\}]^2 \qquad (5.63)$$

the absolute and relative fluctuations can be obtained from the Bhabha-Ramakrishnan solutions. Similar expressions can be found for the photons in the cascade by using the product densities $g_1(E;t)$ and $g_2(E_1,E_2;t)$. In Ref. 83 Ramakrishnan and Mathews give tables of the following: $\mathscr{E}\{X(E;t)\}$, $\mathscr{E}\{X^2(E;t)\}$, $\mathscr{D}^2\{X(E;t)\}$, $\mathscr{D}^2\{X(E;t)\}/\mathscr{E}\{X(E;t)\}$, and $\mathscr{D}^2\{X(E;t)\}/[\{\mathscr{E}\{X(E;t)\}\}^2 - \mathscr{E}\{X(E;t)\}]$. These tables are constructed for various values of the thickness parameter t and $\log(E_0/E)$.

2. *The Method of Regeneration Points. The Jánossy G-Equations.*
In Sec. 2.5 we introduced the concept of a regenerative process and utilized the method of regeneration points to derive the Bellman-Harris equation. In 1950, Jánossy[1] [40] utilized the method of regeneration points to derive the diffusion equations satisfied by the generating function of the probability distribution associated with a cascade process. The diffusion equations obtained by Jánossy are called the *G*-equations. The outstanding advantage of the regeneration-point approach is that it avoids the mathematical difficulties encountered when the energy parameter E is continuous and energy transitions are taken into consideration.

---

[1] Cf. also Jánossy [43].

The G-equations were first derived for a nucleon cascade, and then the form for an electron-photon cascade was considered. Hence, we will first derive the G-equations for the nucleon cascade and then consider the G-equations for the electron-photon cascade.

Before deriving the G-equations, the cross section for nucleon collisions will be discussed.[1] It is assumed that the cross section for nucleon collisions is a homogeneous function of the primary and secondary energies; hence,

$$w(E_0,E',E'')\, dE'\, dE''\, dt = w\left(\frac{E'}{E_0}, \frac{E''}{E_0}\right) \frac{dE'\, dE''}{E_0^2}\, dt \qquad (5.64)$$

is the probability that a primary particle of energy E_0 experiences a collision in the interval $(t,\, t + dt)$ giving rise to secondary particles in the energy intervals $(E',\, E' + dE')$ and $(E'',\, E'' + dE'')$. It is now assumed that

$$w(\epsilon',\epsilon'') = 0 \qquad \text{for } \epsilon' + \epsilon'' > 1,\, \epsilon' < \epsilon'' \qquad (5.65)$$

where $\epsilon' = E'/E_0$ and $\epsilon'' = E''/E_0$; hence, the total cross section per unit length of absorber given by

$$\int_0^1 \int_0^1 w(\epsilon',\epsilon'')\, d\epsilon'\, d\epsilon'' = \alpha$$

is a constant.[2]

Let $\Phi(\epsilon,n_1,n_2;t)$ denote the probability that the cascade will be made up of n_1 particles with energy greater than $\epsilon E_0\ (= E)$ and n_2 particles with energy less than ϵE_0 after the depth t, the cascade being initiated by one primary particle of energy E_0. To derive the functional equation for $\Phi(\epsilon,n_1,n_2;t)$, we proceed as follows, the reasoning being the same as that employed in the derivation of the Bellman-Harris equation.

Consider the first collision the primary particle experiences as being a regeneration point for the stochastic cascade process. The probability that the primary particle traverses a thickness $t - \tau$ without collision is $e^{-\alpha(t-\tau)}$, and the probability that a particle will experience a collision in the next interval of length $d\tau$, resulting in a division of its energy into the intervals $d\epsilon'$ and $d\epsilon''$, is from (5.64)

$$w(\epsilon',\epsilon'')\, d\epsilon'\, d\epsilon''\, d\tau \qquad (5.66)$$

Now, in order to have at depth t a cascade with $n = n_1 + n_2$ particles, it is necessary that the particles with energies ϵ' and ϵ'' develop independent cascades giving rise to $n_1 = n_1' + n_1''$ and $n_2 = n_2' + n_2''$

---

[1] Cf. Heitler and Jánossy [35,36].

[2] For the case of infinite cross sections we refer to Harris [32,33].

particles, respectively. If we now sum over all possible divisions of the cascade, we obtain

$$
\Phi(\epsilon,n;t) = \Phi(\epsilon,n_1,n_2;t)
$$

$$
= \int_0^t e^{-\alpha(t-\tau)}\,d\tau \sum_{\substack{n_1'+n_1''=n_1 \\ n_2'+n_2''=n_2}} \int_0^\infty \int_0^\infty \Phi\left(\frac{\epsilon}{\epsilon'}\,,\,n_1',n_2';\tau\right)
$$

$$
\times\,\Phi\left(\frac{\epsilon}{\epsilon''}\,,\,n_1'',n_2'';\tau\right) w(\epsilon',\epsilon'')\,d\epsilon'\,d\epsilon'' \tag{5.67}
$$

Now let

$$
G(\epsilon,z;t) = \sum_{n=0}^\infty \Phi(\epsilon,n;t)z^n \tag{5.68}
$$

be the generating function of $\Phi(\epsilon,n;t)$. In (5.68) we have put $n = (n_1,n_2)$ and $z = (z_1,z_2)$. Hence from (5.67) and (5.68) we obtain

$$
G(\epsilon,z;t) = \int_0^t e^{-\alpha(t-\tau)} \int_0^\infty \int_0^\infty G\left(\frac{\epsilon}{\epsilon'}\,,\,z;\tau\right) G\left(\frac{\epsilon}{\epsilon''}\,,\,z;\tau\right) w(\epsilon',\epsilon'')\,d\epsilon'\,d\epsilon''\,d\tau \tag{5.69}
$$

If we now multiply (5.69) by $e^{\alpha t}$, differentiate with respect to t, and multiply by $e^{-\alpha t}$, we obtain the Jánossy G-equation

$$
\frac{\partial G(\epsilon,z;t)}{\partial t} + \alpha G(\epsilon,z;t) = \int_0^\infty \int_0^\infty G\left(\frac{\epsilon}{\epsilon'}\,,\,z;t\right) w(\epsilon',\epsilon'')\,d\epsilon'\,d\epsilon'' \tag{5.70}
$$

It is important to note that Eq. (5.69) is valid only for $n > 1$, since in obtaining (5.67) we did not consider the probability that the primary particle experiences no collision.[1] In case of no collision there are two possibilities which must be considered:

1. $E_0(t) > E$. In this case we have only one particle with energy greater than E at depth t, namely, the primary.

2. $E_0(t) \leqslant E$. In this case the primary continuously loses energy, so that no particle of energy greater than E can arrive at the depth t. These possibilities can be taken into consideration by adding to (5.67) the term

$$
[\Delta_n(E, E_0(t)) + \Delta_{n-1}(E, E_0(t))]e^{-\alpha t}
$$

where

$$
\Delta_n(E_1,E_2) = 1 \qquad \text{for } n = 0 \text{ and } E_1 > E_2
$$

$$
= 0 \qquad \text{otherwise}
$$

---

[1] In obtaining the Bellman-Harris equation the probability of the initial "particle" not being transformed was expressed by the term $z[1 - G(t)]$.

Hence Eq. (5.69) becomes

$$G(\epsilon,z;t) = \int_0^t e^{-\alpha(t-\tau)}$$

$$\times \int_0^\infty \int_0^\infty G\left(\frac{\epsilon}{\epsilon'}, z;\tau\right) G\left(\frac{\epsilon}{\epsilon''}, z;\tau\right) w(\epsilon',\epsilon'')\, d\epsilon'\, d\epsilon''\, d\tau$$

$$+ [\Delta_0(E,E_0(t)) + z\Delta_0(E,E_0(t))]e^{-\alpha t} \qquad (5.71)$$

This equation is valid for $n \geqslant 0$. We note, however, that Eq. (5.70) remains unchanged. This is because multiplication of (5.71) by $e^{\alpha t}$ and subsequent differentiation with respect to t eliminates the term added to Eq. (5.71).

To determine the initial condition to be imposed, we assume that the cascade is initiated by a single primary of energy E_0, i.e., with $\epsilon = 1$. Hence

$$\Phi(\epsilon,n_1,n_2;0) = \delta(1-n_1)\delta(n_2) \qquad \text{for } \epsilon < 1$$

$$= \delta(n_1)\delta(1-n_2) \qquad \text{for } \epsilon > 1 \qquad (5.72)$$

Therefore the G-equation must satisfy the initial condition

$$G(\epsilon,z_1,z_2;0) = z_1 \qquad \text{for } \epsilon < 1$$

$$= z_2 \qquad \text{for } \epsilon > 1 \qquad (5.73)$$

We now consider the G-equations for electron-photon cascades. In this case let $\Phi^{(i)}(\epsilon,n;t)$ denote the probability that at thickness t there are n particles in the cascade. The superscript $i = 1$ if the primary particle is an electron, and $i = 2$ if the primary particle is a photon. In the definition of $\Phi^{(i)}(\epsilon,n;t)$ we have put $n = (n_1,n_2,n_3,n_4)$, where n_1 and n_2 represent the number of electrons in the cascade with energies greater than ϵE_0 and less than ϵE_0, respectively, and n_3 and n_4 represent the number of photons in the cascade with energies greater than ϵE_0 and less than ϵE_0, respectively.

We first consider the G-equation for Approximation A. In this case $\Phi^{(i)}(\epsilon,n;t)$ does not depend explicitly on the energy of the primary particle. Let

$$w^{(1)}(\epsilon)\, d\epsilon\, dt$$

denote the probability that in the interval $(t, t + dt)$ an electron of energy E omits a photon of energy ϵE in the $(E\epsilon, E\epsilon + E\, d\epsilon)$ and let

$$w^{(2)}(\epsilon)\, d\epsilon\, dt$$

denote the probability that in the interval $(t, t + dt)$ a photon of energy E produces a pair of electrons, one of which has energy ϵE. The functions $w^{(1)}(\epsilon)$ and $w^{(2)}(\epsilon)$ are the Bethe-Heitler expressions for Bremsstrahlung and pair production, respectively. Let

$$\alpha_i = \int_0^1 w^{(i)}(\epsilon)\, d\epsilon \qquad i = 1, 2$$

denote the total cross section per unit length of absorber. Since the Bremsstrahlung cross section diverges, it is assumed that $w^{(1)}(\epsilon) = 0$ for $0 < \epsilon < k$, where k is small.

By using the method of regeneration points as in the case of the nucleon cascade, we obtain

$$\frac{\partial G^{(i)}(\epsilon,z;t)}{\partial t} + \alpha_i G^{(i)}(\epsilon,z;t)$$
$$= \int_0^1 G^{(1)}\left(\frac{\epsilon}{\epsilon'}, z;t\right) G^{(3-i)}\left(\frac{\epsilon}{1-\epsilon'}, z;t\right) w^{(i)}(\epsilon')\, d\epsilon' \quad (5.74)$$

where

$$G^{(i)}(\epsilon,z;t) = \sum_{n=0}^{\infty} \Phi^{(i)}(\epsilon,n;t) z^n \qquad i = 1, 2 \tag{5.75}$$

is the generating function of $\Phi^{(i)}(\epsilon,n;t)$. In (5.75) we have put $z = (z_1, z_2, z_3, z_4)$. Equations (5.74) are the Jánossy G-equations for electron-photon cascades under Approximation A. The initial conditions to be imposed are

$$G^{(i)}(\epsilon, z_1^*, z_2^*, 0) = z_i^* \qquad \text{for } \epsilon \leqslant 1$$
$$= 1 \qquad \text{for } \epsilon > 1 \tag{5.76}$$

where $z_1^* = (z_1, z_2)$ and $z_2^* = (z_3, z_4)$.

For electron-photon cascades under Approximation B, we put

$$\beta_i = \beta \qquad \text{for } i = 1$$
$$= 0 \qquad \text{for } i = 2 \tag{5.77}$$

where β is the amount of energy lost per unit path by an electron due to collisions. The Jánossy G-equations for electron-photon cascades under Approximation B are

$$\frac{\partial G^{(i)}(\epsilon,z;t)}{\partial t} + \alpha_i G^{(i)}(\epsilon,z;t) + \beta_i \frac{\partial G^{(i)}(\epsilon,z;t)}{\partial \epsilon}$$
$$= \int_0^1 G^{(1)}\left(\frac{\epsilon}{\epsilon'}, z,t\right) G^{(3-i)}\left(\frac{\epsilon}{1-\epsilon'}, z; t\right) w^{(i)}(\epsilon')\, d\epsilon' \quad (5.78)$$

The G-equations were introduced in cascade theory as a means of studying the fluctuation problem; hence, several authors have concentrated on obtaining solutions for the moments, and not on methods of solving the G-equations. Lopuszański [51–53] has developed a direct method of solving the G-equations and has used the method to obtain solutions of the G-equations under Approximations A and B. However, Messel [61] has shown that the solutions obtained by Lopuszański are only formal expansions in terms of factorial moments. We refer to the above papers for details.

In our study of the Bellman-Harris process, we observed that the moments of the random variable could be obtained by differentiation of the functional equation for the generating function and solving the resulting integral equations. The same procedure is applicable to the G-equations.[1] For example, let $S_{ij}(\epsilon,t)$ denote the mean or expected number of particles of type j at thickness $t > 0$ with energy greater than ϵE_0 due to a single incident particle of type i with energy E_0. As before, $i,j = 1, 2$, where 1 denotes electrons and 2 denotes photons. It can be shown that, under Approximation A, $S_{ij}(\epsilon,t)$ satisfies the integral equation

$$S_{ij}(\epsilon,t) = \int_0^t e^{\alpha_i(t-\tau)} \int_0^1 \left[S_{1j}\left(\frac{\epsilon}{\epsilon'}, \tau\right) \right.$$

$$\left. + S_{3-i,j}\left(\frac{\epsilon}{1-\epsilon'}, \tau\right) w^{(i)}(\epsilon')\, d\epsilon' \right] d\tau + \delta_{ij} e^{-\alpha_i t} \quad (5.79)$$

for $t \geqslant 0$, $0 \leqslant \epsilon < 1$.

In a series of papers, Urbanik, Lopuszański, and others have obtained results on the asymptotic behavior of the distribution function $\Phi^{(i)}(\epsilon,n_1,n_2;t)$ and the mean $S_{ij}(\epsilon,t)$ under Approximation A. In particular, Urbanik [103] has shown that, for $0 < \epsilon < 1$, $i,j = 1,2$,

$$\lim_{t\to\infty} e^{\lambda t} S_{ij}(\epsilon,t) = 0 \qquad \text{for } \lambda < \min(\alpha_1,\alpha_2)$$

$$= \infty \qquad \text{for } \lambda \geqslant \min(\alpha_1,\alpha_2) \quad (5.80)$$

where α_1 is the total cross section for Bremsstrahlung and α_2 is the total cross section for pair production. It has also been shown that $S_{ij}(\epsilon,t)$ can be expressed in the form

$$S_{ij}(\epsilon,t) = e^{-\alpha_2 t} W_{ij}(\epsilon,t) \quad (5.81)$$

where the function $W_{ij}(\epsilon,t)$ has the properties:

(a) $\lim_{t\to\infty} e^{-\beta t} W_{ij}(\epsilon,t) = 0$ for each $\beta > 0$

(b) $\lim_{t\to\infty} W_{ij}(\epsilon,t) = \infty$

Let $\Phi_{ij}(\epsilon;t)$ denote the probability of finding at least one particle of type j at thickness t with energy greater than ϵE_0 and an arbitrary number of particles of type $(3 - j)$ due to a single primary particle of type i with energy E_0. Lopuszański [57], by utilizing the results of

[1] Cf. Jánossy and Messel [44].

Urbanik, has shown that for large depth of the absorber the mean number of particles $S_{ij}(\epsilon,t)$ is of the same order of magnitude as the probability distribution $\Phi_{ij}(\epsilon;t)$. This result is based on the inequality

$$\left(\frac{1}{\epsilon}\right)^{-1} S_{ij}(\epsilon,t) \leqslant \Phi_{ij}(\epsilon;t) \leqslant S_{ij}(\epsilon,t)$$

By writing $\Phi_{ij}(\epsilon;t)$ in the form

$$\Phi_{ij}(\epsilon,t) = e^{-\alpha_2 t} V_{ij}(\epsilon,t) \qquad 0 < \epsilon < 1 \tag{5.82}$$

where $V_{ij}(\epsilon,t)$ tends to infinity, as $t \to \infty$, more slowly than an exponential function, the above inequality yields

$$\lim_{t \to \infty} e^{-\beta t} V_{ij}(\epsilon,t) \leqslant \lim_{t \to \infty} e^{-\beta t} W_{ij}(\epsilon,t) = 0 \tag{5.83}$$

Similarly, we obtain

$$\lim_{t \to \infty} V_{ij}(\epsilon,t) \geqslant \left(\frac{1}{\epsilon}\right)^{-1} \lim_{t \to \infty} W_{ij}(\epsilon,t) = \infty \tag{5.84}$$

Eqs. (5.82) to (5.84) represent a generalization of Urbanik's result for the distribution function $\Phi_{ij}(\epsilon,t)$.

Konwent and Lopuszański [49] and H. Stachowiak [100] have also investigated the asymptotic behavior of the ratio of the mean number of photons to the mean number of electrons. In particular, it has been shown that

$$\lim_{t \to \infty} \frac{S_{i,2}(\epsilon',t)}{S_{i,1}(\epsilon,t)} = 0 \qquad \epsilon' > \epsilon \tag{5.85}$$

and

$$\lim_{t \to \infty} \frac{S_{i,2}(\epsilon',t)}{S_{i,1}(\epsilon,t)} = \infty \tag{5.86}$$

The second result can be interpreted as follows: The ratio of the mean number of photons to the mean number of electrons tends to infinity as the absorber depth tends to infinity, provided the threshold energy of the instrument detecting the particles is the same for electrons and photons. That is, if the depth of absorber is sufficiently large, the number of detected photons is greater than the number of detected electrons, and this ratio will increase with increasing depth. The first result can be interpreted similarly: The ratio of the mean number of photons with energy greater than $\epsilon' E_0$ to the mean number of electrons with energy ϵE_0, where $\epsilon' > \epsilon$, tends to zero as the absorber depth tends to infinity. It is of interest to note that the above results are independent of the type of particle initiating the cascade. The dependence of the above limits on the type of particle initiating the cascade is discussed in Ref. 100.

Let $\Phi_i(\epsilon,n_1,n_2,t)$ denote the probability of finding n_1 electrons and n_2 photons with energies greater than ϵE_0 in a cascade at thickness t, if the primary particle was of type i and energy E_0. It has been shown in Ref. 49 that for large depth of absorber

$$\Phi_i(\epsilon,0,0;t) \sim 1$$

$$\Phi_i(\epsilon,0,1;t) \sim S_{i,2}(\epsilon,t) \tag{5.87}$$

$$\Phi_i(\epsilon,n_1,n_2;t) = 0(\Phi_i(\epsilon,0,1;t))$$

where $(n_1,n_2) \neq (0,1)$ and $n_1 + n_2 > 0$. The above results give the asymptotic solution of the distribution function for t tending to infinity. Hence, (5.87) is the first approximation for the distribution function for very large depth of absorber.

The above results also admit an interpretation in terms of conditional probabilities. If we put

$$\Phi_i^*(\epsilon,n_1,n_2;t) = \frac{\Phi_i(\epsilon,n_1,n_2;t)}{\displaystyle\sum_{n_1+n_2>0}\sum \Phi_i(\epsilon,n_1,n_2;t)} \tag{5.88}$$

then (5.87) yields

$$\Phi_i(\epsilon,0,1;t) \sim 1 \qquad \Phi_i^*(\epsilon,n_1,n_2;t) \sim 0 \tag{5.89}$$

This result can be interpreted as follows: If the depth of absorber is sufficiently large and if at least one particle has been detected, we may be sure that only one particle was detected, and that this particle was a photon.

In obtaining the above asymptotic results, (5.85) to (5.89), no assumption was made concerning the manner in which the cascade developed. However, it was assumed that the differential cross sections were homogeneous functions of the energies and possessed no singularities and that in the interval $0 \leqslant E \leqslant E_0$ no finite subintervals exist for which $w^{(i)}$ $(i = 1,2)$ vanishes (cf. [49]). The essential condition is that $\alpha_2 < \alpha_1$; that is, the total cross section for pair production is less than the total cross section for Bremsstrahlung.

In Ref. 25 Czerwonko extends the results obtained by Lopuszanski, Konwent, and Stachowiak by considering the asymptotic behavior of the higher factorial moments of electron-photon cascades. In particular, he shows that the main factor determining the development of electron-photon cascades at large depths of the absorber is the product of the total number of particles and the threshold energy.

We remarked earlier that the G-equations were introduced in cascade theory primarily as a means of solving the fluctuation problem. Therefore, the purpose was to obtain explicit expressions for the first two moments. It has been shown by Messel and Potts [67] that the G-equations are not needed in cascade theory for the solution of the

tion problem, since it is possible to obtain expressions for the ...ments without recourse to these equations. The method developed by Messel and Potts, which utilizes a simple relation between two fundamental equations in cascade theory, is discussed in the section on nucleon cascades.

3. *The Method of Characteristic Functionals.* In the theory of stochastic processes the characteristic functional plays a role similar to that of the characteristic function. In particular, it can be used to study the joint distribution of a continuous infinity of random variables. The characteristic functional, which appears to be a very powerful tool for studying the structure of stochastic processes, was introduced in cascade theory by Bartlett and Kendall [4]. For a rigorous and detailed discussion of the characteristic functional we refer to Bochner [16].

Let the random variable $X(E;t)$ denote the number of particles at thickness t with energy greater than E; then, as before, we denote by $-dX(E;t)$ the random variable representing the number of particles in the interval $(E, E + dE)$. The *characteristic functional* is defined as

$$\mathscr{C}[\alpha(E;t)] = \mathscr{E}\left\{\exp\left\{-i\int_E \alpha(E)\,dX(E;t)\right\}\right\} \qquad (5.90)$$

where $\alpha(E)$ is an arbitrary real function of the energy E, which is the argument of the functional $\mathscr{C}$.

We now consider the application of the method of characteristic functionals in the study of electron-photon cascades. It is clear that in the case of electron-photon cascades the characteristic functional will have two arbitrary argument functions, say $\mu(E)$ and $\eta(E)$, associated with the random variables representing the number of electrons and photons, respectively. Hence, we put

$$\mathscr{C}_i[\mu(E),\eta(E);t] = \mathscr{E}\left\{\exp\left\{-i\int_E [\mu(E)\,dX_i(E;t) + \eta(E)\,dY_i(E;t)]\right\}\right\} \quad (5.91)$$

where $-dX_i(E;t)$ and $-dY_i(E;t)$ are the random variables representing the number of electrons and photons, respectively, in the interval $(E, E + dE)$. The subscript $i = 1$ if the cascade is initiated by an electron, and $i = 2$ if the cascade is initiated by a photon.

Having defined the characteristic functional, application of the method of regeneration points leads to the following integral equation for $\mathscr{C}_i[\mu(E),\eta(E);t]$:

$$\mathscr{C}_i[\mu(E),\eta(E),E_0;t] = \int_0^{E_0}\int_0^t \exp\{-\alpha_i(E_0)\tau\}\mathscr{C}_1[\mu(E), \eta(E), E'; t - \tau]$$

$$\times \mathscr{C}_{3-i}[\mu(E), \eta(E), E_0 - E'; t - \tau]w^{(i)}(E',E_0)\,dE'\,d\tau$$

$$+ \exp\{i\mu(E_0) - \alpha_i(E_0)t\} \quad (5.92)$$

In the above equation, $w^i(E',E_0)$ is the differential cross section, $\alpha_i(E_0)$ is the total cross section. Differentiation of (5.92) yields following integrodifferential equation for the characteristic functional:

$$\frac{\partial \mathscr{C}_i[\mu(E),\eta(E),E_0;t]}{\partial t} = -\alpha_i(E_0)\mathscr{C}_i[\mu(E),\eta(E),E_0;t]$$
$$+ \int_0^{E_0} \mathscr{C}_1[\mu(E),\eta(E),E';t]\mathscr{C}_{(3-i)}[\mu(E),\eta(E),E_0-E';t]$$
$$\times w^{(i)}(E',E_0)\,dE' \qquad i = 1, 2 \tag{5.93}$$

From the definition of the probability generating function and the characteristic functional, it should be clear that it is possible to obtain the former from the latter. To demonstrate this, let $\Phi(E_0,E,n;t)$ denote the probability that there are n electrons at depth t with energies greater than or equal to E due to a primary particle of energy E_0, and let $G(E_0,E,s;t)$ denote the generating function of the probabilities $\Phi(E_0,E,n;t)$. Then

$$G(E_0,E,s;t) = \sum_{n=0}^{\infty} \Phi(E_0,E,n;t)s^n = C[\alpha_s(E'),E_0;t] \tag{5.94}$$

where
$$\alpha_s(E') = \frac{1}{i}\log s \qquad \text{for } E' \geqslant E$$

$$= 0 \qquad \text{for } E' < E$$

Similarly, the kth moments of the distribution are given by

$$S_k(E_0,E;t) = -\frac{\partial^k H(E_0,E,s;t)}{\partial s^k}\bigg]_{s=0} \tag{5.95}$$

where
$$H(E_0,E,s;t) = G(E_0,E,e^{-s};t) = C[\alpha_s(E'),E_0;t] \tag{5.96}$$

with
$$i\alpha_s(E') = -s \qquad \text{for } E' \geqslant E$$

$$= 0 \qquad \text{for } E' < E$$

The method of the characteristic functional will be used in Sec. 5.4, where we discuss the ionization cascade. For other applications in cascade theory we refer to Scott [96].

5.3 Nucleon Cascades

A. Introduction. In Sec. 5.1 we gave a brief discussion of nucleon cascades. It was stated that nucleon cascades result from the penetration of nuclei by high-energy nucleons (neutrons or protons), the nucleons that penetrate the nuclei causing the formation of additional nucleons, which in turn collide with other nuclei of the medium. In this section we consider the stochastic theory of nucleon cascades, with special reference to the one-dimensional case. In the treatment of nucleon cascades only one type of particle is considered, since it is not customary to distinguish between neutrons and protons. We will consider, in particular, the development of nucleon cascades in homogeneous nuclear matter and in a finite absorber. The fluctuation problem for nucleon cascades is also considered.

Before studying the development of a nucleon cascade, we state the assumptions concerning the nucleon-nucleon collisions and the associated cross sections. It is assumed that all high-energy nucleon collisions take place between two and only two nucleons. Also, for energies greater than several Bev (billion electron volts), every nucleon-nucleon collision will be inelastic, leading to the production of a recoil and a secondary nucleon with a certain fractional energy loss into meson production. Hence, in a collision of a nucleon of energy E_0 with a nucleon at rest we can put $E_0 = E_1 + E_2 + E_m$, where E_1 and E_2 are the energies of the secondary and recoil nucleons and E_m is the energy loss of the primary nucleon into a meson or mesons.

It is further assumed that for nucleon-nucleon collisions the cross section is a homogeneous function of the primary and secondary energies only. This assumption, which is based on the Heitler-Jánossy theory [35], is analogous to the assumption made in the theory of electron-photon cascades, since in that case the Bethe-Heitler cross sections for Bremsstrahlung and pair production are homogeneous functions of the primary and secondary energies only. Therefore, we have for the probability that in a nucleon-nucleon collision a primary nucleon of energy E_0 should give rise to secondary and recoil nucleons with energies E_1 and E_2, respectively, and an energy loss E_m into meson production

$$w(E_1, E_2; E_0)\, dE_1\, dE_2 = w\left(\frac{E_1}{E_0}, \frac{E_2}{E_0}\right) E_0^{-2}\, dE_1\, dE_2$$

where $E_0 = E_1 + E_2 + E_m$. From the above we have

$$\int_0^1 \int_0^{1-E_1/E_0} w\left(\frac{E_1}{E_0}, \frac{E_2}{E_0}\right) \frac{dE_1\, dE_2}{E_0^2} = k \text{ (constant)}$$

Hence the total cross section is independent of the primary and secondary energies. The function $w(E_1/E_0, E_2/E_0)$ has the following general properties: (1) Since it is impossible to distinguish between the recoil and secondary nucleons after collision, $w(E_1/E_0, E_2/E_0)$ must be a symmetric function in E_1 and E_2; (2) $w(E_1/E_0, E_2/E_0)$ must vanish whenever either E_1 or E_2 is equal to zero and whenever E_1 or E_2 is equal to the primary energy E_0, since in every nucleon-nucleon collision two nucleons are emitted and the collisions are assumed to be inelastic. Furthermore, $E_1 + E_2 \leqslant E_0$, and $w(E_1/E_0, E_2/E_0)$ must be positive definite.

For a detailed discussion of nucleon cascades we refer to Messel [60].

B. Equations for Nuclecn Cascades in Homogeneous Nuclear Matter and in a Finite Absorber. In this section we consider the fundamental equations for nucleon cascades in homogeneous nuclear matter and in a finite absorber. A nucleon cascade developing within nuclear matter is called a *cascade in homogeneous nuclear matter*, and a nucleon cascade developing in an absorber is called a *cascade in a finite absorber*.

1. *Cascade in Homogeneous Nuclear Matter.* Let $F_n(\epsilon_1, \epsilon_2, \ldots, \epsilon_n; t) \, d\epsilon_1 \, d\epsilon_2, \ldots, d\epsilon_n$ denote the differential probability of finding n nucleons in the intervals $(\epsilon_i, \epsilon_i + d\epsilon_i)$, $\epsilon_i = E_1/E_0$, $i = 1, 2, \ldots, n$, and none elsewhere, at depth t in homogeneous nuclear matter, due to a single incident primary nucleon of unit energy. Messel [59] has shown that $F_n(\epsilon_1, \epsilon_2, \ldots, \epsilon_n; t)$ satisfies the diffusion equation

$$\frac{\partial F_n(\epsilon_1, \epsilon_2, \ldots, \epsilon_n; t)}{\partial t} = -n F_n(\epsilon_1, \epsilon_2, \ldots, \epsilon_n; t)$$

$$+ \sum_{i \neq j} \int_0^1 F_{n-1}(\epsilon_1', \epsilon_2', \ldots, \epsilon_{n-2}', \epsilon; t) w\left(\frac{\epsilon_j}{\epsilon}, \frac{\epsilon_i}{\epsilon}\right)^{-2} d\epsilon \quad (5.97)$$

where the primes on the ϵ_i signify that neither ϵ_i nor ϵ_j is included among the $n - 2$ energy ratios ϵ_i'. It has been shown that $F_n(\epsilon_1, \epsilon_2, \ldots, \epsilon_n; t)$ can be expressed as

$$F_n(\epsilon_1, \epsilon_2, \ldots, \epsilon_n; t) = f_n(t) A_n(\epsilon_1, \epsilon_2, \ldots, \epsilon_n) \quad (5.98)$$

where

$$f_n(t) = \frac{1}{n!} \int_0^1 \cdots \int_0^1 F_n(\epsilon_1, \epsilon_2, \ldots, \epsilon_n; t) \, d\epsilon_1 \, d\epsilon_2 \cdots d\epsilon_n$$

$$= e^{-t}(1 - e^{-t})^{n-1} \quad (5.99)$$

This function is the probability of finding n ($n \geqslant 1$) particles with arbitrary energies at depth t. It is clear that $f_n(t)$ is the Furry distribution with parameter $\lambda = 1$; hence $\sum_{n=0}^{\infty} f_n(t) = 1$, showing that $F_n(\epsilon_1, \epsilon_2, \ldots, \epsilon_n; t)$ is normalized to unity. From (5.97) and (5.98) we see that the function $A(\epsilon_1, \epsilon_2, \ldots, \epsilon_n)$ is given by

$$A_n(\epsilon_1, \epsilon_2, \ldots, \epsilon_n) = \frac{1}{n-1} \sum_{i \neq j} A_{n-1}(\epsilon_1', \epsilon_2', \ldots, \epsilon_{n-2}', \epsilon) w\left(\frac{\epsilon_i}{\epsilon}, \frac{\epsilon_j}{\epsilon}\right) \epsilon^{-2} d\epsilon \quad (5.100)$$

To solve the diffusion equation for F_n, we take advantage of the fact that F_n can be represented in the form (5.98), with $f_n(t)$ known, and introduce the Mellin transform of A_n. Let $M_n(s_1, s_2, \ldots, s_n) = \mathcal{M}\{A_n(\epsilon_1, \epsilon_2, \ldots, \epsilon_n)\}$. From (5.100) we have

$$M_n(s_1, s_2, \ldots, s_n) = \frac{1}{n-1} \sum_{i \neq j} W(s_i, s_j) M_{n-1}(s_1', s_2', \ldots, s_{n-2}', s_i + s_j) \quad (5.101)$$

where

$$W(s_i, s_j) = \int_0^1 \int_0^{1-\epsilon_i} \epsilon_i^{s_i} \epsilon_j^{s_j} w(\epsilon_i, \epsilon_j) \, d\epsilon_i \, d\epsilon_j$$

Assuming that the cascade is initiated by a single nucleon of energy E_0 at $t = 0$, we have the initial condition

$$F_1(\epsilon; 0) = A_1(\epsilon) = \delta(1 - \epsilon) \quad (5.102)$$

hence

$$M_1(s) = 1$$

The solution of (5.101) is

$$M_n(s_1, s_2, \ldots, s_n) = \frac{2^{n-1}}{(n-1)!} \Lambda_n(s_1, s_2, \ldots, s_n) \quad (5.103)$$

where

$$\Lambda_n(s_1, s_2, \ldots, s_n) = \prod_{k=n-1}^{1} \Sigma' W(s_k, s_{k+1} + \cdots + s_n) \quad (5.104)$$

In (5.104) the prime on the summation sign signifies summation over all combinations of the $k + 1$ transform variables $(s_1, \ldots, s_k, s_{k+1} + \cdots + s_n)$ taken two at a time.

From (5.99), (5.103), (5.104), and the application of the inversion theorem for Mellin transforms, we obtain the solution

$$F_n(\epsilon_1, \epsilon_2, \ldots, \epsilon_n; t) = \frac{2^{n-1}}{(n-1)!} L_n \Lambda_n(s_1, s_2, \ldots, s_n) f_n(t) \quad (5.105)$$

where

$$L_n = \frac{1}{(2\pi i)^n} \int_{c_1 - i\infty}^{c_1 + i\infty} \cdots \int_{c_n - i\infty}^{c_n + i\infty} \epsilon_1^{-s_1+1} \cdots \epsilon_n^{-s_n+1} \, ds_1 \cdots ds_n \quad (5.106)$$

is the inverse Mellin transform operator.

From a knowledge of $F_n(\epsilon_1, \epsilon_2, \ldots, \epsilon_n; t)$ it is possible, at least in principle, to obtain all other distribution functions occurring in the theory of nucleon cascades in homogeneous nuclear matter. That this can be done is of great advantage, for it enables one to obtain distribution functions which are of greater interest in experimental situations than $F_n(\epsilon_1, \epsilon_2, \ldots, \epsilon_n; t)$. For example, from $F_n(\epsilon_1, \epsilon_2, \ldots, \epsilon_n; t)$ we can obtain the probability, say $\Phi(\epsilon, n; t)$, of finding n nucleons with energies greater than or equal to ϵE_0 and an arbitrary number with energies less than ϵE_0 at depth t. This probability is given by

$$\Phi(\epsilon, n; t) = \sum_{i=0}^{\infty} (i! n!)^{-1} \int_0^\epsilon d\epsilon_1 \cdots \int_0^\epsilon d\epsilon_i \int_0^1 d\epsilon_{i+1} \cdots \int^1 d\epsilon_{i+n} \, F_{i+n}(\epsilon_1, \epsilon_2, \ldots, \epsilon_{i+n}; t)$$

$$(5.107)$$

2. *Cascade in a Finite Absorber.* Let $H_n(E_0; E_1, \ldots, E_n, t)$ $dE_1\, dE \cdots dE_n$ denote the differential probability that after a depth t a primary nucleon of energy E_0 has given rise to n nucleons with energies in the intervals $(E_i, E_i + dE_i)$, $i = 1, 2, \ldots, n$. Messel and Potts have shown that $H_n(E_0; E_1, \ldots, E_n; t)$ satisfies the diffusion equation

$$\frac{\partial H_n(E_0; E_1, \ldots, E_n; t)}{\partial t} = -n H_n(E_0; E_1, \ldots, E_n; t)$$

$$+ \sum_{i=1}^{n} \sum_{(i)} Q_i(\epsilon; E_1', \ldots, E_i') H_{n-i+1}(E_0; E_{i+1}', \ldots, E_n'; t) \quad (5.108)$$

where $\sum_{(i)}$ indicates summation over all compositions of the energies $E_1, \ldots, E_n$ into two groups $(E_1', \ldots, E_i')$ and $(E_{i+1}', \ldots, E_n')$. The function $Q_i(E_1, \ldots, E_i)$ is given by

$$Q_i(E_1, \ldots, E_i) = N(\gamma_A) F_i(E_1, \ldots, E_i; t) \quad (5.109)$$

where $F_i(E_1, \ldots, E_n; t)$ is given by (5.105), and

$$N(\gamma_A) = \int_0^{\gamma_A} \frac{2x}{\gamma_A^2}\, dx$$

In the above expression γ_A is defined as the average number of collisions which a nucleon experiences on making a diametrical passage through a nucleus of atomic weight A. The cross section $Q_i(E_0; E_1, \ldots, E_i)$ given by (5.109) is the cross section for nucleon-nucleus collisions. For the development of a cascade in a finite absorber this cross section is used in place of the cross section $w(\epsilon_1, \epsilon_2)$ for nucleon-nucleon collisions. If $Q_i(E_0; E_1, \ldots, E_i)$ is normalized to unity, the depth t is measured in interaction mean free paths of the absorber concerned.

Let

$$R_n(s_1, \ldots, s_n; p)$$
$$= \int_0^\infty dE_1 \cdots \int_0^\infty dE_n \int_0^\infty e^{-pt} \epsilon_1^{s_1} \cdots \epsilon_n^{s_n} H_n(E_0; E_1, \ldots, E_n; t)\, dt \quad (5.110)$$

$$N_n(s_1, \ldots, s_n)$$
$$= \int_0^\infty dE_1 \cdots \int_0^\infty dE_n \epsilon_1^{s_1} \cdots \epsilon_n^{s_n} Q_n(E_0; E_1, \ldots, E_n) \quad (5.111)$$

and

$$h = 1 - \frac{2}{\gamma_A^2}[1 - (1 + \gamma_A)e^{-\gamma_A}]$$

Then it has been shown (cf. [66]) that the solution of the diffusion equation (5.108) satisfying the initial condition

$$H_1(E_0; E; 0) = \delta(E_0 - E) \quad (5.112)$$

is given by

$$H_n(E_0; E_1, \ldots, E_n; t) = E_0^{-n} K_n^* R_n(s_1, \ldots, s_n; p) \qquad (5.113)$$

where K_n^* is the inverse Laplace transform operating on the n-fold inverse Mellin transform of the energy ratios. In (5.113) R_{n+1} is given by

$$R_{n+1} = (p + h)^{-1} \sum_c \prod_{j=m}^{1} \sum_{c'} N_{n(j)+1}$$

$$\times (s_{q(j-1)+1}, \ldots, s_{q(j)}, s_{q(j)+1} + \cdots + s_{n+1})[p + (q(j) + 1)h]^{-1} \qquad (5.114)$$

where (1) $\sum_c$ indicates the sum over the 2^{n-1} compositions of n; c is the composition $n(1), \ldots, n(m)$, with $\Sigma_{j=1}^m n(j) = n$, where m is the number of nucleon-nucleus collisions; and $n(j) + 1$ gives the number of nucleons arising from the jth nucleon-nucleus collision; (2) $q(j) = \Sigma_{i=1}^j n(i)$, $q(0) = 0$, and from (1) $q(m) = n$; (3) $\sum_{c'}$ indicates the sum over all combinations $C_{n(j)+1}^{q(j)+1}$ of the $q(j) + 1$ symbols $s_1, \ldots, s_{q(j)}, s_{q(j)+1} + \cdots + s_{n+1}$ taken $n(j) + 1$ at a time.

Needless to say, the expression for R_{n+1} is rather complicated; however, Messel and Potts [66] have given a method of "trees" which, for given n, permits R_{n+1} to be written in a somewhat simpler form.

As in the case of cascades in homogeneous nuclear matter, we can define a function $\Phi(E_0; E, n, t)$ which expresses the probability of finding n nucleons with energies greater than or equal to E and an arbitrary number with energies less than E at depth t. This probability is obtained from $H_n(E_0; E_1, \ldots, E_n; t)$ by the relation

$$\Phi(E_0; E, n; t) = \sum_{i=0}^{\infty} (i! n!)^{-1} \int_0^E dE_1 \cdots \int_0^E dE_i \int_E^{\infty} dE_{i+1}$$

$$\cdots \int_E^{\infty} dE_{i+n} H_{i+n}(E_0; E_1, \ldots, E_n; t) \qquad (5.115)$$

C. The Fluctuation Problem for Nucleon Cascades.[1] In the last section we considered the diffusion equations for certain distribution functions describing the development of nucleon cascades. As in the case of electron-photon cascades, we will not consider the distribution functions but will devote our attention to the moments of the distributions. Therefore, we now consider the various methods that have been developed for obtaining the moments.

1. *Calculation of Moments from the Diffusion Equations.* We first consider the case of a cascade in homogeneous nuclear matter. Let

---

[1] For a treatment of the fluctuation problem for electron-photon cascades by using the methods of this section we refer to Refs. 64 and 65.

$S_k(\epsilon,t)$ denote the kth factorial moment of the distribution function $\Phi(\epsilon,n;t)$ given by (5.107). By definition

$$S_k(\epsilon,t) = \sum_{i=0}^{\infty} \left[\frac{(k+i)!}{i!} \right] \Phi(\epsilon, k+i; t) \tag{5.116}$$

Messel and Potts [66] have shown that

$$S_k(\epsilon,t) = \int_{\epsilon}^{1} d\epsilon_1, \ldots, \int_{\epsilon}^{1} d\epsilon_k C_k(\epsilon_1, \ldots, \epsilon_k; t) \tag{5.117}$$

where $C_k(\epsilon_1, \ldots, \epsilon_k; t)$ satisfies the diffusion equation

$$\frac{\partial C_k(\epsilon_1, \ldots, \epsilon_k; t)}{\partial t} = -k C_k(\epsilon_1, \ldots, \epsilon_k; t)$$

$$+ \int_0^1 \sum_{i=1}^{k} C_k(\epsilon_1', \ldots, \epsilon_{k-1}', \epsilon; t) \bar{w}\left(\frac{\epsilon_i}{\epsilon}\right)^{-1} d\epsilon$$

$$+ \int_0^1 \sum_{i \neq h} C_{k-1}(\epsilon_1', \ldots, \epsilon_{k-2}', \epsilon; t) \bar{w}\left(\frac{\epsilon_i}{\epsilon}, \frac{\epsilon_h}{\epsilon}\right) \epsilon^{-2} d\epsilon \tag{5.118}$$

where

$$\bar{w}(\epsilon) = \int_0^{1-\epsilon} [w(\epsilon_1,\epsilon) + w(\epsilon,\epsilon_1)] \, d\epsilon_1$$

The function $C_k(\epsilon_1, \ldots, \epsilon_k; t) \, d\epsilon_1, \ldots, d\epsilon_k$ gives the probability of finding k nucleons in the intervals $(\epsilon_i; t_i + d\epsilon_i)$, $i = 1, \ldots, k$, and any number with arbitrary energies at depth t. If the function $C_k(\epsilon_1, \ldots, \epsilon_k; t)$ is known, then from (5.117) the moments $S_k(\epsilon,t)$ can be obtained by integrating over the k energy variables of $C_k(\epsilon_1, \ldots, \epsilon_k; t)$. The solution of (5.118) is given by (cf. [59,66])

$$C_k(\epsilon_1, \ldots, \epsilon_k; t) = K_k V_k(s_1, \ldots, s_n; p) \tag{5.119}$$

where

$$V_{k+1}(s_1, \ldots, s_{k+1}; p)$$

$$= [p + \alpha(s_1 + \cdots + s_{k+1})]^{-1} \prod_{j=k}^{1} \Sigma' W(s_j; s_{j+1} + \cdots + s_{k+1})$$

$$\times [p + \alpha(s_1) + \cdots + \alpha(s_j) + \alpha(s_{j+1} + \cdots + s_{k+1})]^{-1} \tag{5.120}$$

In (5.119) K_k is the inverse Laplace transform of the k-fold inverse Mellin transform

$$\frac{1}{(2\pi i)^k} \int_{c_1-i\infty}^{c_1+i\infty} \frac{ds_1}{s_1} \cdots \int_{c_k-i\infty}^{c_k+i\infty} \frac{ds_k}{s_k} e^{-(s_1 + \cdots + s_k)}$$

operating on $V_k(s_1, \ldots, s_k; p)$, in (5.120) the function $W(s_j,s_k)$ is the same as before, and the function $\alpha(s) = 1 - W(0,s) - W(s,0)$. Also, in (5.120) Σ' indicates summation over all combinations C_2^{j+1} of the $j+1$

transform variables $s_1, \ldots, s_j, s_{j+1} + \cdots + s_k$ taken two at a time. Hence, from (5.117) the kth moment $S_k(\epsilon,t)$ is given by

$$S_k(\epsilon,t) = K_k V_k(s_1, \ldots, s_k; p) \tag{5.121}$$

In the case of a nucleon cascade in a finite absorber, the kth factorial moment is given by

$$T_k(E_0;E;t) = \sum_{i=0}^{\infty} \left[\frac{(k+i)!}{i!} \right] \Phi(E_0; E, k+i; t) \tag{5.122}$$

where $\Phi(E_0;E,n;t)$ is given by (5.115). The factorial moments in this case are obtained by using the relation

$$T_k(E_0;E;t) = \int_E^{\infty} dE_1 \cdots \int_E^{\infty} dE_k\, J_k(E_0; E_1, \ldots, E_k; t) \tag{5.123}$$

where $J_k(E_0; E_1, \ldots, E_n; t)\, dE_1 \cdots dE_k$ defines the probability of finding k particles with energies in the intervals $(E_i, E_i + dE_i)$, $i = 1, \ldots, k$, and an arbitrary number of particles with arbitrary energies at depth t due to a single primary nucleon of energy E_0. It has been shown that $J_k(E_0; E_1, \ldots, E_k; t)$ satisfies the diffusion equation

$$\frac{\partial J_k(E_0; E_1, \ldots, E_k; t)}{\partial t} = -kJ_k(E_0; E_1, \ldots, E_k; t)$$
$$+ \sum_{i=1}^{k} \sum_{j=i}^{\infty} \sum_{(i)} [(j-i)!]^{-1} \int_0^{\infty} d\epsilon \int_0^{\infty} d\xi_1 \cdots \int_0^{\infty} d\xi_{j-i}$$
$$\times\, Q_j(\epsilon; E_1', \ldots, E_i', \xi_1, \ldots, \xi_{j-i})$$
$$\times\, J_{k-i+1}(E_0; E_{i+1}', \ldots, E_k', \epsilon; t) \tag{5.124}$$

In (5.124) $\sum_{(i)}$ indicates summation over all compositions of $E_1, \ldots, E_k$ into two groups $(E_1', \ldots, E_{k-i}')$ and $(E_{k-i+1}', \ldots, E_k')$.

The solution of (5.124) is given by

$$J_k(E_0; E_1, \ldots, E_k; t) = E_0^{-k} K_k^* Y_k(s_1, \ldots, s_k; p) \tag{5.125}$$

where

$$Y_{k+1}(s_1, \ldots, s_{k+1}; p) = [p + h(s_1 + \cdots + s_{k+1})]^{-1}$$
$$\times \sum_c \sum_{j=m}^{1} \sum_{c'} B_{k(j)+1}(s_{q(j-1)+1}, \ldots, s_{q(j)}, s_{q(j)+1} + \cdots + s_{k+1})$$
$$\times [p + h(s_1) + \cdots + h(s_{q(j)}) + h(s_{q(j)+1} + \cdots + s_{k+1})]^{-1} \tag{5.126}$$

In (5.126)

$$B_k(s_1, \ldots, s_k) = \int_0^1 d\epsilon_1 \cdots \int_0^1 d\epsilon_k\, \epsilon_1^{s_1} \cdots \epsilon_k^{s_k} N(\gamma_A) C_k$$
$$h(s) = 1 - 2[1 - (1 + \gamma_A \alpha(s))e^{-\gamma_A \alpha(s)}] \cdot [\gamma_A \alpha(s)]^{-2}$$

and the summation convention is the same as that for (5.114).

Hence, from (5.123) and (5.125), the kth factorial moment is given by

$$T_k(E_0;E;t) = K_k^* Y_k(s_1, \ldots, s_k; p) \tag{5.127}$$

2. *The Jánossy G-Equations.* In Sec. 5.2C the Jánossy G-equations were derived for both nucleon and electron-photon cascades. We now consider the G-equations for a nucleon cascade in homogeneous nuclear matter and obtain the moments of the associated distribution function. Let $G(\epsilon,z;t)$ denote the generating function of the probabilities $\Phi(\epsilon,z;t)$, where $\Phi(\epsilon,n;t)$ is the probability that at depth t there are n nucleons with energies greater than ϵE_0 due to a single primary nucleon of energy E_0. It has been shown that $G(\epsilon,z;t)$ satisfies the integral equation

$$G(\epsilon,z;t) = \int_0^t e^{-\alpha(t-\tau)} \int_0^\infty \int_0^\infty G\left(\frac{\epsilon}{\epsilon'}, z; \tau\right) G\left(\frac{\epsilon}{\epsilon''}, z, \tau\right) w(\epsilon',\epsilon'') \, d\epsilon' \, d\epsilon'' \, d\tau \tag{5.128}$$

Now let $S_k(\epsilon,t)$ denote the kth factorial moment of $\Phi(\epsilon,n;t)$. Jánossy has shown that for $0 \leqslant \epsilon < 1$ the factorial moments satisfy the integral equation

$$S_k(\epsilon,t) = \int_0^t e^{-\alpha(t-\tau)} \int_0^1 \int_0^1 \sum_{n=0}^k \binom{k}{n} S_n\left(\frac{\epsilon}{\epsilon'}, \tau\right)$$

$$\times S_{k-n}\left(\frac{\epsilon}{\epsilon''}, \tau\right) w(\epsilon',\epsilon'') \, d\epsilon' \, d\epsilon'' \, d\tau + \delta_{1k} e^{-\alpha t} \tag{5.129}$$

with $S_0(\epsilon,t) = 1$. If we consider the integral equation for the mean, we have

$$m(\epsilon,t) = \int_0^t e^{-\alpha(t-\tau)} \int_0^1 \int_0^1 \left[m\left(\frac{\epsilon}{\epsilon'}, \tau\right) + m\left(\frac{\epsilon}{\epsilon''}, \tau\right) \right] w(\epsilon',\epsilon'') \, d\epsilon' \, d\epsilon'' \, d\tau + e^{-\alpha t} \tag{5.130}$$

where $m(\epsilon,t) = S_1(\epsilon,t)$. Similarly, we can obtain an integral equation for the second factorial moment. From the solutions of these integral equations the mean and variance of the number of nucleons in the cascade can be obtained.

We now consider some results due to Urbanik and Lopuszański concerning the asymptotic behavior of the probability distribution function and the factorial moments. These results are the analogues of those given in Sec. 5.2C for electron-photon cascades.

Urbanik [103,104] has shown that for $0 < \epsilon < 1$ the kth factorial moments can be written as

$$S_k(\epsilon,t) = e^{-\alpha k} W_k(\epsilon,k) \qquad k = 1, 2, \ldots, [\epsilon^{-1}] \tag{5.131}$$

and

$$\lim_{t\to\infty} e^{\lambda t} S_k(\epsilon,t) = 0 \qquad \text{for } \lambda < \alpha$$

$$= \infty \qquad \text{for } \lambda \geqslant \alpha \tag{5.132}$$

where α is the total cross section for nucleon-nucleon collisions. Since (5.132) holds for $k = 1$, the above result gives the asymptotic behavior of the mean number of nucleons and its dependence on the total cross section.

Lopuszański [54–56] has utilized the results of Urbanik in his study of the probability distribution function associated with nucleon cascades. Let

$$\Gamma(\epsilon,s;t) = 1 - G(\epsilon,s;t)$$

It has been shown that

$$\lim_{t\to\infty} \left\{ \frac{\Gamma(\epsilon',0;t)}{\Gamma(\epsilon,0;t)} \right\} = 0 \tag{5.133}$$

for $0 < \epsilon < \epsilon'$. The interpretation of (5.133) is that for very large depth of the absorber, if we detect at least one nucleon of energy greater than ϵ, we may be certain that the energy of the nucleons of the cascade is in the interval $(\epsilon, \epsilon + d\epsilon)$.

Let

$$\Phi^*(\epsilon,n;t) = \frac{\Phi(\epsilon,n;t)}{\displaystyle\sum_{m=1}^{\infty} \Phi(\epsilon,m;t)}$$

that is, $\Phi^*(\epsilon,m;t)$ is the probability of finding n nucleons of energy greater than ϵ at depth t, given that there is at least one nucleon of energy greater than ϵ at depth t. It has been shown that

$$
\begin{aligned}
\lim_{t\to\infty} \Phi^*(\epsilon,1;t) &= 1 \\
\lim_{t\to\infty} \Phi^*(\epsilon,n;t) &= 0 \qquad \text{for } n \geqslant 2
\end{aligned}
\tag{5.134}
$$

The above results obtain for $\epsilon > 0$. The interpretation of (5.134) is that for very large depth of absorber, if we find at least one nucleon with energy greater than ϵ, we may be certain that there is only one nucleon. We can also write the results of (5.134) in terms of the probability distribution function $\Phi(\epsilon,n;t)$ for large t as

$$
\begin{aligned}
\Phi(\epsilon,0;t) &\sim 1 \\
\Phi(\epsilon,1;t) &\sim S_1(\epsilon,t) = m(\epsilon,t) \\
\Phi(\epsilon,n;t) &= 0(\Phi(\epsilon,1;t)) \qquad \text{for } n > 1
\end{aligned}
\tag{5.135}
$$

These results represent the first step in an approximation procedure for large but finite depths. Lopuszański has also shown that

$$
\begin{aligned}
\Phi(\epsilon,2;t) &\sim \tfrac{1}{2}S_2(\epsilon;t) \\
\Phi(\epsilon,n;t) &= 0(\Phi(\epsilon,n;t)) \qquad n \geqslant 3
\end{aligned}
\tag{5.136}
$$

and in general

$$\Phi(\epsilon, n; t) \sim \frac{1}{n!} S_n(\epsilon; t)$$

$$\Phi(\epsilon, n; t) = o(\Phi(\epsilon, m; t)) \qquad \text{for } n > m, \ \epsilon > 0$$

$$n, m = 1, 2, \ldots$$

(5.137)

It was pointed out in Sec. 5.2C that Messel and Potts have shown that the Jánossy G-equations are not needed for the solution of the fluctuation problem and are not, therefore, needed in cascade theory. We now show that the solutions for the moments can be obtained by a direct method which utilizes a simple relation between two fundamental equations in cascade theory. The G-equations are not used in establishing this relationship.

Let $\Phi(E_0; E, n; t)$ denote the probability of finding n particles at depth t with energies greater than E_0 and an arbitrary number of particles with energies less than E_0, and let the functions $H_n(E_0; E_1, \ldots, E_n; t)$, $J_n(E_0; E_1, \ldots, E_n; t)$, and $T_n(E_0; E; t)$ be defined as in Sec. 5.3B. Hence

$$\Phi(E_0; E, n; t) = \sum_{i=0}^{\infty} (i! n!)^{-1} \int_E^{\infty} dE_1 \cdots \int_E^{\infty} dE_n \int_0^E dE_{n+1}$$

$$\cdots \int_0^E dE_{n+i} \, H_{n+i}(E_0; E_1, \ldots, E_{n+i}; t) \qquad (5.138)$$

$$T_n(E_0; E; t) = \sum_{j=0}^{\infty} \frac{(n+j)!}{j!} \, \Phi(E_0; E, n+j; t) \qquad (5.139)$$

and

$$J_n(E_0; E_1, \ldots, E_n; t) = \sum_{k=0}^{\infty} (k!)^{-1} \int_0^{\infty} dE_{n+i} \cdots \int_0^{\infty} dE_{n+k}$$

$$\times H_{n+k}(E_0; E_1, \ldots, E_{n+k}; t) \qquad (5.140)$$

By inserting (5.139) in (5.138) we have

$$T_n(E_0; E; t) = \sum_{i=0}^{\infty} \sum_{j=0}^{\infty} (i! j!)^{-1} \int_E^{\infty} dE_1 \cdots \int_E^{\infty} dE_{n+i}$$

$$\times \int_0^E dE_{n+i+1} \cdots \int_0^E dE_{n+i+j} \, H_{n+i+j}(E_0; E_1, \ldots, E_{n+i+j}; t) \qquad (5.141)$$

By putting $i = k - j$, (5.141) becomes

$$T_n(E_0; E; t) = \int_E^{\infty} dE_1 \cdots \int_E^{\infty} dE_n \sum_{k=0}^{\infty} (k!)^{-1} \int_0^{\infty} dE_{n+1}$$

$$\cdots \int_0^{\infty} dE_{n+k} \, H_{n+k}(E_0; E_1, \ldots, E_{n+k}; t) \qquad (5.142)$$

Now, from (5.140) we obtain

$$T_n(E_0;E;t) = \int_E^\infty dE_1 \cdots \int_E^\infty dE_n J_n(E_0; E_1, \ldots, E_n; t) \qquad (5.143)$$

The above relation gives the nth moment expressed in terms of the distribution function $J_n(E_0; E_1, \ldots, E_n; t)$ by integration over the n energy variables. The function $J_n(E_0; E_1, \ldots, E_n; t)$ is given by (5.125); hence, the moments of the distribution function can be obtained without the use of the Jánossy G-equations.

3. *The Jánossy Functions.* Jánossy [41] has also considered another approach to the fluctuation problem for nucleon cascades. In this approach the function $\Psi_n(E_1, \ldots, E_n)$ is studied, where $\Psi_n(E_1, \ldots, E_n) \, dE_1 \cdots dE_n$ is defined as the probability that there is one particle in the interval $(E_1, E_1 + dE_1)$, one in $(E_2, E_2 + dE_2)$, $\ldots$, one in $(E_n, E_n + dE_n)$, and none in the other energy states. We will not discuss the method of Jánossy functions, since a complete correspondence between the Jánossy functions and the product density functions has been established. Ramakrishnan [80] has shown that

$$\Psi_n(E_1, \ldots, E_n) = P(n)f_n^{(n)}(E_1, \ldots, E_n) \qquad (5.144)$$

where $P(n)$ is the probability that there are n particles in the entire energy range, and $f_n^{(n)}(E_1, \ldots, E_n)$ represents the product density of degree n, given that there are n particles in the entire energy range. Determination of $P(n)$ and $f_n^{(n)}(E_1, \ldots, E_n)$ leads to the relation

$$f_n(E_1, \ldots, E_n) = \sum_{i=n}^\infty \frac{1}{(i-n)!} \int_E \cdots \int_{E_{n+1}} \Psi_i(E_1, \ldots, E_i) \, dE_i \cdots dE_{n+1}$$

$$(5.145)$$

5.4 Ionization Cascades

A. Introduction. We now consider the theory of ionization cascades. An *ionization cascade* is the cascade of knock-on electrons produced by a fast primary particle passing through an absorber. The primary particle is a meson which by ionization gives rise to energetic secondary electrons, called *knock-on electrons*, which in turn give rise to ordinary multiplication showers (cf. [1,39]). The theory of ionization cascades will be studied by first considering the development of the knock-on electron cascade and then considering the ionization electron cascade. The results of this section are due to Moyal [71,73].[1] The method of characteristic functionals is utilized.

---

[1] Messel [61] has shown that Moyal's solution is only a formal expansion of the distribution function in terms of its factorial moments. In particular, he has shown that any procedure utilizing an iteration upon the energy variable will lead to such an expansion.

B. The Knock-on Electron Cascade. A single primary knock-on electron of energy E_0 incident on an absorber produces a *knock-on electron cascade* which is analogous to the nucleon cascade. In the case of a knock-on electron cascade, a fast primary knock-on ionizes an atom, producing thereby secondary electrons. The cascade process develops in this manner, terminating when its end products are too slow to ionize additional atoms, have escaped the absorber, or have been absorbed.

Let the random variable $X(E;t)$ denote the number of electrons in the cascade at time t with energy greater than or equal to E due to a single primary knock-on electron of energy E_0. Denote by $\mathscr{C}[\theta(E),E_0;t]$ the characteristic functional of the distribution of $X(E;t)$, that is,

$$\mathscr{C}[\theta(E),E_0;t] = \mathscr{E}\left\{\exp\left\{i\int_0^E \theta(E)\,dX(E;t)\right\}\right\} \tag{5.146}$$

where $\theta(E)$ is an arbitrary real function of the energy E. Let $f(E_0; E_1, \ldots, E_n; t)\,dE_1 \cdots dE_n$ denote the probability that the primary knock-on produces exactly n electrons with energies in the intervals $(E_i; E_i + dE_i)$, $i = 1, \ldots, n$, with the E_i so ordered that $E_1 \leqslant E_2 \leqslant \cdots \leqslant E_n$. We can now write

$$\mathscr{C}[\theta(E),E_0;t] = \sum_{n=1}^{\infty} \int_0^{E_0} dE_n \int_0^{E_n} dE_{n-1} \cdots \int_0^{E_2} dE_1$$

$$\times \exp\left\{i\sum_{j=1}^n \theta(E_j)\right\} f(E_0; E_1, \ldots, E_n; t) \tag{5.147}$$

Before deriving the equation which the characteristic functional satisfies, we state the assumptions concerning the mechanism by which the cascade develops and give the differential cross section for electron-atom collisions. Concerning the mechanism, it is assumed that:

1. The absorber is homogeneous.

2. The successive ionizing collisions are statistically independent.

3. An electron will ionize an atom and contribute two electrons to the cascade only if its energy is greater than a mean ionization potential I characteristic of the absorber atoms.

4. Escape, absorption, radiation loss, and other extraneous effects, e.g., the increase in ionization due to the acceleration of electrons in electrode fields, and photoelectric effects in the absorber or walls of its container, are neglected.

These assumptions are suitable for applications of the theory to ionization or cloud chambers.

Let $R(E',E)\,dE'$ denote the differential cross section for an electron-atom collision; that is, $R(E',E)\,dE'$ is the probability that an electron of energy E will collide with an atom of the absorber and lose energy in the interval $(E', E' + dE')$. For large E' the Thomson cross section

$$R(E',E)\, dE' = \frac{2\pi e^4 Z}{mv^2} \frac{dE'}{E'^2}$$

is a good approximation. In the above, Z is the atomic number of the absorber atoms, m is the mass, v is the velocity of the electron, and e is the electron charge.

Since every electron-atom collision may not result in ionization, it is necessary to distinguish between ionizing and nonionizing collisions. By assumption (3) an electron will ionize an atom only if its energy is greater than I, the mean ionization potential; hence we let

$$\alpha_i(E) = Nv \int_I^E R(E',E)\, dE' \qquad \text{and} \qquad \alpha_e(E) = Nv \int_0^I R(E',E)\, dE'$$

denote the rates for ionizing and nonionizing collisions, respectively. We put $\alpha(E) = \alpha_i(E) + \alpha_e(E)$. In the above expressions N is the number density of absorber atoms.

The probability densities for the energy loss E' under the condition that an ionizing or nonionizing collision has occurred are given by

$$r_i(E',E) = \frac{NvR(E',E)}{\alpha_i(E)} \qquad \text{for } E' \geqslant I$$

$$= 0 \qquad \text{for } E' < I$$

$$r_e(E',E) = \frac{NvR(E',E)}{\alpha_e(E)} \qquad \text{for } E' < I$$

$$= 0 \qquad \text{for } E' \geqslant I$$

$$\alpha(E)r(E',E) = \alpha_i(E)r_i(E',E) \qquad \text{for } E' \geqslant I$$

$$= \alpha_e(E)r_e(E',E) \qquad \text{for } E' < I$$

The equation satisfied by the characteristic functional is derived by using the method of regeneration points. Hence we consider the change in $\mathscr{C}[\theta(E),E_0;t]$ in the interval $(0,dt)$. In the interval $(0,dt)$ we consider the following mutually exclusive events which can contribute to a change in $C[\theta(E),E_0;t]$:

1. The primary electron will not experience a collision with an atom, the probability of this event being $1 - \alpha(E)\, dt$.

2. The electron collides with an atom and loses energy in the interval $(E', E' + dE')$ owing to a nonionizing collision, the probability of this event being $\alpha_e(E)r_e(E',E)\, dE'\, dt$.

3. The electron collides with an atom and loses energy in the interval $(E', E' + dE')$ owing to an ionizing collision, the probability of this

event being $\alpha_i(E)r_e(E',E)\,dE'\,dt$. The occurrence of this event results in the production of two cascade electrons with energies $(E - E')$ and $(E' - I)$, which now act as primaries for two independent cascades.

By adding the above probabilities, we obtain the functional equation

$$\mathscr{C}[E;t] = [1 - \alpha(E)\,dt]\mathscr{C}[E;\,t - dt]$$
$$+ \alpha_e(E)\,dt\int_0^I \mathscr{C}[E - E';\,t - dt]r_e(E',E)\,dE'$$
$$+ \alpha_i(E)\,dt\int_0^E \mathscr{C}[E - E';\,t - dt]\mathscr{C}[E' - I;\,t - dt]r_i(E',E)\,dE' \quad (5.148)$$

The above equation can be reduced to the integrodifferential equation

$$\frac{\partial\mathscr{C}[E;t]}{\partial t} = \alpha_e(E)\{-\mathscr{C}[E;t] + \int_0^I \mathscr{C}[E - E';\,t]r_e(E',E)\,dE'\}$$
$$+ \alpha_i(E)\{-\mathscr{C}[E;t] + \int_I^E \mathscr{C}[E - E';\,t]\mathscr{C}[E' - I;\,t]r_i(E',E)\,dE'\} \quad (5.149)$$

We first consider the solution of (5.149) when the dissipation of energy due to nonionizing collisions is neglected, i.e., we put $\alpha_e(E) = 0$. This approximation is justified in studying the high-energy part of the cascade, since for high energies ionizing collisions will occur in greater number than nonionizing collisions. Let us now assume that the functional $\mathscr{C}[E;t]$ is known for energies $E < kI$, where k is a positive integer. In this case, the integral over the interval (I,E) in (5.149) is a known function of E, say $H(E;t)$, for $kI \leqslant E \leqslant (k + 1)I$. Hence, in this interval $\mathscr{C}[E;t]$ satisfies the equation

$$\frac{\partial\mathscr{C}[E;t]}{\partial t} = -\alpha_i(E)\mathscr{C}[E;t] + \alpha_i(E)H[E;t]$$
$$= -\alpha_i(E)\mathscr{C}[E;t] + \alpha_i(E)\int_I^E \mathscr{C}[E - E';\,t]\mathscr{C}[E' - I;\,t]r_i(E',E)\,dE' \quad (5.150)$$

If the cascade is initiated by a single particle with energy E, then

$$\mathscr{C}[E;0] = e^{i\theta(E)} \quad (5.151)$$

is the initial condition which the solution of Eq. (5.150) must satisfy. Hence the solution of Eq. (5.150) is

$$\mathscr{C}[E;t] = \exp\{i\theta(E) - \alpha_i(E)t\} + \alpha_i(E)\int_0^t \exp\{-\alpha_i(E)(t - \tau)\}H(E;\tau)\,d\tau \quad (5.152)$$

Since

$$\mathscr{C}[E;t] = e^{i\theta(E)} \quad (5.153)$$

for $E < I$, that is, there is no ionizing collision in this interval of energies, the solution (5.152) can be used as an iteration relation to obtain $\mathscr{C}[E;t]$ in the successive intervals $(kI, (k+1)I)$, $k = 1, 2, \ldots$. For example, in the interval $I \leqslant E < 2I$, we have

$$\mathscr{C}[E;t] = \exp\{i\theta(E) - \alpha_i(E)t\}$$
$$+ \alpha_i(E)\int_I^E \exp\quad \theta(E - E') + \theta(E' - I)]\}r_i(E',E)\,dE' \quad (5.154)$$

To obtain the solution of Eq. (5.149) without neglecting the dissipation of energy due to nonionizing collisions, we proceed as follows: Let $g(E'',E;t)\,dE''$ denote the probability that the primary electron of initial energy E suffers no ionizing collisions, but makes a transition to an energy in the interval $(E'', E'' + dE'')$ by nonionizing collisions in time t. The usual reasoning leads to the equation

$$\frac{\partial g(E'',E;t)}{\partial t} = -\alpha(E)g(E'',E;t)$$
$$+ \alpha_e(E)\int_0^I g(E'', E - E'; t)r_e(E',E)\,dE' \quad (5.155)$$

Messel [60] has shown that the solution of Eq. (5.155) can be written in the form of the infinite series

$$g(E'',E;t) = \sum_{n=0}^{\infty} g_n(E'',E;t)$$

where

$$g_0(E'',E;t) = \delta(E'' - E)e^{-\alpha(E)t}$$

and

$$g_n(E'',E;t) = \int_0^t d\tau \int_0^I dE'\, g_{n-1}(E'', E - E'; t - \tau)$$
$$\times e^{-\alpha(E)}\alpha_e(E)r_e(E',E) \quad \text{for } n \geqslant 1$$

We now have

$$\mathscr{C}[E;t] = \int_0^E e^{i\theta(E'')}g(E'',E;t)\,dE'' + \int_0^t d\tau \int_I^E dE''$$
$$\times \int_I^{E''} dE'\, \mathscr{C}[E'' - E'; t - \tau]\mathscr{C}[E' - I; t - \tau]$$
$$\times r_i(E',E'')\alpha_i(E'')g(E'',E;\tau) \quad (5.156)$$

This equation has the following interpretation: The first term on the right-hand side is the contribution to $\mathscr{C}[E;t]$ if no ionizing collisions occur in the interval $(0,t)$. Now, the probability that the first ionizing collisions occur in the interval $(\tau, \tau + d\tau)$ and give rise to two electrons whose energies are in the intervals $(E' - I, E' + dE' - I)$ and $(E'' - E', E'' - E' + dE')$, respectively, is

$$r_i(E',E'')\alpha_i(E'')g(E'',E;\tau)$$

Hence, if we multiply this probability by the characteristic functionals of the two resultant electron cascades and integrate over all intermediate times τ and energies E', E'', we obtain the second term, which represents the contribution to $\mathscr{C}[E;t]$ from the first ionizing collision in $(0,t)$.

We close this discussion of knock-on electron cascades by remarking that (5.156) can be used to calculate $\mathscr{C}[E;t]$ by iteration for increasing values of E. The iteration procedure starts with

$$\mathscr{C}[E;t] = \int_0^E e^{i\theta(E'')} g(E'',E;t)\, dE \qquad (5.157)$$

for $E < I$. This completes the solution of Eq. (5.149).

C. The Ionization Electron Cascade. In order to complete the description of the ionization cascade, it is necessary to consider electron cascades which are initiated by fast primary ionizing particles which are not electrons but act as a line source of primary knock-on electrons. Each of the primary knock-on electrons produced in this manner initiates an independent cascade of the type discussed in Sec. 5.4B.

To describe cascades of this type we add the following assumptions to those of Sec. 5.4B:

5. The successive ionizing collisions of the primary particles are independent.

6. The loss of energy of the primary particle by all causes (including ionization) is negligible compared with its initial energy in a given time interval of length t.

Let $R_p(E')\, dE'$ denote the differential cross section for the primary to collide with an atom of the absorber and lose energy in the range $(E', E' + dE')$. For E' large, and smaller than the maximum transferable energy, this cross section has the same form as the Thomson cross section, but with velocity v now denoting the velocity of the primary particle. This velocity is assumed to be constant in the interval $(0,t)$. Now let

$$\alpha_p = Nv \int_I^\infty R_p(E')\, dE'$$

and

$$r_p = \frac{NvR_p(E')}{\alpha_p} \qquad \text{for } E' \geqslant I$$

$$= 0 \qquad \text{for } E' < I$$

denote the ionization rate (assumed constant) and probability density per ionizing collision for an energy loss E', respectively.

Let $\mathscr{C}^*[\theta(E);t]$ denote the characteristic functional for the electron cascade initiated by the primary particle. By proceeding as before, i.e., by considering the change in $\mathscr{C}^*[t]$ in the initial time interval $(0,dt)$, we consider the following events and the associated probabilities:

1. The probability of the primary particle not experiencing an ionizing collision in $(0,dt)$ is $1 - \alpha_p\,dt$.

2. The probability that an ionizing collision takes place, resulting in the production of a knock-on electron with energy in the interval $(E - I, E' + dE' - I)$, is $\alpha_p r_p(E')\,dE'\,dt$. This knock-on will then initiate an independent electron cascade whose characteristic functional at time t is $\mathscr{C}[E' - I; t]$. Hence

$$\mathscr{C}^*[t] = (1 - \alpha_p\,dt)\mathscr{C}^*[t - dt]$$
$$+ \alpha_p\,dt\mathscr{C}^*[t - dt]\int_I^\infty \mathscr{C}[E' - I; t - dt]r_p(E')\,dE' \qquad (5.158)$$

On differentiating, we obtain

$$\frac{\partial \mathscr{C}^*[t]}{\partial t} = \alpha_p\{\mathscr{C}_p[t] - 1\}\mathscr{C}^*[t] \qquad (5.159)$$

where
$$\mathscr{C}_p[t] = \int_I^\infty \mathscr{C}[E' - I; t]r_p(E')\,dE' \qquad (5.160)$$

is the characteristic functional of the knock-on cascade produced in a single ionizing collision of the primary particle.

The solution of (5.159) which satisfies the initial condition $\mathscr{C}^*[t] = 1$ is

$$\mathscr{C}^*[t] = \exp\left\{\alpha_p\int_0^t (C_p[\tau] - 1)\,d\tau\right\} \qquad (5.161)$$

The characteristic functional $\mathscr{C}[t]$ is known; hence $\mathscr{C}_p[t]$, which is given by (5.160), is also known. We can, therefore, determine $\mathscr{C}^*[t]$, and this gives a complete solution of the ionization cascade problem.

The fluctuation problem for ionization cascades can be studied by utilizing the relationship between the characteristic functional and the probability generating function (cf. Sec. 5.2C).

5.5 The Ramakrishnan-Srinivasan Approach to Cascade Theory

A. Introduction. Throughout this chapter a fundamental role has been played by the probability distribution of the number of particles at depth t with energies greater than E, say, due to a single particle of energy E_0, and the moments of this distribution. In this section we consider a new approach to the study of cascade theory due to Ramakrishnan and Srinivasan [88]. In the Ramakrishnan-Srinivasan approach there is introduced a function $\Phi_i^*(E_0; n_1, \ldots, n_m; E_1, \ldots, E_m; t)$ which represents the probability that n_i particles of type i, each with *energy greater than E_i at the time of its production* (E_i is called the *primitive energy*), are produced by a single particle of type i with energy E_0 in the interval $(0,t)$. It has been suggested that this function has more elegant mathematical properties than the distribution function usually studied and that it is more suitable for the study of cascade processes in nuclear emulsions and in the bubble chamber.

B. The Product Density Functions. We now define the product density functions in the Ramakrishnan-Srinivasan theory, and in the next section we derive the integral equations which the product density functions satisfy in the case of an electron-photon cascade.

Let $F_n(E_1, \ldots, E_n; t_1, \ldots, t_n) \, dE_1 \cdots dE_n \, dt_1 \cdots dt_n$ denote the joint probability that an electron of energy in the interval $(E_1, E_1 + dE_1)$ is produced in the interval $(t_1, t_1 + dt_1)$, an electron of energy in the interval $(E_2, E_2 + dE_2)$ is produced in the interval $(t_2, t_2 + dt_2)$, etc. Hence, $F_n(E_1, \ldots, E_n; t_1, \ldots, t_n)$ is the product density function of degree n for electrons. Similarly, $G_n(E_1, \ldots, E_n; t_1, \ldots, t_n)$ is the product density function of degree n for photons and $H_{i,n-1}(E_1, \ldots, E_i;$ $E_{i+1}, \ldots, E_n; t_1, \ldots, t_n)$ is the mixed product density function for electrons and photons. For the mixed product density in the case described, we consider i electrons with energies in the intervals $(E_j, E_j + dE_j)$ produced in the intervals $(t_j, t_j + dt_j)$, $j = 1, 2, \ldots, i$, and $n - i$ photons with energies in the intervals $(E_k, E_k + dE_k)$ produced in the intervals $(t_k, t_k + dt_k)$, $k = 1, 2, \ldots, n - i$.

It is important to note that in the new approach the product densities are defined over both the E and t spaces. Previously, the product densities were defined over the E space only. The properties of product density functions defined over both parameter spaces have been established by Ramakrishnan [82]. By utilizing these properties, it can be shown that, if the product density functions are integrated over the range E_c to ∞ with respect to the n energy variables $E_1, \ldots, E_n$, they still remain product densities of the particles produced in the intervals $(t_i, t_i + dt_i)$, $i = 1, 2, \ldots, n$, the primitive energy of each particle being greater than E_c. When this "new" product density is obtained, the moments of number of particles can then be determined by integration over the appropriate range of t. Hence, the kth moment of the number of electrons produced in the cascade in the interval $(0,t)$, the primitive energy of each being greater than E_c, is given by

$$\mathcal{E}\{X^k(E_c;t)\} = \sum_{i=1}^{k} C_i^k \int_{E_c}^{\infty} dE_1 \cdots \int_{E_c}^{\infty} dE_i \int_0^t dt_1$$
$$\cdots \int_0^t F_i(E_1, \ldots, E_i; t_1, \ldots, t_i) \, dt_i \quad (5.162)$$

where the coefficients C_i^k are defined as in Sec. 5.3B.

C. Integral Equations for the Product Densities. In deriving the equations satisfied by the product densities we assume that the differential cross sections for Bremsstrahlung and pair production are given by the Bethe-Heitler theory; hence the cross sections are the same as the cross sections employed in Sec. 5.2B. The notation will also be the same as that employed in the discussion of product density functions given in Sec. 5.2B.

We first consider the equations for product densities of degree 1. Let $F_{i,1}(E_0,E;t)$ and $G_{i,1}(E_0,E;t)$ denote the product densities of degree 1 for electrons and photons, respectively, where the subscript $i = 1$ if the cascade is initiated by an electron and $i = 2$ if the cascade is initiated by a photon, the initiating particle having energy E_0. The equations for the product densities are easily obtained; we have

$$F_{i,1}(E_0,E;t) = 2 \int_E^{E_0} g_{i,1}(E_0,E';t) \rho(E,E') \, dE' \tag{5.163}$$

$$G_{i,1}(E,E;t) = \int_E^{E_0} f_{i,1}(E_0,E',t) R(E' - E,E') \, dE' \tag{5.164}$$

In the above equation $f_{i,1}(E_0,E;t)$ and $g_{i,1}(E_0,E';t)$ are the "old" product densities of degree 1 for electrons and photons, respectively. Hence, by definition, $f_{i,1}(E_0,E;t) \, dE$ represents the probability that an electron is in the interval $(E, E + dE)$ and $g_{i,1}(E_0,E;t)$ represents the probability that a photon is in the interval $(E, E + dE)$. In writing Eqs. (5.163) and (5.164) it was assumed that an electron with energy $E' > E$ should exist at depth t and with probability one radiate a photon in the interval $(t, t + dt)$; similarly, a photon with energy $E' > E$ should exist at depth t and with probability one produce a pair of electrons in the interval $(t, t + dt)$. The factor 2 enters in (5.163) because we do not distinguish between positive and negative electrons.

In order to derive the product density functions of degree 2 and the mixed product density functions, it is necessary to introduce the concepts of parent and ancestor particles. A particle A at depth t is said to be the *parent* of another particle B if the particle B is produced directly by A. And a particle A is said to be the *ancestor* at depth t_1 of another particle B produced at depth t_2, where $t_2 > t_1$, if for the production of B we have to consider the cascade produced by A. From these definitions we see that the parent of a particle is an ancestor, but the converse is not true. The *primitive ancestor* is defined as that particle (electron or photon) which initiates the cascades at $t = 0$.

We consider only the derivation of the integral equation for $F_{i,2}(E_0,E_1,E_2;t_1,t_2)$, since the procedure is the same for the other equations. The events which contribute to the probability $F_{i,2}(E_0,E_1, E_2;t_1,t_2)$ can be classified into two mutually exclusive sets:

1. The two electrons, one electron produced in the interval $(t_1, t_1 + dt_1)$ and the other produced in the interval $(t_2, t_2 + dt_2)$, where $t_2 > t_1$, have a common ancestor at depth t_1.

2. The two electrons have two different ancestors at depth t_1.

The contribution due to (1) can be described as follows: The common ancestor at depth t_1 is the parent of an electron with energy in the interval $(E_1, E_1 + dE_1)$ produced in the interval $(t_1, t_1 + dt_1)$, and hence is a photon with proper energy. Now, the probability for the existence of a photon of energy E' at depth t is $g_{i,1}(E_0, E'; t_1) \, dE'$; hence

$$2g_{i,1}(E_0, E'; t_1)\rho(E_1, E') \, dE_1 \, dE' \, dt_1$$

represents the probability that a photon with energy in the interval $(E', E' + dE')$ is in existence at depth t_1 and produces an electron with energy in the interval $(E_1, E_1 + dE_1)$ in the interval $(t_1, t_1 + dt_1)$. Since it is assumed that an electron which is produced in the interval $(t_2, t_2 + dt_2)$ has the same ancestor as the electron produced in $(t_1, t_1 + dt_1)$, it is equivalent to stating that the electron with energy E_1 or the electron with energy $E' - E_1$ is an ancestor at depth $t_1 + dt_1$. The probability of the above events is given by

$$2\int_{E_1}^{E_0} g_{i,1}(E_0, E'; t)\rho(E_1, E')\{F_{1,1}(E_1, E_2; t_2 - t_1)I(E_1 - E_2)$$

$$+ F_{1,1}(E' - E_1, E_2; t_2 - t_1)I(E' - (E_1 + E_2))\} \, dE' \quad (5.165)$$

In the above $I(x)$ is the Heaviside unit function, i.e.,

$$I(x) = 1 \qquad \text{for } x > 0$$
$$= 0 \qquad \text{otherwise}$$

The contribution due to (2) can be determined as follows: We have assumed that at depth t there are two different ancestors, and since one of the ancestors is the parent in the interval $(t_1, t_1 + dt_1)$ of the electron with energy E_1, then both of the particles can be photons or one of them a photon and the other an electron. The probability of these events is given by

$$2\int_{E_1}^{E_0} dE'' \int_{E_2}^{E_0} dE' \, h_{i,1}(E_0, E', E''; t_1)\rho(E_1, E'')F_{1,1}(E', E_2; t_2 - t_1)$$

$$+ 2\int_{E_2}^{E_0} dE'' \int_{E_1}^{E_0} dE' \, g_{i,2}(E_0, E', E''; t_1)\rho(E_1, E')F_{2,1}(E'', E_2; t_2 - t_1) \quad (5.166)$$

Since the events are mutually exclusive, we add (5.165) and (5.166) to obtain the functional equation for $F_{i,2}(E_0, E_1, E_2; t_1, t_2)$.

By using similar arguments for the product densities $G_{i,2}(E_0,E_1,E_2;t_1,t_2)$ and $H_{i,1}(E_0,E_1,E_2;t_1,t_2)$, we obtain the following functional equations:

$$G_{i,2}(E_0,E_1,E_2;t_1,t_2) = \int_{E_1}^{E_2} f_{i,1}(E_0,E';t_1)R(E'-E_1,E')$$

$$\times \{G_{2,1}(E_1,E_2;t_2-t_1)I(E_1-E_2) + G_{1,1}(E'-E_1,E_2;t_2-t_1)$$

$$\times I(E'-(E_1+E_2))\}\,dE' + \int_{E_1}^{E_0}dE'\int_{E_2}^{E_0}h_{i,1}(E_0,E',E'';t_1)$$

$$\times R(E'-E_1,E')G_{2,1}(E'',E_2;t_2-t_1)\,dE''$$

$$+\int_{E_1}^{E_0}dE\int_{E_2}^{E_0} f_{i,2}(E_0,E,E'';t_1)R(E'-E_1,E')G_{1,1}(E'',E_2;t_2-t_1)\,dE''$$

$$\tag{5.167}$$

$$H_{i,1}(E_0,E_1,E_2;t_1,t_2) = 2\int_{E_1}^{E_0} g_{i,1}(E_1,E';t_1)\rho(E_1,E')$$

$$\times \{G_{1,1}(E_1,E_2;t_2-t_1)I(E_1-E_2) + G_{1,1}(E'-E_1,E_2;t_2-t_1)$$

$$\times I(E'-(E_1+E_2))\}\,dE' + 2\int_{E_1}^{E_0}dE''\int_{E_2}^{E_0}h_{i,1}(E_0,E',E'';t_1)$$

$$\times \rho(E_1,E'')G_{1,1}(E',E_2;t_2-t_1)\,dE'$$

$$+2\int_{E_1}^{E_0}dE''\int_{E_2}^{E_0} g_{i,2}(E_0,E',E'';t_1)\rho(E_1,E'')G_{2,1}(E',E_2;t_2-t_1)\,dE' \tag{5.168}$$

In deriving the above equations we assumed that $t_2 > t_1$. Should $t_1 > t_2$, the product densities for electrons and photons satisfy the same functional equations, but we must interchange t_1,t_2 and E_1,E_2 on the right-hand side of these equations. For the mixed product density we have a new equation:

$$H_{i,1}(E_0,E_1,E_2;t_1,t_2) = \int_{E_1}^{E_0} f_{i,1}(E_0,E';t_2)R(E'-E_2,E')$$

$$\times dE' \{F_{1,1}(E',E_1;t_1-t_2)I(E'-(E_2+E_1)) + F_{2,1}(E_2,E_1;t_1-t_2)$$

$$\times I(E_2-E_1)\} + \int_{E_2}^{E_0}dE'\int_{E_1}^{E_0}dE'' h_{1,1}(E_0,E',E'',t_2)R(E'-E_2,E')$$

$$\times F_{2,1}(E'',E_2;t_1-t_2) + \int_{E_2}^{E_0}dE'\int_{E_1}^{E_0}dE'' f_{i,2}(E_0,E',E'';t_2)$$

$$\times R(E'-E_2,E')F_{1,1}(E'',E_1;t_1-t_2) \qquad t_1 > t_2 \tag{5.169}$$

The equations for the new product density functions of degrees 1 and 2, as well as the equations for the mixed product density, have been solved by Ramakrishnan and Srinivasan by using the Mellin transformation. By using these solutions the moments can be obtained, and the fluctuation problem can be studied on the basis of the Ramakrishnan-Srinivasan theory.

D. Other Studies Based on the New Approach. 1. Srinivasan and Ranganathan [99] have given numerical calculations of the mean number of electrons in a cascade with energies greater than E at depth t. In particular, they have obtained the following result: Let the random variable $X(E;t)$ denote the number of electrons produced in the interval $(0,T)$ with energy greater than E at the point of production and let $Y(E;t)$ denote the number of electrons at depth t with energies greater than E. Then $\mathscr{E}\{X(E;t)\}$ approaches a finite value as t approaches infinity for any finite E, and $\mathscr{E}\{Y(E;t)\}$ approaches zero as t approaches infinity. Calculations have also been given for the mean number of electrons produced by a single incident electron for small depth of absorber.

2. Ramakrishnan and Srinivasan [89] have used the new approach to study the development of an electron-photon cascade with ionization loss. In particular, the mean number of particles produced for infinite depth is determined.

5.6 Some Additional Studies on Cascade Processes

A. Three-dimensional Theory of Cascade Processes. In the course of this chapter we have restricted our attention to one-dimensional cascade processes, i.e., we have studied the longitudinal development of the cascade only. A three-dimensional theory of cascade processes has been developed by Messel and Green [62,63] and Chartres and Messel [20,21] (cf. also [60]). In this theory the angular and radial development as well as the longitudinal development of the cascade is considered. We refer also to the paper of Kamata and Nishimera [45] for the development and applications of a three-dimensional theory of cascade processes.

B. N-component Cascades. In addition to restricting our attention to one-dimensional cascades, we have considered only those cascades made up of one (nucleon cascades) or two (electron-photon cascades) types of particles. In many situations it is necessary to consider cascades made up of N types of particles, $N = 1, 2, \ldots$. Olbert and Stora [75] have developed a theory of high-energy N-component cascades of cosmic rays. They obtain a system of simultaneous integrodifferential equations the solution of which gives the number of particles of type $i(i = 1, 2, \ldots, N)$ in the cascade at atmospheric depth t within the energy interval $(E, E + dE)$.

In Ref. 42 Jánossy gives a generalization of the G-equations which permits the study of N-component cascades.

C. Age-dependent Branching Stochastic Processes in Cascade Theory. The theory of age-dependent branching stochastic processes[1] has been applied in cascade theory to study several simple models of electron-photon cascades. Three cases have been considered: (1) the transformation probabilities are constants [15], (2) the transformation probabilities are functions of the depth of absorber [105], and (3) the transformation probabilities are functions of both the energy and the depth of absorber.[2]

D. Straggling of the Range of Fast Particles. The term *straggling* is used to denote the random decrease in energy of a particle as it passes through matter. The theory of stochastic processes has been utilized to obtain the probability distribution of the *range of the particle*, the range being defined as the thickness of matter traversed by the particle before it is absorbed. Hence, given the initial energy of the particle, the problem is to determine the distribution of the thickness of the matter traversed by the particle before it drops below a given critical energy E_c.

Jánossy [39] has used the Poisson process as a simple stochastic model of straggling. A more realistic treatment of this problem has been given by Ramakrishnan and Mathews [85]. These authors obtain an expression for the probability of absorption and the probability density function of the range under the assumptions that the particle can lose energy deterministically by ionization and randomly by Bremsstrahlung.

E. Multiple Processes in Electron-Photon Cascades. The experimental study of high-energy electron-photon cascades in cosmic rays points up the need for theoretical studies which take into consideration third-order processes of the following types: (1) Electron scattering with pair production, i.e., trident production; (2) electron scattering with emission of two photons; and (3) photon scattering with pair production. Ramakrishnan et al. [90] have considered this problem and have obtained differential cross sections for processes 2 and 3. At present these results have not been used in the stochastic theory of cascade processes. For a stochastic treatment of the trident process utilizing the Messel-Potts theory [65] we refer to the paper of Gardner [29].

F. Numerical Studies on the Fluctuation Problem. From the complexity of the fluctuation problem in cascade theory it is clear that the analytical studies must be supplemented by numerical

---

[1] Cf. Sec. 2.5.
[2] P. D. Kapadia, unpublished, 1956.

studies in order to make the available probability distribution and expressions for moments of greater value. The use of electronic computers greatly facilitates the formulation of computational procedures for the study of cascade processes. In a very important paper, Butcher and Messel [17] give numerical results for the number distribution of electrons above a given energy due to primary electrons and photons. Their results essentially complete the fluctuation problem for electron-photon cascades. For additional numerical studies we refer to Butcher et al. [18], Gardner et al. [30], and Ramakrishnan and Mathews [83,84].

Bibliography

1 Arley, N.: "On the Theory of Stochastic Processes and Their Applications to the Theory of Cosmic Radiation," John Wiley & Sons, New York, 1949.

2 Bartlett, M. S.: Processus stochastiques ponctuels, *Ann. inst. H. Poincaré*, vol. 14, pp. 35–60, 1954.

3 Bartlett, M. S.: "An Introduction to Stochastic Processes," Cambridge University Press, New York, 1955.

4 Bartlett, M. S., and D. G. Kendall: On the Use of the Characteristic Functional in the Analysis of Some Stochastic Processes Occurring in Physics and Biology, *Proc. Cambridge Phil. Soc.*, vol. 47, pp. 65–78, 1951.

5 Bass, L.: On the Stochastic Equation for the Energy Loss of Fast Electrons in Matter, *Proc. Indian Acad. Sci.*, sec. A, vol. 43, pp. 423–427, 1956.

6 Belenkiǐ, S. Z.: "Cascade Processes in Cosmic Rays" (in Russian), Gosudarstv. Izdat. Tehn.-Teor. Lit., Moscow, 1948.

7 Bernstein, I. B.: Improved Calculations on Cascade Shower Theory, *Phys. Rev.*, vol. 80, pp. 995–1005, 1950.

8 Bethe, H., and W. Heitler: On the Stopping of Fast Particles and on the Creation of Positive Electrons, *Proc. Roy. Soc. (London)*, ser. A, vol. 146, pp. 83–112, 1934.

9 Bhabha, H. J.: On the Stochastic Theory of Continuous Parametric Systems and Its Applications to Electron Cascades, *Proc. Roy. Soc. (London)*, ser. A, vol. 202, pp. 301–322, 1950.

10 Bhabha, H. J.: Note on the Complete Stochastic Treatment of Electron Cascades, *Proc. Indian Acad. Sci.*, sec. A, vol. 32, pp. 154–161, 1951.

11 Bhabha, H. J., and S. K. Chakrabarty: The Cascade Theory with Collision Loss, *Proc. Roy. Soc. (London)*, ser., A vol. 181, pp. 267–303, 1943.

12 Bhabha, H. J., and S. K. Chakrabarty: Further Calculations on the Cascade Theory, *Phys. Rev.*, vol. 74, pp. 1352–1363, 1948.

13 Bhabha, H. J., and W. Heitler: The Passage of Fast Electrons and the Theory of Cosmic Showers, *Proc. Roy. Soc. (London)*, ser. A, vol. 159, pp. 432–458, 1937.

14 Bhabha, H. J., and A. Ramakrishnan: The Mean Square Deviation of the Number of Electrons and Quanta in the Cascade Theory, *Proc. Indian Acad. Sci.*, sec. A, vol. 32, pp. 141–153, 1950.

15 Bharucha-Reid, A. T.: Age-dependent Branching Stochastic Processes in Cascade Theory, *Phys. Rev.*, vol. 96, pp. 751–753, 1954.

16 Bochner, S.: "Harmonic Analysis and the Theory of Probability," University of California Press, Berkeley, 1955.

17   Butcher, J. C., and H. Messel: Electron Number Distribution in Electron-Photon Showers, *Phys. Rev.*, vol. 112, pp. 2096–2106, 1958.
18   Butcher, J. C., B. A. Chartres, and H. Messel: Tables of Average Numbers for Electron-Photon Showers at Small Depths of Absorber, *Nuclear Phys.*, vol. 6, pp. 271–281, 1958.
19   Carlson, J. F., and J. R. Oppenheimer: On Multiplicative Showers, *Phys. Rev.*, vol. 51, pp. 210–231, 1937.
20   Chartres, B. A., and H. Messel: Moments of the Angular Distribution for High Energy Nuclear Collisions, *Phys. Rev.*, vol. 87, pp. 748–749, 1952.
21   Chartres, B. A., and H. Messel: Three Dimensional Theory of Electron-Photon Showers, *Proc. Phys. Soc. (London)*, ser. A, vol. 67, pp. 158–166, 1954.
22   Chartres, B. A., and H. Messel: New Formulation of a General Three-dimensional Cascade Theory, *Phys. Rev.*, vol. 96, pp. 1651–1654, 1954.
23   Chartres, B. A., and H. Messel: Angular Distribution in Electron-Photon Showers without the Landau Approximation, *Phys. Rev.*, vol. 104, pp. 517–525, 1956.
24   Clementel, E., and G. Puppi: Sulla componente nucleonica nell'atomosfera, *Nuovo Cimento*, ser. 9, vol. 8, pp. 936–951, 1951.
25   Czerwonko, J.: Asymptotic Behaviour of Higher Factorial Moments of Electron-Photon Cascades at Large Depths, *Acta Phys. Polon.*, vol. 16, pp. 305–326, 1957.
26   Fay, H.: Statistical Fluctuations Occurring in Electron-Photon Cascades of 1000 GeV Primary Energy, *Nuovo Cimento*, ser. 10, vol. 6, pp. 1516–1519, 1957.
27   Furry, W. H.: On Fluctuation Phenomena in the Passage of High Energy Electrons through Lead, *Phys. Rev.*, vol. 52, pp. 569–581, 1937.
28   Ganguly, S.: Mixed Nucleon-Meson Cascades in Finite Absorbers, *Proc. Natl. Sci. India*, pt. A, vol. 22, pp. 40–53, 1956.
29   Gardner, J. W.: The Electromagnetic Cascade with Tridents, *Nuovo Cimento*, ser. 10, vol. 7, pp. 10–22, 1958.
30   Gardner, J. W., H. Gellman, and H. Messel: Numerical Calculations on the Fluctuation Problem in Cascade Theory, *Nuovo Cimento*, ser. 10, vol. 2, pp. 58–74, 1955.
31   Gupta, M. R.: Analysis of Bursts Produced by Mesons, *Nuovo Cimento*, ser. 10, vol. 7, pp. 39–52, 1958.
32   Harris, T. E.: The Random Fluctuations of Cosmic-ray Cascades, *Proc. Natl. Acad. Sci. U.S.*, vol. 43, pp. 509–512, 1957.
33   Harris, T. E.: "Branching Processes," Springer-Verlag, Berlin, Vienna, in press.
34   Heitler, W.: "The Quantum Theory of Radiation," Oxford University Press, New York, 1954.
35   Heitler, W., and L. Jánossy: On the Absorption of Meson-producing Nucleons, *Proc. Phys. Soc. (London)*, ser. A, vol. 62, pp. 374–385, 1949.
36   Heitler, W., and L. Jánossy: On the Size-Frequency Distribution of Penetrating Showers, *Proc. Phys. Soc. (London)*, ser. A, vol. 62, pp. 669–683, 1949.
37   Ivanenko, I. P.: Cascade Showers Produced by Electrons and Photons in Light and Heavy Elements (in Russian), *Ž. Eksper. Teoret. Fiz.*, vol. 31, pp. 86–104, 1956. (English translation: *Soviet Physics JETP*, vol. 4, pp. 115–130, 1957.)
38   Ivanenko, I. P.: Particle Angular Distribution Functions at the Cascade

Shower Maximum (in Russian), Ž. Eksper. Teoret. Fiz., vol. 33, pp. 825–827, 1957. (English translation: Soviet Physics JETP, vol. 6, pp. 637–639, 1958.)

39 Jánossy, L.: "Cosmic Rays," Oxford University Press, New York, 1950.

40 Jánossy, L.: Note on the Fluctuation Problem of Cascades, Proc. Phys. Soc. (London), ser. A, vol. 63, pp. 241–249, 1950.

41 Jánossy, L.: On the Absorption of a Nucleon Cascade, Proc. Roy. Irish Acad., ser. A, vol. 53, pp. 181–188, 1950.

42 Jánossy, L.: Studies on the Theory of Cascades: I, II, Acta Phys. Acad. Sci. Hung., vol. 2, pp. 289–333, 1952. (Also in Ž. Eksper. Teoret. Fiz., vol. 26, pp. 386–404, 518–538, 1954.)

43 Jánossy, L.: Note on the Fluctuation Problem of Cascades, Proc. Phys. Soc. (London), ser. A, vol. 66, pp. 117–118, 1953.

44 Jánossy, L., and H. Messel: Fluctuations of the Electron-Photon Cascade— Moments of the Distribution, Proc. Phys. Soc. (London), ser. A, vol. 63, pp. 1101–1115, 1950.

45 Kamata, K., and J. Nishimera: The Lateral Spread and the Angular Structure Functions of Electron Showers, Progr. Theoret. Phys. (Kyoto), Supplement, no. 6, pp. 93–155, 1958.

46 Khristov, Kh. Ia.: On the Green's Function of the Kinetic Equations of a Cascade (in Russian), Ž. Eksper. Teoret. Fiz., vol. 33, pp. 683–695, 1957. (English translation in Soviet Physics JETP, vol. 6, pp. 522–534, 1958.)

47 ˈ Khristov, Kh. Ia.: Propagation of Cascades in a Multi-layer Medium (in Russian), Ž. Eksper. Teoret. Fiz., vol. 33, pp. 877–882, 1957. (English translation: Soviet Physics JETP, vol. 6, pp. 676–680, 1958.)

48 Khristov, Kh. Ia.: Correlations in the Distribution of Cascade Particles (in Russian), Ž. Eksper. Teoret. Fiz., vol. 33, pp. 883–888, 1957. (English translation in Soviet Physics JETP, vol. 6, pp. 680–684, 1956.)

49 Konwent, H., and J. Lopuszański: Some Remarks on the Asymptotic Behaviour of the Electron-Photon Cascade for Large Depths of the Absorber, Acta Phys. Polon., vol. 15, pp. 191–203, 1956.

50 Landau, L., and G. Rumer: The Cascade Theory of Electronic Showers, Proc. Roy. Soc. (London), ser. A, vol. 166, pp. 213–228, 1938.

51 Lopuszański, J.: Lösung der G-Gleichungen von Jánossy für die kosmischen Schauer, Acta Phys. Polon., vol. 15, pp. 191–203, 1953.

52 Lopuszański, J.: Solution of the Fluctuation Problem in the Cascade by Means of the G-equations of Jánossy without Ionization Loss, Acta Phys. Polon., vol. 14, pp. 191–196, 1955.

53 Lopuszański, J.: The Solution of the G-equations of Jánossy with Ionization Loss, Acta Phys. Polon., vol. 14, pp. 251–253, 1955.

54 Lopuszański, J.: The Asymptotic Behaviour of the Probability Distribution Function of Cosmic Cascades, Acta Phys. Polon., vol. 14, pp. 265–267, 1955.

55 Lopuszański, J.: Evaluation of the Probability Distribution Function for Large Depth of the Absorber, Acta Phys. Polon., vol. 14, pp. 269–271, 1955.

56 Lopuszański, J.: Some Remarks on the Asymptotic Behaviour of the Cosmic Ray Cascade for Large Depth of the Absorber: II. Asymptotic Behaviour of the Probability Distributions, III. Evaluation of the Distribution Function, Nuovo Cimento, ser. 10, vol. 2, suppl. 4, pp. 1150–1160, 1161–1167, 1955.

57 Lopuszański, J.: Some Remarks on the Theory of the Electron-Photon Cascade, Acta Phys. Polon., vol. 15, pp. 177–180, 1956.

58 Messel, H.: Further Results on the Fluctuation Problem in Electron-Photon Cascade Theory and the Probability Distribution Function, *Proc. Phys. Soc. (London)*, ser. A, vol. 64, pp. 807–813, 1951.

59 Messel, H.: The Solution of the Fluctuation Problem in Nucleon Cascade Theory: Homogeneous Nuclear Matter, *Proc. Phys. Soc. (London)*, ser. A, vol. 65, pp. 465–472, 1952.

60 Messel, H.: The Development of a Nucleon Cascade, in J. G. Wilson (ed.), "Progress in Cosmic Ray Physics," vol. 2, pp. 135–216, North-Holland Publishing Company, Amsterdam, 1954.

61 Messel, H.: On the Solutions of the Fluctuation Problem in Cascade Showers, *Nuovo Cimento*, ser. 10, vol. 4, pp. 1339–1348, 1956.

62 Messel, H., and H. S. Green: The Angular and Lateral Distribution Functions for the Nucleon Component of the Cosmic Radiation, *Phys. Rev.*, vol. 87, pp. 738–747, 1952.

63 Messel, H., and H. S. Green: The General Three-dimensional Theory of Cascade Processes, *Proc. Phys. Soc. (London)*, ser. A, vol. 66, pp. 1009–1018, 1953.

64 Messel, H., and R. B. Potts: Cascade Theories with Ionisation Loss, *Phys. Rev.*, vol. 87, pp. 759–767, 1952.

65 Messel, H., and R. B. Potts: The Solution of the Fluctuation Problem in Electron-Photon Shower Theory, *Phys. Rev.*, vol. 86, pp. 847–851, 1952.

66 Messel, H., and R. B. Potts: The Solution of the Fluctuation Problem in a Finite Absorber for Nucleon Cascades, *Proc. Phys. Soc. (London)*, ser. A, vol. 65, pp. 473–480, 1952.

67 Messel, H., and R. B. Potts: Note on the Fluctuation Problem in Cascade Theory, *Proc. Phys. Soc. (London)*, ser. A, vol. 65, pp. 854–856, 1952.

68 Messel, H., and R. B. Potts: Longitudinal Development of Extensive Air Showers, *Nuovo Cimento*, ser. 9, vol. 10, pp. 754–777, 1953.

69 Mitra, A. N.: The Fluctuation Problem in μ-meson Bursts, *Nuclear Phys.*, vol. 3, pp. 262–272, 1957.

70 Moyal, J. E.: Stochastic Processes and Statistical Physics, *J. Roy. Statist. Soc.*, ser. B, vol. 11, pp. 150–210, 1949.

71 Moyal, J. E.: Theory of Ionization Fluctuations, *Phil. Mag.*, vol. 46, pp. 263–280, 1955.

72 Moyal, J. E.: Statistical Problems in Nuclear and Cosmic Ray Physics, *Proc. Intern. Statist. Inst. Conf.*, Rio de Janeiro, pp. 1–12, 1955.

73 Moyal, J. E.: Theory of the Ionization Cascade, *Nuclear Phys.*, vol. 1, pp. 180–194, 1956.

74 Nordsieck, A., W. E. Lamb, and G. E. Uhlenbeck: On the Theory of Cosmic-ray Showers: I. The Furry Model and the Fluctuation Problem, *Physica*, vol. 7, pp. 344–360, 1940.

75 Olbert, S., and R. Stora: Theory of High-energy N-component Cascades, *Ann. Phys.*, vol. 1, pp. 247–269, 1957.

76 Pál, L.: On the Theory of Stochastic Processes in Cosmic Radiation (in Russian), *Ž. Eksper. Teoret. Fiz.*, vol. 30, pp. 362–366, 1956. (English translation: *Soviet Physics JETP*, vol. 3, pp. 233–237, 1956.)

77 Popov, Iu. A.: Solution of the Fundamental Diffusion Equation for Cosmic Ray Particles Emitted by a Constant Energy Concentrated Pulsed Source (in Russian), *Ž. Eksper. Teoret. Fiz.*, vol. 31, pp. 86–104, 1956. (English translation in *Soviet Physics JETP*, vol. 4, pp. 85–90, 1957.)

78 Ramakrishnan, A.: A Note on the Size-Frequency Distribution of

Penetrating Showers, *Proc. Phys. Soc. (London)*, ser. A, vol. 63, pp. 861–863, 1950.

79 Ramakrishnan, A.: Stochastic Processes Relating to Particles Distributed in a Continuous Infinity of States, *Proc. Cambridge Phil. Soc.*, vol. 46, pp. 596–602, 1950.

80 Ramakrishnan, A.: A Note on Jánossy's Mathematical Model of a Nucleon Cascade, *Proc. Cambridge Phil. Soc.*, vol. 48, pp. 451–456, 1951.

81 Ramakrishnan, A.: Some Simple Stochastic Processes, *J. Roy. Statist. Soc.*, ser. B, vol. 13, pp. 131–140, 1951.

82 Ramakrishnan, A.: On the Molecular Distribution Functions of a One-dimensional Fluid: I, *Phil. Mag.*, vol. 45, pp. 401–409, 1954.

83 Ramakrishnan, A., and P. M. Mathews: Numerical Work on the Fluctuation Problem of Electron Cascades, *Progr. Theoret. Phys. (Kyoto)*, vol. 9, pp. 679–681, 1953.

84 Ramakrishnan, A., and P. M. Mathews: Studies in the Stochastic Problem of Electron-Photon Cascades, *Progr. Theoret. Phys. (Kyoto)*, vol. 11, pp. 95–177, 1954.

85 Ramakrishnan, A., and P. M. Mathews: Straggling of the Range of Fast Particles as a Stochastic Process, *Proc. Indian Acad. Sci.*, sec. A, vol. 41, pp. 202–209, 1955.

86 Ramakrishnan, A., and S. K. Srinivasan: Two Simple Stochastic Models of Cascade Multiplication, *Progr. Theoret. Phys. (Kyoto)*, vol. 11, pp. 595–603, 1954.

87 Ramakrishnan, A., and S. K. Srinivasan: Fluctuations in the Number of Photons in an Electron-Photon Cascade, *Progr. Theoret. Phys. (Kyoto)*, vol. 13, pp. 93–99, 1955.

88 Ramakrishnan, A., and S. K. Srinivasan: A New Approach to the Cascade Theory, *Proc. Indian Acad. Sci.*, sec. A, vol. 44, pp. 263–273, 1956.

89 Ramakrishnan, A., and S. K. Srinivasan: A Note on Cascade Theory with Ionization Loss, *Proc. Indian Acad. Sci.*, sec. A, vol. 45, pp. 133–138, 1957.

90 Ramakrishnan, A., S. K. Srinivasan, N. R. Ranganathan, and R. Vasudevan: Multiple Processes in Electron-Photon Cascades, *Proc. Indian Acad. Sci.*, sec. A, vol. 45, pp. 311–326, 1957.

91 Rankin, B.: The Distribution in Energy and in Number of Electrons and Photons in Cascade, *University of California Radiation Laboratory Report* 2322, 1953.

92 Rankin, B.: The Concept of Sets Enchained by a Stochastic Process and Its Use in Cascade Shower Theory, *University of California Radiation Laboratory Report* 2652, 1954.

93 Rossi, B.: "High-energy Particles," Prentice-Hall, Inc., Englewood Cliffs, N.J., 1952.

94 Rossi, B., and K. Greisen: Cosmic-ray Theory, *Revs. Modern Phys.*, vol. 13, pp. 240–309, 1941.

95 Scott, W. T.: On a Difference Equation Method in Cosmic-ray Shower Theory, *Phys. Rev.*, vol. 80, pp. 611–615, 1950.

96 Scott, W. T.: Characteristic Functionals in Cascade Theory (Abstract), *Phys. Rev.*, vol. 86, pp. 145–146, 1952.

97 Scott, W. T., and G. E. Uhlenbeck: On the Theory of Cosmic-ray Showers: II. Further Contributions to the Fluctuation Problem, *Phys. Rev.*, vol. 62, pp. 497–508, 1942.

98 Snyder, H. S.: Comparison of Calculations on Cascade Theory, *Phys. Rev.*, vol. 76, pp. 1563–1571, 1949.

99   Srinivasan, S. K., and N. R. Ranganathan: Numerical Calculations on the New Approach to the Cascade Theory: I, II, *Proc. Indian Acad. Sci.*, sec. A, vol. 45, pp. 69–73, 268–272, 1957.

100   Stachowiak, H.: Some Properties of Distributions of Electron-Photon Cascades at Large Absorber Depth, *Acta Phys. Polon.*, vol. 15, pp. 181–190, 1956.

101   Tamm, I., and S. Belenkiĭ: On the Soft Component of Cosmic Rays at Sea Level, *J. Phys. U.S.S.R.*, vol. 1, pp. 177–198, 1939.

102   Tamm, I., and S. Belenkiĭ: The Energy Spectrum of Cascade Electrons, *Phys. Rev.*, vol. 70, pp. 660–664, 1946.

103   Urbanik, K.: Bemerkungen über die mittlere Anzahl von Partikeln in gewissen stochastischen Schauern, *Studia Math.*, vol. 15, pp. 34–42, 1955.

104   Urbanik, K.: Some Remarks on the Asymptotic Behaviour of the Cosmic Ray Cascade for Large Depth of the Absorber: I. Evaluation of the Factorial Moments, *Nuovo Cimento*, ser. 10, vol. 2, suppl. 4, pp. 1147–1149, 1955.

105   Woods, W. M., and A. T. Bharucha-Reid: Age-dependent Branching Stochastic Processes in Cascade Theory: II, *Nuovo Cimento*, ser. 10, vol. 10, pp. 569–578, 1958.

6

Applications in Physics: Additional Applications

6.1 Introduction

In addition to the theory of cascade processes, Markov chains and processes have found many applications in other branches of physics. In this chapter we present some of them. Section 6.2 is devoted to the equations of radioactive decay, these equations being of fundamental importance in the theory of radioactive transformations. In Sec. 6.3 we consider the theory of particle counters, and in Sec. 6.4 we consider a problem concerning nuclear fission detectors. The results of these sections are of interest to research workers concerned with the design and construction of counting devices, and of importance for the interpretation of counter registrations. Section 6.5 is devoted to the theory of tracks in nuclear research emulsions. Photographic emulsions are used in the fields of cosmic-ray and nuclear physics, and a stochastic theory of tracks is of value in interpreting the results obtained from developed photographic plates. Finally, in Sec. 6.6 we consider some problems in the theory of nuclear reactors.

6.2 Theory of Radioactive Transformations

In 1902 E. Rutherford and F. Soddy developed the general theory of radioactive transformations. This theory was based on the assumption that radioactive atoms are unstable and disintegrate according to the laws of chance. Following disintegration the resulting "daughter" atom has physical and chemical properties different from the preceding, or "parent," atom. The newly formed atom may in turn be unstable and pass through a number of transformations, each of which is characterized by the emission of an α or β particle (cf. Rutherford, Chadwick, and Ellis [34]).

To describe the development of a radioactive family, let $N_1(t), N_2(t)$, ..., $N_n(t)$ represent the number of atoms of the successive elements $E_1, E_2, \ldots, E_n$ at time t. Let λ_i denote the disintegration rate or constant associated with the element E_i, $i = 1, 2, \ldots, n$. The law of radioactive transformation states that the number of atoms, say $-\Delta N_1$, which decay to the daughter element E_2 in the interval $(t, t + \Delta t)$ is proportional to both Δt and $N_1(t)$. Hence

$$\Delta N_1(t) = -\lambda_1 N_1(t) \, \Delta t \tag{6.1}$$

If the daughter element E_2 is radioactive, similar reasoning gives

$$\Delta N_2(t) = \lambda_1 \, {}_1(t) \, \Delta t - \lambda_2 N_2(t) \, \Delta t \tag{6.2}$$

The radioactive family develops in this way, with E_{i-1} decaying into E_i according to the rule

$$\Delta N_{i+1}(t) = \lambda_i N_i(t) \, \Delta t - \lambda_{i+1} N_{i+1}(t) \, \Delta t \tag{6.3}$$

Finally, should E_n be a stable element, it will increase according to the rule

$$\Delta N_{n+1}(t) = \lambda_n N_n(t) \, \Delta t \tag{6.4}$$

It is clear that the above equations can be replaced by the system of differential equations

$$\frac{dN_1(t)}{dt} = -\lambda_1 N_1(t)$$

$$\frac{dN_{i+1}(t)}{dt} = \lambda_i N_i(t) - \lambda_{i+1} N_{i+1}(t) \qquad i = 1, 2, \ldots, n-1 \tag{6.5}$$

$$\frac{dN_{n+1}(t)}{dt} = \lambda_n N_n(t)$$

The above system of equations describes the radioactive decay or transformation process (cf. Bateman [2]). It is of interest to obtain a solution to the above system when the initial conditions $N_i(0)$, $i = 1, 2, \ldots, n$, are given. We now treat several cases which are of interest in applications.

1. Only E_1 present initially. Let $N_1(0) = n_{10}$, $N_i(0) = 0$, $i = 2, 3, \ldots, n$. By applying the Laplace transformation to the system (6.5) we obtain the subsidiary equations

$$(s + \lambda_1) n_1(s) = n_{10}$$

$$(s + \lambda_i) n_i(s) - \lambda_{i-1} n_{i-1}(s) = 0 \qquad i = 2, 3, \ldots, n \tag{6.6}$$

where $n_i(s)$ is the Laplace transform of $N_i(t)$. By solving the above equation successively, we obtain

$$n_1(s) = \frac{n_{10}}{s + \lambda_1} \tag{6.7}$$

$$n_i(s) = \frac{n_{10}\prod_{j=1}^{i-1}\lambda_j}{\prod_{j=1}^{i}(s + \lambda_j)} \qquad i = 2, \ldots, n-1 \tag{6.8}$$

$$n_n(s) = \frac{n_{10}\prod_{j=1}^{n-1}\lambda_j}{s\prod_{j=1}^{n-1}(s + \lambda_j)} \tag{6.9}$$

Application of the inversion theorem yields

$$N_i(t) = n_{10}\sum_{j=1}^{i}\alpha_{ij}e^{-\lambda_j t} \qquad i = 1, 2, \ldots, n-1$$

$$N_n(t) = n_{10}\left[1 - \sum_{j=1}^{n-1}\alpha_{n-1,j}\left(\frac{\lambda_{n-1}}{\lambda_j}\right)e^{-\lambda_j t}\right] \tag{6.10}$$

where

$$\alpha_{ij} = \frac{\prod_{k=1}^{i-1}\lambda_k}{\prod_{\substack{k=1 \\ k \neq j}}^{i}(\lambda_k - \lambda_j)} \tag{6.11}$$

2. Only E_1 present initially; all disintegration constants equal, i.e., $\lambda_1 = \lambda_2 = \cdots = \lambda_n = \lambda$. From Eq. (6.6) the subsidiary equation in this case is

$$n_i(s) = \frac{\lambda^{i-1}n_{10}}{(s + \lambda)^i} \tag{6.12}$$

By inverting, we obtain

$$N_i(t) = \frac{n_{10}(\lambda t)^{i-1}}{(i-1)!}e^{-\lambda t} \qquad i = 1, 2, \ldots, n-1 \tag{6.13}$$

3. Element E_1 is supplied from a primary source at a constant rate μ per unit time for $0 < t < T$, and then suddenly removed; initial conditions $N_i(0) = 0$, $i = 1, 2, \ldots, n$. In this case the first equation of (6.5) is replaced by

$$\frac{dN_1(t)}{dt} = -\lambda_1 N_1(t) + Q(t) \tag{6.14}$$

where

$$Q(t) = \mu \qquad \text{for } 0 < t < T$$
$$= 0 \qquad \text{for } t > T$$

The subsidiary equation for (6.14) is

$$(s + \lambda_1)n_1(s) = \frac{\mu}{s}(1 - e^{-sT}) \qquad (6.15)$$

Hence, the solutions of the subsidiary equations obtained for the first case must be multiplied by the factor $(1 - e^{-sT})s^{-1}$. Therefore, upon inversion, we obtain the solutions

$$N_i(t) = \mu\left(\frac{1}{\lambda_i} - \sum_{j=1}^{i} \frac{\alpha_{ii}}{\lambda_j} e^{-\lambda_j t}\right) \qquad i = 1, 2, \ldots, n-1$$

$$N_n(t) = \mu\left(t - \sum_{j=1}^{n-1}\left\{\frac{1}{\lambda_j} - \left(\frac{\lambda_{n-1}}{\lambda_j^2}\right)\alpha_{n-1,j}e^{-\lambda_j t}\right\}\right) \qquad (6.16)$$

for the interval $0 < t < T$. For $t > T$, we have

$$N_i(t) = \mu \sum_{j=1}^{i} \frac{\alpha_{ij}}{\lambda_j} e^{-\lambda_j t}(e^{\lambda_j T} - 1) \qquad i = 1, 2, \ldots, n-1$$

$$N_n(t) = \mu\left(T - \sum_{j=1}^{n-1}\left(\frac{\lambda_{n-1}}{\lambda_j^2}\right)\alpha_{n-1,j}e^{-\lambda_j t}(e^{\lambda_j T} - 1)\right) \qquad (6.17)$$

We close this section by remarking that the differential equations for radioactive transformations can be given a stochastic interpretation as follows: The disintegration of a radioactive atom can be considered as the transition from one physical state to another. Hence, we can interpret the radioactive disintegration process as a stochastic process with a finite number of states. Now, let us assume that in the interval $(t, t + \Delta t)$ the probability of an atom in state i passing to state $i + 1$ is $\lambda_i \Delta t + o(\Delta t)$. We are, therefore, dealing with a stochastic birth process, and the reasoning employed in Chap. 2 leads to the relation

$$P_i(t + \Delta t) = (1 - \lambda_i \Delta t)P_i(t) + \lambda_{i-1}P_{i-1}(t) \Delta t + o(\Delta t) \qquad (6.18)$$

where $P_i(t)$ is the probability that the system is in state i at time t. Equation (6.18) leads to the system of equations

$$\frac{dP_0(t)}{dt} = -\lambda_0 P_0(t)$$

$$\frac{dP_i(t)}{dt} = -\lambda_i P_i(t) + \lambda_{i-1}P_{i-1}(t) \qquad i = 1, 2, \ldots, n \qquad (6.19)$$

where $\lambda_n = 0$. This stochastic system of equations is formally the same as the deterministic system (6.5). In fact, (6.19) can be obtained from (6.5) by replacing $N_i(t)$ by $P_{i-1}(t)$, with the initial condition $P_0(0) = 1$. We see, therefore, that in the second case treated above the associated stochastic process is the Poisson process.

6.3 Theory of Particle Counters

A. Introduction. The theory of particle counters is concerned with the formulation and study of stochastic processes associated with the registration of impulses, due to radioactive substances, by a counting device designed to detect and record the impulses. The general problem can be described as follows: Consider a counter placed within range of a radioactive substance. We first consider the sequence of random events consisting of the arrival of radioactive impulses at the counter. This sequence is called the *primary sequence of events*. Owing to the inertia of the counter, all impulses arriving at the counter will probably not be registered or counted. The interval of time during which the counter is unable to register an impulse is called the *resolving time*, or *dead time*. Hence the recorded events, i.e., the registrations by the counter, form a *secondary sequence of events* the distribution of which depends upon the distribution of the primary sequence of events and upon the mathematical model used for the action of the counter. The recorded events can be considered as a sequence of secondary events selected successively from the primary sequence of events. The basic problem in the theory of particle counters can now be stated as follows: Determine the distribution function of the registered events, if the distribution function of the sequence of primary events and the selection rule (mathematical model for the action of the counter) are known.

Let $t_1, t_2, \ldots$ denote the times at which particles arrive at the counter; let $\chi_1, \chi_2, \ldots$ denote the duration of the successive pulses initiated by the particles arriving at $t_1, t_2, \ldots$; and let $t_1', t_2', \ldots$ denote the times of the successive registrations by the counter. The theory of particle counters usually considers two selection rules or types of models for the action of the counter. These models, referred to as Type I and Type II counters, can be described as follows:[1]

Type I counter. In a Type I counter there is a fixed resolving time, or dead time, $\tau > 0$ such that an event of the primary sequence $\{t_i\}$ is selected if and only if no event of the secondary sequence $\{t_i'\}$, i.e., a registration, has taken place during the preceding time interval of length τ.

Type II counter. In a Type II counter the resolving time is random and specified by the rule that an event of the primary sequence $\{t_i\}$ is selected if and only if *no event of the primary sequence* has taken place during the preceding time interval of length τ.

The difference between Type I and Type II counters is illustrated in Fig. 6.1. Type I counters are used in the theory of Geiger-Müller

[1] These definitions of Type I and Type II counter hold only in the case where the dead time has a constant length.

counters and mechanical recorders, and Type II counters are used in the theory of electron multipliers, scintillation or crystal counters, and electronic amplifiers.

In addition to assuming that the counter is either Type I or Type II, it is usually assumed that the sequence of primary events is a Poisson

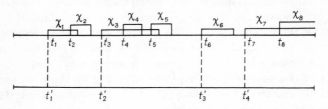

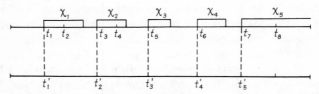

Figure 6.1   Type I counter, above;   Type II counter, below.

process, i.e., the probability of exactly k primary events in the interval $(0,t)$ is

$$\frac{(\lambda t)^k}{k!} e^{-\lambda t}$$

where $\lambda \Delta t + o(\Delta t)$ is the probability of a primary event in the interval $(t, t + \Delta t)$. In addition to making the counter problem easier to handle mathematically, the assumption of a Poisson process for the sequence of primary events makes the stochastic model more realistic, since the impulses from radioactive substances tend to behave randomly in time, at least in time intervals which are short relative to the half-life of the substance.[1]

In this section we present the theory of particle counters as developed by W. Feller and L. Takács. In Sec. 6.3B we present the simple and elegant theory due to Feller [10]. This theory centers around the integral equations of renewal theory and the application of the theory of the Laplace transformation. In Sec. 6.3C we present the theory as developed by Takács [45,46]. This theory is based on a general

---

[1] The half-life of a radioactive substance is defined as the time required for one-half of the substance to decay.

selection rule which includes the Type I and Type II counters as special cases.

B. General Theory. Applications to Type I and Type II Counters. Let $\{t_i'\}$ denote the sequence of registration times, and let $T_i' = t_i' - t_{i-1}' \ (i \geqslant 1)$ denote the time between the $(i-1)$st and ith registration. Denote by T_0' the time up to the first registration. Following the $(i-1)$st registration, the counter is inoperative for a period of time equal to its resolving time, after which the ith registration can take place. Once the counter is free, the situation is the same for all registrations.[1] Hence the sequence $\{t_i'\}$ forms a *recurrent process*, i.e., the time differences $T_i' = t_i' - t_{i-1}'$ are equidistributed independent positive random variables. From the definition of the T_i', the time up to the kth observation is given by

$$S_k = \sum_{i=0}^{k-1} T_i' \tag{6.20}$$

that is, S_k is the sum of k independent random variables.

Let $Y(t)$ denote the number of registrations during the interval $[0,t)$ and let $R(y,t) = \mathscr{P}\{Y(t) = y\}$. From (6.20) we have

$$R(y,t) = \mathscr{P}\{Y(t) = y\} = \mathscr{P}\{S_{y-1} \leqslant t\} - \mathscr{P}\{S_y \leqslant t\} \tag{6.21}$$

If we now assume that the probability distribution of registration times is Poisson, then

$$F_0(t) = 1 - e^{-\lambda t} \qquad \text{for } t \geqslant 0$$
$$= 0 \qquad \text{for } t < 0 \tag{6.22}$$

is the probability that the time T_0' from an arbitrary moment to the next event will not exceed t. The random variables T_i', $i \geqslant 1$, have the common distribution function

$$F(t) = \mathscr{P}\{T_i' \leqslant t\} \tag{6.23}$$

and
$$F_0(t) = \mathscr{P}\{T_0' \leqslant t\} \tag{6.24}$$

The distribution of T_0' is given by (6.22).

Let $F_k(t)$ denote the k-fold convolution of the distribution function $F(t)$ with itself. $F_k(t)$ can be determined, starting from $F_1(t) = F(t)$, by the recurrence formula

$$F_{k+1}(t) = \int_0^t F_k(t-u)\, dF(u) \qquad k = 1, 2, \ldots \tag{6.25}$$

[1] The times $\{t_i'\}$ are also the regeneration points of the process.

Hence, from (6.20), (6.23), and (6.25), we have

$$\mathscr{P}\{S_y \leqslant t\} = F_y(t) \tag{6.26}$$

and from (6.21)

$$R(y,t) = F_{y-1}(t) - F_y(t) \tag{6.27}$$

Let $M(t)$ denote the expected or mean number of registrations up to time t. From (6.21) and (6.26) we obtain

$$M(t) = \mathscr{E}\{Y(t)\} = \sum_{k=0}^{\infty} k R(k,t) = \sum_{k=0}^{\infty} F_k(t) \tag{6.28}$$

It is of interest to note that $M(t)$ satisfies an integral equation of the renewal type, namely,

$$M(t) = F_0(t) + \int_0^t M(t - u)\, dF(u) \tag{6.29}$$

This equation is obtained from (6.28) by using the recurrence relation (6.25).

In order to obtain $M(t)$, and an asymptotic relation for large t, we utilize the theory of the Laplace transformation. Let

$$\varphi(s) = \int_0^{\infty} e^{-st}\, dF(t) \tag{6.30}$$

$$\varphi_k(s) = \int_0^{\infty} e^{-st}\, dF_k(t) \tag{6.31}$$

and

$$\mu(s) = \int_0^{\infty} e^{-st} M(t)\, dt \tag{6.32}$$

denote the Laplace-Stieltjes and Laplace transforms of $F(t)$, $F_k(t)$, and $M(t)$, respectively. From the recurrence relation (6.25) we obtain

$$\varphi_k(s) = \varphi_0(s)[\varphi(s)]^{k-1} \tag{6.33}$$

Hence, from (6.27) and (6.28) application of the Laplace transformation yields

$$\mu(s) = \mathscr{L}\{M(t)\} = \frac{\varphi_0(s)}{s[1 - \varphi(s)]} \tag{6.34}$$

In order to obtain the asymptotic relation that $M(t)$ satisfies, we must expand $\varphi_0(s)$ and $\varphi(s)$ as power series in s and determine in (6.34) the coefficient of s^{-2}. Let

$$m = \int_0^{\infty} t\, dF(t) \tag{6.35}$$

denote the average or mean time between two registrations and let

$$\sigma^2 = \int_0^\infty (t-m)^2 \, dF(t)$$

$$= \int_0^\infty t^2 \, dF(t) - m^2 \tag{6.36}$$

denote the variance of the time between two registrations.

If we now expand e^{-st} as a power series, (6.30) yields

$$\varphi(s) = 1 - ms + \tfrac{1}{2}(\sigma^2 + m^2)s^2 + \cdots \tag{6.37}$$

and for $\varphi_0(s)$ we obtain

$$\varphi_0(s) = 1 - m_0 s + \cdots \tag{6.38}$$

In (6.38) m_0 is the average of T_0', i.e., m_0 is the average length of time before the first registration. In the Poisson case m_0 is λ^{-1}, where λ is the Poisson parameter. By introducing (6.27) and (6.38) into (6.34), we obtain

$$\mu(s) = \frac{1}{ms^2} + \left(\frac{1}{2} + \frac{\sigma^2}{2m^2} - \frac{m_0}{m}\right)\frac{1}{s} + \cdots \tag{6.39}$$

Inversion of (6.39) shows that $M(t)$ satisfies the asymptotic relation[1]

$$M(t) \sim \frac{t}{m} + \frac{1}{2} + \frac{1}{2}\frac{\sigma^2}{m^2} - \frac{m_0}{m} + o(t) \tag{6.40}$$

It is of interest to note that for $m < \infty$

$$\lim_{t\to\infty}\frac{M(t)}{t} = \frac{1}{m} \tag{6.41}$$

Let

$$D(t) = 2\sum_{k=1}^\infty k F_{k-1}(t) \tag{6.42}$$

then from (6.27) the variance of the number of registrations up to time t is given by

$$\mathscr{D}^2\{Y(t)\} = B(t) = \sum_{k=1}^\infty k^2 R(k,t) - [M(t)]^2$$

$$= D(t) + M(t) - [M(t)]^2 \tag{6.43}$$

[1] We refer to Bellman [5] and Feller [9] for a discussion of the asymptotic properties of integral equations of the renewal type. We also remark that (6.40) does not hold in general. If $F(t)$ is not a lattice distribution, then (6.39) holds.

If we denote by $\nu(s)$ the Laplace transform of $D(t)$, we obtain

$$\nu(s) = \frac{2\varphi_0(s)}{s[1 - \varphi(s)]^2} \tag{6.44}$$

By proceeding as before, it can be shown that $B(t)$, the variance of the number of registrations up to time T, satisfies the asymptotic relation

$$B(t) \sim \frac{\sigma^2 t}{m^3} \tag{6.45}$$

Analogously to (6.41), we have for $m < \infty$

$$\lim_{t \to \infty} \frac{B(t)}{t} = \frac{\sigma^2}{m^3} \tag{6.46}$$

We also note from (6.25) and (6.42) that $D(t)$ also satisfies an integral equation of the renewal type, namely,

$$D(t) = 2M(t) + \int_0^t D(t - u)\, dF(u) \tag{6.47}$$

We now consider the application of the above results to Type I and Type II counters.

Type I Counters. For Type I counters $F_0(t) = \mathscr{P}\{T_0' \leqslant t\} = 1 - e^{-\lambda t}$, $t > 0$. By definition,

$$F(t) = \mathscr{P}\{T_i' \leqslant t\} = 1 - e^{-\lambda(t-\tau)} \qquad \text{for } t \geqslant \tau$$
$$= 0 \qquad \text{for } t \leqslant \tau \tag{6.48}$$

where τ is the resolving time. Hence $F(t)$ differs from $F_0(t)$ only by a change of origin, the new origin being a regeneration point of the process. From (6.25) we observe that, if $F(t)$ is replaced by $F_0(t)$, the derivative of $F_k(t)$ would give the Poisson distribution. Therefore

$$\frac{dF_k(t)}{dt} = \frac{\lambda^{k+1}(t - k\tau)^k}{k!} \qquad \text{for } t \geqslant k\tau$$
$$= 0 \qquad \text{for } t \leqslant k\tau \tag{6.49}$$

Hence, from (6.28) and (6.49) we have for the *mean number of registrations for a Type I counter*

$$M(t) = 1 + \left[\frac{t}{\tau}\right] - \sum_{i=0}^{[t/\tau]} e^{-\lambda(t-i\tau)} \sum_{j=0}^{i} \frac{\lambda^j(t - i\tau)^j}{j!}$$
$$= \sum_{i=0}^{[t/\tau]} e^{-\lambda(t-i\tau)} \sum_{j=i+1}^{\infty} \frac{\lambda^j(t - i\tau)^j}{j!} \tag{6.50}$$

In the above expressions $[t/\tau]$ denotes the greatest positive integer not exceeding the ratio t/τ.

Finally, from (6.40) and (6.45), the asymptotic expressions for the mean and variance of the number of registrations in the interval $[0,t)$ for a Type I counter are given by

$$M(t) \sim \frac{\lambda t}{1 + \lambda\tau} + \frac{\lambda^2\tau^2}{2(1 + \lambda\tau)^2} \tag{6.51}$$

and

$$B(t) \sim \frac{\lambda t}{(1 + \lambda\tau)^3} \tag{6.52}$$

Type II Counters. In applying the general theory to Type II counters we first observe that the distribution function $F(t)$ is unknown, since by definition the Type II counter may be inoperative for an arbitrarily long period following the first registration. This condition will obtain if a large number of impulses arrive at the counter in rapid succession following the first registration. Of these impulses only the first will be registered. Let

$$p = e^{-\lambda\tau} \qquad q = 1 - p = 1 - e^{-\lambda\tau}$$

where τ is the resolving time. Hence, the probability that, once the counter is inoperative, exactly k impulses will extend the inoperative period is pq^k. Now let $T_i = t_i - t_{i-1}$ denote the time between the arrival of the $(i-1)$st and ith impulses at the counter. Hence the counter is inoperative for a total time given by

$$S_k^* = \sum_{i=1}^{k} T_i$$

In this case we have a sum of k random variables, the number k being a random variable.[1]

From (6.22) and the definition of q we see that

$$Q(t) = \frac{\mathscr{P}\{T_0' \leqslant t\}}{q} = \frac{1 - e^{-\lambda t}}{1 - e^{-\lambda\tau}} \qquad 0 < t < \tau \tag{6.53}$$

is the conditional probability that an impulse arriving at the counter will succeed another impulse by less than t, given that the interval between them does not exceed the resolving time τ. Therefore, if it is known that k impulses have arrived at the counter during the inoperative period, we have

$$Q(t) = \mathscr{P}\{T_i \leqslant t\} \qquad 0 < t < \tau \quad i = 1, 2, \ldots, k$$

[1] Cf. the remarks in Appendix A on compound distributions.

Let $\eta(s)$ denote the Laplace-Stieltjes transform of $Q(t)$. We have

$$\eta(s) = \mathscr{L}\{Q(t)\} = \int_0^\tau e^{-st}\, dQ(t)$$

$$= \frac{\lambda}{\lambda + s}\left(\frac{1 - e^{-(\lambda+s)\tau}}{1 - e^{-\lambda\tau}}\right)$$

hence $$\mathscr{L}\{\mathscr{P}\{S_k^* \leqslant t\}\} = [\eta(s)]^k$$

In order to obtain the transform of the conditional distribution of S_k^*, under the assumption that k impulses arrived during the inoperative period, it is necessary to take into consideration the *constant resolving time* τ following the last registration. This requires a multiplication of $[\eta(s)]^k$ by $e^{-\tau s}$. Hence, the conditional distribution has the transform $e^{-\tau s}[\eta(s)]^k$. In view of the above, the transform of the absolute distribution $\mathscr{P}\{S_k^* \leqslant t\}$ is

$$\mathscr{L}\{\mathscr{P}\{S_k^* \leqslant t\}\} = pe^{-s\tau}\sum_{k=0}^\infty [q\eta(s)]^k$$

$$= \frac{(\lambda + s)e^{-(\lambda+s)\tau}}{s + \lambda e^{-(\lambda+s)\tau}} \tag{6.54}$$

Now, the time T_i' between the $(i-1)$st and the ith registrations is made up of a time whose distribution is the inverse distribution of (6.54) and the time from the moment the counter is inoperative to the next impulse (cf. the figure illustrating Type II counters). The distribution of the latter time is $1 - e^{-\lambda t}$, whose Laplace transform is $\lambda/(\lambda + s)$. Hence the Laplace-Stieltjes transform of the distribution of T_i', i.e., $F(t)$, is given by

$$\varphi(s) = \frac{\lambda e^{-(\lambda+s)\tau}}{s + \lambda e^{-(\lambda+s)\tau}} \tag{6.55}$$

We can now utilize the general theory to determine the mean and variance of the number of registrations in the interval $[0,t)$. If we now introduce (6.55) into (6.34), we obtain

$$\mu(s) = \mathscr{L}\{M(t)\} = \frac{\lambda}{(\lambda + s)s^2}\,(s + \lambda e^{-(\lambda+s)\tau}) \tag{6.56}$$

Inversion of (6.56) gives

$$M(t) = 1 - e^{-\lambda t} \qquad\qquad \text{for } t \leqslant \tau$$

$$= 1 - e^{-\lambda\tau} + \lambda(t - \tau)e^{-\lambda} \quad \text{for } t \geqslant \tau \tag{6.57}$$

The above expression for $M(t)$ is the *exact* expression for the *mean number of registrations for Type II counters.*

If we now introduce (6.55) into (6.44), invert $\nu(s)$, and use the relation $B(t) = D(t) + M(t) - [M(t)]^2$, we obtain

$$\mathscr{D}^2\{Y(t)\} = B(t) = \lambda(t - \tau)e^{-\lambda\tau}(1 - 2\lambda\tau e^{-\lambda\tau})$$
$$- e^{-\lambda\tau} + (1 + \lambda\tau)^2 e^{-2\lambda} \qquad \text{for } t > 2\tau \quad (6.58)$$

The above expression is the *variance of the number of registrations* in the interval $[0,t)$ for Type II counters.

C. General Model of a Particle Counter. In Sec. 6.3B we considered the general theory of particle counters and the application of this theory to Type I and Type II counters. We now consider a general model of a particle counter which includes the Type I and Type II counters as special cases. The generalization is achieved by introducing a general model for the action of the counter, i.e., a general selection rule, which includes the usual types studied as special cases.

The results of this section contain as a special case most of the earlier work on Type I and Type II counters. We refer to the following papers: 10, 12, 13, 19–22, 30, 37–39, 43, and 46.

Consider a primary sequence of impulses arriving at a counter and let $\{t_i\}$ denote the sequence of arrival times, where, without loss of generality, we put $t_0 = 0$. Let $T_i = t_i - t_{i-1}, i = 0, 1, \ldots$, denote the difference between the times of arrival of the $(i - 1)$st and ith impulses. The arrival times form a recurrent process; hence, the T_i are equidistributed independent positive random variables. Let

$$G(t) = \mathscr{P}\{T_i \leqslant t\}$$

denote the distribution of the random variables T_i and let

$$\mu = \int_0^\infty t \, dG(t) \qquad \text{and} \qquad s^2 = \int_0^\infty (t - \mu)^2 \, dG(t)$$

denote the mean and variance of T_i', respectively.

Let $\{t_i'\}$ denote the sequence of registration times. The sequence $\{t_i'\}$ can be considered as those events in the primary sequence $\{t_i\}$ which are selected by the counter. Since $\{t_i\}$ is a recurrent process, the secondary sequence $\{t_i'\}$ is a recurrent process also. In view of the above, we can consider a counting device to be an instrument for transforming a recurrent process into another recurrent process. As before, let

$$F(t) = \mathscr{P}\{T_i' \leqslant t\}$$

denote the distribution function of the random variable T_i' and let

$$m = \int_0^\infty t \, dF(t) \qquad \text{and} \qquad \sigma^2 = \int_0^\infty (t - m)^2 \, dF(t)$$

denote the mean and variance of T_i', respectively.

We now consider the mathematical model for the action of the counter. Put $t_0 = t_0'$ and assume that at time t_0 an impulse, the duration of which is χ_0, arrives at the counter. The impulse arriving at time t_i $(i \geqslant 1)$ results in a registration with probability p, $0 \leqslant p \leqslant 1$, if at time t_i a registration is in course, and with probability one if at time t_i there is no registration in course. We see, therefore, that we are dealing with a model for Type I counters if $p = 0$ and a model for Type II counters if $p = 1$. This general selection rule is due to Albert and Nelson [1].

The problem we first consider is that of determining the number of impulses arriving at the counter in the interval $[0,t)$. Let the random variable $X(t)$ denote the number of impulses arriving at the counter in the interval $[0,t)$ and let

$$P(x,t) = \mathscr{P}\{X(t) \leqslant x\}$$

In addition to determining the above distribution function, we also determine the following probabilities:

(a) $\mathscr{P}\{X(T + t) - X(T) \leqslant x\}$, where $T > 0$ is fixed
(b) $\lim\limits_{t \to \infty} \mathscr{P}\{X(T + t) - X(T) \leqslant x\}$
(c) $\mathscr{P}\{X(\tau + t) - X(\tau) \leqslant x\}$, where τ is a random variable uniformly distributed over the interval $(0,T)$

Let us now assume that $G(t) = \mathscr{P}\{T_i \leqslant t\}$ is known. The probability $P(x,t)$ can be obtained as follows: The probability of at most x impulses arriving at the counter in the interval $[0,t)$ is the same as the probability that the time of arrival of the $(x + 1)$st impulse is greater than t. Since the probability of the latter event is one minus the probability that the time of arrival of the $(x + 1)$st impulse is less than or equal to t, we have

$$P(x,t) = \mathscr{P}\{X(t) \leqslant x\} = \mathscr{P}\{t < t_{x+1}\}$$
$$= 1 - \mathscr{P}\{t_{x+1} \leqslant t\} = 1 - G_{x+1}(t)$$

where $G_n(t)$ is the n-fold convolution of $G(t)$ with itself, with

$$G_0(t) = 1 \qquad \text{if } t \geqslant 0$$
$$= 0 \qquad \text{if } t < 0$$

Let $\beta(t)$ denote the mean or expected number of impulses in the interval $[0,t)$, whence

$$\beta(t) = \mathscr{E}\{X(t)\} = \sum_{n=1}^{\infty} G_n(t) \tag{6.59}$$

Similarly, the variance of the number of impulses in $[0,t)$ is given by

$$\gamma(t) = \mathscr{D}^2\{X(t)\} = \sum_{n=1}^{\infty} (2n - 1)G_n(t) - [\beta(t)]^2 \tag{6.60}$$

Let

$$P_T(x,t) = \mathscr{P}\{X(T+t) - X(T) \leqslant x\}$$

We now show that

$$P_T(x,t) = 1 - K_T(t)*G_x(t) \tag{6.61}$$

where

$$K_T(t) = \int_T^{T+t} [1 - G(T+t-u)d]\,d\beta(t) \tag{6.62}$$

In (6.61) $K_T(t)*G_x(t)$ is the convolution of $K_T(t)$ with $G_x(t)$. Let the random variable Z_T represent the distance between the time T and the time of arrival of the next impulse—if the counter is free. Now, the probability of at most x impulses arriving in the interval $(T, T+t)$ is $1 - \mathscr{P}\{(Z_T \leqslant t)* G_x(t)\}$. We observe that the event $\{Z_T \leqslant t\}$ occurs when there is at least one arrival in the interval $(T, T+t)$. This event can occur in several mutually exclusive ways. In the interval $(T, T+t)$ the arrival of the last impulse can occur at $t_1, t_2, \ldots, t_x, \ldots$; hence,

$$\mathscr{P}\{Z_T \leqslant t\} = K_T(t) = \sum_{x=1}^{\infty} \int_T^{T+t} [1 - G(T+t-u)]\,dG_x(t)$$

$$= \int_T^{T+t} [1 - G(T+t-u)]\,d\beta(t) \tag{6.63}$$

since $\beta(t) = \sum_{x=1}^{\infty} G_x(t)$. Therefore, $K_T(t)$ can be interpreted as the probability that the next arrival after time T occurs in the interval $(T, T+t)$.

We now consider the limiting behavior of $P_T(x,t)$ as T approaches infinity. Let

$$P^*(x,t) = \lim_{T \to \infty} P_T(x,t) = \lim_{T \to \infty} \mathscr{P}\{X(T+t) - X(T) \leqslant x\}$$

To obtain $P^*(x,t)$, we assume that $G(t)$ is not a lattice distribution[1] and that its expected value μ is finite. Since $K_T(t)$ is the only expression in (6.61) that depends on T, we need only consider $\lim_{T \to \infty} K_T(t)$. In view of the above assumptions on $G(t)$, application of a result due to Blackwell [6] and Doob [8] shows that $\lim_{T \to \infty} K_T(t) = K(t)$ exists. From (6.62) we have

$$\lim_{T \to \infty} K_T(t) = K(t) = \frac{1}{\mu} \int_0^t [1 - G(u)]\,du \tag{6.64}$$

Hence from (6.61) we have

$$P^*(x,t) = \lim_{T \to \infty} P_T(x,t) = 1 - K(t)*G_x(t) \tag{6.65}$$

[1] A discrete distribution of a random variable X is called a *lattice distribution* if there exist numbers α and $\beta > 0$ such that every possible value of X can be written in the form $\alpha + \kappa\beta$, where k assumes integer values.

This limiting distribution is the stationary distribution of $X(t)$ in the strict sense. That is, $P^*(x,t)$ is the probability that, after an infinitely long time, x impulses at most arrive at the counter in an interval of length t. In case $G(t)$ is a lattice distribution, $P^*(x,t)$ cannot be considered as the limit of $P_T(x,t)$. In this case we say that a stationary distribution of $X(t)$ exists in the *weak sense*.

Finally, we consider the case when τ is a random variable uniformly distributed over the interval $(0,T)$. Let

$$Q_T(x,t) = \mathscr{P}\{X(\tau + t) - X(\tau) \leqslant x\}$$

From (6.61) we have

$$Q_T(x,t) = 1 - K_T^*(t) * G_x(t) \tag{6.66}$$

where

$$K_T^*(t) = \int_0^t [1 - G(t - u)] \left\{ \frac{\beta(T + u) - \beta(u)}{T} \right\} du \tag{6.67}$$

Now if μ, the expected value of $G(t)$, is finite, we have [cf. (6.41)]

$$\lim_{t \to \infty} \frac{\beta(t)}{t} = \frac{1}{\mu} \tag{6.68}$$

hence $\lim_{T \to \infty} K_T^*(t) = K^*(t) = K(t)$ exists. Therefore, for $\mu < \infty$,

$$\lim_{T \to \infty} Q_T^*(x,t) = P^*(x,t) \tag{6.69}$$

exists. In this case $P^*(x,t)$ is called the stationary distribution of $X(t)$ in the *wide sense*.

We now turn to the problem of determining the distribution of the number of registrations in the interval $[0,t)$. As in Sec. 6.3B, let the random variable $Y(t)$ denote the number of registrations in the interval $[0,t)$ and let

$$R(y,t) = \mathscr{P}\{Y(t) \leqslant y\}$$

As we remarked earlier, both $\{t_i\}$ and $\{t_i'\}$ are recurrent processes; hence, the problem of determining $R(y,t)$ and the associated distributions is the same as that considered for the distribution $P(x,t)$. We have only to obtain the distribution function $F(t)$, where $F(t) = \mathscr{P}\{T_i' \leqslant t\}$. Let

$$\varphi(s) = \int_0^\infty e^{-st} \, dF(t) \tag{6.70}$$

and

$$\mu(s) = \int_0^\infty e^{-st} \, dM(t) \tag{6.71}$$

denote the Laplace-Stieltjes transforms of $F(t)$ and $M(t)$, respectively, where [cf. (6.28)]

$$M(t) = \mathscr{E}\{Y(t)\} = \sum_{n=1}^{\infty} F_n(t) \qquad (6.72)$$

denotes the expected or mean number of registrations up to time t.

In order to determine $F(t)$, it is necessary to decide on the selection rule and make some assumption concerning the distribution of the duration of the impulses. We now consider the case of the general selection rule, and we assume that χ_i, the duration of the ith impulse, is a constant for every i, say $\chi_i = \alpha$, $\alpha > 0$.

The distribution function $F(t)$ can be determined by utilizing the relationship between $M(t)$ and $F_n(t)$ and the method of the Laplace transformation. From (6.71) we have

$$\mu(s) = \frac{\varphi(s)}{1 - \varphi(s)} \qquad (6.73)$$

hence

$$\varphi(s) = \frac{\mu(s)}{1 + \mu(s)} \qquad (6.74)$$

In view of the above, if the Laplace transform of $M(t)$ is known, we know the Laplace transform of $F(t)$ and can, therefore, obtain $F(t)$ by inversion. We first show that

$$M(t) = \Phi(t) - \Phi(\alpha) + \int_0^t M(t - u) \, d\Phi(u) \qquad \text{for } t \geqslant \alpha$$

$$= 0 \qquad \text{for } t < \alpha \qquad (6.75)$$

The distribution function $\Phi(t)$ is given by

$$\Phi(t) = pm^*(t) \qquad \text{for } 0 < t < \alpha$$

$$= m_\alpha^*(t) - m_\alpha^*(\alpha) + pm^*(\alpha) \qquad \text{for } \alpha \leqslant t \leqslant \infty$$

where

$$m_\alpha^*(t) = G(t) + (1 - p)\int_0^\alpha G(t - u) \, dm^*(u)$$

and

$$m^*(t) = \sum_{n=1}^{\infty} (1 - p)^{n-1} G_n(t)$$

That $M(t) = 0$ for $t < \alpha$ is clear. The integral equation for $M(t)$, $t \geqslant \alpha$, can be derived as follows: Let t_y denote the first arrival time in the primary sequence $\{t_i\}$ which gives rise to an impulse. Then the conditional expectation of $Y(t)$, given that $t_y = u$, is given by

$$\mathscr{E}\{Y(t) \mid t_y = u\} = M(t - u) \qquad \text{for } 0 < u \leqslant \alpha$$

$$= 1 + M(t - u) \qquad \text{for } \alpha < u \leqslant t$$

$$= 0 \qquad \text{for } t < u < \infty \qquad (6.76)$$

We also observe that $\Phi(t) = \mathscr{P}\{t_y \leqslant t\}$. This follows from the definition of $\mathscr{P}\{t_y \leqslant t\}$, namely,

$$\mathscr{P}\{t_y \leqslant t\} = \sum_{n=1}^{\infty} \mathscr{P}\{y = n, t_n \leqslant t\} \tag{6.77}$$

In particular, if $\{t_i\}$ is a Poisson process with parameter λ, we have

$$\begin{aligned}\Phi(t) &= 1 - e^{-\lambda p t} && \text{for } 0 \leqslant t \leqslant \alpha \\ &= 1 - e^{-\lambda t + \lambda \alpha (1-p)} && \text{for } \alpha \leqslant t \leqslant \infty \end{aligned} \tag{6.78}$$

For Type I counters ($p = 0$) we have

$$\begin{aligned}\Phi(t) &= 0 && \text{for } 0 \leqslant t \leqslant \alpha \\ &= 1 - e^{-\lambda(t-\alpha)} && \text{for } \alpha \leqslant t \leqslant \infty \end{aligned} \tag{6.79}$$

and for Type II counters ($p = 1$) we have

$$\Phi(t) = 1 - e^{-\lambda t} \qquad \text{for } 0 \leqslant t \leqslant \infty \tag{6.80}$$

From (6.73) and (6.75) we have

$$\mu(s) = \frac{\displaystyle\int_{\alpha}^{\infty} e^{-st}\, d\Phi(t)}{1 - \displaystyle\int_{0}^{\infty} e^{-st}\, d\Phi(t)} \tag{6.81}$$

and from (6.74)

$$\varphi(s) = \frac{\displaystyle\int_{\alpha}^{\infty} e^{-st}\, d\Phi(t)}{1 - \displaystyle\int_{0}^{\alpha} e^{-st}\, d\Phi(t)} \tag{6.82}$$

From (6.82) $F(t)$, for the general selection rule, can be uniquely determined by inversion. Once $F(t)$ is determined, we can obtain $P_T(y,t)$ and $P^*(y,t)$ from (6.61) and (6.65), respectively, with $G(t)$ replaced by $F(t)$.

As a final problem we derive an expression for the probability, say $S(t)$, that all impulses initiated in the interval $(0,t)$ are completed at time t. Hence, $S(t)$ is the probability that at time t there is no impulse in course. For $t < \alpha$ it is clear that $S(t) = 0$. For $t \geqslant \alpha$ the event that there is no impulse in course can occur in the following mutually exclusive ways:

1. $t_y > t$; i.e., the time of arrival of the yth impulse is greater than t. The contribution of this event to $S(t)$ is $1 - \Phi(t)$.

2. $t_y = u$, where $0 < u \leqslant t$, but all impulses initiated in $[0,t)$ are completed at time t. The contribution of this event to $S(t)$ is

$$\int_{0}^{t} S(t - u)\, d\Phi(u)$$

Therefore, for $t \geqslant \alpha$, $S(t)$ satisfies the integral equation

$$S(t) = 1 - \Phi(t) + \int_0^t S(t - u) \, d\Phi(u) \tag{6.83}$$

Application of the Laplace transformation to (6.83) gives

$$\omega(s) = \mathscr{L}\{S(t)\} = \frac{\displaystyle\int_\alpha^\infty [1 - \Phi(t)]e^{-st} \, dt}{1 - \displaystyle\int_0^\infty e^{-st} \, d\Phi(t)} \tag{6.84}$$

Given $\Phi(t)$, the Laplace transform of $S(t)$ can be determined; the inversion of it gives $S(t)$. We note, however, that the limit of $S(t)$ as t approaches infinity, if it exists, can be determined without recourse to inversion. By using Theorem B.5, we have

$$S = \lim_{t \to \infty} S(t) = \lim_{s \to 0} s\omega(s) \tag{6.85}$$

hence
$$S = \frac{\displaystyle\int_\alpha^\infty [1 - \Phi(t)] \, dt}{\displaystyle\int_0^\infty [1 - \Phi(t)] \, dt} \tag{6.86}$$

From (6.86) we see that, in order to obtain S, it is only necessary to know $\Phi(t)$.

6.4   A Problem Concerning Nuclear Fission Detectors

A.   Introduction. Instruments known as *fission detectors* are used in the determination of the energy distribution of the fission fragments resulting from nuclear fission processes. Two fission fragments are produced from every fission process; hence the number of fission pulses observed within a solid angle of two radians is equal to the number of fissions produced in the material.

In the design and construction of ionization chambers for the study of nuclear fission processes it is necessary to consider the problem of excluding (or minimizing) the ionizing action of α particles. The energy of the fission fragments can vary over a wide range, but the expected or mean energy of the fragments is in each case an order of magnitude larger than the energy of the α particles. Therefore it is possible for pulses due to α particles to "pile up" and produce a pulse which is registered as a pulse created by fission fragments. In order to correct for the pulses that are due to the α particles, it is necessary to have at least an approximate estimate of the number of such cases.

The problem of correcting for the number of fissions has been considered by Rossi and Staub [33], in whose treatment it was assumed that the pulses have rectangular shape and duration τ. Under these assumptions the following approximate estimate of the number of registrations per unit time containing n unwanted pulses due to α particles was obtained:

$$C(n) = \frac{n_0}{1 + n_0\tau} \frac{(n_0\tau)^{n-1}}{(n-1)!} e^{-n_0\tau} \tag{6.87}$$

where n_0 is the true number of pulses per unit time. Since in most cases the pulses have an exponential shape, the above expression holds only approximately. However, it gives the right order of magnitude if the duration of the pulse τ is taken equal to the resolving time.

Pál [26] has considered the case when pulses have an exponential rather than rectangular shape. In the next section we present Pál's treatment of this problem.

B. Pulses with an Exponential Shape. Let A denote the amplitude of the pulse produced by the α particles and let $H(x) = \mathscr{P}\{A < x\}$. Let the random variable $X(t)$ denote the sum of the pulses produced by the α particles in the interval $(0,t)$. It is assumed throughout that the number of pulses satisfies a Poisson distribution. Now let $R(x,t) = \mathscr{P}\{X(t) < x\}$. A functional equation for $R(x,t)$ can be derived, in the usual manner, by considering that $R(x, t + \Delta t)$ is made up of the following mutually exclusive events:

1. The sum of the pulses is less than $xe^{\Delta t/\tau}$ at time t, and no new pulse appears in $(t, t + \Delta t)$.

2. The sum of the pulses is less than $(x - y)e^{\Delta t/\tau}$ at time t, and a new pulse with amplitude in the interval $(y, y + dy)$ appears in $(t, t + \Delta t)$. Therefore

$$R(x, t + \Delta t) = (1 - N\,\Delta t)R(xe^{\Delta t/\tau}; t)$$

$$+ N\,\Delta t \int_0^x R((x - y)e^{\Delta t/\tau}, t)\,dH(y) + o(\Delta t)$$

In the above, N is the pulse density and the term $o(\Delta t)$ represents the probability of more than one pulse entering in the interval $(t, t + \Delta t)$. In the limit we obtain from the above the functional equation[1]

$$\frac{\partial R(x,t)}{\partial t} = \frac{x}{\tau}\frac{\partial R(x,t)}{\partial x} - N\left\{R(x,t) - \int_0^x R(x - y, t)\,dH(y)\right\} \tag{6.88}$$

[1] For the proof of the existence of $\dfrac{\partial R(x,t)}{\partial x}$ and $\lim\limits_{t\to\infty} R(x,t) = R(x)$ we refer to [42] or [44].

If $R(x,t) \to R(x)$ as $t \to \infty$, we obtain from Eq. (6.88) the steady-state equation

$$\frac{x}{\tau} \frac{\partial R(x)}{\partial x} = N\left\{ R(x) - \int_0^x R(x-y)\, dH(y) \right\} \qquad (6.89)$$

Since the pulses due to the α particles and the pulse due to the fission fragments are independent, the steady-state distribution function of the number of pulses produced by the fission fragments, say $S(x)$, satisfies

$$\frac{x}{\tau} \frac{\partial S(x)}{\partial x} = M\left\{ S(x) - \int_0^x S(x-y)\, dK(y) \right\} \qquad (6.90)$$

where M is the pulse density and $K(y)$ is the distribution function of the pulse amplitude. From $R(x)$ and $S(x)$ the *distribution function of the output voltage of the ionization chamber*, say $P(x)$, can be obtained, since

$$P(x) = \int_0^x R(x-y)\, dS(y) \qquad (6.91)$$

Let V denote the recording or registration level of the discriminator which follows the ionization chamber. Should the voltage at the input of the discriminator increase above the recording level V, a "nuclear fission" is recorded. Hence in this case, the pulses connected with the pile-up of pulses from α particles are also recorded. Now let Q_1 denote the density of pure fissions and Q_2 denote the density associated with the pile-up of pulses from α particles. Therefore, the total number of recorded pulses is given by

$$Q = Q_1 + Q_2 \qquad (6.92)$$

The basic problem is to determine the function $Q(V)$. From the definitions of $H(x)$, $K(x)$, and $P(x)$, we have

$$Q_1(V) = M \int_0^V [1 - K(y)]\, dP(V - y) \qquad (6.93)$$

and

$$Q_2(V) = N \int_0^V [1 - H(y)]\, dP(V - y) \qquad (6.94)$$

We now assume that

$$H(x) = 1 - e^{-\lambda x} \qquad (6.95)$$

where λ^{-1} is the mean value of the amplitude of the pulses. Since the ionizing effects of the fission fragments are canceled out, the distribution function of the pulse amplitude is given by

$$K(x) = \mu \int_0^x (1 - e^{-\mu(x-y)}) e^{-\mu y}\, dy \qquad (6.96)$$

where $2\mu^{-1}$ is the mean value of the amplitude of the fission fragments.

Let $r(x)$ and $s(x)$ denote the density functions associated with $R(x)$ and $S(x)$ and let $\rho(p)$ and $\sigma(p)$ denote the Laplace-Stieltjes transforms of $r(x)$ and $s(x)$, respectively. Then from (6.89) and (6.95) we have

$$\frac{d \log \rho(p)}{dp} = -\frac{N\tau}{p + \lambda} \tag{6.97}$$

Therefore

$$\rho(p) = \frac{\lambda^{N\tau}}{(p + \lambda)^{-N}} \tag{6.98}$$

and

$$r(x) = \frac{\lambda(\lambda x)^{N\tau-1}e^{-\lambda x}}{\Gamma(N\tau)} \tag{6.99}$$

Hence, the distribution function $R(x)$ is given by

$$R(x) = \frac{\Gamma(\lambda x, N\tau)}{\Gamma(N\tau)} \tag{6.100}$$

In (6.100) $\Gamma(\lambda x, N\tau)$ is the incomplete gamma function.

Similarly, from (6.90) and (6.96) we have

$$\frac{d \log \sigma(p)}{dp} = -M\tau[(p + \mu)^{-1} + \mu(p + \mu)^{-2}] \tag{6.101}$$

Therefore

$$\sigma(p) = \mu^{M\tau}e^{-M\tau p(p+\mu)^{-1}}(p + \mu)^{-M\tau} \tag{6.102}$$

and

$$s(x) = \mu\left(\frac{x\mu}{M\tau}\right)^{(M\tau-1)/2}e^{-(M\tau+\mu x)}I_{M\tau-1}(2\sqrt{M\tau\mu x}) \tag{6.103}$$

where $I_k(x) = i^{-k}J_k(ix)$ is the Bessel function of order k with imaginary argument.

Let $q_1(p)$ and $q_2(p)$ denote the Laplace-Stieltjes transforms of $Q_1(x)$ and $Q_2(x)$, respectively. From (6.93), (6.94), and the above results, we have

$$q_1(p) = \left(\frac{M}{N}\right)(p + \lambda)(p + \mu)^{-1}[1 + \mu(p + \mu)^{-1}]q_2(p) \tag{6.104}$$

and

$$q_2(p) = N\lambda^{N\tau}\mu^{M\tau}e^{-M\tau p(p+\lambda)^{-1}}(p + \lambda)^{-(N\tau+1)}(p + \mu)^{-M\tau} \tag{6.105}$$

For values of V of practical importance $VM\mu \ll 1$. Inverting (6.104) and (6.105) under this assumption yields the approximate expressions

$$Q_1(V) \sim Me^{-(M\tau+\mu V)}(V\lambda)^{N\tau}(V\mu)^{M\tau}(1 + \mu V) \tag{6.106}$$

$$Q_2(V) \sim Ne^{-(M\tau+\lambda V)}(V\lambda)^{N\tau}(V\mu)^{M\tau} \tag{6.107}$$

$Q_1(V)$ gives the number of recordings per unit time of true fission, and $Q_2(V)$ gives the number of false fissions per unit time associated with the pile-up of particles from α particles.

6.5 Theory of Tracks in Nuclear Research Emulsions

A. Introduction. The use of sensitive photographic emulsions to record the tracks of particles is of great importance in cosmic-ray and nuclear physics. A photographic emulsion is made up of crystals of roughly spherical shape suspended in gelatin. When a charged particle passes through the emulsion, it activates some of the crystals through which it travels. When the emulsion is developed, each "activated" crystal grows into a "grain" of larger radius, while the crystals that were not activated dissolve away. Clearly, if two activated crystals are close enough to one another, the developed, or final, grains may overlap and form what is termed a "blob." Because microscopic methods are employed in examining the developed plates and because the size of the grains is close to the limit of optical resolution, it is usually difficult to tell whether a blob is made up of one or more grains. Therefore, the standard procedure is to do "grain counting" by counting blobs or by counting the gaps between the blobs.

The experimental procedure used points up the need for theoretical expressions which relate the observed number of gaps per unit length of track, or some other measurable quantity which is characteristic of the track, to the physical characteristics of the emulsion (e.g., initial crystal size, spatial distribution of initial crystals, and final grain size). In the theoretical study of track formation, probability methods are indicated, since the process of track formation is statistical in character. In addition, in experimental studies it is necessary to know the parameters (e.g., mean and variance) associated with the measured track characteristics.

We now consider the theory of track formation developed by Blatt [7]. This theory utilizes the method of regeneration points, which has been considered in Sec. 2.5 and in connection with other physical applications.

For a discussion of the applications of photographic emulsions in cosmic-ray and nuclear physics, and the techniques employed, we refer to Shapiro [35] and Voyvodic [52].

B. Fundamental Equations and Their Solutions. Moments. In developing a stochastic theory of track formation it is necessary to take into consideration the spatial distribution of the initial grains. In the model we consider it is assumed that the initial grains traversed by the charged particles all have their centers along the path of the particle. This "linear approximation" simplifies the theory considerably.

Under the above assumption, the visible characteristics of the particle track depend only on the following probability distributions:

1. The distribution in final size after development. Let ρ denote the

radius of the developed particle; then $G(\rho)\,d\rho$ represents the probability that the radius is in the interval $(\rho,\ \rho + d\rho)$.

2. The "survival" probability $Q(t)$. The survival probability is defined as the probability that the distance between the center points of adjacent activated grains exceeds t; that is, $Q(t)$ is the probability that the charged particle, after activating a crystal along its path, "survives" for a distance greater than t before activating the next crystal.[1] If we assume that $Q(t)$ admits a density function, we then denote by $q(\tau)\,d\tau$ the probability that the center of the next activated crystal (regeneration point of the process) is in the interval $(\tau,\ \tau + d\tau)$. Hence, $q(\tau) = -dQ/d\tau$.

For arbitrary functions $G(\rho)$ and $Q(t)$ we now consider the derivation of the functional equations for the probability distributions associated with the random variables of interest in the theory of tracks.

Let $X_r(t)$ denote the number of countable gaps,[2] each one of length equal to or greater than r, in a cell (track) length t, and let $Y_r(t)$ denote the total length of these X_r gaps, counting for each gap only the excess of its length over r.

Let

$$P_n(r;t) = \mathscr{P}\{X_r(t) = n\} \qquad n = 0, 1, \ldots$$

In order to derive the functional equation for $P_n(r,t)$ by using the method of regeneration points, we consider the position τ of the next activated crystal. Considering the position τ and the final radii ρ_1 and ρ_2 of the initial and next grains, there are three mutually exclusive events that must be considered:

1. The grains growing from the initial and the next crystal merge to form a blob, or are so close that the gap between them is less than r, the minimum gap length. Clearly, this takes place when $\rho_1 + \rho_2 + r > \tau$. In this case the distance $t - \tau$ must contain n gaps of length greater than r. The probability of this event is expressed by the integral

$$\int_0^\infty d\rho_1 \int_0^\infty d\rho_2 \int_0^{\rho_1+\rho_2+r} P_n(r; t - \tau)G(\rho_1)G(\rho_2)q(\tau)\,d\tau$$

2. The next crystal is far enough away to form a gap of length greater than r, but its center is still within the cell length t. Hence $\rho_1 + \rho_2 + r < \tau < t$. In this case the distance $t - \tau$ must contain $n - 1$ gaps, each

---

[1] We remark that $Q(t) = 1 - G(t)$, where $G(t)$ is the distribution function for the length of time between "events" in the Bellman-Harris theory of age-dependent processes.

[2] The definition of a countable gap used in Blatt's theory is not the same as the usual one. Here the definition of a countable gap consists of two parts: (1) the initial point $t = 0$ of each cell length must be inside a blob, and (2) the grain terminating each countable gap must have its center point within the cell length t.

with length greater than r. The probability of this event is expressed by the integral

$$\int_0^\infty d\rho_1 \int_0^\infty d\rho_2 \int_{\rho_1+\rho_2+r}^t P_{n-1}(r; t-\tau)G(\rho_1)G(\rho_2)q(\tau)\, d\tau$$

3. The next activated crystal is at a distance greater than t; hence, there is no gap within the cell length t. The probability of this event is expressed by the integral

$$\int_t^\infty \delta_n q(\tau)\, d\tau = Q(t)\delta_n$$

where δ_n is the Dirac delta function.

Since the above events are mutually exclusive, we add the respective probabilities to obtain the functional equation

$$P_n(r;t) = \int_0^\infty d\rho_1 \int_0^\infty d\rho_2 \int_0^{\rho_1+\rho_2+r} P(r; t-\tau)G(\rho_1)G(\rho_2)q(\tau)\, d\tau$$

$$+ \int_0^\infty d\rho_1 \int_0^\infty d\rho_2 \int_{\rho_1+\rho_2+r}^t P_{n-1}(r; t-\tau)G(\rho_1)G(\rho_2)q(\tau)\, d\tau + Q(t)\delta_n$$

$$n = 0, 1, \ldots, r > 0, t \geqslant 0 \quad (6.108)$$

The formal solution of (6.108) can be obtained by means of the Laplace transformation. Let $p_n(r;s)$ denote the Laplace transform of $P_n(r;t)$. The subsidiary equation is given by the difference equation

$$p_n(r;s) = [1 - h(s) - g(r,s)]p_n(r;s) + g(r,s)p_{n-1}(r;s) + \frac{1}{s}h(s)\delta_n \quad (6.109)$$

where

$$h(s) = s\int_0^\infty e^{-st}Q(t) = 1 - \int_0^\infty e^{-st}q(t)\, dt \quad (6.110)$$

$$g(r,s) = \int_0^\infty d\rho_1 \int_0^\infty d\rho_2 \int_{\rho_1+\rho_2+r}^\infty e^{-st}G(\rho_1)G(\rho_2)q(t)\, dt \quad (6.111)$$

For $n = 0$, the solution of (6.108) is

$$p_0(r;s) = \frac{h(s)}{s[h(s) + g(r,s)]} \quad (6.112)$$

Hence, by iteration we obtain

$$p_n(r;s) = [K(r,s)]^n p_0(r;s) \quad (6.113)$$

where

$$K(r,s) = \frac{g(r,s)}{h(s) + g(r,s)} \quad (6.114)$$

Equation (6.113) is the complete formal solution of the functional equation (6.108). In order to obtain the probabilities $P_n(r;t)$, it is necessary to invert (6.113). For given functions $G(\rho)$ and $Q(t)$ that are of interest in applications, the evaluation of the integrals (6.110) and (6.111) should not be difficult; however, the inversion of (6.113) is another matter, and an expression for $P_n(r;t)$ in closed form in terms of tabulated functions is probably difficult to obtain.

In experimental situations it is usually the mean and variance of $X_r(t)$ that are of primary interest, and not the probabilities $P_n(r;t)$. We now show that expressions for the moments can be obtained in terms of $h(s)$ and $g(r,s)$; hence, the inversion problem can be ignored if the moments are of interest. Let

$$F_k(r;t) = \sum_{n=0}^{\infty} n(n-1) \cdots (n-k+1) P_n(r;t) \qquad (6.115)$$

denote the kth factorial moment of the distribution $P_n(r;t)$ and let $f_k(r;s)$ denote the Laplace transform of $F_k(r;t)$. From (6.113) and (6.115) we obtain

$$f_k(r;s) = k! p_0(r;s) \frac{[K(r,s)]^k}{[1-K(r,s)]^{k+1}}$$

$$= \frac{k!}{s} \left[\frac{g(r,s)}{h(s)} \right]^k \qquad (6.116)$$

The inversion of (6.116) can be carried out by the method of residues. In particular, if we only consider the contribution of the pole $s = 0$, inversion of (6.116) yields

$$F_k(r;t) = \frac{d^k}{ds^k} \left\{ e^{st} \left[\frac{sg(r,s)}{h(s)} \right]^k \right\}_{s=0} \qquad (6.117)$$

$$= \sum_{j=0}^{k} \frac{k!}{j!(k-j)!} t^{k-j} \left\{ \frac{d^j}{ds^j} \left[\frac{sg(r,s)}{h(s)} \right]^k \right\}_{s=0} \qquad (6.118)$$

Therefore, the moments of the distribution can be obtained by the evaluation of the integrals (6.110) and (6.111) and differentiation. From (6.117) or (6.118), the mean and variance are easily obtained, since

$$m(t) = \mathscr{E}\{X_r(t)\} = F_1(r;t) \qquad (6.119)$$

and $$\mathscr{D}^2\{X_r(t)\} = F_2(r;t) + F_1(r;t) - [F_1(r;t)]^2 \qquad (6.120)$$

We now consider the functional equation for the distribution of the total gap length. Let $P(r;y,t)\, dy$ denote the probability that the total gap length $Y_r(t)$ is in the interval $(y, y+dy)$. By using the same procedure as before, the method of regeneration points leads to the functional equation

$$P(r;y,t) = \int_0^\infty d\rho_1 \int_0^\infty d\rho_2 \int_0^{\rho_1+\rho_2+r} P(r; y, t-\tau)G(\rho_1)G(\rho_2)q(\tau) \, d\tau$$

$$+ \int_0^\infty d\rho_1 \int_0^\infty d\rho_2 \int_{\rho_1+\rho_2+r}^t P(r; y-\tau+\rho_1+\rho_2+r, t-\tau)G(\rho_1)G(\rho_2)$$

$$\times q(\tau) \, d\tau + Q(t)\delta(y) \qquad r \geqslant 0, y \geqslant 0, t \geqslant 0 \qquad (6.121)$$

where $\delta(y)$ is the Dirac delta function. To solve (6.121), we introduce the double Laplace transform of $P(r;y,t)$, say $p(r;u,s)$. The subsidiary equation is

$$p(r;u,s) = [1 - \{h(s) + g(r,s) - v(r,u,s)\}]p(r;u,s) + \frac{1}{s}h(s) \qquad (6.122)$$

where $h(s)$ and $g(r,s)$ are the same as before, and

$$v(r,u,s) = \int_0^\infty d\rho_1 \int_0^\infty d\rho_2 \int_{\rho_1+\rho_2+r}^\infty \exp\{(\rho_1 + \rho_2 + r)u - (u+s)t\}G(\rho_1)G(\rho_2)q(t) \, dt$$

$$(6.123)$$

with

$$v(r,0,s) = g(r,s) \qquad (6.124)$$

and

$$\lim_{u\to\infty} v(r,u,s) = 0 \qquad (6.125)$$

From (6.122) we obtain the formal solution

$$p(r;u,s) = \frac{h(s)}{s[h(s) + g(r,s) - v(r,u,s)]} \qquad (6.126)$$

As in the case considered earlier, the inversion of $p(r;u,s)$ to obtain the probabilities $P(r;y,t)$ is difficult; hence, we turn our attention to the problem of determining the moments of $P(r;y,t)$.

If we now multiply $P(r;y,t)$ by y^k, integrate over y, and apply the Laplace transformation, we have

$$p_k(r;s) = (-1)^k \left[\frac{\partial^k p(r;u,s)}{\partial u^k}\right]_{u=0} \qquad (6.127)$$

Hence, $p_k(r;s)$ is the Laplace transform of the kth factorial moment. From (6.126) we see that the only term that depends on u is $v(r,u,s)$. Let

$$g_k(r,s) = (-1)^k \left[\frac{\partial^k v(r,u,s)}{\partial u^k}\right]_{u=0}$$

$$= \int_0^\infty d\rho_1 \int_0^\infty d\rho_2 \int_{\rho_1+\rho_2+r}^\infty e^{-st}(t - \rho_1 - \rho_2 - r)^k G(\rho_1)G(\rho_2)q(t) \, dt$$

where $g_0(r,s) = g(r,s)$. From (6.126), (6.127), and by using the fact that $v(r,o,s) = g(r,s)$, we have, for example,

$$p_1(r,s) = \frac{g_1(r,s)}{sh(s)} \tag{6.128}$$

and

$$p_2(r,s) = \frac{g_2(r,s)}{sh(s)} + \frac{2}{s}\left[\frac{g_1(r,s)}{h(s)}\right]^2 \tag{6.129}$$

If we put

$$w_{kj}(r) = \frac{d^k}{ds^k}\left[\frac{sg_j(r,s)}{h(s)}\right]_{s=0}$$

and again consider only the contribution from the pole at $s = 0$, we obtain the following expression for the *mean and variance of $Y_r t$*:

$$m(t) = \mathscr{E}\{Y_r(t)\} = w_{01}t + w_{11} \tag{6.130}$$

and

$$\mathscr{D}^2\{Y_r(t)\} = (2w_{01}w_{11} + w_{02})t + (w_{11}^2 + 2w_{01}w_{21} + w_{12}) \tag{6.131}$$

For details of approximation procedures and evaluation of the integrals occurring in the above theory, we refer to Ref. 7.

C. Remarks on Some Specific Models. Throughout the above discussion we have not stipulated the functions $G(\rho)$ and $Q(t)$. In practical situations it is necessary to make certain assumptions concerning the functional forms of $G(\rho)$ and $Q(t)$ and then compare the theoretical predictions with the observations that actually obtain.

Several models based on different $G(\rho)$ and $Q(t)$ have been studied. The first model we consider was introduced by O'Ceallaigh.[1] In this model it is assumed that $G(\rho)$ is a delta function, i.e., the radius of the developed grain is assumed to have a unique value. The "survival" probability distribution is assumed to be the negative exponential, that is,

$$Q(t) = e^{-\lambda t} \qquad \lambda > 0 \tag{6.132}$$

Hence, there is a constant probability $\lambda\,\Delta t + o(\Delta t)$ for activating the next crystal in the interval $(t,\, t + \Delta t)$. This model, which is Markovian because of the form of $Q(t)$, neglects the correlations between positions of crystal centers in the photographic emulsion. While this assumption is unrealistic, it is a good approximation if the emulsion is very dilute, which is the same as saying that the probability of a charged particle activating any crystal it traverses is very small.

Another model based on (6.132) has been considered by Happ, Hull, and Moorish [14]. In this model, however, it is assumed that the grain radius ρ has a parabolic distribution.

[1] Cf. [7].

In order to take into consideration the arrangement of the crystals in the solution, Herz and Davis [15] have postulated a model in which $Q(t)$ is the step function

$$
\begin{aligned}
Q(t) &= 1 && \text{for } 0 < t < \beta \\
&= 1 - p && \text{for } \beta < t < 2\beta \\
&= (1 - p)^2 && \text{for } 2\beta < t < 3\beta \\
&\quad\vdots \\
&= (1 - p)^{n-1} && \text{for } (n - 1)\beta < t < n\beta
\end{aligned} \tag{6.133}
$$

In this case it is assumed that the crystals are arranged along the path of the charged particle in a regular lattice. In (6.133) β is the distance between crystals and p is the probability that a charged particle will activate each crystal it traverses.

Since the assumptions on $Q(t)$ made by O'Ceallaigh and by Herz and Davis represent rather extreme and unrealistic situations, a model which modifies the O'Ceallaigh model by taking into consideration the size of the initial crystal has been considered by Blatt. This model assumes

$$
\begin{aligned}
Q(t) &= 1 && \text{for } 0 < t < \alpha \\
&= e^{-\lambda(t-\alpha)} && \text{for } t > \alpha
\end{aligned} \tag{6.134}
$$

where α is the diameter of each crystal.

For a detailed discussion of these models and an analysis of some experimental data in each case, we refer to the paper of Blatt. On the basis of the analysis presented by Blatt, it seems that it is necessary to formulate stochastic models of track formation which consider the approximate lattice structure of the spatial distribution of the undeveloped crystals in the photographic emulsions, since it has been shown that models which do not consider the lattice are in disagreement with experimental data.

6.6 Some Problems in the Theory of Nuclear Reactors

A. Introduction. The theory and design of nuclear reactors presents many mathematical problems, a large number of which involve probability or statistical consideration. In this section we consider a stochastic model for the process of the slowing down of neutrons which takes place in the moderators of nuclear reactors. The problem of the slowing down of neutrons is fundamental in the theory of nuclear chain reactions. Consider a chain reaction taking place in a nuclear pile. The uranium nuclei undergo fission, which results in the liberation of neutrons. The initial velocity of the liberated neutrons is high but is soon reduced as the neutrons pass through the moderator. After the slowing-down process, the neutrons diffuse through the material before being finally absorbed. For a detailed discussion of this problem and related problems, we refer to the text of Glasstone and Edlund [11].

The model we consider was formulated and studied by Takács [48]. In Sec. 6.6B we consider the stochastic process associated with the lethargy of the neutron, and in Sec. 6.6C we determine the probability distribution of the time required for the lethargy of the neutron to reach a given value. Section 6.6D is concerned with the problem of determining the probability distribution of the number of collisions which the neutron experiences before its lethargy reaches a given value. Finally, in Sec. 6.6E we determine the expected or mean time required for the lethargy of the neutron to reach a given value.

B. The Stochastic Process Associated with the Lethargy of the Neutron. Consider a source neutron with a fixed initial energy E_0 at $t = 0$. Let $E(t)$ and $L(t)$ respectively denote the energy and lethargy of the neutron at time t. The *lethargy* of the neutron is defined as the logarithmic energy decrement, that is,

$$L(t) = \log \frac{E_0}{E(t)} \tag{6.135}$$

or

$$E(t) = E_0 e^{-L(t)} \tag{6.136}$$

We now assume that the neutron moves in a homogeneous medium of infinite extent consisting of atomic nuclei of r different nuclei. Let $N_1, N_2, \ldots, N_r$ denote the *spatial density* (i.e., the average number per unit volume) of the r types of atoms. Also, let $\sigma_{s1}, \sigma_{s2}, \ldots, \sigma_{sr}$ and $\sigma_{a1}, \sigma_{a2}, \ldots, \sigma_{ar}$ denote the scattering and absorption cross sections, respectively, of the neutron with respect to the r different types of nuclei. Put

$$\sigma_i = \sigma_{si} + \sigma_{ai} \qquad \gamma_i = N_i \sigma_i \qquad \gamma_{si} = N_i \sigma_{si}$$

where $i = 1, 2, \ldots, r$, and

$$C = \sum_{i=1}^{r} \gamma_i \qquad C_s = \sum_{i=1}^{r} \gamma_{si}$$

In general, the quantities γ_i, γ_{si}, C, and C_s are functions of the lethargy of the colliding neutron, rather than constants as we shall assume.

In formulating the model for the slowing down of neutrons, it is assumed that:

1. The probability of a neutron of lethargy l colliding with a nucleus of type i ($i = 1, 2, \ldots, r$) in the interval $(t, t + \Delta t)$ is

$$a \gamma_i(l) e^{-l/2} \Delta t + o(\Delta t)$$

2. The probability of scattering is

$$a\gamma_{si}(l)e^{-l/2}\,\Delta t + o(\Delta t)$$

In the above, $a = \sqrt{2E_0/m}$ and m is the mass of the neutron. It is also assumed that the collisions are such that, in the case of scattering, the increase of the lethargy of the neutron is independent of its value before collision.

Let $H_i(l)$ denote the distribution function of the increase in lethargy experienced by the neutron upon scattering on a nucleus of type i. It is assumed that

$$\begin{aligned} H_i(l) &= 0 && \text{for } l \leqslant 0 \\ &= \frac{1 - e^{-l}}{1 - \alpha_i} && \text{for } 0 \leqslant l \leqslant \log\frac{1}{\alpha_i} \\ &= 1 && \text{for } l > \log\frac{1}{\alpha_i} \end{aligned} \quad (6.137)$$

In (6.137)

$$\alpha_i = \left(\frac{A_i - 1}{A_i + 1}\right)^2 \quad (6.138)$$

where A_i is the mass number of the ith stationary nucleus.

Denote by $A(t)$ the event that the neutron was not absorbed in the interval $(0,t)$. Let

$$F(l,t) = \mathscr{P}\{L(t) \leqslant l,\, A(t)\}$$

that is, $F(l,t)$ is the probability that, during the interval $(0,t)$, the neutron will not be absorbed and at time t its lethargy is at most l. In order to derive the functional equation for $F(l,t)$, we consider the changes in the system during the interval $(t,\, t + \Delta t)$. If at time t the neutron has not been absorbed and $L(t) = y$, $0 \leqslant y \leqslant l$, then in the interval $(t,\, t + \Delta t)$ the probability of the lethargy increasing to l due to scattering on the ith nucleus is

$$a\int_0^l \gamma_{si}(y)e^{-y/2}\,\Delta t\, H_i(l - y)\, d_y F(y,t) + o(\Delta t)$$

Since we must consider all r nuclei, the total probability is given by

$$a\sum_{i=1}^r \int_0^l \gamma_{si}(y)e^{-y/2}\,\Delta t\, H_i(l - y)\, d_y F(y,t) + o(\Delta t)$$

Similarly, the probability of no increase in lethargy due to collision with any of the r nuclei is

$$\int_0^l [1 - aC(y)e^{-y/2}\,\Delta t]\, d_y F(y,t) + o(\Delta t)$$

where $C(y) = \sum_{i=1}^{r} \gamma_i(y)$. Since the above events are mutually exclusive, we have

$$F(l, t + \Delta t) = \int_0^l [1 - aC(y)e^{-y/2}\,\Delta t]\,d_y F(y,t)$$
$$+ a\sum_{i=1}^{r} \gamma_{si}(y)e^{-y/2}\,\Delta t\, H_i(l - y)\,d_y F(y,t) + o(\Delta t)$$

Since

$$\int_0^l d_y F(y,t) = F(y,t)$$

we obtain, after transposing and passing to the limit, the integro-differential equation

$$\frac{\partial F(l,t)}{\partial t} = -a\int_0^l e^{-y/2}\left[C(y) - \sum_{i=1}^{r}\gamma_{si}(y)H_i(l - y)\right]d_y F(y,t) \quad (6.139)$$

This equation is to be solved with the initial condition

$$F(l,0) = 0 \quad l \geqslant 0$$
$$= 1 \quad l < 0 \qquad (6.140)$$

The solution of (6.139) is

$$F(l,t) = \sum_{n=0}^{\infty} \frac{(-1)^n (at)^n}{n!} G_n(l) \qquad (6.141)$$

where

$$G_0(l) = 1 \quad l \geqslant 0$$
$$= 0 \quad l < 0$$

and

$$G_{n+1}(l) = \int_0^l e^{-y/2}\left[C(y) - \sum_{i=1}^{r}\gamma_{si}(y)H_i(l - y)\right]dG_n(y)$$

We now consider the process $\{L(t),\, t \geqslant 0\}$ under the following assumptions: (1) the absorption and scattering cross sections are constants; hence

$$C(l) = C \qquad \gamma_{si}(l) = \gamma_{si} \qquad C_s = \sum_{i=1}^{r} \gamma_{si}$$

and (2) the distribution function of the total increase in lethargy is given by

$$H(l) = \frac{\sum_{i=1}^{r} \gamma_{si} H_i(l)}{\sum_{i=1}^{r} \gamma_{si}} \qquad (6.142)$$

Under the above assumptions Eq. (6.139) becomes

$$\frac{\partial F(l,t)}{\partial t} = -w \int_0^l e^{-y/2}[1 - \rho H(l-y)] \, d_y F(y,t) \tag{6.143}$$

where $w = aC$ and $\rho = C_s/C$. Let

$$\psi(s,t) = \int_0^\infty e^{-sl} \, d_l F(l,t)$$

denote the Laplace-Stieltjes transform of $F(l,t)$. From (6.143) we obtain the differential-difference equation

$$\frac{\partial \psi(s,t)}{\partial t} = -w[1 - \rho\varphi(s)]\psi(s + \tfrac{1}{2}, t) \tag{6.144}$$

where

$$\varphi(s) = \int_0^\infty e^{-sl} \, dH(l)$$

is the Laplace-Stieltjes transform of $H(l)$. Since $L(0) = 0$, (6.144) is to be solved with the initial condition $\psi(s,0) = 1$. The solution of (6.144) is

$$\psi(s,t) = 1 - w[1 - \rho\varphi(s)] \int_0^t \psi(s + \tfrac{1}{2}, u) \, du$$

$$= \sum_{n=0}^\infty \frac{(-1)^n (wt)^n}{n!} \prod_{j=0}^{n-1} \left[1 - \rho\varphi\left(s + \frac{j}{2}\right) \right] \tag{6.145}$$

By using the above solution, we now obtain several expressions that are of interest in applications of the theory. Let $\mathscr{P}\{A(t)\}$ denote the *probability that the neutron will not be absorbed in the interval* $(0,t)$. Since $\mathscr{P}\{A(t)\} = F(\infty,t)$, and $F(\infty,t) = \psi(0,t)$, we have from (6.145)

$$\mathscr{P}\{A(t)\} = F(\infty,t) = \psi(0,t)$$

$$= \sum_{n=0}^\infty \frac{(-1)^n (wt)^n}{n!} \prod_{j=0}^{n-1} \left[1 - \rho\varphi\left(\frac{j}{2}\right) \right] \tag{6.146}$$

Let $\mathscr{E}\{L(t) \mid A(t)\}$ denote the *expected or mean lethargy of the neutron at time t, given that absorption has not taken place.* Clearly,

$$\mathscr{E}\{L(t) \mid A(t)\} = \frac{-\left. \dfrac{\partial \psi(s,t)}{\partial s} \right]_{s=0}}{\psi(0,t)} \tag{6.147}$$

hence

$$\mathscr{E}\{L(t) \mid A(t)\} = \sum_{n=1}^\infty \frac{(-1)^n (wt)^n}{n!} \left\{ \prod_{j=0}^{n-1} \left[1 - \rho\varphi\left(\frac{j}{2}\right) \right] \sum_{k=0}^{n-1} \left(\frac{-\rho\varphi'(k/2)}{1 - \rho\varphi(k/2)} \right) \right\}$$

$$\div \sum_{n=0}^\infty \frac{(-1)^n (wt)^n}{n!} \prod_{j=0}^{n-1} \left[1 - \rho\varphi\left(\frac{j}{2}\right) \right] \tag{6.148}$$

By using (6.136), the above expression becomes

$$\mathscr{E}\{L(t) \mid A(t)\} = E_0\left\{\frac{\psi(1,t)}{\psi(0,t)}\right\}$$

$$= E_0\left\{\frac{\displaystyle\sum_{n=0}^{\infty}[(-1)^n(wt)^n/n!]\prod_{j=0}^{n-1}[1 - \rho\varphi(1+j/2)]}{\displaystyle\sum_{n=1}^{\infty}[(-1)^n(wt)^n/n!]\prod_{j=0}^{n-1}[1 - \rho\varphi(j/2)]}\right\} \qquad (6.149)$$

From the above we see that explicit expression can be obtained for $\mathscr{P}\{A(t)\}$ and $\mathscr{E}\{L(t) \mid A(t)\}$ in terms of the scattering and absorption cross section and the Laplace-Stieltjes transform of $H(l)$.

C. Distribution of the Time Required for the Lethargy of the Neutron to Reach a Given Value. Let the random variable T_l represent the time at which the lethargy of the neutron reaches the value l; that is,

$$T_l = \inf_{L(t)>l} t$$

Let $\mathscr{P}\{T_l \leqslant t, A(T_l)\}$ denote the probability that T_l is at most t and absorption has not taken place in the interval $(0, T_l)$. Let $g(l,t)$ denote the probability density of T_l. To be more precise, $g(l,t)\,\Delta t + o(\Delta t)$ is the probability that the neutron will not be absorbed until T_l, and T_l is in the interval $(t, t + \Delta t)$.

By employing the reasoning used to derive the functional equation for $F(l,t)$, we have

$$g(l,t) = a\sum_{i=1}^{r}\int_0^l e^{-y/2}\gamma_{si}(y)[1 - H_i(l-y)]\,d_yF(y,t) \qquad (6.150)$$

Therefore, the probability of the event $A(T_l)$ is

$$\mathscr{P}\{A(T_l)\} = \int_0^{\infty} g(l,t)\,dt \qquad (6.151)$$

Similarly, the conditional expectation of T_l is given by

$$\mathscr{E}\{T_l \mid A(T_l)\} = \frac{\displaystyle\int_0^t tg(l,t)\,dt}{\displaystyle\int_0^{\infty} g(l,t)\,dt} \qquad (6.152)$$

For the special case considered in Sec. 6.6B explicit expressions for $\mathscr{P}\{A(T_l)\}$ and $\mathscr{E}\{T_l \mid A(T_l)\}$ can be obtained.

D. Distribution of the Number of Collisions before T_l. Let the random variable M_l denote the number of collisions experienced by the neutron in the interval $(0, T_l)$ and let

$$Q_n(l) = \mathscr{P}\{L(n) \leqslant l, A(n)\} \qquad n = 0, 1, \ldots$$

Since $L(0) = 0$, we have

$$Q_0(l) = 1 \qquad \text{for } l \geqslant 0$$
$$= 0 \qquad \text{for } l < 0 \tag{6.153}$$

and
$$Q_n(l) = \sum_{i=1}^{r} \int_0^l \frac{\gamma_{si}(y) H_i(l-y)}{C(y)} \, dQ_{n-1}(y) \qquad n = 1, 2, \ldots \tag{6.154}$$

The probability distribution of M_l is given in terms of the $Q_n(l)$ by

$$\mathscr{P}\{M_l = n, A(T_l)\} = \sum_{i=1}^{r} \int_0^l \frac{\gamma_{si}(y)[1 - H_i(l-y)]}{C(y)} \, dQ_{n-1}(y)$$

$$= \int_0^l \frac{C_s(y)}{C(y)} \, dQ_{n-1}(y) - Q_n(l) \tag{6.155}$$

Since $\mathscr{P}\{A(T_l)\} = \sum_{n=1}^{\infty} \mathscr{P}\{M_l = n, A(T_l)\}$, we have from (6.155)

$$\mathscr{P}\{A(T_l)\} = 1 - \int_0^l \left[1 - \frac{C_s(y)}{C(y)}\right] dK(y) \tag{6.156}$$

where $K(l) = \sum_{n=0}^{\infty} Q_n(l)$. This expression can be used in place of (6.151). From (6.154) we obtain

$$K(l) = 1 + \sum_{i=1}^{r} \int_0^l \frac{\gamma_{si}(y) H_i(l-y)}{C(y)} \, dK(y) \tag{6.157}$$

E. Expected Time Required for the Lethargy of the Neutron to Reach a Given Value. We now use the results of Secs. 6.6C and D to obtain an expression for the expected time required for the lethargy of the neutron to reach a given value. Let

$$K(l;z) = \sum_{n=0}^{\infty} Q_n(l;z) \qquad z \geqslant 0$$

where
$$Q_0(l;z) = 1 \qquad \text{for } l \geqslant z$$
$$= 0 \qquad \text{for } l < z$$

and
$$Q_n(l;z) = \sum_{i=1}^{r} \int_0^l \frac{\gamma_{si}(y) H_i(l-y)}{C(y)} \, d_y Q_{n-1}(y;z)$$

When $z = 0$, $K(l;z)$ and $Q_n(l;z)$ are given by (6.157) and (6.154), respectively.

Now, by using (6.152) and the results of Sec. 6.6D, we have

$$\mathscr{E}\{T_l \mid A(T_l)\} = \frac{1}{\mathscr{P}\{A(T_l)\}} \int_0^l \left[1 - \int_y^l \left(1 - \frac{C_s(u)}{C(u)}\right) d_u K(u;y)\right] \frac{e^{y/2}}{aC_s(y)}\, dK(y)$$

$$(6.158)$$

For the special case considered in Sec. 6.6B, we have

$$Q_n(l) = \rho^n H_n(l) \qquad \text{and} \qquad K(l) = \sum_{n=0}^{\infty} \rho^n H_n(l)$$

We assume $\qquad\qquad K(l;z) = K(l-z)$

From (6.156)

$$\mathscr{P}\{A(T_l)\} = 1 - (1 - \rho)K(l) \qquad (6.159)$$

and from (6.158)

$$\mathscr{E}\{T_l \mid A(T_l)\} = \frac{1}{w\rho\,\mathscr{P}\{A(T_l)\}} \int_0^l e^{y/2}[1 - (1-\rho)K(l-y)]\, dK(y) \quad (6.160)$$

We close this section with the calculation of $\mathscr{P}\{A(T_l)\}$ and $\mathscr{E}\{T_l \mid A(T_l)\}$ for the hydrogen atom. For the hydrogen atom

$$H(l) = 1 - e^{-l} \qquad l \geqslant 0$$

hence $\qquad\qquad K(l) = 1 + \frac{\rho}{1-\rho}(1 - e^{-(1-\rho)l})$

and from (6.159)

$$\mathscr{P}\{A(T_l)\} = \rho e^{-(1-\rho)l} \qquad (6.161)$$

Similarly, from (6.160) and (6.161) we have

$$\mathscr{E}\{T_l \mid A(T_l)\} = \frac{2}{w}(e^{l/2} - 1) + \frac{1}{w\rho} \qquad (6.162)$$

Since $E = E_0 e^{-l}$, $\rho = C_s/C$, and $w = aC = \sqrt{2E_0/m}\,C$, the above expression becomes

$$\mathscr{E}\{T_l \mid A(T_l)\} = \frac{\sqrt{2m}}{C}\left(\frac{1}{\sqrt{E}} - \frac{1}{\sqrt{E_0}}\right) + \frac{\sqrt{2m}}{2C_s\sqrt{E_0}} \qquad (6.163)$$

The above expression can be compared with the expression for the total time, say T, required for the energy of a source neutron to be decreased from the fission energy E_0 to the thermal energy E_{th} obtained by Fermi (cf. [11]):

$$T = \frac{\sqrt{2m}\,\bar{\lambda}_s}{\xi}\left(\frac{1}{\sqrt{E_{\text{th}}}} - \frac{1}{\sqrt{E_0}}\right) \qquad (6.164)$$

where m is the actual mass of the neutron, ξ is the average logarithmic energy change per collision, and $\bar{\lambda}_s$ is the average of the scattering mean free path, i.e., the average distance the neutron travels between collisions.

F. Other Studies. Other studies concerned with the slowing down of neutrons are those of Mogyoródi and Németh [23] and Pál [27]. In Ref. 23 the number of collisions that take place in a given time interval is determined, and in Ref. 27 there are obtained expressions for the collision density and resonance escape probability which are more accurate than those previously obtained by other methods.

In Ref. 28 L. Pál formulates a general theory of stochastic processes in nuclear reactors. The fundamental equations describing the stochastic process are obtained, and from these equations are derived the generalized equations for the average neutron density and neutron importance and the fluctuation of neutron density.

Klahr [18] has used the theory of diffusion processes to calculate steady-state neutron distributions in moderator materials. In this approach the probability distribution of the neutron in phase space, as a function of the time from its birth, is obtained by solving a Fokker-Planck (forward Kolmogorov) equation whose coefficients depend on the one-collision probability distributions.

Bibliography

1 Albert, G. E., and L. Nelson: Contributions to the Theory of Counter Data, *Ann. Math. Statist.*, vol. 24, pp. 9–22, 1953.
2 Bateman, H.: The Solution of a System of Differential Equations Occurring in the Theory of Radio-active Transformations, *Proc. Cambridge Phil. Soc.*, vol. 15, pp. 423–427, 1910.
3 Bay, Z.: Electron Multiplier as an Electron Counting Device, *Rev. Sci. Instr.*, vol. 12, pp. 127–133, 1941.
4 Békéssy, A.: Über der Wahrscheinlichkeitsverteilung der Impulsanzahl bei fehlerhaft arbeitenden Untersetzern, *Magyar Tud. Akad. Alkalm. Mat. Int. Közl.*, vol. 3, pp. 171–181, 1955.
5 Bellman, R.: "A Survey of the Mathematical Theory of Time Lag, Retarded Control and Hereditary Processes," RAND Monograph R-256, 1954.
6 Blackwell, D.: A Renewal Theorem, *Duke Math. J.*, vol. 15, pp. 145–150, 1948.
7 Blatt, J. M.: Theory of Tracks in Nuclear Research Emulsions, *Australian J. Phys.*, vol. 8, pp. 248–272, 1955.
8 Doob, J. L.: Renewal Theory from the Point of View of the Theory of Probability, *Trans. Am. Math. Soc.*, vol. 63, pp. 422–438, 1948.
9 Feller, W.: On the Integral Equation of Renewal Theory, *Ann. Math. Statist.*, vol. 12, pp. 243–267, 1941.
10 Feller, W.: On Probability Problems in the Theory of Counters. In "Courant Anniversary Volume," pp. 105–115, Interscience Publishers, New York, 1948.
11 Glasstone, S., and M. C. Edlund: "The Elements of Nuclear Reactor Theory," D. Van Nostrand Company, Inc., Princeton, N.J., 1952.
12 Gnedenko, B. V.: On the Theory of Geiger-Müller Counters (in Russian), *Ž. Eksper. Teoret. Fiz.*, vol. 11, pp. 101–106, 1941.

13 Hammersley, J. M.: On Counters with Random Dead Time: I, *Proc. Cambridge Phil. Soc.*, vol. 49, pp. 623–637, 1953.

14 Happ, W. W., T. E. Hull, and A. H. Moorish: A Statistical Analysis of Grain Counting in Photographic Emulsions, *Can. J. Phys.*, vol. 30, pp. 699–714, 1952.

15 Herz, A. J., and G. Davis: The Characteristics of Tracks in Nuclear Research Emulsions, *Australian J. Phys.*, vol. 8, pp. 129–135, 1955.

16 Jánossy, L.: A Statistical Problem of Cascade Processes (in Hungarian), *Magyar Tud. Akad. Mat. Fiz. Oszt. Közl.*, vol. 1, pp. 213–217, 1951.

17 Jánossy, L.: A Stochastic Process Occurring in the Theory of the Multiplier Tube (in Hungarian), *Magyar Tud. Akad. Fiz. Oszt. Közl.*, vol. 10, pp. 357–367, 1951.

18 Klahr, C. N.: Calculations of Neutron Distributions by the Methods of Stochastic Processes, *Nuclear Sci. and Eng.*, vol. 3, pp. 269–285, 1958.

19 Kosten, L.: On the Frequency Distribution of the Number of Discharges Counted by a Geiger-Müller Counter in a Constant Interval, *Physica*, vol. 10, pp. 749–756, 1943.

20 Kurbatov, J. D., and H. B. Mann: Correction of G-M Counter Data, *Phys. Rev.*, vol. 68, pp. 40–43, 1945.

21 Levert, C., and W. L. Scheen: Probability Fluctuations of Discharges in a Geiger-Müller Counter Produced by Cosmic Radiation, *Physica*, vol. 10, pp. 225–238, 1943.

22 Mann, H. B.: A Note on the Correction of Geiger-Müller Counter Data, *Quart. Appl. Math.*, vol. 4, pp. 307–309, 1946.

23 Mogyoródi, J., and G. Németh: On Probabilistic Problems Connected with the Process of Slowing Down of Neutrons in Nuclear Reactors (in Hungarian), *Publ. Math. Inst. Hung. Acad. Sci.*, vol. 1, pp. 337–348, 1956.

24 Mycielski, J.: On the Distance between Signals in the Nonhomogeneous Poisson Stochastic Process, *Studia Math.*, vol. 15, pp. 303–313, 1956.

25 Olsson, O.: A Theoretical Study of the Time Energy Distribution of Slowed-down Neutrons, *Arkiv Fysik*, vol. 10, pp. 129–144, 1955.

26 Pál, L.: Application of the Theory of Stochastic Processes to the Investigation of Nuclear Fission (in Russian), *Ž. Eksper. Teoret. Fiz.*, vol. 30, pp. 367–373, 1956. (English translation in *Soviet Phys. JETP*, vol. 3, pp. 264–268, 1956.)

27 Pál, L.: Some Problems Concerning the Slowing Down of Neutrons (in Hungarian), *Publ. Math. Inst. Hung. Acad. Sci.*, vol. 1, pp. 41–54, 1956.

28 Pál, L.: On the Theory of Stochastic Processes in Nuclear Reactors, *Nuovo Cimento*, ser. 10, vol. 7, suppl. 1, pp. 25–42, 1958.

29 Pyke, R.: On Renewal Processes Related to Type I and Type II Counter Models, *Ann. Math. Statist.*, vol. 29, pp. 737–754, 1958.

30 Ramakrishnan, A.: Counters with Random Dead Time, *Phil. Mag.*, vol. 45, pp. 1050–1052, 1954.

31 Ramakrishnan, A., and P. M. Mathews: A Stochastic Problem Relating to Counters, *Phil. Mag.*, vol. 44, pp. 1122–1128, 1953.

32 Rényi, A.: "Calculus of Probabilities" (in Hungarian), Tankönyvkiado, Budapest, 1954.

33 Rossi, B. B., and H. H. Staub: "Ionization Chambers and Counters," McGraw-Hill Book Company, Inc., New York, 1949.

34 Rutherford, E., J. Chadwick, and C. D. Ellis: "Radiations from Radioactive Substances," Cambridge University Press, New York, 1930.

35 Shapiro, M. M.: Tracks of Nuclear Particles in Photographic Emulsions, *Revs. Modern Phys.*, vol. 13, pp. 58–71, 1941.

36 Smith, W. L.: On Renewal Theory, Counter Problems and Quasi-Poisson Processes, *Proc. Cambridge Phil. Soc.*, vol. 53, pp. 175–193, 1957.

37 Takács, L.: Wahrscheinlichkeitstheoretische Behandlung von Koinzidenz-Erscheinungen, mit Ereignissen gleicher Zeitdauer, *Comptes rendus du premier congrès des mathématiciens Hongrois*, pp. 731–740, 1950.

38 Takács, L.: Occurrence and Coincidence Phenomena in Case of Happenings with Arbitrary Distribution Law of Duration, *Acta Math. Acad. Sci. Hung.*, vol. 2, pp. 275–298, 1951.

39 Takács, L.: A New Method for Discussing Recurrent Stochastic Processes (in Hungarian), *Magyar Tud. Akad. Alkalm. Mat. Int. Közl.*, vol. 2, pp. 135–151, 1953.

40 Takács, L.: Coincidence Problems Arising in the Theory of Counters (in Hungarian), *Magyar Tud. Akad. Alkalm. Mat. Int. Közl.*, vol. 2, pp. 153–163, 1953.

41 Takács, L.: Some Investigations Concerning Recurrent Stochastic Processes of a Certain Type (in Hungarian), *Magyar Tud. Akad. Alkalm. Mat. Int. Közl.*, vol. 3, pp. 115–128, 1954.

42 Takács, L.: On Secondary Processes Generated by a Poisson Process and Their Applications in Physics, *Acta Math. Acad. Sci. Hung.*, vol. 5, pp. 203–236, 1954.

43 Takács, L.: On Processes of Happenings Generated by Means of a Poisson Process, *Acta Math. Acad. Sci. Hung.*, vol. 6, pp. 81–99, 1955.

44 Takács, L.: On Stochastic Processes Connected with Certain Physical Recording Apparatuses, *Acta Math. Acad. Sci. Hung.*, vol. 6, pp. 363–380, 1955.

45 Takács, L.: On the Sequence of Events, Selected by a Counter from a Recurrent Process of Events, *Teor. Veroyatnost. i Primenen.*, vol. 1, pp. 90–102, 1956.

46 Takács, L.: On a Probability Problem Arising in the Theory of Counters, *Proc. Cambridge Phil. Soc.*, vol. 52, pp. 483–498, 1956.

47 Takács, L.: On Secondary Stochastic Processes Generated by Recurrent Stochastic Processes, *Acta Math. Acad. Sci. Hung.*, vol. 7, pp. 17–29, 1956.

48 Takács, L.: On Some Probabilistic Problems in the Theory of Nuclear Reactors (in Hungarian), *Publ. Math. Inst. Hung. Acad. Sci.*, vol. 1, pp. 55–66, 1956.

49 Takács, L.: Some Probabilistic Problems Concerning the Counting of Particles. A Note on a Paper of A. Békéssy (in Hungarian), *Publ. Math. Inst. Hung. Acad. Sci.*, vol. 1, pp. 93–98, 1956.

50 Takács, L.: Über die wahrscheinlichkeitstheoretische Behandlung der Anodenstromschwankungen von Elektronenröhren, *Acta Phys. Sci. Hung.*, vol. 7, pp. 25–50, 1957.

51 Takács, L.: On Some Probability Problems Concerning the Theory of Counters, *Acta Math. Acad. Sci. Hung.*, vol. 8, pp. 127–138, 1957.

52 Voyvodic, L.: Particle Identification with Photographic Emulsions, and Related Problems, in J. G. Wilson (ed.), "Progress in Cosmic Ray Physics," vol. 2, pp. 217–288, North-Holland Publishing Company, 1954.

7

Applications in Astronomy and Astrophysics

7.1 Introduction

In this chapter we consider some studies that deal with the application of the theory of stochastic processes to certain problems in astronomy and astrophysics. Statistical and probabilistic methods have for many years been invaluable tools in the study of stellar systems (cf. [2]); however, it is only recently that problems in astronomy and astrophysics have been considered within the framework of the theory of stochastic processes. It is reasonable to assume that stochastic processes will play an important role in the study of stellar systems and that attempts will be made to reexamine the deterministic approach to many astrophysical problems by utilizing the theory of stochastic processes.

In Sec. 7.2 we consider the theory of fluctuations in brightness of the Milky Way as developed by S. Chandrasekar and G. Münch. The development of this theory centers around an integral equation for the probability distribution of the observed brightness of the Milky Way. The theory we present is based on the work of A. Ramakrishnan, who develops the theory of fluctuations by assuming that the stochastic process associated with the observed brightness has the Markov property.

A discussion of the Neyman-Scott theory of the spatial distribution of galaxies is given in Sec. 7.3. In particular, we consider the postulates underlying a stochastic model of the distribution of galaxies and the generating function of the joint probability distribution of the numbers of galaxies visible on photographs of arbitrary regions in the stellar space.

In Sec. 7.4 we consider the stochastic approach to the theory of radiative transfer. The theory is not only of fundamental importance in the study of radiation fields in the atmosphere, but also plays an outstanding role in the study of neutron transport problems which arise in the theory and design of nuclear reactors. The stochastic approach to the equations of radiative transfer is due to S. Ueno.

7.2 Theory of Fluctuations in Brightness of the Milky Way

A. Introduction. In a series of papers Chandrasekar and Münch [4] and Münch [8] have developed a theory of fluctuations in brightness of the Milky Way.[1] The statistical problem involved can be stated as follows: Let stars and interstellar clouds occur with a uniform distribution in a plane of infinite extent, and let the system extend to a linear distance L in the direction of a line of sight. Also, let the following be given:

1. That there is a deterministic contribution of amount $\beta\, d\tau$ from the element of length $d\tau$, at $t = \tau$, to the intensity measured at $t = 0$. This contribution is due to stars occurring with a uniform distribution along t.

2. That clouds, characterized by an average optical thickness τ_*, occur with a Poisson distribution

$$\frac{(\lambda t)^n e^{-\lambda t}}{n!}$$

in any interval of length t, where λ is the probability per unit distance t that a cloud occurs in any interval.

3. That a cloud has transparency factor q, with probability density $\psi(q)$. Hence $\alpha\psi(q)\, dq$ is the probability per unit distance t that radiation of a given intensity I changes to an interval between Iq and $I(q + dq)$. To say that a cloud has transparency q is to mean that the cloud reduces the intensity of radiation of the light of the stars immediately behind it by the factor q.

Given the above, the problem is to determine the probability density $g(I,L)$ of the observed brightness I. Chandrasekar and Münch derived the following integral equation for $g(I,L)$:

$$g(u,\xi) + \frac{\partial g(u,\xi)}{\partial u} + \frac{\partial g(u,\xi)}{\partial \xi} = \int_{u/\xi}^{1} g\left(\frac{u}{q}, \xi\right) \psi(q)\, \frac{dq}{q} \tag{7.1}$$

where u is the observed brightness and ξ is the extent of the system along

---

[1] Cf. also Münch [9].

the line of sight. In the above, u and ξ are measured in suitable astrophysical units.

In the first four papers of their series, Chandrasekar and Münch solved Eq. (7.1) for several cases that are of astrophysical interest. In the fifth paper, the assumption that the interstellar matter occurs in the form of discrete clouds was replaced by the assumption that the distribution of the cloud density is continuous but exhibits fluctuations of a statistical character.

In this section we treat the astrophysical problem of fluctuations in brightness of the Milky Way within the framework of the theory of Markov processes. This treatment is due to Ramakrishnan [17–20] and Ramakrishnan and Mathews [21,22]. In Sec. 7.2B we derive the Chandrasekar-Münch integral equation by using the fact that the stochastic process defined by $g(u,\xi)$ is a Markov process. A solution of the Chandrasekar-Münch equation utilizing the Mellin transform is also given. In Sec. 7.2C a stochastic model of a fluctuating density field is given and its application to the fluctuation problem is considered.

B. Derivation and Solution of the Chandrasekar-Münch Integral Equation. In order to derive the Chandrasekar-Münch integral equation, we assume that the stochastic process defined by the probability density $g(u,\xi)$ is Markovian. Let the state of the system be specified by the pair (u,ξ) and consider the change in $g(u,\xi)$ induced by a change in ξ of $d\xi$. We consider, therefore, the contributions to the probability $g(u, \xi + d\xi)$. We have the following:

1. The system is in the state (u',ξ) and in the interval $(\xi, \xi + d\xi)$ moves to the state (u,ξ), $u > u'$. The probability of the system being in the state u' is $g(u',\xi)\,du'$; therefore, the measure of this contribution is

$$\alpha g(u',\xi)\psi\left(\frac{u}{u'}\right)\frac{du}{u}\,du'\,d\xi \tag{7.2}$$

where $u/u' = q$. Hence the measure of the total contribution is obtained by integrating (7.2) over u'; thus

$$\alpha\,du\,d\xi\int_{u'} g(u',\xi)\psi\left(\frac{u}{u'}\right)\frac{du'}{u'} \tag{7.3}$$

2. The system is in the state (u,ξ), and in the interval $(\xi, \xi + d\xi)$ the intensity u drops to u'. The measure of this decrease is given by

$$\alpha g(u,\xi)\,d\xi\,du\int_0^u \psi\left(\frac{u'}{u}\right)\frac{du'}{u} = \alpha g(u,\xi)\,du\,d\xi \tag{7.4}$$

since

$$\int_0^1 \psi(q)\,dq = 1 \tag{7.5}$$

3. The system is in the state $(u - \beta \, d\xi, \xi)$, and the deterministic contribution $\beta \, d\xi$ in the interval $(\xi, \xi + d\xi)$ results in the transition of the system from $(u - \beta \, d\xi, \xi)$ to $(u, \xi + d\xi)$.

Since the above events are mutually exclusive, we have

$$g(u, \xi + d\xi) \, du = -\alpha g(u,\xi) \, du \, d\xi + g(u - \beta \, d\xi, \xi) \, du$$
$$+ \, du \, d\xi \alpha \int_{u'} g(u',\xi)\psi\left(\frac{u}{u'}\right)\frac{du'}{u} \quad (7.6)$$

On passing to the limit, we obtain

$$\frac{\partial g(u,\xi)}{\partial \xi} = -\alpha g(u,\xi) - \beta \, \frac{\partial g(u,\xi)}{\partial u} + \alpha \int_u^\beta g(u',\xi)\psi\left(\frac{u}{u'}\right)\frac{du'}{u} \quad (7.7)$$

In (7.7) the integration over u' is over the range u to $\beta\xi$, since $g(u,\xi) = 0$ for $u > \beta\xi$. If we now put $\alpha = \beta = 1$ in (7.7), we obtain the Chandrasekar-Münch integral equation (7.1), since for $\beta = 1$

$$\int_u^\xi g(u',\xi)\psi\left(\frac{u}{u'}\right)\frac{du'}{u} = \int_{u/\xi}^1 g\left(\frac{u}{q}, \xi\right)\psi(q)\frac{dq}{q}$$

We now consider the solution of Eq. (7.1). The method of solution is based on the Mellin transformation, and the physical conditions of the problem require that

$$u \geqslant 0 \quad (7.8)$$

$$\int_u g(u,\xi) \, du = 1 \qquad \text{for all } \xi \quad (7.9)$$

and
$$g(u,0) = \delta(u) \qquad 0 \leqslant q \leqslant 1 \quad (7.10)$$

For complex s, let

$$p(s,\xi) = \int_0^\infty g(u,\xi)u^s \, du$$

and
$$\varphi(s) = \int_0^\infty \psi(q)q^s \, dq = \int_0^1 \psi(q)q^s \, dq$$

denote the Mellin transforms[1] of $g(u,\xi)$ and $\psi(q)$, respectively. Application of the Mellin transformation, as defined above, to (7.1) yields the differential-difference equation

$$\frac{\partial p(s,\xi)}{\partial \xi} = -p(s,\xi) + \varphi(s)p(s,\xi) + sp(s-1, \xi) \quad (7.11)$$

[1] We remark that the above definition of the Mellin transform is not the usual one. The above form, however, is of value in comparing integral moments.

The initial conditions for the subsidiary equation (7.11) are

$$p(s,0) = 0 \qquad \text{for } s \neq 0 \tag{7.12}$$

and
$$p(0,0) = 1 \tag{7.13}$$

since $g(u,0) = \delta(u)$, and $p(0,\xi) = 1$ for all ξ.

We now note that since the transform variable s is complex, Eq. (7.11) cannot be solved by iteration. In order to obtain the solution of Eq. (7.11) by iteration, the following technique is used:[1] Replace s by n, where n assumes nonnegative-integer values. In this case $p(n,\xi)$ represents the nth moment of u and $p(n,\xi)$ satisfies (7.11) and the initial conditions given above. If conditions can be found such that the substitution of the continuous variable s in the solution $p(n,\xi)$ is meaningful and if the resulting function $p(s,\xi)$ satisfies (7.11) and the initial conditions, then $p(s,\xi)$ is the required solution of Eq. (7.11).

Solutions of Eq. (7.11) have been obtained in several cases of astrophysical interest [22]. In this section we shall consider only one case. We take the function $\psi(q')$ to be defined as follows:

$$\psi(q') = \delta(q' - q) \tag{7.14}$$

The above definition of $\psi(q')$ is an expression of the assumption that the clouds are equally transparent. From the definition of $\varphi(s)$ we see that (7.14) is equivalent to the assumption that $\varphi(s) = q^s$.

By iteration, the solution of Eq. (7.11), when $s = n$, is

$$p(n,\xi) = n! \sum_{k=0}^{n} \frac{e^{-(1-\varphi(k))\xi}}{\displaystyle\prod_{\substack{j=0 \\ j \neq k}}^{n} [\varphi(k) - \varphi(j)]}$$

$$= n! \sum_{k=0}^{n} \frac{e^{-(1-q^k)\xi}}{\displaystyle\prod_{\substack{j=0 \\ j \neq k}}^{n} (q^k - q^j)} \qquad n = 0, 1, \ldots \tag{7.15}$$

If we put

$$(-1)^n A_k^n = \frac{1}{\displaystyle\prod_{\substack{j=0 \\ j \neq k}}^{n} (q^k - q^j)}$$

$p(n,\xi)$ can be written in the form

$$p(n,\xi) = \xi^n e^{-\xi} + n! e^{-\xi} \sum_{m=1}^{\infty} \frac{\xi^{n+m}}{(n+m)!} \sum_{k=0}^{m} (-1)^m A_k^m q^{(n+m)k} \tag{7.16}$$

If we now replace n by s, it can be easily verified that (7.16) satisfies (7.11); hence, (7.16) is the correct solution when $\varphi(s) = q^s$.

---

[1] Cf. Ramakrishnan and Mathews [22].

If we now apply the inversion theorem to (7.16), we obtain

$$g(u,\xi) = \delta(u - \xi)e^{-\xi} \sum_{m=l}^{\infty} \sum_{k=0}^{l-1} \frac{A_k^m (u - \xi q^k)^{m-1}}{(m-1)!} \tag{7.17}$$

for $q^l < u \leqslant q^{l-1}$. Hence, (7.17) is the solution[1] of the Chandrasekar-Münch equation in the case when all clouds are equally transparent.

C. Stochastic Model of a Fluctuating Density Field. In the fifth paper of their series, Chandrasekar and Münch introduced the concept of a fluctuating density field. That is, the assumption that the interstellar matter occurs in the form of discrete clouds was replaced by the assumption that the probability distribution of the cloud density is continuous but exhibits statistical fluctuations. In this section we consider a stochastic model of a fluctuating density field, due to Ramakrishnan [19], and its application to the fluctuation problem. Before considering the model, we derive an expression for the correlation between the densities at n points in the field, since it is this correlation function which is of primary interest in the astrophysical problem.

Let the vector $\mathbf{t} = (t^{(1)}, \ldots, t^{(n)})$ denote the position of a point in n space and let the random variable $\rho(\mathbf{t})$ denote the density associated with the point denoted by $\mathbf{t}$. Also, let $P_\rho(\mathbf{t})$ denote the probability distribution function of $\mathbf{t}$. We now introduce the random variable

$$X(V) = \int_V \rho \, dv \tag{7.18}$$

where V is the domain of integration and $dv = dt^{(1)} \cdots dt^{(n)}$. From the definition of $X(V)$ we can write $\rho \, dv_1 = dX(v_1)$, where $dv_1 = dt_1^{(1)} \cdots dt_1^{(n)}$, etc. We now partition V into k parts, $dv_1, \ldots, dv_k$, and let

$$X(V) = \sum_{i=1}^{k} dX(v_i) \qquad k \to \infty \tag{7.19}$$

If we denote the nth moment of $X(V)$ by $\mathscr{E}\{X^n(V)\}$, then from (7.19)

$$\mathscr{E}\{X^n(V)\} = \mathscr{E}\{dX(v_1) + dX(v_2) + \cdots + dX(v_k)\}^n \tag{7.20}$$

and

$$\mathscr{E}\{X(V)\} = \mathscr{E}\{dX(v_1)\} + \cdots + \mathscr{E}\{dX(v_k)\} \tag{7.21}$$

as $k \to \infty$. That is,

$$\mathscr{E}\{X(V)\} = \int_V \mathscr{E}\{\rho(\mathbf{t}_1)\} \, dv_1 \tag{7.22}$$

since

$$\mathscr{E}\{dX(v_i)\} = \int_V \mathscr{E}\{\rho(\mathbf{t}_i)\} \, dv_i \tag{7.23}$$

[1] In Ref. 18 the same solution is obtained by using the method of regeneration points.

Let us now assume that for all i and j

$$\mathscr{E}\{dX(v_i)\,dX(v_j)\} = 0(dv_i\,dv_j) \tag{7.24}$$

and let

$$\mathscr{E}\{\rho(\mathbf{t}_i)\rho(\mathbf{t}_j)\}\,dv_i\,dv_j \tag{7.25}$$

denote the expectation defined by (7.24). In view of the above, $\mathscr{E}\{\rho(\mathbf{t}_i)\rho(\mathbf{t}_j)\}$ expresses the density correlation between any two points $\mathbf{t}_i$ and $\mathbf{t}_j$. If we now consider n points $\mathbf{t}_1, \ldots, \mathbf{t}_n$, we obtain

$$\mathscr{E}\{dX(v_1)\cdots dX(v_n)\} = \mathscr{E}\{\rho(\mathbf{t}_1)\cdots\rho(\mathbf{t}_n)\}\,dv_1\cdots dv_n \tag{7.26}$$

Hence, as $k \to \infty$ we have the following result:

$$\mathscr{E}\{X^n(V)\} = \int_V\cdots\int_V \mathscr{E}\{\rho(\mathbf{t}_1)\cdots\rho(\mathbf{t}_n)\}\,dv_1\cdots dv_n \tag{7.27}$$

This result is of interest because it states that correlations of degree less than n do not enter into the expression for the nth moment of $X(V)$.

Let

$$P_{\rho_1\rho_2}(\mathbf{t}_1,\mathbf{t}_2)\,d\rho_2 = \mathscr{P}\{\rho(\mathbf{t}_2) = \rho_2 \mid \rho(\mathbf{t}_1) = \rho_1\}$$

that is, $P_{\rho_1\rho_2}(\mathbf{t}_1,\mathbf{t}_2)$ is the conditional probability that $\rho(\mathbf{t}_2) = \rho_2$, given that $\rho(\mathbf{t}_1) = \rho_1$. If $\mathbf{t}$ is a continuous parameter, the function $P_{\rho_1\rho_2}(\mathbf{t}_1,\mathbf{t}_2)$ is determined only in the case when $\mathbf{t}$ is one-dimensional (i.e., $\mathbf{t} = t$), and the associated stochastic process is homogeneous with respect to t and has the Markov property. If the process is homogeneous, $P_{\rho_1\rho_2}(t_1,t_2) = P_{\rho_1\rho_2}(t_2 - t_1)$ for $t_2 > t_1$, and if there exists a function $Q_{\rho_1\rho_2}$ such that

$$\lim_{t_2 \to t_1} P_{\rho_1\rho_2}(t_2 - t_1) = Q_{\rho_1\rho_2} \cdot (t_2 - t_1) \qquad \rho_1 \neq \rho_2$$

then the conditional probability $P_{\rho_1\rho_2}(t_2 - t_1)$ can be expressed in terms of $Q_{\rho_1\rho_2}$. We remark that in the above we have assumed that the parameter $\mathbf{t}$ is one-dimensional. This assumption is essential, because if $\mathbf{t}$ is n-dimensional, the $\mathbf{t}_1, \ldots,$ etc., cannot be ordered. Hence, for the remainder of this section we assume that $\mathbf{t}$ is one-dimensional.

From the theory presented in Chap. 2 it can be shown that under appropriate regularity conditions,

$$\lim_{t_2 \to t_1} P_{\rho_1\rho_2}(t_2 - t_1) = Q_{\rho_1\rho_2} \cdot T + \delta(\rho_2 - \rho_1)\phi(T)$$

where $T = t_2 - t_1$, and

$$\phi(T) = 1 - T\int_{\rho_2} Q_{\rho_1\rho_2}\,d\rho_2$$

If the above limit exists, then $P_{\rho_1\rho_2}(T)$ satisfies the forward equation

$$\frac{\partial P_{\rho_1\rho_2}(T)}{\partial t_2} = -P_{\rho_1\rho_2}(T)\int_{\rho'} Q_{\rho_2\rho'}\,d\rho' + \int_{\rho'} P_{\rho_1\rho'}(T)Q_{\rho'\rho_2}\,d\rho' \tag{7.28}$$

In order to calculate the correlation function, it is necessary to consider the joint probability that $\rho(t_n) = \rho_n$, $\rho(t_{n-1}) = \rho_{n-1}, \ldots$, and $\rho(t_1) = \rho_1$, given that $\rho(t_0) = \rho_0$. If the stochastic process is homogeneous and Markovian with respect to t, we have

$$P_{\rho_0\rho_1\cdots\rho_n}(t_0, t_1, \ldots, t_n) = P_{\rho_0\rho_1}(t_1 - t_0)P_{\rho_1\rho_2}(t_2 - t_1) \cdots P_{\rho_{n-1}\rho_n}(t_n - t_{n-1})$$

In view of the above,

$$\mathscr{E}\{\rho(t_1) \cdots \rho(t_n)\} = \int_{\rho_n}\int_{\rho_{n-1}} \cdots \int_{\rho_1} \rho_1\rho_2 \cdots \rho_n P_{\rho_0\rho_1}(t_1 - t_0)$$

$$\times P_{\rho_1\rho_2}(t_2 - t_1) \cdots P_{\rho_{n-1}\rho_n}(t_n - t_{n-1})\, d\rho_1 \cdots d\rho_n \quad (7.29)$$

Also, if ρ_1 admits a stationary distribution as $t_1 - t_0 \to \infty$, that is

$$\lim_{t_1-t_0\to\infty} P_{\rho_0\rho_1}(t_1 - t_0) = \psi(\rho_1)$$

we have

$$\mathscr{E}\{\rho(t_1) \cdots \rho(t_n)\} = \int_{\rho_n} \cdots \int_{\rho_1} \rho_1 \cdots \rho_n \psi(\rho_1)$$

$$\times P_{\rho_1\rho_2}(t_2 - t_1) \cdots P_{\rho_{n-1}\rho_n}(t_n - t_{n-1})\, d\rho_1 \cdots d\rho_n \quad (7.30)$$

The above is a formal expression of the correlation function of degree n (where n is a positive integer) in terms of the conditional probabilities $P_{\rho_i\rho_{i-1}}(t_i - t_{i-1})$, $i = 2, \ldots, n$, of the associated stochastic process.

In order to obtain an explicit expression for the correlation function of degree n associated with the Markovian density field, we now assume that

$$\lim_{T\to0} P_{\rho\rho'}(T) = Q_{\rho\rho'} \cdot T$$

where we now put $T = |t' - t|$. We next assume that $Q_{\rho\rho'}$ depends only on ρ'. Hence

$$Q_{\rho\rho'} = R_{\rho'}$$

In view of this assumption, we have

$$\int_{\rho'} Q_{\rho\rho'}\, d\rho' = \int_{\rho'} R_{\rho'}\, d\rho' = \gamma \quad (7.31)$$

that is, the total probability of a transition from the state ρ to any other state is independent of ρ. We denote this total probability by γ. From (7.31) and (7.28) we see that $P_{\rho\rho'}(T)$ satisfies

$$\frac{\partial P_{\rho\rho'}(T)}{\partial T} = -\gamma P_{\rho\rho'}(T) + \int_{\rho''} P_{\rho\rho''}(T)R_{\rho'}\, d\rho''$$

$$= -\gamma P_{\rho\rho'}(T) + R_{\rho'} \quad (7.32)$$

since $\int_{\rho''} P_{\rho\rho''}(T)\,d\rho'' = 1$. The solution of Eq. (7.32) is

$$P_{\rho\rho'}(T) = \frac{R_{\rho'}}{\gamma}\,[1 - e^{-\gamma T}] + \delta(\rho' - \rho)e^{-\gamma T} \qquad (7.33)$$

From the definition of $\psi(\rho)$ we have

$$\lim_{T\to\infty} P_{\rho\rho'}(T) = \frac{R_{\rho'}}{\gamma} = \psi(\rho')$$

Now, from (7.33) and (7.30) we obtain

$$\mathscr{E}\{\rho(t)\rho(t')\} - (\mathscr{E}\{\rho\})^2 = [\mathscr{E}\{\rho^2\} - (\mathscr{E}\{\rho\})^2]e^{-\gamma T}$$

where

$$\mathscr{E}\{\rho^n\} = \int_0^\infty \rho^n \psi(\rho)\,d\rho$$

Let us now assume that γ, defined by (7.31), is very large. In this case it is clear that $P_{\rho\rho'}(T)$ tends to a stationary state very rapidly. It is also clear that, for $T \gg 1/\gamma$, γ is large:

$$P_{\rho\rho'}(T) \to \psi(\rho')$$

hence, if the above condition obtains, the distribution at the position t' is independent of the distribution at the position t. It is important to note that regardless of how large γ may be, the densities at the positions t and t' are not independent for T in the interval $(0, 1/\gamma)$. This property of the Markovian density field is of great interest in astrophysical applications, for it reflects the fact that regardless of how "wild" or erratic the density fluctuation may be (i.e., however large γ may be) densities at two positions separated by a distance less than $1/\gamma$ are correlated. We also note that the above property enables us to use a Markovian density field as an approximation to a wildly fluctuating density field when the correlation between two points tends to zero as the distance between the two points increases.

We now calculate the correlation function of degree n. This is obtained by introducing (7.33) into (7.30). We first observe that, if γ is assumed to be large, the densities $\rho(t_1), \ldots, \rho(t_n)$ are independent, provided $t_i - t_{i-1} \gg 1/\gamma$. The approximation, which assumes that γ is large, is equivalent to replacing $e^{-\gamma t'}$ by $1/\gamma\,\delta(t')$, where $\delta(t')$ is the Dirac δ-function, when integrating with respect to t' over any domain $0 \leqslant t' \leqslant t$ for which $t \gg 1/\gamma$. That this is true, we have only to note that

$$\int_0^t e^{-\gamma t'}\,dt' = \frac{1}{\gamma}\,(1 - e^{-\gamma t}) = \frac{1}{\gamma} \qquad \text{for } t \gg \frac{1}{\gamma}$$

In view of the above, the distribution function (7.33) can be rewritten as

$$P_{\rho\rho'}(T) = \psi(\rho')\left[1 - \frac{1}{\gamma}\,\delta(t'-t)\right] + \frac{\delta(\rho'-\rho)\delta(t'-t)}{\gamma} \quad (7.34)$$

From (7.27), (7.29), and (7.34), we have, neglecting terms of order $(1/\gamma)^2$,

$$\mathscr{E}\{X^n(t)\} = \int_0^t dt_1 \cdots \int_0^t dt_n \int_{\rho_n} d\rho_n \cdots \int_{\rho_1} d\rho_1 \cdots \rho_n$$

$$\times\, \psi(\rho_1)\cdots\psi(\rho_n)\left\{1 + \frac{1}{\gamma}\left[-\sum_{i=1}^n \delta(t_i - t_{i-1})\right.\right.$$

$$\left.\left.+\sum_{i=1}^n \frac{\delta(\rho_i - \rho_{i-1})\,\delta(t_i - t_{i-1})}{\psi(\rho_i)}\right]\right\} + 0\!\left(\frac{1}{\gamma^2}\right) \quad (7.35)$$

The above yields

$$\mathscr{E}\{X^n(t)\} = (\mathscr{E}\{X(t)\})^n + \frac{1}{\gamma}\,n(n-1)(\mathscr{E}\{X(t)\})^{n-1}\mathscr{E}\{\rho\}\frac{\mathscr{E}\{\rho^2\} - (\mathscr{E}\{\rho\})^2}{(\mathscr{E}\{\rho\})^2}$$

$$= (\mathscr{E}\{X(t)\})^n + \alpha^2\tau n(n-1)(\mathscr{E}\{X(t)\})^{n-1} \quad (7.36)$$

where

$$\alpha^2 = \frac{\mathscr{E}\{\rho^2\} - (\mathscr{E}\{\rho\})^2}{(\mathscr{E}\{\rho\})^2} \qquad \tau = \frac{\mathscr{E}\{\rho\}}{\gamma}$$

It is of interest to note that

$$\mathscr{E}\{X^2(t)\} - (\mathscr{E}\{X(t)\})^2 = 2\alpha^2\tau\mathscr{E}\{X(t)\}$$
$$= 2\alpha^2\tau\mathscr{E}\{\rho\}t \quad (7.37)$$

since $\mathscr{E}\{X(t)\} = \mathscr{E}\{\rho\}t$. From (7.37) we observe that the variance of $X(t)$ is proportional to t if $t \gg 1/\gamma$. In other words, the random variables $X(t_1)$ and $X(t_2)$ are independent if the domains t_1 and t_2 are disjoint.

We now apply the results obtained above to the problem of fluctuations in brightness of the Milky Way. As in Sec. 7.2B we will assume (1) that the observer is at the origin $t = 0$, (2) that along t there is a continuous, but fluctuating, distribution of interstellar matter in a stationary state, and (3) that the stars are uniformly distributed over the interval $[0,t]$ and in the interval $(\tau, \tau + d\tau)$ give a deterministic contribution of radiation of intensity equal to $d\tau$.

Now let κ denote the absorption coefficient and let ρ denote the density of interstellar matter. We then have that

$$X(t) = \kappa\int_0^t \rho\, d\tau$$

is the optical thickness of matter corresponding to the distance t. By a suitable choice of units, we can put $\kappa = 1/\mathscr{E}\{\rho\}$. This means that radiation of unit intensity, on passing through matter of extension t, is reduced to an intensity $e^{-X(t)}$, where

$$X(t) = \int_0^t \rho'\, d\tau \qquad \rho' = \frac{\rho}{\mathscr{E}\{\rho\}}$$

In this case $\mathscr{E}\{X(t)\} = t$, but α^2 remains the same.

Let us now consider what happens to the radiation in the interval $(0,t]$ before it reaches the observer at the origin $t = 0$. The amount of radiation emanating from the interval $(\tau, \tau + d\tau)$ traverses a distance τ, or an optical path $X(\tau)$, before reaching the observer, and in doing so it is reduced to $e^{-X(\tau)}\, d\tau$. Therefore, with each point τ we can associate a random variable $Y(\tau) = e^{-X(\tau)}$. If we integrate $Y(\tau)$ over the interval $[0,t]$, we obtain the intensity of radiation reaching the observer at the origin. Let

$$I(t) = \int_0^t Y(\tau)\, d\tau$$

denote the intensity of radiation reaching the observer at the origin. We are particularly interested in calculating the integral moments of $I(t)$, and they can be determined if we know the correlation function $\mathscr{E}\{Y(t_1) \cdots Y(t_n)\}$. Now let $P_X(t)\, dX$ denote the probability that the optical length of extension t is in the interval $(X, X + dX)$. We then have

$$\mathscr{E}\{Y(t_1) \cdots Y(t_n)\} = \int_0^\infty \cdots \int_0^\infty e^{-X_1} e^{-(X_1 + X_2)} \cdots e^{-(X_1 + \cdots + X_n)}$$
$$\times\ P_{X_1}(t_1) P_{X_2}(t_2 - t_1) \cdots P_{X_n}(t_n - t_{n-1})\, dX_1 \cdots dX_n$$
$$(7.38)$$

The equation given above can be obtained as follows: We first compute the joint probability that optical lengths corresponding to the intervals $(0,t_i)$ are in the intervals $(\mu_i, \mu_i + d\mu_i)$, respectively, where $i = 1, \ldots, n$. This probability is given by

$$P_{\mu_1}(t_1) P_{\mu_2 - \mu_1}(t_2 - t_1) \cdots P_{\mu_n - \mu_{n-1}}(t_n - t_{n-1})\, d\mu_1 \cdots d\mu_n$$

We next introduce the transformation

$$\mu_1 = X_1$$
$$\mu_2 - \mu_1 = X_2$$
$$\cdot$$
$$\cdot$$
$$\cdot$$
$$\mu_n - \mu_{n-1} = X_n$$

The Jacobian of this transformation is 1; hence, we obtain (7.38). **We** now observe that (7.38) can be written as

$$\mathscr{E}\{Y(t_1) \cdots Y(t_n)\} = \int_0^\infty \cdots \int_0^\infty \exp\{-[nX_1 + (n-1)X_2 + \cdots + X_n]\}$$
$$\times P_{X_1}(t_1) P_{X_2}(t_2 - t_1) \cdots P_{X_n}(t_n - t_{n-1}) \, dX_1 \cdots dX_n$$
$$(7.39)$$

Consider the integral

$$\int_0^\infty e^{-nX} P_X(t) \, dt$$

By definition, the above integral is equal to $\mathscr{E}\{e^{-nX(t)}\}$. If we expand $e^{-nX(t)}$ as a power series, we have

$$\mathscr{E}\{e^{-nX(t)}\} = \left\{1 - nX(t) + \frac{n^2}{2} X^2(t) \cdots\right\}$$

From (7.36) we obtain

$$\mathscr{E}\{e^{-nX(t)}\} = \exp\{-n\mathscr{E}\{X(t)\}\}[1 + n^2\alpha^2\tau\mathscr{E}\{X(t)\}]$$
$$= e^{-nt}(1 + n^2\alpha^2\tau t) \tag{7.40}$$

since $\mathscr{E}\{X(t)\} = t$.

If we now introduce (7.40) into (7.38) and neglect terms of the order of τ^2, we have

$$\mathscr{E}\{Y(t_1) \cdots Y(t_n)\} = \exp\{-[nt_1 + (n-1)(t_2 - t_1) + \cdots + (t_n - t_{n-1})]\}$$
$$\times \{1 + [n^2t_1 + (n-1)^2(t_2 - t_1) + \cdots + (t_n - t_{n-1})]\alpha^2\tau\}$$
$$(7.41)$$

for $t_1 < t_2 < \cdots < t_n$. In computing $\int_0^t Y(\tau) \, d\tau$ we integrate $\mathscr{E}\{Y(t_1) \cdots Y(t_n)\}$ with respect to $t_1, \ldots, t_n$ over the whole domain t. Therefore, the nth moment of the intensity of radiation reaching the observer is given by

$$\mathscr{E}\{I^n(t)\} = \int_0^t \cdots \int_0^t \mathscr{E}\{Y(t_1) \cdots Y(t_n)\} \, dt_1 \cdots dt_n$$
$$= n! \int_0^t dt_1 \int_{t_1}^t dt_2 \cdots \int_{t_{n-1}}^t \mathscr{E}\{Y(t_1) \cdots Y(t_n)\} \tag{7.42}$$

for $t_1 < t_2 < \cdots < t_n$.

In astrophysical problems it is of interest to obtain the moments for a system of infinite extent. In this case, (7.42) yields

$$\mathscr{E}\{I^n(\infty)\} = \mathscr{E}\{I^n\} = 1 + \frac{n(n+1)}{2}\alpha^2\tau + 0(\tau^2) \tag{7.43}$$

Since
$$\mathscr{E}\{I\} = 1 + \alpha^2 \tau + 0(\tau^2)$$

we have
$$\frac{\mathscr{E}\{I^n\}}{(\mathscr{E}\{I\})^n} = 1 + \frac{n(n-1)}{2}\,\alpha^2\tau$$

Ramakrishnan has also shown that for t finite the nth moments of $I(t)$ are given by

$$\mathscr{E}\{I^n(t)\} = n!\sum_{k=0}^{n} \frac{e^{-a_k t}}{\prod_{\substack{j=0 \\ j \neq k}}^{n} (a_j - a_k)} \tag{7.44}$$

where
$$a_n = n(1 - n\alpha^2\tau)$$

We close this discussion of a stochastic model of a fluctuating density field by stating the following result, which enables us to compare the model based on a fluctuating density field with the model based on discrete clouds: A discrete cloud system with a Poisson distribution and with optical thickness $\tau_* = 2\alpha^2\tau$ for each cloud can always replace a fluctuating density field with critical length $\mathscr{E}\{\rho\}/\gamma = \tau$.

D. Other Studies. We close this section on the problem of fluctuations in brightness of the Milky Way with brief comments on some additional studies that are of interest.

Agekyan[1] has recently obtained a generalization of the Chandrasekar-Münch integral equation. This generalization is obtained by assuming that the emission per unit volume and the probability of occurrence of a cloud per unit length along the line of sight are both functions of position.

Ramakrishnan[20] has generalized the Chandrasekar-Münch formulation of the fluctuation problem by dropping the assumption that in the interval $(\tau, \tau + d\tau)$ there is a deterministic contribution $\beta\,d\tau$ to the intensity u, and in its place assuming that in $(\tau, \tau + d\tau)$ a star occurs with probability $\mu\,d\tau$ and a given star has intensity in the interval $(\eta, \eta + d\eta)$ with probability $\phi(\eta)\,d\eta$. In view of the above, the mean or expected contribution to the intensity u by the element $d\tau$ is

$$\mu\mathscr{E}\{\eta\}\,d\tau$$

where
$$\mathscr{E}\{\eta\} = \int_{\eta} \eta\phi(\eta)\,d\eta$$

Hence, $\mu\mathscr{E}\{\eta\}$ corresponds to β in the previous formulation.

This new formulation suggests the following problems: (1) What is the probability distribution of the total intensity u, that is, the sum of the intensities of the stars? (2) What is the probability of observing n stars, each with intensity greater than η? These problems have been studied by Ramakrishnan by using the method of product density functions.

7.3 Theory of the Spatial Distribution of Galaxies

A. Introduction. In this section we discuss the theory of the spatial distribution of galaxies due to J. Neyman and E. L. Scott. Contributions to this theory are reported in a series of papers by Neyman and Scott and their associates [10–15,24]. For an excellent summary of this work and for a discussion of numerical studies we refer to the paper by Neyman, Scott, and Shane [15].

The theory of the spatial distribution of galaxies presents problems which are in general similar to those encountered in other areas of applied stochastic processes. For example, the astronomer obtains photographs of galaxies by using telescopic instruments, and these photographs, in turn, must be analyzed with the aid of a model of the distribution of galaxies in order to infer the distribution that actually obtains in stellar space.

In Sec. 7.3B we consider a stochastic model of the spatial distribution of galaxies which is termed a "model of simple clustering." In particular, we consider the postulates on which the model is based and discuss the functions which characterize the theoretical distribution of galaxies.

B. Stochastic Model of Simple Clustering of Galaxies. We now consider the stochastic model of the spatial distribution of galaxies developed in Ref. 11. In order to develop the model, it is first necessary to state the postulates underlying it, since the postulates reflect the structural properties of the galactic space and are formulated in terms of the functions which characterize the distribution of galaxies.

The postulates are as follows:

1. Galaxies occur only in clusters.

2. Associated with every cluster is a random variable X which denotes the number of galaxies belonging to the cluster. The random variables X associated with the different clusters are assumed to be mutually independent and identically distributed. Let $G_X(s)$ denote the generating function of the probability distribution of X, $X = 1, 2, \ldots$. The variables X are independent of the other variables occurring in the formulation of the model.

3. The distribution of positions of galaxies within a cluster is random and is governed by a probabilistic law which is the same for all clusters. To be more precise, given a time t and given the coordinates (u',v',w') of the cluster center C at time t, there exists a function $f(\eta,t)$ which denotes the conditional probability density of the coordinates (x,y,z) of

the galaxy.[1] For t fixed, $f(\eta,t)$ is assumed to be a continuous function of the distance η between any given point (x',y',z') and the cluster center C. The distance η is given by

$$\eta = [(x' - u')^2 + (y' - v')^2 + (z' - w')]^{\frac{1}{2}}$$

i.e., η is the distance function, or metric, in Euclidean 3-space. It is also assumed that the coordinate axes are orthogonal and that the observer is at the origin.

4. At each instant t, to every region R in the stellar space there is associated a random variable $Y(R,t)$ which denotes the number of cluster centers in R at time t. The distribution of Y, for t fixed, depends only on the volume V of the region R and is independent of the geometry or location of R. Let $G_Y(s \mid V)$ denote the generating function of the probability distribution of Y.

Let R be a region of finite volume, and let R_0 be a subregion or subset of R. If it is known that there are exactly n $(n > 1)$ cluster centers, say $C_1, \ldots, C_n$, in R and if $a_1, \ldots, a_m$ is an arbitrary combination of $m < n$ numbers selected out of $1, \ldots, n$, then the probability that there are exactly m cluster centers in R_0 and that they are $C_{a_1}, \ldots, C_{a_m}$ is independent of the combination $a_1, \ldots, a_m$ and is equal to the conditional probability that $Y(R_0) = m$, given that $Y(R) = n$ divided by the number of combinations of m objects out of the given n.

If we use the notation $C_i \in R_0$ to denote that the ith cluster center is in the region R_0, let $\prod_{i=1}^{n}(C_{a_i} \in R_0)$ denote the logical product of the m propositions $C_{a_i} \in R_0$, $i = 1, \ldots, m$; i.e., the product symbol represents the proposition that all the cluster centers $C_{a_1}, \ldots, C_{a_m}$ are in R_0. With this notation the conditional probability can be written as

$$\mathscr{P}\{[Y(R_0) = m] \cdot \prod_{i=1}^{m}(C_{a_i} \in R_0 \mid Y(R) = n)\} = \frac{1}{\binom{n}{m}} \mathscr{P}\{Y(R_0) = m \mid Y(R) = n\}$$

In addition to the above postulates, it is necessary to introduce a function which denotes the probability that the photograph will contain a noticeable image of the galaxy concerned. The introduction of such a function is necessary in order to establish a relation between the distribution of galaxies in space and the distribution of images on the photographic plate. Let us consider a galaxy with coordinates (x,y,z) that emitted a light signal that reaches the photographic plate at the time the photograph of the sky is taken. Denote by T the time at

---

[1] In Ref. 11 the time variable t was not introduced, because a static universe was assumed.

which the light signal was emitted and by ξ the distance which at time T separated the galaxy from the observer. Since the observer is at the origin of the coordinate system, and since we are in Euclidean 3-space, the distance ξ is simply

$$\xi = (x^2 + y^2 + z^2)^{\frac{1}{2}}$$

We denote by $\theta(\xi)$ the probability that the photograph will contain a noticeable image of the galaxy. It is assumed that the function $\theta(\xi)$ is defined for every galaxy, and that θ is a continuous function for all values of ξ.

We now consider an expression for the generating function of the joint distribution of the numbers of galaxies visible within two arbitrary regions. Let ω_1 and ω_2 denote two arbitrary regions (not necessarily disjoint) in space, which will be photographed on idealized plates. We will also let ω_1 and ω_2 denote the solid angles, with vertices at the origin of the coordinate system, corresponding to these two regions. Also, let $m_i, i = 1, 2$, denote the limiting magnitude of the ith region. We assume that $m_1 \leqslant m_2$. Finally, let the random variables X_1 and X_2 denote the numbers of galaxies visible on photographs taken of the regions ω_1 and ω_2, respectively.

Let $G_{X_1,X_2}(s_1,s_2)$ denote the generating function of the joint probability distribution of X_1 and X_2. It has been shown in Ref. 11 that

$$G_{X_1,X_2}(s_1,s_2) = e^{\psi(s_1,s_2)} \tag{7.45}$$

where $\quad \psi(s_1,s_2) = \int\int\int_{-\infty}^{\infty} h[G_X(1 - P_1(1 - s_1) - P_2(1 - s_2)$

$$- P_3(1 - s_1 s_2))] \, du \, dv \, dw \tag{7.46}$$

In (7.46) $\qquad h(s) = -h_0 + \sum_{i=1}^{\infty} h_i s^i$

where $h(1) = 0$, and the $h_i, i = 0, 1, \ldots$, are nonnegative constants. We have also introduced the following functions: $P_1(u,v,w)$ denotes the probability that a galaxy from a cluster with its center at (u,v,w) will be visible on the plate taken over the regions ω_1 but not on the plate taken over ω_2. Since we have assumed $m_1 \leqslant m_2$, we have

$$P_1(u,v,w) = \underset{\omega_1-(\omega_1\cap\omega_2)}{\int\int\int} f(\eta)\theta_1(\xi) \, dx \, dy \, dz \tag{7.47}$$

Similarly, $P_2(u,v,w)$ denotes the probability that a galaxy from a cluster with its center at (u,v,w) will be visible on the plate taken over ω_2 but not on the plate taken over ω_1. Hence we have

$$P_2(u,v,w)= \underbrace{\int\int\int}_{\omega_1-(\omega_1\cap\omega_2)} f(\eta)\theta_2(\xi)\,dx\,dy\,dz + \underbrace{\int\int\int}_{\omega_1\cap\omega_2} f(\eta)[\theta_2(\xi) - \theta_1(\xi)]\,dx\,dy\,dz$$

(7.48)

Finally, $P_3(u,v,w)$ denotes the probability that a galaxy from a cluster with its center at (u,v,w) will be visible on both plates. We have

$$P_3(u,v,w) = \underbrace{\int\int\int}_{\omega_1\cap\omega_2} f(\eta)\theta_1(\xi)\,dx\,dy\,dz$$

(7.49)

In the above expressions we have used the set-theoretic notation $\omega_1 \cap \omega_2$ to denote the intersection of ω_1 and ω_2, i.e., the volume common to both regions.

From the above formulas we see that, once the functions $h(s)$, $G_X(s)$, $f(\eta)$, and $\theta(\xi)$ are specified, the generating function (7.45) of the joint probability distribution of X_1 and X_2 is completely determined; hence, the joint probability density can be obtained.

For a detailed study of a specific model, based on certain assumptions concerning the functions $h(s)$, $G_X(s)$, $f(\eta)$, and $\theta(\xi)$, we refer to Ref. 14.

We close this section by remarking that the moments of X_1 and X_2 can be obtained from $G_{X_1,X_2}(s_1,s_2)$ by differentiation. Let σ_{ij} denote the central product moment of order i with respect to X_1 and of order j with respect to X_2; hence,

$$\sigma_{ij} = \mathscr{E}\{[X_1 - \mathscr{E}\{X_1\}]^i[X_2 - \mathscr{E}\{X_2\}]^j\}$$

(7.50)

for $i, j = 0, 1, \ldots$. By differentiating (7.45) we obtain, for example,

$$\mathscr{E}\{X_i\} = \frac{\partial\psi(s_1,s_2)}{\partial s_i}\bigg]_{s_i=1} \qquad i = 1, 2$$

(7.51)

$$\sigma_{20} = \mathscr{D}^2\{X_1\} = \frac{\partial^2\psi(s_1,s_2)}{\partial^2 s_1}\bigg]_{s_1=1} + \mathscr{E}\{X_1\}$$

(7.52)

$$\sigma_{02} = \mathscr{D}^2\{X_2\} = \frac{\partial^2\psi(s_1,s_2)}{\partial^2 s_2}\bigg]_{s=1} + \mathscr{E}\{X_2\}$$

(7.53)

7.4   Stochastic Theory of Radiative Transfer

A.   Introduction. The theory of radiative transfer is concerned with the study of the transfer of radiant energy through media that absorb, scatter, or emit radiant energy. As a branch of mathematical physics its most important applications are to astrophysical and neutron transport problems. In astrophysics the carrier of radiant energy is generally assumed to be a photon, while in neutron transport theory the neutron is considered to be the carrier.

The mathematical problems in radiative transfer theory are, in the main, concerned with the solution of an integrodifferential equation for the specific intensity of radiation. The particular equation involved will depend, of course, on the nature of the radiation phenomenon being considered and on the postulated mechanism for the transfer of radiant energy. For a detailed discussion of radiative transfer theory we refer to the books of Chandrasekar [3] and Kourganoff and Busbridge [6], and for a rigorous treatment of the theory based on measure theory and functional analysis, we refer to the work of Preisendorfer [16].

In this section we discuss the work of Ueno [25–29], which is concerned with the development of a stochastic theory of radiative transfer. The fundamental problem in this theory is the determination of the emergent intensity and the source function in a radiation field by utilizing the probability distribution of emission. In Sec. 7.4B we derive the probability distribution of emission by utilizing the theory of Markov processes, and in Sec. 7.4C we consider, within this framework, the multiple scattering of photons for Milne's problem in semi-infinite space with isotropic scattering. For additional applications we refer to Refs. 25 to 29.

We close this introductory section with a brief discussion of the specific intensity and the source function associated with a radiation field. For a detailed discussion we refer to Refs. 3 and 6. In analyzing a radiation field it is often necessary to consider the amount of radiant energy, say dE_ν, in a specified frequency interval $(\nu, \nu + d\nu)$ which is transported across an element of area $d\sigma$. Let 0 be a fixed point and let L be a fixed line passing through 0 (Fig. 7.1). Let the point 0 be considered in the element of area $d\sigma$ and let θ be the angle between the line L and the line N normal to $d\sigma$. If we now draw a line parallel to L through each point $0'$ contained in $d\sigma$ such that the line is the axis and $0'$ is the vertex, we obtain an elementary cone of solid angle $d\omega$. The collection of all cones constructed in this manner defines a semi-infinite truncated cone with the element of area $d\sigma$ as the finite end.

Now let dE_ν denote the amount of energy, in the frequency interval $(\nu, \nu + d\nu)$, which is transmitted across $d\sigma$ in time dt in directions lying within the truncated cone. It is known that for radiation fields occurring in nature the ratio

$$\frac{dE_\nu}{\cos\theta \, d\sigma \, d\omega \, d\nu \, dt}$$

approaches a limit, as $d\sigma$, $d\omega$, $d\nu$, and dt approach zero, which is a function of the point 0 and the line L but is independent of the angle θ. The limit, which is denoted by I_ν, defines the *specific intensity of radiation* at the point 0 along the line L. Hence from the ratio given above, the amount of energy transported across an element $d\sigma$, in a specified frequency interval $(\nu, \nu + d\nu)$, in a direction making an angle θ with the normal to $d\sigma$, within an elementary solid angle $d\omega$, and in time dt, is given by

$$dE_\nu = I_\nu \cos \theta \, d\sigma \, d\omega \, d\nu \, dt$$

The construction given above also defines what is called a *pencil of radiation.*

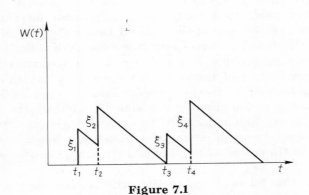

Figure 7.1

We now define the absorption and emission coefficients, since the interaction between radiation and matter is usually expressed in terms of these coefficients, which are used to define the source function associated with a radiation field.

A pencil of radiation in passing through a medium will be measured by its interaction with matter. Let us assume that a pencil of radiation of intensity I_ν upon passing through a thickness of matter ds in the direction of its propagation becomes $I_\nu + dI_\nu$. We can then write

$$dI_\nu = -\kappa_\nu \rho I_\nu \, ds$$

where ρ is the density of the matter. The quantity κ_ν defines the *mass absorption coefficient* for radiation of frequency ν. Hence

$$\kappa_\nu = -\frac{1}{\rho I_\nu} \frac{dI_\nu}{ds}$$

The *emission coefficient* j_ν is defined in such a way that the amount of radiant energy emitted by an element of mass dm confined to an element of solid angle $d\omega$, in the frequency interval $(\nu, \nu + d\nu)$ and in time dt, is given by

$$j_\nu \, dm \, d\omega \, d\nu \, dt$$

The ratio of the emission coefficient to the absorption coefficient at a given point in the radiation field plays a fundamental role in the theory of radiative transfer. The ratio is called the *source function*, and we denote it by J_ν. Thus, by definition

$$J_\nu = \frac{j_\nu}{\kappa_\nu}$$

B.   Stochastic Equation for the Probability Distribution of Emission. In this section we derive the stochastic equation for the probability distribution of emission of radiant energy in a semi-infinite plane-parallel atmosphere. Let $M(t)$ denote a random variable associated with a point in space, depending on the one-dimensional parameter t, and let $P_\mu(t) = \mathscr{P}\{M(t) = \mu\}$ denote the probability distribution of $M(t)$. In the above μ is the cosine of the angle between the direction of the light beam and the normal to the atmosphere and t is the optical depth measured in total absorption. By definition, $P_\mu(t) \, d\mu$ is the probability that $M(t)$ is in the interval $(\mu, \mu + d\mu)$ at depth t. Physically, $P_\mu(t) \, d\mu$ can be interpreted as the emission at the point t in the direction μ. Let

$$P_{\mu_1 \mu_2}(t_1, t_2) \, d\mu_2 = \mathscr{P}\{M(t_2) = \mu_2 \mid M(t_1) = \mu_1\} \qquad \text{for } t_2 > t_1$$

denote the probability that $M(t_2)$ is in the interval $(\mu_2, \mu_2 + d\mu_2)$, given that $M(t_1) = \mu_1$. If we now assume that $P_\mu(t)$ is homogeneous with respect to time and that the process $\{M(t), t \geqslant 0\}$ is Markovian, then $P_\mu(t)$ satisfies the functional equation[1]

$$P_\mu(t) = \int_0^1 P_{\mu'}(t - \tau) P_{\mu' \mu}(\tau) \, d\mu' \tag{7.54}$$

for $0 < \tau < t$. Let us now assume that there exists a function $Q_{\mu' \mu}$ such that

$$\lim_{\tau \to 0} P_{\mu' \mu}(\tau) = Q_{\mu' \mu} \cdot \tau + \delta(\mu - \mu')\phi(\tau) \tag{7.55}$$

where $\phi(\tau)$ is given by $\phi(\tau) = 1 - \int_0^1 Q_{\mu' \mu} \, d\mu.$

We now assume $$Q_{\mu' \mu} = \frac{P_\mu(0)}{2\mu'} \tag{7.56}$$

and require that $$\int_0^1 Q_{\mu \mu'} \, d\mu' = \frac{1}{\mu} \tag{7.57}$$

---

[1] Equation (7.54) is a modification of the Chapman-Kolmogorov equation.

In (7.57) $1/\mu$ can be interpreted as the total cross section. By inserting (7.55) in (7.54), using (7.56), and passing to the limit, we obtain

$$\frac{\partial P_\mu(t)}{\partial t} = -\frac{P_\mu(t)}{\mu} + \tfrac{1}{2}P_\mu(0)\int_0^1 P_{\mu'}(t)\frac{d\mu'}{\mu'} \tag{7.58}$$

Equation (7.58) admits the following interpretation: The left-hand side represents the change in the probability of the system being in the state μ as a function of the optical depth t. The first term on the right-hand side represents the decrease in the probability of the state (μ,t) if a parallel beam of radiation emerging in the direction μ was transmitted along the interval $(t, t + dt)$ without being scattered or absorbed, and the second term is a measure of the total contribution to the probability $P_\mu(t + dt)$ if the system in the state (μ',t), with probability $P_{\mu'}(t)\,d\mu'$, changed to the state μ. This transition takes place as a consequence of one or more scattering processes.

The formal solution of Eq. (7.58) is given by

$$P_\mu(t) = P_\mu(0)e^{-t/\mu} + \tfrac{1}{2}P_\mu(0)\int_0^t d\tau \int_0^1 e^{-(t-s)/\mu}P_{\mu'}(t)\frac{d\mu'}{\mu'} \tag{7.59}$$

We remark that Eq. (7.58) is similar in form to the Kolmogorov-Feller integrodifferential equation considered in Chap. 2.

It has been shown by Ueno that the probability distribution $P_\mu(t)$ derived on the assumption that the multiple scattering process is Markovian is equivalent to the *probability of emission* $\Phi_\mu(t)$ considered by Sobolev.[1] Here $\Phi_\mu(t)$ is the probability that a photon absorbed at a given point t will be reemitted in a given direction μ in the radiation emerging from the surface of the atmosphere.

C. The H-Equation. The Law of Darkening. For a diffuse radiation field, i.e., a radiation field which arises as a consequence of one or more scattering processes, it has been shown (Ref. 6, pp. 165–166) that the source function $J(\mu,t)$ satisfies the integrodifferential equation

$$\frac{\partial J(\mu,t)}{\partial t} = -\frac{J(\mu,t)}{\mu} + \tfrac{1}{2}J(\mu,0)\int_0^1 J(\mu',t)\frac{d\mu'}{\mu'} \tag{7.60}$$

Hence, a comparison of Eqs. (7.58) and (7.60) implies the equivalence of $P_\mu(t)$ and $J(\mu,t)$, when $J(\mu,0)$ satisfies condition (7.57).

Let us now put $J(\mu,0) = H(\mu)$. The functional equation for $H(\mu)$ is given by

$$H(\mu) = 1 + \tfrac{1}{2}\int_0^1 \rho(\mu',\mu)\,d\mu' \tag{7.61}$$

where
$$\rho(\mu',\mu) = \int_0^\infty P_\mu(t)e^{-t/\mu'}\frac{dt}{\mu'} \tag{7.62}$$

[1] Cf. S. Ueno [26].

that is, $\rho(\mu',\mu)$ is the Laplace transform of $P_\mu(t)$, where the Laplace transform is defined as

$$\mathscr{L}_{1/\mu}\{f(x)\} = \int_0^\infty f(x)e^{-x/\mu}\,\frac{dx}{\mu}$$

The function $\rho(\mu',\mu)$ satisfies the relation

$$\rho(\mu',\mu)\mu' = \rho(\mu,\mu')\mu = S(\mu,\mu')$$

where $S(\mu,\mu')$ is the standard scattering function.

By applying the Laplace transformation as defined above to Eq. (7.58) and performing the integrations, we obtain

$$\rho(\mu',\mu)\left(\frac{1}{\mu'} + \frac{1}{\mu}\right) = \frac{H(\mu)}{\mu'} + \tfrac{1}{2}H(\mu)\int_0^1 \rho(\mu',\mu'')\,\frac{d\mu''}{\mu''} \qquad (7.63)$$

By use of (7.61), the above becomes

$$\rho(\mu',\mu) = \frac{\mu}{\mu + \mu'}\,H(\mu)H(\mu') \qquad (7.64)$$

If we now substitute (7.64) in (7.61), we have

$$H(\mu) = 1 + \tfrac{1}{2}\mu H(\mu)\int_0^1 \frac{H(\mu')\,d\mu'}{\mu + \mu'} \qquad (7.65)$$

The above equation is the functional equation for the H-function introduced by Chandrasekar [3]. In this case the characteristic function $\Psi(\mu') = \tfrac{1}{2}$, which obtains in the case of conservative isotropic scattering. It can be shown that $H(\mu)$ satisfies condition (7.57); hence, we have justified the assumption that the multiple scattering process of photons in Milne's problem in semi-infinite space with isotropic scattering has the Markov property.

We now consider the problem of determining the angular distribution of the emergent radiation in a semi-infinite atmosphere in the case of a constant net flux.[1] The angular distribution is called the *law of darkening*, and it is denoted by the function

$$I(\mu,0) \qquad 0 \leqslant \mu \leqslant 1$$

The Milne first integral equation for the source function in this case can be written in the form

$$J(t) = B(t) + G(t) \qquad (7.66)$$

where $$G(t) = -B(t) + \Lambda_t[B(\tau)] + \Lambda_t[G(\tau)] \qquad (7.67)$$

---

[1] The net flux πF represents the flux of radiation normal to the plane of stratification.

and Λ_t is the Hopf operator

$$\Lambda_t[y(\tau)] = \tfrac{1}{2} \int_0^\infty y(\tau) E_1(|t - \tau|) \, d\tau$$

In the above $E_1(z)$ is the first exponential integral

$$E_1(z) = \int_0^1 e^{-z/x} \frac{dx}{x}$$

In (7.66) the function $B(t)$ is a particular solution of the Milne problem with a constant net flux πF.

Ueno has shown that the angular distribution $I(\mu,0)$ can be expressed in the form

$$I(\mu,0) = \mathscr{L}_{1/\mu}\{B(t)\} + \mathfrak{S}_\mu[f(t)] \tag{7.68}$$

where $f(t) = -B(t) + \Lambda_t[B(\tau)]$ and $\mathfrak{S}_\mu$ is the scattering operator

$$\mathfrak{S}_\mu[f(t)] = \int_0^\infty P_\mu(t) f(t) \frac{dt}{\mu} \tag{7.69}$$

In (7.68) the function $B(t)$ is given by

$$B(t) = \tfrac{3}{4} F t \tag{7.70}$$

By using the fact that (Ref. 6, p. 45)

$$\Lambda_t[\tau] = t + \tfrac{1}{2} E_3(t)$$

where

$$E_3(z) = \int_0^1 e^{-z/x} \frac{dx}{x^3}$$

is the third exponential integral, together with Eqs. (7.62), (7.69), and (7.70), Eq. (7.68) can be written in the form

$$I(\mu,0) = \tfrac{3}{4} F \left[\mu + \tfrac{1}{2} \int_0^\infty E_3(t) P_\mu(t) \frac{dt}{\mu} \right]$$

$$= \tfrac{3}{4} F \left[\mu + \frac{1}{2\mu} \int_0^1 \rho(\mu',\mu)(\mu')^2 \, d\mu' \right] \tag{7.71}$$

If we now use the relationship between $\rho(\mu',\mu)$ and $H(\mu)$ given by (7.64) and use (7.65), Eq. (7.71) becomes

$$I(\mu,0) = \tfrac{3}{4} F H(\mu)[\mu(1 - \tfrac{1}{2}\alpha_0) + \tfrac{1}{2}\alpha_1] \tag{7.72}$$

where α_0 and α_1 represent the (zero)th and first moments of the function $H(\mu)$. The ith moment of $H(\mu)$ is given by

$$\alpha_i = \int_0^1 \mu^i H(\mu) \, d\mu$$

hence (cf. [3])

$$\alpha_0 = 2 \qquad \alpha_1 = \frac{2}{\sqrt{3}}$$

Therefore, the angular distribution of the emergent radiation is given by

$$I(\mu,0) = \frac{\sqrt{3}}{4} FH(\mu) \qquad (7.73)$$

Bibliography

1 Agekyan, T. A.: On the Theory of Fluctuations in Brightness of the Milky Way and Metagalaxy (in Russian), *Vestnik Leningrad Univ.*, vol. 11, pp. 145–154, 1956.

2 Chandrasekar, S.: Stochastic Problems in Physics and Astronomy, *Revs. Modern Phys.*, vol. 15, pp. 1–89, 1943.

3 Chandrasekar, S.: "Radiative Transfer," Oxford University Press, New York, 1950.

4 Chandrasekar, S., and G. Münch: The Theory of Fluctuations in Brightness of the Milky Way: I–V, *Astrophys. J.*, vol. 112, pp. 380–392, 393–398, 1950; vol. 114, pp. 110–122, 1951; vol. 115, pp. 94–102, 103–123, 1952.

5 Chandrasekar, S., and G. Münch: On Stellar Statistics, *Astrophys. J.*, vol. 113, pp. 150–165, 1951.

6 Kourganoff, V., and I. W. Busbridge: "Basic Methods in Transfer Problems," Oxford University Press, New York, 1952.

7 McVittie, G. C.: Relativity and the Statistical Theory of the Distribution of Galaxies, *Astron. J.*, vol. 60, pp. 105–115, 1955.

8 Münch, G.: The Theory of Fluctuations in Brightness of the Milky Way: VI, *Astrophys. J.*, vol. 121, pp. 291–299, 1955.

9 Münch, G.: Stochastic Processes of Astronomical Interest, *Proceedings of the Symposia in Applied Mathematics*, vol. 7: Applied Probability, pp. 51–66, McGraw-Hill Book Company, Inc., New York, 1957.

10 Neyman, J.: Sur la théorie probabiliste dans amas de galaxies et le problème de vérification de l'hypothèse de l'expansion de l'univers, *Ann. inst. H. Poincaré*, vol. 14, pp. 201–244, 1955.

11 Neyman, J., and E. L. Scott: A Theory of the Spatial Distribution of Galaxies, *Astrophys. J.*, vol. 116, pp. 144–163, 1952.

12 Neyman, J., and E. L. Scott: Spatial Distribution of Galaxies—Analysis of the Theory of Fluctuations, *Proc. Natl. Acad. Sci. U.S.*, vol. 40, pp. 873–881, 1954.

13 Neyman, J., and E. L. Scott: Statistical Approach to Problems of Cosmology, *J. Roy. Statist. Soc.*, ser. B, vol. 20, pp. 1–43, 1958.

14 Neyman, J., E. L. Scott, and C. D. Shane: On the Spatial Distribution of Galaxies: A Specific Model, *Astrophys. J.*, vol. 117, pp. 92–133, 1953.

15 Neyman, J., E. L. Scott, and C. D. Shane: Statistics of Images of Galaxies with Particular Reference to Clustering, *Proceedings of the Third Berkeley Symposium on Mathematical Statistics and Probability*, vol. 3, pp. 75–111, 1956.

16 Preisendorfer, R. W.: A Mathematical Foundation of Radiative Transfer Theory, *J. Math. Mech.*, vol. 6, pp. 695–730, 1957.

17 Ramakrishnan, A.: On an Integral Equation of Chandrasekar and Münch, *Astrophys. J.*, vol. 115, pp. 141–144, 1952.

18 Ramakrishnan, A.: Stochastic Processes Associated with Random Divisions of a Line, *Proc. Cambridge Phil. Soc.*, vol. 49, pp. 473–485, 1953.

19 Ramakrishnan, A.: A Stochastic Model of a Fluctuating Density Field: I, II, *Astrophys. J.*, vol. 119, pp. 443–455, 682–685, 1954.

20   Ramakrishnan, A.: On Stellar Statistics, *Astrophys. J.*, vol. 122, pp. 24–31, 1955.

21   Ramakrishnan, A., and P. M. Mathews: On a Class of Stochastic Integro-differential Equations, *Proc. Indian Acad. Sci.*, sec. A, vol. 38, pp. 450–466, 1953.

22   Ramakrishnan, A., and P. M. Mathews: The Solution of an Integral Equation of Chandrasekar and Münch, *Astrophys. J.*, vol. 119, pp. 81–90, 1954.

23   Ramakrishnan, A., and R. Vasudevan: On the Distribution of Visible Stars, *Astrophys. J.*, vol. 126, pp. 513–518, 1957.

24   Scott, E. L., C. D. Shane, and M. D. Swanson: Comparison of the Synthetic and Actual Distribution of Galaxies on a Photographic Plate, *Astrophys. J.*, vol. 119, pp. 91–112, 1954.

25   Ueno, S.: The Formation of Absorption Lines by Coherent and Non-coherent Scattering: IV. The Solution of the Equation of Transfer by the Probabilistic Method, *Contrib. Inst. Astrophys. (Kyoto)*, no. 64, 1956.

26   Ueno, S.: The Probabilistic Method for Problems of Radiative Transfer: II. Milne's Problem, *Astrophys. J.*, vol. 126, pp. 413–417, 1957.

27   Ueno, S.: The Probabilistic Method for Problems of Radiative Transfer: III. Line Formation by Coherent Scattering, *J. Math. Mech.*, vol. 7, pp. 629–641, 1958.

28   Ueno, S.: La méthode probabiliste pour les problèmes de transfert du rayonnement. La réflexion diffuse et la transmission dans l'atmosphère finie, *Compt. rend. acad. sci. Paris*, vol. 247, pp. 1443–1445, 1958.

29   Ueno, S.: La méthode probabiliste pour les problèmes de transfert du rayonnement. La réflexion diffuse et la transmission dans l'atmosphère finie avec la diffusion non cohérente, *Compt. rend. acad. sci. Paris*, vol. 247, pp. 1557–1559, 1958.

8

Applications in Chemistry

8.1 Introduction

The theory of Markov processes has many applications in the field of chemistry. In particular, Markov chain methods are of great importance in the statistical theory of polymer chains; and the theory developed in Chap. 2 is applicable to the study of chemical reaction kinetics. Chemical kinetics can be defined as the study of the time rates of chemical reactions and of the factors which influence those rates (cf. [9]). The mathematical theory of chemical reaction kinetics is concerned with the prediction of the concentrations of the reactant species as a function of time, the concentration of a reactant being defined as the number of molecules per constant volume. In the classical deterministic theory the concentration is expressed as a continuous real-valued function of time and the mechanism of the reaction is expressed by a system of differential (or integral) equations, the solution of which gives the concentrations of the various reactant species. In the stochastic theory the fundamental random variables are the concentrations of the reactant species at time t, and the problem is to determine the probability distribution of these concentrations.

A stochastic approach to the study of chemical reaction kinetics is especially useful in the initial stages of a chemical reaction, for it is during this period that the probability of two molecules of the same species reacting with one another is small compared with the probability that a molecule of one species will react with molecules belonging to a number of different species. Stochastic considerations are also required in the analysis of irreproducible chemical reactions, i.e., reactions which exhibit random fluctuations about the theoretical concentration curves

359

obtained on the basis of the deterministic theory, the pattern of these fluctuations varying from experiment to experiment.

In this chapter we shall restrict our attention to some stochastic models for chemical reaction kinetics. The models we consider in Sec. 8.2 are for autocatalytic reactions, unimolecular reactions, bimolecular reactions, and some chain reactions. For the stochastic treatment of other reactions we refer to Bartholomay [1]. In Sec. 8.3 we refer to some additional applications of stochastic methods in chemistry.

8.2   Some Stochastic Models for Chemical Reaction Kinetics

A.   Simple Model for an Autocatalytic Reaction. We now consider a kinetic model for an autocatalytic reaction proposed by Delbrück [6]. This model is applicable to the study of the production of enzymes from precursor substances by an autocatalytic reaction, the rate of production of the enzyme being proportional both to the concentration of the precursor and to the concentration of the enzyme.

The model is formulated with reference to the following experimental setup: Consider a large number of experimental samples. At $t = 0$ each sample contains one molecule of enzyme and a large amount of precursor. Now assume that the fraction of samples which form their second enzyme molecule in a given interval of time is always proportional to the fraction of samples which have not formed their second enzyme molecule at the beginning of the interval of time and is independent of time. If k $(k > 0)$ is the reaction rate constant, let $k\,\Delta t + o(\Delta t)$ denote the probability that any particle will double in the interval $(t, t + \Delta t)$.

Now let the random variable $X(t)$ represent the number of enzyme molecules in the system at time t and let $P_x(t) = \mathscr{P}\{X(t) = x\}$, $x = 1$, $2, \ldots$. The samples in which doubling takes place leave the state x and enter the state $x + 1$. The probability that more than one enzyme will double in $(t, t + \Delta t)$ is $o(\Delta t)$. Hence, if at time t there are x enzyme molecules in the system, the probability $P_x(t)$ can change as follows:

1. It can increase by $k(x - 1)P_{x-1}(t)\,\Delta t$ owing to the number of samples that contain $x - 1$ enzymes and which will produce one additional enzyme.

2. It can decrease by $kxP_x(t)\,\Delta t$ owing to the number of samples that contain x enzymes.

From the above we can immediately write the differential-difference equation which $P_x(t)$ satisfies. We have

$$\frac{dP_x(t)}{dt} = k(x - 1)P_{x-1}(t) - kxP_x(t) \tag{8.1}$$

The above equation is the differential equation for the simple birth process discussed in Chap. 2. In this case the birth rate $\lambda_x = kx$. If Eq. (8.1) is solved with initial conditions

$$P_x(0) = 1 \qquad x = x_0$$
$$= 0 \qquad x \neq x_0 \tag{8.2}$$

we have

$$P_x(t) = \binom{x-1}{x-x_0} e^{-kx_0 t}(1 - e^{-kt})^{x-x_0} \qquad x \geqslant x_0$$
$$= 0 \qquad\qquad\qquad\qquad\qquad\qquad x < x_0 \tag{8.3}$$

All of the properties of this distribution discussed in Chap. 2 are applicable to this case. In particular, the mean and variance of the number of enzyme molecules in the system are

$$\mathscr{E}\{X(t)\} = x_0 e^{\lambda t} \tag{8.4}$$

$$\mathscr{D}^2\{X(t)\} = x_0 e^{\lambda t}(e^{\lambda t} - 1) \tag{8.5}$$

B. The Unimolecular Reaction. In this section we consider a stochastic model for the unimolecular reaction

$$A \xrightarrow{k} B$$

The results of this section are due to Bartholomay [1,2]. That is, we consider the case of two species, A (the reactant) and B (the product), in which A is converted irreversibly into B, the rate of reaction being given by the constant $k > 0$. To construct a stochastic model for the unimolecular reaction, we proceed as follows: Let the random variable $X(t)$ denote the concentration of A at time t and let $X(0) = x_0 > 0$. On the basis of the physical nature of the unimolecular reaction we make the following assumptions:

1. The probability of a single transformation of x molecules in the interval $(t, t + \Delta t)$, given that in $(0,t)$ exactly $(x - x_0)$ transformations occurred, is $kx\,\Delta t + o(\Delta t)$.

2. The probability of more than one transformation in $(t, t + \Delta t)$ is $o(\Delta t)$.

3. The species A and B are statistically independent.

4. The reverse reaction, i.e., B → A, occurs with probability zero.

On the basis of the above assumptions we can immediately write down the differential-difference equation for $P_x(t)$, where $P_x(t) = \mathscr{P}\{X(t) = x\}$, $x = 0, 1, 2, \ldots, x_0$. We have

$$\frac{dP_x(t)}{dt} = -kxP_x(t) + k(x+1)P_{x+1}(t) \qquad x = 0, 1, \ldots, x_0 \tag{8.6}$$

This system of equations is to be solved with the initial conditions

$$P_x(0) = 1 \qquad x = x_0$$
$$= 0 \qquad \text{otherwise} \tag{8.7}$$

Equation (8.6) is the equation for the simple death process considered in Chap. 2; hence we have

$$P_x(t) = \binom{x_0}{x} e^{-x_0 kt}(e^{kt} - 1)^{x_0 - x} \qquad 0 \leqslant x \leqslant x_0 \tag{8.8}$$

Similarly, we have that the mean and variance of the concentration of A are

$$\mathscr{E}\{X(t)\} = x_0 e^{-kt} \tag{8.9}$$
$$\mathscr{D}^2\{X(t)\} = x_0 e^{-kt}(1 - e^{-kt}) \tag{8.10}$$

That the stochastic treatment of the unimolecular reaction should lead to a pure death process is to be expected on physical grounds, since the concentration of A can only decrease from its initial value x_0 to zero as the reaction proceeds. It is also of interest to remark that the expression for the expected concentration of A [Eq. (8.9)] is the same as the expression for the concentration of A predicted by the deterministic theory of the unimolecular reaction.

C. The Bimolecular Reactions and the Law of Mass Action. A chemical reaction in which two molecules combine, resulting in the formation of a new molecule, is termed *bimolecular*. In this section we derive the stochastic differential equations that describe the change in the number of newly formed molecules in a bimolecular reaction. The law of mass action for bimolecular reactions is also derived. This section is based on the studies of Rényi [17]. Bimolecular reactions have also been considered by Bartholomay [1].

Consider a liquid that consists of molecules of two species, say A and B. Upon collision of an A molecule with a B molecule, a new molecule, say C, is formed:

$$A + B \rightarrow O$$

with probability one. Let the random variables $X_1(t)$, $X_2(t)$, $X_3(t)$ represent the numbers of A, B, and C molecules in the liquid at time t, respectively, and let x_1, x_2, x_3 ($x_i \geqslant 0$) denote the values that these random variables can assume. If at $t = 0$, $X_1(0) = x_{10}$ and $X_2(0) = x_{20}$, then $X_1(t) = x_{10} - X_3(t)$ and $X_2(t) = x_{20} - X_3(t)$.

Since we are primarily interested in $X_3(t)$, let $P_{x_3}(t) = \mathscr{P}\{X_3(t) = x_3\}$, where $x_3 = 0, 1, \ldots, C^*$ and $C^* = \min(x_{10}, x_{20})$. Now let $\lambda \, \Delta t + o(\Delta t)$ denote the probability of a collision between any A molecule and B molecule in the interval $(t, t + \Delta t)$. Define

$$h(x_3) = (x_{10} - x_3)(x_{20} - x_3)$$

The usual reasoning leads to the relation

$$P_{x_3}(t + \Delta t) = [1 - \lambda h(x_3)] \Delta t \, P_{x_3}(t) + \lambda h(x_3 - 1) \Delta t \, P_{x_3-1}(t) + o(\Delta t)$$

$$(8.11)$$

As $\Delta t \to 0$ we obtain the differential-difference equations

$$\frac{dP_{x_3}(t)}{dt} = \lambda[h(x_3 - 1)P_{x_3-1}(t) - h(x_3)P_{x_3}(t)] \qquad x_3 = 1, 2, \ldots, C*$$

$$\frac{dP_0(t)}{dt} = -\lambda h(x_3)P_0(t)$$

$$(8.12)$$

To solve this system, we use the Laplace transformation. By applying the transformation to Eqs. (8.12), we obtain

$$Q_{x_3}(s) = \mathcal{L}\{P_{x_3}(t)\} = \frac{h(x_3 - 1)Q_{x_3-1}(s)}{s + \lambda h(x_3)} \qquad x_3 = 1, 2, \ldots, C* \quad (8.13)$$

From (8.12) we have

$$\frac{dP_0(t)}{dt} = -\lambda x_{10}x_{20}P_0(t) \tag{8.14}$$

with the initial condition $P_0(0) = 1$; hence,

$$P_0(t) = e^{-\lambda x_{10}x_{20}t} \tag{8.15}$$

The Laplace transform of $P_0(t)$ is

$$Q_0(s) = \frac{1}{s + \lambda x_{10}x_{20}} \tag{8.16}$$

By using (8.16) in (8.13), we can obtain $Q_{x_3}(s)$ by iteration; hence,

$$Q_{x_3}(s) = \frac{\lambda^{x_3} \prod_{i=1}^{x_3} h(i-1)}{\prod_{i=0}^{x_3}[s + \lambda h(i)]} \qquad x_3 = 0, 1, \ldots, C* \tag{8.17}$$

By employing the partial-fraction representation, we have

$$Q_{x_3}(s) = \sum_{i=0}^{x_3} \frac{\gamma_{x_3 i}}{s + \lambda h(i)} \tag{8.18}$$

where

$$\gamma_{x_3 i} = \frac{h(0)h(1) \cdots h(x_3 - 1)}{[h(0) - h(i)][h(1) - h(i)] \cdots [h(i-1) - h(i)]}$$
$$\times [h(i+1) - h(i)] \cdots [h(x_3) - h(i)]$$

Hence

$$Q_{x_3}(s) = h(0)h(1) \cdots h(x_3 - 1) \sum_{i=1}^{x_3} \{[s + \lambda h(i)] \prod_{\substack{i \neq j \\ 0 \leqslant j \leqslant x_3}} [h(j) - h(i)]\}^{-1} \tag{8.19}$$

By applying the inverse transformation, we have

$$P_{x_3}(t) = \sum_{i=0}^{x_3} e^{-\lambda h(i)t} \left[\prod_{i=0}^{x_3-1} h(i) \right] \left[\prod_{\substack{i \neq j \\ 0 \leqslant j \leqslant x_3}} \frac{1}{h(j) - h(i)} \right] \tag{8.20}$$

for $x_3 = 0, 1, \ldots, C^*$.

In applications it is of interest to determine the expected number of C molecules in the mixture at time t. The expected number is given by

$$m(t) = \mathcal{E}\{X_3(t)\} = \sum_{x_3=0}^{C^*} x_3 P_{x_3}(t) \tag{8.21}$$

From (8.12) we see that $m(t)$ satisfies the differential equation

$$\frac{dm(t)}{dt} = \lambda\{[x_{10} - m(t)][x_{20} - m(t)]\} + \mathcal{D}^2\{X_3(t)\}$$

or

$$\frac{dm(t)}{dt} = \lambda[\mathcal{E}\{X_1(t)\}\mathcal{E}\{X_2(t)\}] + \mathcal{D}^2\{X_3(t)\} \tag{8.22}$$

where $\mathcal{E}\{X_1(t)\}$ and $\mathcal{E}\{X_2(t)\}$ are the expected number of A and B molecules, respectively, and $\mathcal{D}^2\{X_3(t)\}$ is the variance of the number of C molecules. Equation (8.22) can be written as

$$\frac{dm(t)}{dt} = \lambda\mathcal{E}\{X_1(t)X_2(t)\}$$

since the mean of the product $X_1(t)X_2(t)$ is

$$\begin{aligned} \mathcal{E}\{X_1(t)X_2(t)\} &= \mathcal{E}\{[x_{10} - X_3(t)][x_{20} - X_3(t)]\} \\ &= x_{10}x_{20} - (x_{10} + x_{20})\mathcal{E}\{X_3(t)\} + \mathcal{E}\{X_3^2(t)\} \\ &= x_{10}x_{20} - (x_{10} + x_{20})\mathcal{E}\{X_3(t)\} + [\mathcal{E}\{X_3(t)\}]^2 + \mathcal{D}^2\{X_3(t)\} \\ &= \mathcal{E}\{X_1(t)\}\mathcal{E}\{X_2(t)\} + \mathcal{D}^2\{X_3(t)\} \end{aligned}$$

Now $\mathcal{D}^2\{X_3(t)\} \ll \mathcal{E}\{X_1(t)\}\mathcal{E}\{X_2(t)\}$; hence, it can be neglected. Therefore, as a first approximation we have

$$\frac{dm(t)}{dt} = \lambda\{[x_{10} - m(t)][x_{20} - m(t)]\} \tag{8.23}$$

This is the so-called *law of mass action*,[1] which is only approximately valid. The solution of (8.23) is

$$m(t) = \frac{e^{-\lambda x_{10}t} - e^{-\lambda x_{20}t}}{(1/x_{10})e^{\lambda - x_{10}t} - (1/x_{20})e^{\lambda - x_{20}t}} \tag{8.24}$$

[1] The law of mass action, which is a basic axiom in the deterministic theory, states that the rate of chemical reaction is proportional to the active masses of the reacting substances.

Hence for small values of t, $m(t) \sim x_{10}x_{20}t$, and

$$\lim_{t \to \infty} m(t) = C^* = \min (x_{10}, x_{20}) \tag{8.25}$$

D. Sequence of Monomolecular Reactions. We now consider a finite sequence of monomolecular reactions, viz.:

$$A \underset{k_2'}{\overset{k_1}{\rightleftarrows}} B \underset{k_3'}{\overset{k_2}{\rightleftarrows}} C$$

where k_1 and k_2 are the forward reaction rates and k_2' and k_3' are the reverse reaction rates. Let $P_{x_1}(t)$, $P_{x_2}(t)$, and $P_{x_3}(t)$ denote the amounts of the substances A, B, and C in the system at time t, respectively. We require $\sum_{i=1}^{3} P_{x_i}(t) = M$ (conservation of mass). The $P_{x_i}(t)$ are not probabilities, but they can be interpreted as probabilities, especially if we require $M = 1$.

It is clear that the differential equations describing the course of the reaction are

$$\frac{dP_{x_1}(t)}{dt} = k_2' P_{x_2}(t) - k_1 P_{x_1}(t)$$

$$\frac{dP_{x_2}(t)}{dt} = k_1 P_{x_1}(t) - (k_2 + k_2') P_{x_2}(t) + k_3' P_{x_3}(t) \tag{8.26}$$

$$\frac{dP_{x_3}(t)}{dt} = k_2 P_{x_2}(t) - k_3' P_{x_3}(t)$$

This system of equations is to be solved with the initial conditions

$$P_{x_i}(0) = x_{i0} \qquad i = 1, 2, 3 \tag{8.27}$$

Let $Q_{x_i}(s) = \mathscr{L}\{P_{x_i}(t)\}$ denote the Laplace transform of $P_{x_i}(t)$. Hence the subsidiary system of equations is given by

$$(k_1 + s)Q_{x_1}(s) - k_2' Q_{x_2}(s) = x_{10}$$
$$-k_1 Q_{x_1}(s) + (k_2 + k_2' + s)Q_{x_2}(s) - k_3 Q_{x_3}(s) = x_{20} \tag{8.28}$$
$$-k_2 Q_{x_2}(s) + (k_3' + s)Q_{x_3}(s) = x_{30}$$

Denote by D the determinant

$$D = \begin{vmatrix} k_1 + s & -k_2' & 0 \\ -k_1 & k_2 + k_2' + s & -k_3 \\ 0 & -k_2 & k_3' + s \end{vmatrix}$$

By solving the system of algebraic equations (8.28), we have

$$Q_{x_1}(s) = \frac{\begin{vmatrix} x_{10} & -k_1' & 0 \\ x_{20} & k_2 + k_2' + s & -k_3 \\ x_{30} & -k_2 & k_3' + s \end{vmatrix}}{D}$$

$$Q_{x_2}(s) = \frac{\begin{vmatrix} k_1 + s & k_{10} & 0 \\ -k_1 & k_{20} & -k_3 \\ 0 & k_{30} & k_3' + s \end{vmatrix}}{D}$$

$$Q_{x_3}(s) = \frac{\begin{vmatrix} k_1 + s & -k_2' & x_{10} \\ -k_1 & k_2 + k_2' + s & x_{20} \\ 0 & -k_2 & x_{30} \end{vmatrix}}{D}$$

In each of the above the degree of the numerator is less than the degree of the denominator and hence we can expand in partial fractions to obtain a solution of the form

$$Q_{x_i}(s) = \sum_{j=1}^{3} \frac{c_j}{s - s_j} \qquad i = 1, 2, 3 \tag{8.29}$$

where s_1, s_2, and s_3 are the roots of the determinant D and the c_j are constants. Clearly $s = 0$ is a root; call this root s_3. By applying the inverse transform, we obtain

$$P_{x_i}(t) = \sum_{j=1}^{2} c_{ij} e^{s_j t} + c_{i3} \qquad i = 1, 2, 3 \tag{8.30}$$

E. Some Chain Reactions. In this section we shall treat two chain reactions which are more involved than the reactions treated in the previous sections. These examples, due to Singer [18], will illustrate the application of stochastic methods to relatively complex reaction schemes.

1. *A Chain Reaction without Branching.* We now consider a chain reaction without branching which can be described by the following scheme:

$$S \xrightarrow{k_1} A \qquad (1)\} \quad \text{chain initiation}$$

$$A + R \xrightarrow{k_2} B + Pr \qquad (2)$$
$$\left. \right\} \quad \text{chain propagation}$$
$$B + S \xrightarrow{k_3} A + I \qquad (3)$$

This type of reaction mechanism has been shown to operate in the reaction between gaseous hydrogen and chlorine. In the above reaction step 1 represents the formation of an active molecule A from the substance S by a first-order reaction, step 2 represents the second-order reaction between A and the reactant R which yields the product Pr and an active molecule B, and step 3 represents the second-order reaction between B and S which yields A and an inert substance I. The k_i are the reaction rate constants associated with the ith step.

Denote by $P(x_1, x_2, x_3, t)$ the probability that the numbers of A, B, and Pr molecules in the system at time t are x_1, x_2, and x_3, respectively. At $t = 0$ let s and r denote the number of S and R molecules present in the system. In order to derive the differential equation describing the course of the reaction, it is necessary to consider the possible transitions in the interval of time $(t, t + \Delta t)$ which lead from the state (x_1, x_2, x_3) into other states and the transitions which bring the system from other states into (x_1, x_2, x_3). The transitions from (x_1, x_2, x_3) into other states and the associated probabilities are

(1) $\quad \mathscr{P}\{(x_1, x_2, x_3) \to (x_1 + 1, x_2, x_3)\} = k_1(s - x_1 - x_2)\, \Delta t + o(\Delta t)$

(2) $\quad \mathscr{P}\{(x_1, x_2, x_3) \to (x_1 - 1, x_2 + 1, x_3 + 1)\}$
$$= k_2(r - x_3)x_1\, \Delta t + o(\Delta t) \qquad (8.31)$$

(3) $\quad \mathscr{P}\{(x_1, x_2, x_3) \to (x_1 + 1, x_2 - 1, x_3)\} = k_2(s - x_1 - x_2)x_2\, \Delta t + o(\Delta t)$

The transitions leading to the state (x_1, x_2, x_3) from other states and the associated probabilities are

(1) $\quad \mathscr{P}\{(x_1 - 1, x_2, x_3) \to (x_1, x_2, x_3)\} = k_1(s - x_1 + 1 - x_2)\, \Delta t + o(\Delta t)$

(2) $\quad \mathscr{P}\{(x_1 + 1, x_2 - 1, x_3 - 1) \to (x_1, x_2, x_3)\}$
$$= k_2(r - x_3 + 1)(x_1 + 1)\, \Delta t + o(\Delta t) \qquad (8.32)$$

(3) $\quad \mathscr{P}\{(x_1 - 1, x_2 + 1, x_3) \to (x_1, x_2, x_3)\}$
$$= k_3(s - x_1 - x_2)(x_2 + 1)\, \Delta t + o(\Delta t)$$

By using (8.31) and (8.32), we can write the differential equation describing the reaction:

$$\frac{dP(x_1, x_2, x_3, t)}{dt} = k_1(s - x_1 + 1 - x_2)P(x_1 - 1, x_2, x_3, t)$$

$$- k_1(s - x_1 - x_2)P(x_1, x_2, x_3, t)$$

$$+ k_2(r - x_3 + 1)(x_1 + 1)P(x_1 + 1, x_2 - 1, x_3 - 1, t)$$

$$- k_2(r - x_3)x_1 P(x_1, x_2, x_3, t)$$

$$+ k_3(s - x_1 - x_2)(x_2 + 1)P(x_1 - 1, x_2 + 1, x_3, t)$$

$$- k_3(s - x_1 - x_2)x_2 P(x_1, x_2, x_3, t) \qquad (8.33)$$

Now let

$$F(s_1,s_2,s_3,t) = \sum_{x_1,x_2,x_3=0}^{\infty} P(x_1,x_2,x_3,t)s_1^{x_1}s_2^{x_2}s_3^{x_3}$$

be the generating function of the probabilities $P(x_1,x_2,x_3,t)$. From (8.33) we obtain the differential equation for the generating function:

$$\frac{\partial F(s_1,s_2,s_3,t)}{\partial t} = k_1(s_1-1)\left(s-s_1\frac{\partial}{\partial s_1}-s_2\frac{\partial}{\partial s_2}\right)F$$

$$+ k_2(s_2s_3-s_1)\frac{\partial}{\partial s_1}\left[\left(r-s_3\frac{\partial}{\partial s_3}\right)F\right]$$

$$+ k_3(s_1-s_2)\frac{\partial}{\partial s_2}\left[\left(s-s_1\frac{\partial}{\partial s_1}-s_2\frac{\partial}{\partial s_2}\right)F\right] \qquad (8.34)$$

Equation (8.34) can be simplified if we use the fact that during the initial stages of the reaction (which are of primary interest), $x_1, x_2, x_3 \ll r, s$. We can, therefore, introduce the following approximations for (8.31) and (8.32):

$$k_1(s-x_1-x_2) \sim k_1s$$
$$k_2(r-x_3)x_1 \sim k_2rx_1 \qquad (8.35)$$
$$k_3(s-x_1-x_2)x_2 \sim k_3sx_2$$

and

$$k_1(s-x_1+1-x_2) \sim k_1s$$
$$k_2(r-x_3+1)(x_1+1) \sim k_2r(x_1+1) \qquad (8.36)$$
$$k_3(s-x_1-x_2)(x_2+1) \sim k_3s(x_2+1)$$

By introducing these approximations into (8.34), we obtain

$$\frac{\partial F}{\partial t} = k_1s(s_1-1)F + k_2r(s_2s_3-s_1)\frac{\partial F}{\partial s_1} + k_3s(s_1-s_2)\frac{\partial F}{\partial s_2} \qquad (8.37)$$

This equation is to be solved with boundary conditions

$$F(0,0,0,0) = 1 \qquad F(1,1,1,t) = 1 \qquad (8.38)$$

since $x_i = 0$ ($i = 1,2,3$) when $t = 0$, and $\sum_{x_1,x_2,x_3=0}^{\infty} P(x_1,x_2,x,t) = 1$.

By employing standard methods in the theory of partial differential equations of the first order, we find

$$\log F = \frac{\alpha k_1s[k_3ss_1+(\alpha k_3s+k_2r+k_3s)s_2]}{k_3s(\alpha-1)(2\alpha k_3s+k_2r-k_3s)}(\exp\{k_3s(\alpha-1)t\}-1)$$

$$+ \frac{k_1s(s_1-\alpha s_2)(k_3s\alpha+k_2r-k_3s)}{(2\alpha k_3s+k_2r-k_3s)(\alpha k_3s+k_2r)}(1-\exp\{-(k_2r+\alpha k_3s)t\}) - k_1st$$

$$(8.39)$$

In (8.39) we have put α equal to the positive root of $\xi^2 + (k_2r - k_3s/k_3s)\xi - (k_2r/k_3s)s_3 = 0$.

Needless to say, the expression for the generating function is rather difficult to work with in order to obtain an explicit expression for $P(x_1, x_2, x_3, t)$. However, the moments, which provide valuable information, can be obtained in the usual way. Let us now consider the expected number of Pr molecules in the system. We have, if $m(x_3, t)$ is the expected number,

$$m(x_3, t) = \frac{\partial \log F}{\partial s_3} \bigg]_{s_1 = s_2 = s_3 = 1}$$
$$= A_1 t + A_2 t^2 - A_3(1 - \exp\{-(k_2r + k_3s)t\}) \qquad (8.40)$$

where $A_1 = (k_1 s(k_2 r)^2)/(k_2 r + k_3 s)^2$, $A_2 = (k_1 k_2 k_3 r s^2)/2(k_2 r + k_3 s)$, and $A_3 = (k_1 s(k_2 r)^2)/(k_2 r + k_3 s)^3$. Similar expressions can be found for $m(x_1, t)$ and $m(x_2, t)$.

2. *A Chain Reaction with Branching.* We now consider the reaction scheme

$$S \xrightarrow{k_1} A \qquad (1)$$
$$A + R \xrightarrow{k_2} B + Pr \qquad (2)$$
chain initiation

$$B + S \begin{cases} \nearrow^{k_3} A + I_1 \qquad (3a) \\ \searrow_{k_3'} 2A + I_2 \qquad (3b) \end{cases}$$
chain propagations

$$A \xrightarrow{k_4} I_3 \qquad (4)$$
chain termination

This reaction scheme differs from the one considered in the last section by the presence of the chain-branching step 3b and the termination step 4. A complete analysis of this reaction would be rather involved; hence, we here study a special case which is obtained by assuming that steps 3a and 3b are fast compared to the other steps. In this case the amount of B present does not affect the over-all reaction rate. For this special case the reaction scheme becomes

$$S \xrightarrow{k_1} A \qquad (1)$$

$$A + R + S \begin{cases} \nearrow^{k_2} Pr + A \qquad (2a) \\ \searrow_{k_3} Pr + 2A \qquad (2b) \end{cases}$$

$$A \xrightarrow{k_4} I \qquad (3)$$

Denote by $P(x_1,x_2,t)$ the probability that the numbers of A and Pr molecules in the system at time t are x_1 and x_2, respectively. I is an inert substance and does not enter into the reaction. As before, let s and r denote the numbers of S and R molecules present in the systems at $t = 0$.

The transitions which lead from the state (x_1,x_2) into other states, and the associated probabilities, are:

(1) $\mathscr{P}\{(x_1,x_2) \to (x_1 + 1, x_2)\} = k_1(s - x_1 - x_2)\,\Delta t + o(\Delta t)$

(2a) $\mathscr{P}\{(x_1,x_2) \to (x_1, x_2 + 1)\} = k_2(r - x_2)x_1\,\Delta t + o(\Delta t)$

$$(8.41)$$

(2b) $\mathscr{P}\{(x_1,x_2) \to (x_1 + 1, x_2 + 1)\} = k_3(r - x_2)x_1\,\Delta t + o(\Delta t)$

(3) $\mathscr{P}\{(x_1,x_2) \to (x_1 - 1, x_2)\} = k_4 x_1\,\Delta t + o(\Delta t)$

For the transitions into the state (x_1,x_2) we have

(1) $\mathscr{P}\{(x_1 - 1, x_2) \to (x_1,x_2)\} = k_1(s - x_1 + 1 - x_2)\,\Delta t + o(\Delta t)$

(2a) $\mathscr{P}\{(x_1, x_2 - 1) \to (x_1,x_2)\} = k_2(r - x_2 + 1)x_1\,\Delta t + o(\Delta t)$

$$(8.42)$$

(2b) $\mathscr{P}\{(x_1 - 1, x_2 - 1) \to (x_1,x_2)\} = k_3(r - x_2 + 1)(x_1 - 1)\,\Delta t + o(\Delta t)$

(3) $\mathscr{P}\{(x_1 + 1, x_2) \to (x_1,x_2)\} = k_4(x_1 + 1)\,\Delta t + o(\Delta t)$

If we make the assumption that $x_1, x_2 \ll r,s$ during the initial stages of the reaction, the probabilities given above can be approximated as follows:

$$k_1(s - x_1 - x_2) \sim k_1 s \qquad k_3(r - x_2)x_1 \sim k_3 r x_1$$
$$k(_2 r - x_2)x_1 \sim k_2 r x_1 \qquad k_4 x_1 = k_4 x_1 \qquad (8.43)$$

and

$$k_1(s - x_1 + 1 - x_2) \sim k_1 s \qquad k_3(r - x_2 + 1)(x_1 - 1) \sim k_3 r(x_1 - 1) \quad (8.44)$$
$$k_2(r - x_2 + 1)x_1 \sim k_2 r x_1 \qquad k_4(x_1 + 1) = k_4(x_1 + 1)$$

The differential-difference equation describing the reaction scheme is

$$\frac{dP(x_1,x_2,t)}{dt} = k_1 s P(x_1 - 1, x_2, t) - k_1 s P(x_1,x_2,t) + k_2 r x_1 P(x_1, x_2 - 1, t)$$

$$- k_2 r x_1 P(x_1,x_2,t) + k_3 r(x_1 - 1)P(x_1 - 1, x_2 - 1, t)$$

$$- k_3 r x_1 P(x_1,x_2,t) + k_4(x_1 + 1)P(x_1 + 1, x_2, t)$$

$$- k_4 x_1 P(x_1,x_2,t) \qquad (8.45)$$

By putting

$$F(s_1,s_2,t) = \sum_{x_1,x_2=0}^{\infty} P(x_1,x_2,t)s_1^{x_1}s_2^{x_2}$$

Eq. (8.45) becomes

$$\frac{\partial F}{\partial t} = k_1 s(s_1 - 1)F + [k_2 r(s_2 - 1)s_1 + k_3 r(s_1 s_2 - 1)s_1 + k_4(1 - s_1)]\frac{\partial F}{\partial s_1}$$

(8.46)

The solution of (8.46) when $F(0,0,0) = F(1,1,1) = 1$ is

$$\log F = k_1 s\{(\alpha - 1)t + (k_3 r s_2)^{-1} \log (\alpha - \beta)$$
$$- (k_3 r s_2)^{-1} \log [s_1 - \beta - (s_1 - \alpha) \exp \{-(\alpha - \beta)k_3 r s_2 t\}]\}$$ (8.47)

In the above solution α and β are the roots of

$$\xi^2 + \left(\frac{k_2}{k_3} - \frac{k_2 r + k_3 r + k_4}{k_3 r s_2}\right)\xi + \frac{k_4}{k_3 r s_2} = 0$$

As in the previous case, (8.47) is difficult to handle, but the moments can be obtained. Let $m(x_2,t)$ denote the expected number of Pr molecules in the system. We have

$$m(x_2,t) = \frac{k_1 s(k_2 r + k_3 r)}{k_3 r - k_4} \frac{\exp\{(k_3 r - k_4)t - 1\}}{(k_3 r - k_4) - t}$$ (8.48)

8.3 Remarks on Other Applications

The applications given thus far have been concerned with certain problems in reaction kinetics. There are, however, other problems in chemistry that have been treated by using stochastic methods. In this section we refer to some of them and give references to the literature.

Montroll and Shuler [13] have developed a general theory of reaction kinetics which is based on the distribution theory of first-passage times. The model can be described as follows: Consider the case in which the attainment of the $(n + 1)$st reactant level represents the completion of the reaction. Assume that reaction occurs only by a molecule passing into the $(n + 1)$st level and that any molecules which reach this level are absorbed. The reaction rate is determined by the rate at which molecules, executing a random walk from level to level, reach the $(n + 1)$st level for the first time. Hence, the mean time required for level $n + 1$ to be reached is the mean first-passage time for the nth level. By utilizing the transport equations for the system, expressions can be obtained for the mean first-passage time and the distribution of first-passage times for transitions past level n. For an application of this theory we refer to the paper of Kim [7].

Stochastic methods also enter in the theory of diffusion-controlled reactions, i.e., reactions which occur immediately upon collision of two reactant particles. For a stochastic approach to the kinetics of diffusion-controlled reactions we refer to the paper of Waite [21]. And for some applications of the theory of diffusion-controlled reactions to biological systems and the quenching of fluorescence we refer to Refs. 5 and 11. For another application of diffusion and random-walk methods we refer to the paper of Medgyessy et al. [10].

The use of statistical theory in the study of polymer chains is well known. Markov chain methods have been used to study the stochastic formation of chain configurations on a square lattice and the excluded-volume effect in polymer chains (cf., for example, [8] and [12]). We refer also to the work of Tchen [19,20] on the statistical properties of high polymers by using Markov chain methods and to the work of Prékopa [14] on the degradation of long-chain polymers. In a highly interesting series of papers Prékopa et al. [15,16] have used stochastic methods to study the rearrangement of linear methylsilicone oils.

Bibliography

1   Bartholomay, A. F.: A Stochastic Approach to Chemical Reaction Kinetics, Harvard University thesis, 1957.

2   Bartholomay, A. F.: Stochastic Models for Chemical Reactions: I. Theory of the Unimolecular Reaction Process, *Bull. Math. Biophys.*, vol. 20, pp. 175–190, 1958.

3   Bartholomay, A. F.: Stochastic Models for Chemical Reactions: II. The Unimolecular Rate Constant, *Bull. Math. Biophys.*, vol. 21, pp. 175–190, 1959.

4   Bharucha-Reid, A. T.: Note on Diffusion-controlled Reactions, *J. Chem. Phys.*, vol. 20, pp. 915–916, 1952.

5   Bharucha-Reid, A. T.: Diffusion-controlled Reactions in Metabolizing Systems, *Arch. Biochem. Biophys.*, vol. 43, pp. 416–423, 1952.

6   Delbrück, M.: Statistical Fluctuations in Autocatalytic Reactions, *J. Chem. Phys.*, vol. 8, pp. 120–124, 1940.

7   Kim, S. K.: Mean First Passage Time for a Random Walker and Its Application to Chemical Kinetics, *J. Chem. Phys.*, vol. 28, pp. 1057–1067, 1958.

8   King, G. W.: Stochastic Methods in Statistical Mechanics, pp. 12–18, "Monte Carlo Method," National Bureau of Standards, 1951.

9   Laidler, K. J.: "Chemical Kinetics," McGraw-Hill Book Company, Inc., New York, 1950.

10 Medgyessy, P., A. Rényi, K. Tettamanti, and I. Vincze: Mathematical Investigations of Chemical Counter Current Distribution in the Case of Non-complete Diffusion (in Hungarian), *Magyar Tud. Akad. Alkalm. Mat. Int. Közl.*, vol. 3, pp. 81–97, 1955.

11 Montroll, E. W.: A Note on the Theory of Diffusion-controlled Reactions with Application to the Quenching of Fluorescence, *J. Chem. Phys.*, vol. 14, pp. 202–211, 1946.

12 Montroll, E. W.: Markoff Chains and the Excluded Volume Effect in Polymer Chains, *J. Chem. Phys.*, vol. 18, pp. 734–743, 1950.

13 Montroll, E. W., and K. E. Shuler: The Application of the Theory of Stochastic Processes to Chemical Kinetics, *Advances in Chemical Physics,* vol. 1, pp. 361–399, 1958.

14 Prékopa, A.: Statistical Treatment of the Degradation Process of Long Chain Polymers (in Hungarian), *Magyar Tud. Akad. Alkalm. Mat. Int. Közl.,* vol. 2, pp. 103–123, 1954.

15 Prékopa, A., and P. Révész: Mathematical Treatment of the Rearrangement of Linear Methylsilicone Oils: II (in Hungarian), *Publ. Math. Inst. Hung. Acad. Sci.,* vol. 1, pp. 349–356, 1956.

16 Prékopa, A., and F. Török: Mathematical Treatment of the Rearrangement of Linear Methylsilicone Oils: I (in Hungarian), *Publ. Math. Inst. Hung. Acad. Sci.,* vol. 1, pp. 67–81, 1956.

17 Rényi, A.: Betrachtung chemischer Reaktionen mit Hilfe der Theorie der stochastischen Prozesse (in Hungarian), *Magyar Tud. Akad. Alkalm. Mat. Int. Közl.,* vol. 2, pp. 93–101, 1954.

18 Singer, K.: Application of the Theory of Stochastic Processes to the Study of Irreproducible Chemical Reactions and Nucleation Processes, *J. Roy. Statist. Soc.,* ser. B, vol. 15, pp. 92–106, 1953.

19 Tchen, C. M.: Stochastic Processes and Dispersion of Configurations of Linked Events, *J. Research Natl. Bur. Standards,* vol. 46, pp. 480–488, 1951.

20 Tchen, C. M.: Random Flight with Multiple Partial Correlations, *J. Chem. Phys.,* vol. 20, pp. 214–217, 1952.

21 Waite, T. R.: Theoretical Treatment of the Kinetics of Diffusion-limited Reactions, *Phys. Rev.,* vol. 107, pp. 463–470, 1957.

9

Applications in Operations Research:
The Theory of Queues

9.1 Introduction

The theory of queues is concerned with the development of mathematical models to predict the behavior of systems that provide services for randomly arising demands. Since the demands for service are assumed to be governed by some probability law, the theory of queues has been developed within the framework of the theory of stochastic processes. It is of great interest to note that the work of A. K. Erlang[1] on telephone traffic problems, carried out as early as 1908, constitutes the first major contribution to the theory of queues. Erlang was primarily concerned with the equilibrium behavior of telephone exchanges, and in studying this problem he derived what we now call the equilibrium form of the Kolmogorov equations for Markov processes with a denumerable number of states.

The theory of queues has been applied to an embarrassingly large number of problems. This is due to the existence of many types of systems that provide service for randomly arising demands. We mention a few of the areas and problems that have been studied by using the theory of queues: (1) telephone traffic, (2) the landing of aircraft, (3) the loading and unloading of ships, (4) machine breakdown and repair, (5) the scheduling of patients in clinics, (6) the timing of traffic lights, (7) restaurant service, (8) check-out stands in supermarkets,

---

[1] See Brockmayer, Halstrøm, and Jensen [11] for the complete works of A. K. Erlang.

(9) inventory control, and (10) the theory of dams and provisioning.

In spite of their seemingly diverse nature all of the problems mentioned above deal with the following type of situation: A "customer" arrives at the "counter" and demands service. If the server is busy with another customer, the newly arrived customer must wait until the server is free. In the meantime other customers may arrive at the service. If customers arrive when the server is busy, they must form a *queue* or *waiting line* until service is available. Here we use the terms "customer" and "counter" in a generic sense. For a particular application these terms will be defined more explicitly. As an example, in the study of telephone traffic the counter is the telephone exchange, the customer is the incoming call, and the server is the trunk line or channel.

In order to describe a given queueing system, it is necessary to specify the following components of the system: (1) *the input process*, (2) *the queue discipline*, and (3) *the service mechanism*.

The *input process* expresses the probability law governing the arrival of "customers" at the "counter" where service is provided. Suppose customers arrive at the counter at times $t_1, t_2, \ldots, t_n$ ($0 < t_1 < t_2 < \cdots < t_n < \infty$) and let $\tau_n = t_{n+1} - t_n$ denote the difference between the time of arrival of the $(n + 1)$st and nth customers. The input process is given by the probability law governing the *sequence of arrival times* $\{t_n\}$ and the *sequence of interarrival* times $\{\tau_n\}$. The simplest hypothesis about the input is that the arrival times follow a Poisson process with parameter λ. In this case the interarrival times τ_n will have the negative-exponential distribution

$$A(\tau) = 1 - e^{-\lambda\tau} \qquad \text{for } \tau \geqslant 0$$

$$= 0 \qquad\qquad \text{for } \tau < 0 \qquad\qquad (9.1)$$

and the random variables $\tau_1, \tau_2, \ldots$ will be statistically independent.

The *queue discipline* is the rule or moral code determining the manner in which customers form a queue and the manner in which they behave while waiting. Throughout this chapter we will assume that the queue discipline can be expressed as "first come, first served."

The *service mechanism* can be described as follows: Let the random variable ξ_n denote the time required to serve the nth customer; hence, the probability law governing the *sequence of service times* $\{\xi_n\}$ expresses the service mechanism of the queueing system. It is natural to assume that the successive service times, $\xi_1, \xi_2, \ldots, \xi_n, \ldots$, are statistically independent of one another and of the sequence of interarrival times $\{\tau_n\}$ and that they have the same distribution function $B(\xi)$, $0 < \xi < \infty$.

There are two distribution functions which are of special interest:

$$B(\xi) = 1 \qquad \text{for } \xi > \frac{1}{\mu}$$

$$= 0 \qquad \text{for } \xi \leqslant \frac{1}{\mu} \qquad (9.2)$$

$$B = 1 - e^{-\mu\xi} \qquad \text{for } \xi \geqslant 0$$

$$= 0 \qquad \text{for } \xi < 0 \qquad (9.3)$$

The first is called deterministic or regular, and the second, of course, is the negative exponential. In the above μ is the *mean* or *expected service time*. Another distribution function, which was introduced by Erlang and which includes the above two as a special case, is

$$dB(\xi) = \frac{(\mu n)^n}{n!} \xi^{n-1} e^{-\mu n\xi} d\xi \qquad (9.4)$$

That is, the service times are distributed as χ^2 with $2n$ degrees of freedom. In this case the customer must pass through n "phases" or "stages" of service. When $n = 1$, we obtain (9.3); when $n \to \infty$, we obtain (9.2).

Throughout this chapter we shall use the description of a queueing system proposed by Kendall [38]. In this description the following symbols are used to specify the input process and service mechanism of the queueing system:

D deterministic or regular
M random or Poisson
E_n χ^2 with $2n$ degrees of freedom
G no special assumption made about $B(\xi)$
GI general independent input, i.e., the random variables τ_n are statistically independent and have the same distribution function $A(\tau)$

With the above symbols a particular type of queueing system can be identified by giving it a label such as $M/G/1$ (Poisson arrivals, no special assumption about the service-time distribution, one server). Hence, the symbol on the left designates the input process, the symbol in the middle designates the service-time distribution, and the symbol on the right specifies the number of servers.

In Sec. 9.2 we consider the stochastic representation of queueing processes. The stochastic processes arising in the theory of queues are in general non-Markovian, and it is only for systems of the type $M/M/s$ that the associated stochastic processes are Markovian. Hence, it is necessary to consider various ways of representing queueing systems so that their stochastic properties can be ascertained.

In Sec. 9.3 we consider queueing problems arising in the theory of telephone traffic, and in Sec. 9.4 we consider queueing problems associated with the servicing of machines. In the last section, we consider two special queueing processes.

The bibliography given at the end of this chapter is not a complete bibliography on the theory of queues, but there are several excellent bibliographies that are readily available. We refer, in particular, to Refs. 19, 37, and 73.

9.2 Representation of Queueing Processes. General Theory

A. Introduction. In order to investigate the stochastic properties of a particular queueing system, it is necessary to formulate a model (or mathematical representation) which is based on (1) the input process, (2) the queue discipline, and (3) the service mechanism which characterize the queueing system. We have observed that the stochastic processes occurring in the theory of queues are in general non-Markovian; hence, special methods are needed in order to study their properties.

In this section we shall discuss several different methods of representing queueing processes. We first consider the Markov chain representation of queueing systems, this representation being based on the concept of an imbedded Markov chain. Next we consider the representation of queueing systems as Markov processes. In this case the Kolmogorov equations are employed. The third method of representing queueing systems leads to mixed Markov processes. Finally, we consider the integral equation representation.

B. Markov Chain Representation. The Method of the Imbedded Markov Chain. A method of reducing non-Markovian queueing processes to Markov chains which is of great importance in queueing theory has been developed by Kendall [37,38]. This method, called the *method of the imbedded Markov chain*, can be described abstractly as follows: Let the state of the queueing system at time t be denoted by the random variable $Y(t)$, so that, in any realization of the process $\{Y(t), t \geqslant 0\}$, the history of the system can be represented as a function $Y(\cdot)$ of time with domain $(-\infty,\infty)$. Let the set Ω_t denote the collection of functions with domain $(-\infty,t]$ and having the same range as $Y(\cdot)$. For each $t \in (-\infty,\infty)$ let θ_t be a specified subset of Ω_t and corresponding to any actual realization of the process let T be the set of those values of $t \in (-\infty,\infty)$ for which θ_t contains as an element the contraction of $Y(\cdot)$ to the reduced domain $(-\infty,t]$.

We next define the random variable

$$X(t) = f_t\{Y(\tau): \tau \leqslant t, t \in T\}$$

where f_t is some specified functional with domain θ_t. If we can select a domain θ_t and a functional f_t, for $t \in (-\infty, \infty)$, such that

1. The set T almost certainly has no finite point of accumulation. (Hence the elements t can be ordered as follows: $\ldots, t_{n-1}, t, t_{n+1}, \ldots$)
2. If $X_n = X(t_n)$ for each $t_n \in T$, then

$$\mathscr{P}\{X_{n+1} = x_{n+1} \mid X_n = x_n, X_{n-1} = x_{n-1}, \ldots\} = \mathscr{P}\{X_{n+1} = x_{n+1} \mid X_n = x_n\}$$

for all n;

the process $\{X_n, n = 0, 1, \ldots\}$ will then be said to be an *imbedded Markov chain*.

In applications of the above method three conditions must be satisfied if it is to be of any value. First, the queueing system must be simple enough to permit a mathematical formulation of the above procedure. Second, for the random variable $X(t)$ to be useful in describing the state of the system, the functional f_t must be sufficiently and suitably sensitive to variations in its argument $Y(\tau)$. Third, the stochastic mechanism governing the transition from one instant in T to the next must be such that the transition probabilities $p_{ij} = \mathscr{P}\{X_{n+1} = j \mid X_n = i\}$ of the imbedded chain can be calculated.

It should be clear that the main advantage of working with the imbedded Markov chain is that it enables us to utilize the theory of Markov chains to obtain some information about the associated non-Markovian process. Of particular importance is the study of the limiting probability distribution associated with the queueing system. Hence, before illustrating the method of the imbedded Markov chain, we state some general theorems which provide criteria for determining whether the queueing system, as described by the Markov chain, is ergodic, transient, or recurrent (cf. Foster [24]). In this chapter we shall show that the Markov chains occurring in the theory of queues are irreducible, aperiodic, and characterized by a matrix of transition probabilities $P = (p_{ij})$, $i, j = 0, 1, 2, \ldots$.

Theorem 9.1: The system is *ergodic* if there exists a non-negative solution of the inequalities

$$\sum_{j=0}^{\infty} p_{ij} x_j \leqslant x_i - 1 \qquad i \neq 0 \tag{9.5}$$

such that

$$\sum_{j=0}^{\infty} p_{0j} x_j < \infty$$

Theorem 9.2: If the system is ergodic, then the finite mean first-passage times d_i from the ith to the zero state satisfy the equations

$$\sum_{j=1}^{\infty} p_{ij}d_j = d_i - 1 \qquad i \neq 0 \tag{9.6}$$

and

$$\sum_{j=1}^{\infty} p_{0j}d_j < \infty$$

In the next two theorems we do not distinguish between recurrent-nonnull and recurrent-null systems.

Theorem 9.3: The system is *transient* if and only if there exists a bounded nonconstant solution of the equations

$$\sum_{j=0}^{\infty} p_{ij}x_j = x_i \qquad i \neq 0 \tag{9.7}$$

Theorem 9.4: The system is *recurrent* if there exists a solution x_i of the inequalities

$$\sum_{j=0}^{\infty} p_{ij}x_j \leqslant x_i \qquad i \neq 0 \tag{9.8}$$

such that $x_i \to \infty$ as $i \to \infty$.

To illustrate the method of the imbedded Markov chain and the application of the above theorems, we consider the queueing system $M/G/1$, that is, the queueing system characterized by a single server, a Poisson input, and arbitrary service-time distribution.

Let y denote the number of customers that are in the queue or are being served at time t. Except in the special case $M/M/1$, the process $\{Y(t), t \geqslant 0\}$ is non-Markovian. We define $Y(\cdot)$ to be continuous to the right at its points of discontinuity and say that $Y(\cdot)$ is in the set θ_t if $Y(t) = Y(t - 0) - 1$, i.e., the value of y has just decreased by unity. Therefore, the set

$$T = \{t_n : n = 1, 2, \dots\}$$

where $t_{n+1} > t_n$ for all n; that is, T is the set of departure times. When $t \in T$, let the random variable $X_n = X(t_n - 0)$ denote the number of customers in the queue after the nth customer has been served. The fact that the input is Poisson ensures that the imbedded chain is a Markov chain with a denumerable number of states.

Now let i and j denote the numbers of customers left behind by two consecutively departing customers, and let

$$p_{ij} = \mathscr{P}\{X_{n+1} = j \mid X_n = i\}$$

Let $P = (p_{ij})$ be the matrix of transition probabilities characterizing the imbedded Markov chain; then

$$P = \begin{bmatrix} k_0 & k_1 & k_2 & k_3 & \cdots \\ k_0 & k_1 & k_2 & k_3 & \cdots \\ 0 & k_0 & k_1 & k_2 & \cdots \\ 0 & 0 & k_0 & k_1 & \cdots \\ \cdots & \cdots & \cdots & \cdots \end{bmatrix} \tag{9.9}$$

where $k_r > 0$ for $r = 0, 1, 2, \ldots$ is given by

$$k_r = \frac{1}{r!} \int_0^\infty (\lambda \xi)^r e^{-\lambda \xi} \, dB(\xi) \tag{9.10}$$

and where $\lambda > 0$ is the parameter characterizing the Poisson input.

We now consider the application of the theorems stated above to the system $M/G/1$. Let

$$\rho = \sum_{n=1}^\infty n k_n \tag{9.11}$$

We shall prove that (1) the system is ergodic if and only if $\rho < 1$, and (2) the system is recurrent if and only if $\rho \leqslant 1$.

We first assume that $\rho < 1$. If we define

$$x_j = \frac{j}{1 - \rho}$$

we find that $\{x_j\}$ satisfies the conditions of Theorem 9.1, and so the system is ergodic. Conversely, assume that the system is ergodic. From the structure of the matrix P we find that, if M_{ij} is the mean first-passage time from the state i to the state j, then

$$M_{i,i-1} = M_{10} \qquad i \neq 0$$

Moreover, we see that from the ith state the zero state can be attained only by passing through the $(i - 1)$st state. Therefore, we have the relation

$$M_{i0} = M_{i,i-1} + M_{i-1,0}$$

By induction, we obtain

$$M_{i0} = i M_{10} \qquad i \neq 0$$

By applying Theorem 9.2, we have

$$M_{10} \sum_{j=1}^\infty p_{ij} j = M_{10} - 1$$

Therefore, from (9.11),

$$M_{10}\rho = M_{10} - 1 \qquad \text{so that} \qquad \rho = 1 - \frac{1}{M_{10}} < 1$$

Now assume that $\rho \leqslant 1$. If we put $x_j = j$, we find that the inequalities (9.8) are satisfied, and $x_j \to \infty$ as $j \to \infty$. Therefore, by Theorem 9.4 the system is recurrent.

Finally, we assume $\rho > 1$. Since $\rho = \sum_{n=0}^{\infty} nk_n > 1$, application of the fundamental theorem for branching processes[1] establishes the existence of a root $\xi \in (0,1)$ of the equation

$$F(s) = \sum_{n=0}^{\infty} k_n s^k = s$$

If we define $x_j = \xi^j$, we see that the Eqs. (9.7) are satisfied and $x_j \to 0$ as $j \to \infty$, with $x_0 = 1$. Therefore, by Theorem 9.3 the system is transient.

For the analysis of other queueing systems by the method of the imbedded Markov chain we refer to Foster [24], Homma [30–32], Kawata [36], Kendall [37,38], and Wishart [89]. In subsequent sections of this chapter we will utilize the method of the imbedded Markov chain to study queueing processes that arise in the theory of telephone traffic and in connection with the servicing of machines.

C. Markov Process Representation. The Kolmogorov Equations. In Sec. 9.1 we pointed out that it is only for queueing systems of type $M/M/s$ that the stochastic process associated with the fluctuations in queue size are Markovian. Hence, in these cases if we let the random variable $X(t)$ denote the number of customers in the queue at time t, i.e., the queue size at time t, the stochastic process $\{X(t), t \geqslant 0\}$ is a Markov process with a denumerable number of states, and its stochastic properties can be obtained from the solutions of the Kolmogorov differential equations representing the process.

Let $P(t) = (p_{ij}(t))$ denote the matrix of transition probabilities associated with the process $\{X(t)\}$. In Chap. 2 we saw that $P(t)$ satisfies the system of Kolmogorov equations

$$\frac{dP(t)}{dt} = P(t)A(t) \qquad \frac{dP(t)}{dt} = A(t)P(t) \tag{9.12}$$

with
$$P(0) = I \qquad \text{the identity matrix} \tag{9.13}$$

---

[1] Cf. Sec. 1.4.

In (9.12) $A(t) = (a_{ij}(t))$ is the matrix of infinitesimal transition probabilities. Therefore, in terms of the matrix elements the above equations become

$$\frac{dP_{ij}(t)}{dt} = \sum_{k=0}^{\infty} P_{ik}(t) a_{kj}(t)$$

(9.14)

$$\frac{dP_{ij}(t)}{dt} = \sum_{k=0}^{\infty} a_{ik}(t) P_{kj}(t) \qquad i,j = 0, 1, \ldots$$

with
$$P_{ij}(0) = \delta_{ij} = 0 \qquad \text{for } i \neq j$$
$$= 1 \qquad \text{for } i = j$$

(9.15)

In order to solve the Kolmogorov equations, it is necessary to specify the functions $a_{ij}(t)$, which in the study of queueing systems will (in general) be functions of time, and the parameters characterizing the interarrival time and service-time distribution functions. Once the $a_{ij}(t)$ are specified, an attempt can be made to solve the Kolmogorov equations by any of the methods used in Chap. 2.

The theory of queues based on the Kolmogorov equations can be divided into two areas or parts, one of which is called the *nonequilibrium theory* and the other the *equilibrium theory*. In the nonequilibrium theory interest centers on the probabilities $P_x(t) = \mathscr{P}\{X(t) = x\}$, $x = 0, 1, \ldots$. That is, one seeks the probability that at time t the queue is of length x. These probabilities are obtained by solving the system of forward Kolmogorov equations, since if the initial state, say x_0, is known, $\{P_{x_0,x}(t)\}$ is the absolute probability distribution at time t. In the equilibrium theory interest centers on the probabilities $\pi_x = \lim_{t \to \infty} P_x(t)$. That is, one is interested in finding the limiting or stationary probability distribution $\{\pi_x\}$ for the states x. These probabilities, if they exist, are obtained by solving the forward Kolmogorov equations when the time derivative is put equal to zero. One of the important problems in queueing theory is to determine under what conditions the probabilities π_x exist.

To a large extent workers in queueing theory have been concerned with the equilibrium properties of queueing systems, and because of that, we will in this chapter restrict our attention to the equilibrium theory of queues. There are, however, several studies that are concerned with the nonequilibrium behavior of queueing systems. We refer, in particular, to the studies of Bailey [5,6], Champernowne [12], Clarke [14], Karlin and McGregor [35], Ledermann and Reuter [51], Luchak [53,54], and Morse [62].

Since the remainder of this chapter will be devoted to the study of the equilibrium behavior of queueing systems, we now consider the non-equilibrium behavior of a simple queueing system.[1] The system we consider is of type $M/M/1$; hence, the system is a single-server queue characterized by negative-exponential interarrival and service-time distribution functions.

Since the queueing system is of type $M/M/1$, we have:

1. In the interval $(t, t + \Delta t)$ the probability of one customer arriving at the service is $\lambda \Delta t + o(\Delta t)$, $\lambda > 0$.

2. In the interval $(t, t + \Delta t)$ the probability of one customer departing (service completed) is $\mu \Delta t + o(\Delta t)$, $\mu > 0$.

3. In the interval $(t, t + \Delta t)$ the probability that there are no arrivals or departures is $1 - (\lambda + \mu) \Delta t + o(\Delta t)$.

4. In the interval $(t, t + \Delta t)$ the probability of more than one arrival or departure is $o(\Delta t)$.

By following the procedure used in Chap. 2, we obtain the differential-difference equations [2]

$$\frac{dP_x(t)}{dt} = \lambda P_{x-1}(t) - (\lambda + \mu)P_x(t) + \mu P_{x+1}(t) \qquad x = 1, 2, \ldots$$

$$\frac{dP_0(t)}{dt} = -\lambda P_0(t) + \mu P_1(t)$$

(9.16)

To solve this system of equations, we use the methods of generating functions and the Laplace transformation. Let

$$F(s,t) = \sum_{x=0}^{\infty} P_x(t)s^x \qquad |s| \leqslant 1$$

denote the generating function of the probabilities $P_x(t)$. Therefore, the equation for the generating function is the partial differential equation

$$s\frac{\partial F(s,t)}{\partial t} = (1 - s)\{(\mu - \lambda s)F(s,t) - \mu p_0(t)\}$$

(9.17)

If the number of customers in the queue at time $t = 0$ is x_0, i.e., $X(0) = x_0$, Eq. (9.17) is to be solved with the initial condition

$$F(s,0) = s^{x_0}$$

(9.18)

---

[1] Cf. Bailey [5,6]. The equilibrium behavior of this system is discussed in Sec. 9.3B.

[2] These are the equations of a birth-and-death process, where a "birth" corresponds to an arrival at the counter and a "death" corresponds to the departure of a customer that has been served.

Now let $f(s,z) = \mathscr{L}\{F(s,t)\}$ be the Laplace transform of $F(s,t)$. From (9.17) and (9.18) we obtain

$$f(s,z) = \frac{s^{x_0+1} - \mu(1-s)p_0(z)}{zs - (1-s)(\mu - \lambda s)} \tag{9.19}$$

where $p_0(z) = \mathscr{L}\{P_0(t)\}$. From the definition of the Laplace transform, $f(s,z)$ must converge everywhere within the unit circle $|s| = 1$, provided $\mathscr{R}(z) > 0$. Therefore, in this region zeros of both the numerator and denominator of the expression for $f(s,z)$ must coincide. Let the zeros of the denominator be

$$\xi_{1,2}(z) = \left\{ \frac{(\lambda + \mu + z) \mp [(\lambda + \mu + z)^2 - 4\lambda\mu]^{\frac{1}{2}}}{2\lambda} \right\} \tag{9.20}$$

so that
$$\xi_1 + \xi_2 = \frac{\lambda + \mu + z}{\lambda}$$

$$\xi_1\xi_2 = \frac{\mu}{\lambda} \tag{9.21}$$

$$z = \lambda(1 - \xi_1)(1 - \xi_2)$$

In (9.20) we select that value of the square root for which the real part is positive. It is easy to verify that $\xi_1(z)$ is the only zero within the unit circle when $\mathscr{R}(z) > 0$. For on the boundary, $|s| = 1$,

$$|(\lambda + \mu + z)s| = |(\lambda + \mu + z)| > \lambda + \mu \geqslant |\lambda s^2 + \mu|$$

Therefore, use of Rouché's theorem[1] shows that $zs - (1-s)(\mu - \lambda s)$ has only one zero within the unit circle, and it can be seen from (9.20) that $|\xi_1(z)| < |\xi_2(z)|$. Hence it is seen that the numerator of (9.19) vanishes when $s = \xi_1(z)$. Therefore, we have

$$p_0(z) = \frac{\xi_1^{x_0+1}}{\mu(1 - \xi_1)} \tag{9.22}$$

By inserting this in (9.19), dividing through by $(s - \xi_1)$, and expanding in powers of s, we can rewrite (9.19) as

$$\begin{aligned}
f(s,z) &= \frac{s^{x_0+1} - (1-s)\xi_1^{x_0+1}/(1-\xi_1)}{-\lambda(s - \xi_1)(s - \xi_2)} \\
&= \frac{1}{\lambda\xi_2} \sum_{i=0}^{x_0-1} \left\{ \frac{(\xi_1/\xi_2)^i - 1}{(\xi_1/\xi_2) - 1} + \frac{(\xi_1/\xi_2)^i}{1 - \xi_1} \right\} \xi_1^{x_0-i} s^i \\
&\quad + \frac{1}{\lambda\xi_2} \left\{ \frac{(\xi_1/\xi_2)^{x_0} - 1}{(\xi_1/\xi_2) - 1} + \frac{(\xi_1/\xi_2)^{x_0}}{1 - \xi_1} \right\} \sum_{i=x_0}^{\infty} \xi_2^{x_0-i} s^i
\end{aligned} \tag{9.23}$$

---

[1] Rouché's theorem states that, if $f(x)$ and $g(x)$ are analytic inside and on a closed contour C and if $|g(x)| < |f(x)|$ on C, then $f(x)$ and $f(x) + g(x)$ have the same number of zeros inside C.

If we now put

$$\xi_1 = \alpha^{\frac{1}{2}}\theta^{-1} \qquad \xi_2 = \alpha^{\frac{1}{2}}\theta \qquad \alpha = \mu/\lambda \qquad (9.24)$$

(9.23) can be expressed as

$$f(s,z) = \lambda^{-1} \sum_{i=0}^{\infty} \left\{ \frac{1}{\theta^{x_0+i}(\theta - \alpha)^{\frac{1}{2}}} - \frac{1}{\theta^{x_0+i+1} - \theta^{x_0+i-1}} \right.$$
$$\left. + \frac{1}{\theta^{|x_0-i|+1} - \theta^{|x_0-i|-1}} \right\} \alpha^{\frac{1}{2}(x_0-i-1)} s^i \qquad (9.25)$$

In order to simplify matters, we consider the Laplace transform of the derivative of $F(s,t)$ with respect to time. From (9.17) we obtain

$$\mathscr{L}\left\{ \frac{\partial F(s,t)}{\partial t} \right\} = zf(s,z) - s^{x_0} = -\lambda(1 - \xi_1)(1 - \xi_2)f(s,z) - s^{x_0} \qquad (9.26)$$

If we now substitute (9.25) in (9.26) and use (9.24), we see that the coefficient of s^i is $\mathscr{L}\{dP_i(t)/dt\}$, the Laplace transform of $dP_i(t)/dt$. It can be shown that this coefficient is the sum of six terms, each proportional to an expression of the form

$$\frac{1}{\theta^{n+1} - \theta^{n-1}} \qquad n \geqslant 0$$

the inverse Laplace transform of which is

$$\frac{1}{2\pi i} \int_{c-i\infty}^{c+i\infty} \frac{e^{zt}}{\theta^{n+1} - \theta^{n-1}} \, dz = (\lambda\mu)^{\frac{1}{2}} e^{\frac{1}{2}(\lambda+\mu)t} I_n(2(\lambda\mu)^{\frac{1}{2}}t) \qquad (9.27)$$

where $I_n(2(\lambda\mu)^{1/2}t)$ is the Bessel function of order n. Hence, for any value of i we have[1]

$$\frac{dP_i(t)}{dt} = \left(\frac{\mu}{\lambda} \right)^{\frac{1}{2}(x_0-i)} e^{-(\lambda+\mu)t} \{ -(\lambda + \mu)I_{x_0-i}(\cdot)$$
$$+ (\lambda\mu)^{\frac{1}{2}}I_{x_0-i-1}(\cdot) + (\lambda\mu)^{\frac{1}{2}}I_{x_0-i+1}(\cdot) + I_{x_0+i+2}(\cdot)$$
$$+ 2(\lambda + \mu)I_{x_0+i+1}(\cdot) + \mu I_{x_0+i}(\cdot) \} \qquad (9.28)$$

where $I_n(\cdot) = I_n(2(\lambda\mu)^{1/2}t)$. From (9.28) the probability distribution of the queue length for a system of type $M/M/1$ can be obtained by integration.

For additional properties of this queueing system we refer to Refs. 5 and 6.

D. Integrodifferential Equation Representation. The Theory of Takács. We now consider another approach to the study of non-Markovian queueing systems of the type $GI/M/1$. In this approach, due to Takács [81], the state of the queueing system is

---

[1] This result has also been obtained by Ledermann and Reuter [51] by utilizing the spectral theory of birth-and-death processes.

characterized by a single random variable which denotes the waiting time of a customer joining the queue at time t. In Sec. 9.2C the state of the queueing system was characterized by the number of customers in the queue at time t; hence, it was necessary to consider an infinite system of differential-difference equations in order to represent the process. In the Takács theory it is only necessary to consider a single integrodifferential equation for the distribution function of the waiting time.

Let the random variable $W(t)$ denote the waiting time of a customer joining the queue at time t. If at time $t = 0$ the server is not busy,

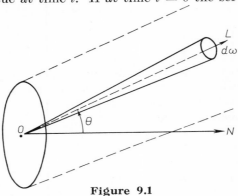

Figure 9.1

$W(0) = 0$. Should the server be busy at $t = 0$, we denote by $W(0) = w_0$ the random variable that denotes the time when the server ceases to be busy. If at some time t', $W(t') = 0$, then it remains zero until a customer joins the queue.

Let $\{t_n\}$ and $\{\xi_n\}$ denote the sequence of arrival times and service times, respectively. The random variable $W(t)$ changes as follows: It jumps upward discontinuously every time a customer with nonzero service time joins the queue. Otherwise $W(t)$ approaches zero with slope -1 until it jumps again or reaches 0. At 0 it remains equal to 0 until another jump occurs. The magnitudes of the jumps $W(t)$ experiences are the service times of the customers, $\xi_1, \xi_2, \ldots$, arriving at times $t_1, t_2, \ldots$. Hence, $W(t)$ changes both discontinuously and continuously. The changes in $W(t)$ are illustrated in Fig. 9.1.

In view of the above, $W(t)$ can be defined as follows: Let $W(0) = w_0$; then for $t_n < t < t_{n+1}$, $n = 0, 1, \ldots$, with $t_0 = 0$,

$$W(t) = W(t_n) - (t - t_n) \qquad \text{for } W(t_n) > t - t_n$$
$$= 0 \qquad \text{for } W(t_n) \leqslant t - t_n \qquad (9.29)$$

and if $t = t_n$,

$$W(t_n) = W(t_n - 0) + \xi_n \qquad (9.30)$$

Hence, if the probability laws governing the sequences of arrival times and service times are known, $W(t)$ can be determined for all $t \geqslant 0$ from the initial condition $W(0) = w_0$ and Eqs. (9.29) and (9.30).

We know that in the case when the arrival times are Poisson and the service times are mutually independent random variables, the process $\{W(t), t \geqslant 0\}$ is Markovian. However, if the arrival times are not Poisson but the differences $\tau_n = t_n - t_{n-1}$ form a recurrent process, the process is non-Markovian. In this case the arrival times t_n are the regeneration points of the process. From the nature of the changes which $W(t)$ experiences we can conclude that the process $\{W(t), t \geqslant 0\}$ is a Markov process of the mixed type; i.e., it is a Markov process characterized by both discontinuous and continuous changes of state.

Within the framework of the Takács theory we shall determine the following: (1) the distribution function of the waiting time, (2) the distribution function of the length of the service (busy) period, and (3) the number of customers served in a service period.

Let $F(t,w) = \mathscr{P}\{W(t) \leqslant w\}$ denote the distribution function of $W(t)$. Hence, $F(t,w)$ gives the probability that the waiting time of a customer joining the queue at time t does not exceed w. We now assume that:

1. The process $\{t_n\}$ is Poisson with parameter $\lambda(t)$; i.e., the probability of a customer arriving in the interval $(t, t + \Delta t)$ is $\lambda(t) \Delta t + o(\Delta t)$, where $\lambda(t)$ is a real-valued, nonnegative, continuous, and bounded function of t.

2. The service times ξ_n are equidistributed mutually independent random variables with common distribution function $B(w)$.

3. The initial condition $W(0) = w_0$ is in general arbitrary, with distribution function $F_0(w) = \mathscr{P}\{w_0 \leqslant w\}$; or, in particular, $w_0 = 0$ with

$$F_0(w) = 0 \qquad \text{for } w < 0$$
$$= 1 \qquad \text{for } w \geqslant 0$$

We now state and prove the following theorem for the distribution function $F(t,w)$:

Theorem 9.5: The distribution function $F(t,w)$ of the waiting time $W(t)$ satisfies the integrodifferential equation

$$\frac{\partial F(t,w)}{\partial t} = \frac{\partial F(t,w)}{\partial w} - \lambda(t)F(t,w) + \lambda(t)\int_0^w B(w-z)\,d_z F(t,z) \qquad (9.31)$$

where all the derivatives occurring in (9.31) are either right-hand derivatives $(w \geqslant 0)$ or left-hand derivatives $(w < 0)$. $F(t,w)$ has a jump of magnitude $F(t,0)$ at $w = 0$, and for $w > 0$ it is continuous for all t. By imposition of the initial condition $F(0,w) = 1$ $(w > 0)$, Eq. (9.31) has a unique solution.

Proof: In order to derive Eq. (9.31), we consider the changes in the interval $(t, t + \Delta t)$ that can bring about the occurrence of the event $\{W(t + \Delta t) \leqslant w\}$. We have the following:

1. In the interval $(t, t + \Delta t)$, the probability that no customer joins the queue is $1 - \lambda(t) \Delta t + o(\Delta t)$. In this case $P\{W(t) \leqslant w + \Delta t\} = F(t, w + \Delta t)$.

2. In the interval $(t, t + \Delta t)$, the probability that one customer joins the queue is $\lambda(t) \Delta t + o(\Delta t)$. In this case if $W(t) = z$ $(0 \leqslant z < w)$, we must have $\xi \leqslant w - z$, the probability that it obtains being $B(w - z)$. The distribution function of z is $F(t,z)$.

3. In the interval $(t, t + \Delta t)$, the probability that more than one customer joins the queue is $o(\Delta t)$.

Since these events are mutually exclusive, we obtain the equation

$$F(t + \Delta t, w) = [1 - \lambda(t) \Delta t]F(t, w + \Delta t)$$

$$+ \lambda(t) \Delta t \int_0^w B(w - z) \, d_z F(t,z) + o(\Delta t) \qquad (9.32)$$

We now have to determine $F(t, w + \Delta t)$. It can be shown that for $w > 0$, $F(t,w)$ is a continuous function of w. For $w = 0$, $F(t,w)$ has in general a jump of magnitude $F(t,0)$. Hence, for $\Delta t > 0$ we have

$$F(t, w + \Delta t) = F(t,w) + \frac{\partial F(t,w)}{\partial w} \Delta t + o(\Delta t) \qquad (9.33)$$

where the partial derivative with respect to w is a right-hand derivative which exists for $w \geqslant 0$.

If we now insert (9.33) in (9.32) and let $\Delta t \to 0$, we obtain Eq. (9.31), the *Takács integrodifferential equation*.[1]

Now, for $\mathscr{R}(s) \geqslant 0$, let

$$\varphi(t,s) = \int_0^\infty e^{-sw} \, dF(t,w) \qquad \text{and} \qquad \psi(s) = \int_0^\infty e^{-sw} \, dB(w)$$

denote the Laplace-Stieltjes transforms of $F(t,w)$ and $B(w)$, respectively. From (9.31) we obtain the partial differential equation

$$\frac{\partial \varphi(t,s)}{\partial t} = [s - \lambda(t) + \lambda(t)\psi(s)]\varphi(t,s) - sF(t,0) \qquad (9.34)$$

---

[1] For a detailed discussion of the Takács integrodifferential equation and its properties we refer to Reich [72]; see also E. Reich, On the Integrodifferential Equation of Takács: II, *Ann. Math Statist.*, vol. 30, pp. 143–148, 1959.

which is to be solved with the initial condition $\varphi(0,s) = 1$. The unique solution of Eq. (9.34) that satisfies the above initial condition is

$$\varphi(t,s) = \exp\{st - [1 - \psi(s)]\Lambda(t)\}$$

$$\times \left[1 - s\int_0^t \exp\{-s\tau + [1 - \psi(s)]\Lambda(\tau)\}F(\tau,0)\,d\tau\right] \quad (9.35)$$

where $F(\tau,0)$ is the probability that at time τ the server is not busy and

$$\Lambda(t) = \int_0^t \lambda(\tau)\,d\tau$$

Hence, the distribution function of the waiting time can be determined uniquely from (9.35) by the inversion theorem for Laplace transforms.

We now determine the limiting distribution

$$F^*(w) = \lim_{t\to\infty} F(t,w)$$

Let

$$\mu = \int_0^\infty w\,dB(w)$$

denote the expected value of the random variable ξ_n and assume that the Poisson density is asymptotically constant, i.e.,

$$\lim_{t\to\infty} \lambda(t) = \lambda > 0$$

We now state the following two theorems.

Theorem 9.6: If $\lambda\mu < 1$, the limiting distribution $F^*(w)$ exists, is independent of the initial distribution $F_0(w)$, and is uniquely determined by the equations

$$F^*(0) = 1 - \lambda\mu \quad (9.36)$$

and

$$\frac{dF^*(w)}{dw} = \lambda\left[F^*(w) - \int_0^w B(w - z)\,dF^*(z)\right] \quad (9.37)$$

where the derivative with respect to w is the right-hand derivative.

If $\lambda\mu \geqslant 1$, the limiting distribution $F^*(w)$ does not exist; however, $\lim_{t\to\infty} F(t,w) = 0$ for all w.

Theorem 9.7: If $\lambda\mu < 1$, the limiting distribution $F^*(w)$ exists, is independent of the initial distribution $F_0(w)$, and is uniquely determined by the Laplace-Stieltjes transform

$$\varphi^*(s) = \int_0^\infty e^{-sw}\, dF^*(w) \qquad \mathscr{R}(s) \geqslant 0$$

satisfying

$$\varphi^*(s) = \frac{1 - \lambda\mu}{1 - \lambda\{[1 - \psi(s)]/s\}} \tag{9.38}$$

We omit the proofs of these two theorems; however, the following remarks are in order. The expression for the Laplace-Stieltjes transform of $F^*(w)$ is equivalent to the solution obtained by Khintchine [41]. If, following Khintchine, we assume that for $\lambda\mu < 1$ the limiting distribution function $F^*(w)$ exists, then Eq. (9.38) can be obtained from the fundamental integrodifferential equation (9.31). For as $t \to \infty$ (9.31) becomes (9.37). Hence, the Laplace-Stieltjes transform is given by

$$\varphi^*(s) = \frac{F^*(0)}{1 - \lambda\{[1 - \psi(s)]/s\}} \tag{9.39}$$

Since $F^*(w)$ is a distribution function, we must have $\lim_{s \to 0} \varphi^*(s) = 1$, and since $\lim_{s \to 0} [1 - \psi(s)]/s = \mu$, we have $F^*(0) = 1 - \lambda\mu$.

As an example we consider the case when ξ_n has an exponential distribution, i.e.,

$$B(w) = 1 - e^{-w/\mu} \qquad \text{for } w \geqslant 0$$
$$= 0 \qquad \text{for } w < 0$$

Hence $\psi(s) = \dfrac{1}{1 + \mu s}$ and $\varphi^*(s) = \dfrac{(1 - \lambda\mu)(1 + \mu s)}{(1 - \lambda\mu) + \mu s}$

Inversion yields the limiting distribution

$$F^*(w) = 1 - \lambda\mu \exp\left\{-\left(\frac{1 - \lambda\mu}{\mu}\right)w\right\} \tag{9.40}$$

Let us now consider the determination of the *distribution function of the length of the service period*. We restrict our attention to the homogeneous case; i.e., we assume that the Poisson parameter $\lambda(t) = \lambda$. To further simplify matters, we assume that $W(0) = w_0 = 0$. Let the random variable θ_n denote the duration of the nth service period and let $G(x) = \mathscr{P}\{\theta_n \leqslant x\}$ denote the distribution function of the service period, which is assumed to be the same for all n.

To determine $G(x)$, we proceed as follows: We assume that a customer joins the queue and is served without waiting. If the duration of his service time is τ, then the probability that in the interval $(0,\tau)$ n customers arrive at the counter is

$$\frac{(\lambda\tau)^n}{n!}\, e^{-\lambda\tau} \qquad\qquad (9.41)$$

If $n = 0$, only one customer is served, and the associated distribution function is $H(x)$. If $n \geqslant 1$, the server, after serving the first customer, starts to attend one of the customers waiting in the queue.

Let $G_n(x)$ denote the n-fold convolution of $G(x)$, i.e., $G_n(x)$ is the distribution of n mutually independent random variables each of which has the distribution function $G(x)$. Then

$$G_n(x) = \int_0^x G_{n-1}(x - \tau)\, dG(\tau) \qquad n = 2, 3, \ldots \qquad (9.42)$$

where $G_1(x) = G(x)$. Now, the length of the service period does not exceed x if the service of the first customer lasts a time τ $(0 < \tau \leqslant x)$, the distribution function of which is $B(\tau)$. During this service time, the probability that n customers $n = 0, 1, \ldots$ arrive is given by (9.41). Also, the probability that the service time of these n customers, and those arriving in the meantime, does not exceed $x - \tau$ is $G_n(x - \tau)$. Hence, we have

$$G(x) = \int_0^x \sum_{n=0}^\infty \frac{(\lambda\tau)^n}{n!}\, e^{-\lambda\tau} G_n(x - \tau)\, dB(\tau) \qquad (9.43)$$

Let $\Gamma(s)$ denote the Laplace-Stieltjes transform of $G(x)$. From Eq. (9.43) we obtain the functional equation

$$\Gamma(s) = \sum_{n=0}^\infty \frac{[\Gamma(s)]^n}{n!} \int_0^\infty (\lambda x)^n e^{(s+\lambda)x}\, dB(x)$$

$$= \sum_{n=0}^\infty (-1)^n \frac{\lambda^n [\Gamma(s)]^n \psi^{(n)}(s + \lambda)}{n!}$$

$$= \psi[s + \lambda - \lambda\Gamma(s)] \qquad (9.44)$$

This functional equation for $\Gamma(s)$ was first obtained by Kendall [37]. The distribution function $G(x)$ can be determined uniquely from $\Gamma(s)$ by inversion. The following theorem is due to Takács.

Theorem 9.8: The Laplace transform $\Gamma(s)$ of the distribution function $G(x)$ is the uniquely defined analytic solution of the functional equation

$$\Gamma(s) = \psi[s + \lambda - \lambda\Gamma(s)] \qquad (9.45)$$

valid for $\mathscr{R}(s) \geqslant 0$, where $\Gamma(s)$ is subject to the condition $\Gamma(\infty) = 0$.

Let p denote the smallest positive number for which

$$\psi[\lambda(1-p)] = p \tag{9.46}$$

Then
$$\lim_{x\to\infty} G(x) = p \tag{9.47}$$

If $\lambda\mu \leqslant 1$, $p = 1$ and $G(x)$ is an honest distribution function, while if $\lambda\mu > 1$, $p < 1$ and $G(x)$ is a dishonest distribution function, i.e., in such cases the service period can be infinite with probability $(1 - p)$.

As an example, we again consider the case when $B(x)$ is the exponential distribution function. Hence $\psi(s) = (1 + \mu s)^{-1}$ and

$$\Gamma(s) = \frac{1 + \mu(s + \lambda) - \sqrt{[1 + \mu(s + \lambda)]^2 - 4\lambda\mu}}{2\lambda\mu} \tag{9.48}$$

Since $\Gamma(\infty) = 0$, we take the root of (9.48) with positive sign. By inverting (9.48), we obtain

$$\frac{dG(x)}{dx} = \exp\left\{-\frac{(1 + \lambda\mu)x}{\mu}\right\} I_1\left(\frac{2\sqrt{\lambda\mu x}}{\mu}\right) \frac{1}{\sqrt{\lambda\mu x}} \tag{9.49}$$

where $I_1(x) = J_1(ix)/i$ and $J_1(x)$ denotes the Bessel function of order 1 with imaginary argument. Here

$$G(\infty) = 1 \qquad \text{for } \lambda\mu \leqslant 1$$

$$= \frac{1}{\lambda\mu} \qquad \text{for } \lambda\mu > 1$$

The moments of $G(x)$ can be determined from the functional equation (9.45). However, the *expected duration of the service period* can be obtained directly from (9.43). By definition,

$$\mathscr{E}\{\theta\} = \int_0^\infty x \, dG(x)$$

hence
$$\mathscr{E}\{\theta\} = \mu + \sum_{n=1}^\infty n\mu \int_0^\infty \frac{(\lambda\tau)^n}{n!} e^{-\lambda\tau} \, dH(\tau) \tag{9.50}$$

As a final problem we consider the determination of the *number of customers that will be served in a service period*. Let P_i denote the probability that i ($i = 1, 2, \dots$) customers are served in a service period. Now the probability that n customers arrive at the counter while one customer is being served is

$$q_n = \int_0^\infty \frac{(\lambda x)^n}{n!} e^{-\lambda x} \, dH(x) \qquad n = 0, 1, 2, \dots \tag{9.51}$$

Now it is clear that

$$P_i = \sum_{\substack{n_1 + \cdots + n_i = i-1 \\ n_1 + \cdots + n_k \geqslant k \ (k=1,2,\ldots,i-1)}} q_{n_1} q_{n_2} \cdots q_{n_i} \tag{9.52}$$

We now introduce the generating function

$$F(s) = \sum_{i=1}^{\infty} P_i s^i \qquad |s| \leqslant 1$$

and the recurrence relations

$$P_1 = q_0$$

$$P_i = \sum_{n=1}^{i-1} q_n \sum_{i_1 + \cdots + i_n = i-1} P_{i_1} P_{i_2} \cdots P_{i_n}$$

From the above recurrence relations, forming the generating function, we obtain the functional equation

$$F(s) = s\psi[\lambda - \lambda F(s)] \tag{9.53}$$

where we have used the relation

$$\sum_{i=0}^{\infty} q_i s^i = \psi[\lambda(1-s)] \tag{9.54}$$

The results are expressed by the following theorem.

Theorem 9.9: The generating function $F(s)$ of the probabilities P_i is the uniquely determined analytic solution of the functional equation

$$F(s) = s\psi[\lambda - \lambda F(s)]$$

subject to the condition $F(0) = 0$. If $x = p$ is the smallest positive real root of $\psi[\lambda(1-x)] = x$, then $\lim_{s \to 1} F(s) = p$.

For applications of the Takács theory to queueing problems we refer to the papers of Beneš [8] and McMillan and Riordan [58]. Beneš considers the properties of queueing systems with Poisson arrivals, and McMillan and Riordan utilize Theorem 9.9 to study a moving-single-server queueing problem.

E. Integral Equation Representation. The Theory of Lindley. We now consider the integral equation representation of queueing system due to Lindley [52].[1] It is assumed that:

1. The process $\{\tau_n\}$ is a recurrent process, and the distribution function $A(t)$ is arbitrary.

---

[1] Cf. also Saaty [76,77].

2. The service times $\xi_1, \xi_2, \ldots$ are equidistributed mutually independent random variables with common distribution function $B(\xi)$.

3. There is only one server at the counter.

In view of the above assumptions, the queueing system we consider is
of the type $GI/G/1$.

Let the random variable $W(t)$ denote the waiting time of the customer
arriving at the service at time t. In general, the process $\{W(t), t \geqslant 0\}$ is
non-Markovian. Now let $W(t_n - 0) = W_n$ denote the waiting time of
the nth customer to arrive at the counter. Hence, the process $\{W_n,
n = 0, 1, \ldots\}$ is the imbedded Markov chain associated with the
queueing system. If at time $t = 0$ the server is free, then $W_0 = 0$;
however, if the server is not free, W_0 denotes the time that elapses
before the server is free. If W_0 is known, the random variables W_n may
be determined successively from the following equation:

$$W_{n+1} = W_n + \xi_n - \tau_{n+1} \qquad \text{if } \tau_{n+1} - \xi_n < W_n$$
$$= 0 \qquad\qquad\qquad \text{if } \tau_{n+1} - \xi_n \geqslant W_n \qquad (9.55)$$

The interpretation of this equation should be clear.

Let $F_n(t) = \mathscr{P}\{W_n \leqslant t\}$ denote the distribution function of W_n; that
is, $F_n(t)$ is the distribution function of the waiting time of the nth
customer to arrive at the service. We remark that, since the random
variables W_n are nonnegative, $F_n(t) = 0$ for $t < 0$, and that $F_n(0)$ is the
probability that the nth customer will not have to wait.

If we start with $F_0(t)$, the sequence of distribution functions $\{F_n(t)\}$
may be determined by the recurrence formula

$$F_{n+1}(t) = \int_0^\infty K(t,u) \, dF_n(u) \qquad n = 0, 1, 2, \ldots \qquad (9.56)$$

where the kernel function $K(t,u)$ expresses the transition probabilities
of the imbedded Markov chain $\{W_n\}$; that is,

$$K(t,u) = P\{W_{n+1} \leqslant t \mid W_n = u\}$$
$$= \int_0^\infty [1 - A(w + u - t)] \, dB(w) \qquad (9.57)$$

The computations indicated may be carried out by utilizing the
recurrence formulas

$$K_n(t) = \int_0^\infty B(t - u) \, dF_n(u) \qquad (9.58)$$

and
$$F_{n+1}(t) = \int_0^\infty [1 - A(u - t)] \, dK_n(u) \qquad (9.59)$$

It is of interest to determine under what conditions an equilibrium distribution of waiting times exists; i.e., when does the limiting distribution function $\lim_{n \to \infty} F_n(t) = F(t)$ exist? This question is answered by the following theorem.

Theorem 9.10: A necessary and sufficient condition that $\lim_{n \to \infty} F_n(t) = F(t)$ exists is that either $\mathscr{E}\{\xi_n\} \leqslant \mathscr{E}\{\tau_n\}$ or $\tau_n - \xi_n = 0$. If $\mathscr{E}\{\xi_n\} \geqslant \mathscr{E}\{\tau_n\}$ and $\tau_n - \xi_n \neq 0$, then $\lim_{n \to \infty} F_n(t) = 0$ for every $t \geqslant 0$.

The limiting distribution $F(t)$, if it exists, is independent of the initial distribution[1] $F_0(t)$ and is the unique solution of the integral equation

$$F(t) = \int_0^\infty K(t,u) \, dF(u) \qquad (9.60)$$

The proof of this theorem, which we omit, is given in Ref. 52. Theorem 9.10 can also be proved by utilizing general theorems in the theory of Markov chains, or it can be proved by establishing the ergodicity of the state 0 (the system is in state 0 if, as $t \to \infty$, there is no customer waiting or being served) with finite mean recurrence time if $\mathscr{E}\{\xi_n\} < \mathscr{E}\{\tau_n\}$.

Equation (9.60), which is an integral equation of the Wiener-Hopf type, has been solved by Lindley for the case of a queueing system of type $D/E_K/1$. Its solution has been considered by Smith [79] for cases in which more general arrival-time distributions are assumed. In particular, Smith has shown that, if the distribution function of the service times is exponential, so is that of the (nonzero) waiting time, regardless of the assumptions made concerning the distribution function of the arrival times.

The integral equation representation has been employed by Gurk[28] to study time-limited queues, i.e., queueing systems for which a customer will wait no longer than a fixed time.

9.3 Applications to Telephone Traffic Theory

A. Introduction. In this section we consider the application of the theory of queues to some problems associated with telephone traffic. The general queueing process considered in telephone traffic theory can be described as follows: Consider a telephone exchange (or center) with m (m finite or infinite) available trunks or channels, and assume that calls arrive at the exchange at times $t_1, t_2, \ldots, t_n, \ldots$ ($0 < t_1 < t_2 < \cdots < t_n < \infty$). The following assumptions are made regarding the handling of calls arriving at the exchange:

[1] If $\tau_n - \xi_n = 0$, then $F(t) = F_0(t)$. This case is to be excluded.

1. A connection (conversation) is realized if the incoming call finds an idle channel, an *idle channel* being defined as a channel which is not engaged or busy.

2. Any idle channel can be utilized by an incoming call. In this case we say that the group of channels is a *full-availability group*.

3. As soon as a conversation is finished, the channel being utilized becomes immediately available for a new call.

4. If incoming calls find that all channels are busy, each new call joins a queue and waits until a channel is available. Hence all channels have a common queue.

We now consider the characterization of the queueing process arising in telephone traffic theory. It is clear from the above remarks that the queue we consider is an m-server queue; hence, we must now consider the distribution functions for the interarrival and service times. It is assumed that the interarrival times τ_n form a sequence of recurrent events; i.e., the interarrival times or differences $t_{n+1} - t_n$ ($n = 1, 2, \dots$) are identically distributed independent positive random variables. Let $A(t)$ denote the common distribution functions of the interarrival times and assume that mean interarrival time

$$\lambda = \int_0^\infty t \, dA(t) < \infty$$

Let ξ_n denote the *holding time* of the call arriving at time t_n; that is, ξ_n is the duration of the connection which is realized by a call arriving at time t_n. It is assumed that the ξ_n ($n = 1, 2, \dots$) are identically distributed independent positive random variables and that they are independent of the process $\{\tau_n\}$. It is further assumed that the distribution function of the holding times is negative-exponential, i.e.,

$$B(\xi) = 1 - e^{-\mu\xi} \quad \text{for } \xi \geqslant 0$$
$$= 0 \qquad \text{for } \xi < 0$$

In view of the above assumptions the queueing processes arising in telephone traffic theory are, in general, of the type $GI/M/m$, where m may be finite or infinite.

One of the main problems in telephone traffic theory is to determine the probability, say $P_x(t)$, that out of a group of m channels a given number x are busy at time t. In the main, emphasis is placed on the development of the equilibrium theory of telephone traffic; i.e., the determination of the limiting probabilities $\pi_x = \lim_{t \to \infty} P_x(t)$.

In Sec. 9.3B we consider the equilibrium theory for telephone exchanges with an infinite and finite number of channels when the queueing process is Markovian. This section is based on the results of

Erlang [11], Kolmogorov [47], Molina [61], and several other authors. In Sec. 9.3C we consider the equilibrium theory when the number of channels is finite and the queueing process is non-Markovian. This section is based on the results of Takács [82,85,86]. For a detailed discussion of telephone traffic theory we refer to the book of Syski [80].

B. Equilibrium Theory for Telephone Exchanges with an Infinite and Finite Number of Channels. The Markovian Case: Negative-exponential Arrival and Holding Times. In this section we consider the equilibrium theory for telephone exchanges with an infinite and finite number of channels. It is assumed throughout that the distribution functions of both arrival and holding times are negative-exponential. Hence, the queueing process is of the type $M/M/m$ and can, therefore, be represented by the Kolmogorov equations.

The first case we consider is that of an exchange with an *infinite number of channels*. Let the random variable $X(t)$ denote the number of channels that are busy at time t, and let $P_x(t) = \mathscr{P}\{X(t) = x\}$. Since the queueing system is of type $M/M/\infty$ and x can assume the values $0, 1, \ldots$, the process $\{X(t), t \geqslant 0\}$ is a Markov process with a denumerable number of states. Hence, the system being in state x means that there are x busy channels. In view of the assumptions concerning the arrival and holding times, we have the following:

1. In the interval $(t, t + \Delta t)$ the probability of a call arriving at the exchange is $\lambda \Delta t + o(\Delta t)$, $\lambda > 0$.

2. If at time t there are x busy channels, the probability that one of them will become available in the interval $(t, t + \Delta t)$ is $x\mu \Delta t + o(\Delta t)$, $\mu > 0$.

3. If at time t there are x busy channels, the probability that none of the channels will become available and a call will not arrive at the exchange in the interval $(t, t + \Delta t)$ is $1 - (\lambda + x\mu) \Delta t + o(\Delta t)$.

4. In the interval $(t, t + \Delta t)$ the probability of any events other than those listed above is $o(\Delta t)$.

From the above assumptions the only transitions possible in $(t, t + \Delta t)$ are $(x \to x + 1)$, $(x \to x - 1)$, and $(x \to x)$; hence, the Markov process is a birth-and-death process with birth and death rates given by

$$\lambda_x = \lambda \qquad \mu_x = x\mu$$

Therefore, the Kolmogorov equations for the probabilities $P_x(t)$ are

$$\frac{dP_0(t)}{dt} = -\lambda P_0(t) + \mu P_1(t)$$

$$\frac{dP_x(t)}{dt} = \lambda P_{x-1}(t) - (\lambda + x\mu)P_x(t) + (x+1)\mu P_{x-1}(t)$$

(9.61)

Since we are primarily concerned with the equilibrium theory, we **have** from Eqs. (9.61) that the limiting probabilities $\pi_x = \lim_{t \to \infty} P_x(t)$ satisfy the following equations:

$$\lambda \pi_0 = \mu \pi_1 \qquad (\lambda + x\mu)\pi_x = \lambda \pi_{x-1} + (x+1)\mu \pi_{x+1} \qquad (9.62)$$

By induction we obtain

$$\pi_x = \frac{(\lambda/\mu)^x}{x!}\, \pi_0 \qquad (9.63)$$

and therefore

$$\pi_x = \frac{(\lambda/\mu)^x}{x!}\, e^{-\lambda/\mu} = \frac{\rho^x}{x!}\, e^{-\rho} \qquad x = 0, 1, \ldots \qquad (9.64)$$

Hence, the limiting or equilibrium distribution is the Poisson distribution with parameter $\rho = \lambda/\mu$—the *traffic intensity*. From (9.64) it is clear that the limiting probabilities are independent of initial state and that $\sum_{x=0}^{\infty} \pi_x = 1$.

It is of interest to note that the limiting distribution (9.64) can also be obtained by the method of generating functions. Let

$$F(s,t) = \sum_{x=0}^{\infty} P_x(t)s^x \qquad |s| \leqslant 1$$

denote the generating function of the distribution $\{P_x(t)\}$. From (9.61) we have that $F(s,t)$ satisfies the partial differential equation

$$\frac{\partial F(s,t)}{\partial t} = (1-s)\left[\mu\, \frac{\partial F(s,t)}{\partial s} - \lambda F(s,t)\right] \qquad (9.65)$$

If we assume $X(0) = x_0$, then the solution of Eq. (9.65) satisfying the initial condition $F(s,0) = s^{x_0}$ is

$$F(s,t) = [1 - (1-s)e^{-\mu t}]^{x_0} \exp\{-\rho(1-s)(1 - e^{-\mu t})\} \qquad (9.66)$$

We observe, however, that for $x_0 = 0$, (9.66) is the generating function of the Poisson distribution with parameter $(1 - e^{-\mu t})$. Hence, for $x_0 = 0$ (no busy channels at $t = 0$),

$$\lim_{t \to \infty} F(s,t) = F(s) = \sum_{x=0}^{\infty} \pi_x s^x = e^{-\rho(1-s)} \qquad (9.67)$$

which is the generating function of the distribution $\{\pi_x\}$ given by (9.64).

We now consider the case of an exchange with a *finite number of channels*. This case is essentially the same as the previous one; however, it is clear that a queue is present only when $x > m$, and then there are $x - m$ calls waiting for channels to become available. As long as at least one of the m channels is idle, there is no difference between the infinite and finite channel cases. However, if the system is in state x with $x > m$, only m calls are being handled, and we have in this case $\mu_x = m\mu$, for $x \geqslant m$. Hence, for $x < m$ the Kolmogorov equations in the finite channel case are given by (9.61), but for $x \geqslant m$ we have

$$\frac{dP_x(t)}{dt} = \lambda P_{x-1}(t) - (\lambda + m\mu)P_x(t) + m\mu P_{x+1}(t) \qquad (9.68)$$

For $x < m$ the limiting probabilities π_x exist and are given by (9.64), but for $x \geqslant m$ the π_x satisfy

$$(\lambda + m\mu)\pi_x = \lambda\pi_{x-1} + m\mu\pi_{x+1} \qquad (9.69)$$

By induction we find

$$\pi_x = \frac{\rho^x}{x!}\,\pi_0 \qquad \text{for } x \leqslant m$$

$$= \frac{\rho^x}{m!\,m^{x-m}}\,\pi_0 \qquad \text{for } x \geqslant m \qquad (9.70)$$

In order for a limiting distribution $\{\pi_x\}$ to exist, the series of ratios π_x/π_0 must converge. From (9.70) we see that we must have $\rho < m$ in order for a limiting distribution to exist; i.e., the traffic intensity must be less than the number of channels. In this case the sum of the limiting probabilities can be normalized to unity by the proper choice of π_0. If $\rho > m$, a limiting distribution does not exist, and in this case $\pi_x = 0$ for all x, which means that gradually the queue becomes unbounded.

C. **Equilibrium Theory for Telephone Exchanges with a Finite Number of Channels. The Non-Markovian Case: Arbitrary Arrival Times and Negative-exponential Holding Times.** Let us now consider the equilibrium theory for telephone exchanges with a finite number of states when it is assumed that the distribution function of the arrival times is no longer exponential but is an arbitrary function. However, we still assume that the distribution function of the service times is exponential. Under these assumptions, the queueing process is no longer Markovian; hence, the limiting probabilities cannot be obtained by recourse to the Kolmogorov equations.

Let the random variable $X(t)$ denote the total number of busy channels and the calls in the queue at time t and let $X_n = X(t_n - 0)$, $n = 1, 2, \ldots$. If at time t, $X(t) = x$, we shall say that the system is in the state x.

We first consider the problem of determining the limiting distributions

$$\pi_x^* = \lim_{n \to \infty} \mathscr{P}\{X_n = x\} \qquad \pi_x = \lim_{t \to \infty} \mathscr{P}\{X(t) = x\}$$

The conditions under which these limiting probabilities exist, and their exact form, are given in Theorems 9.11 and 9.12.

Theorem 9.11: If $m\lambda\mu > 1$, the limiting distribution

$$\pi_x^* = \lim_{n\to\infty} \mathscr{P}\{X_n = x\} \qquad x = 0, 1, 2, \ldots \qquad (9.71)$$

exists and is independent of the initial state of the system. The π_x^* are given by

$$\pi_x^* = \sum_{i=x}^{m-1} (-1)^{i-x} \binom{i}{x} U_i \qquad \text{for } x = 0, 1, \ldots, m-1$$

$$= H\omega^{x-m} \qquad \text{for } x = m, m+1 \qquad (9.72)$$

where ω is the only root of the equation

$$\int_0^\infty \exp\{-m\mu(1-\omega)t\}\, dA(t) = \omega$$

in the interval $(0,1)$, and

$$U_i = HC_i \sum_{k=i+1}^{m} \binom{m}{k} \left[\frac{m(1-\varphi_k)-k}{m(1-\omega)-k}\right][C_k(1-\varphi_k)]^{-1} \qquad (9.73)$$

where

$$H = \left\{\frac{1}{1-\omega} + \sum_{k=1}^{m} \binom{m}{k}\left[\frac{m(1-\varphi_k)-k}{m(1-\omega)-k}\right][C_k(1-\varphi_k)]\right\}^{-1} \qquad (9.74)$$

$$C_k = \prod_{l=1}^{k} \frac{\varphi_l}{1-\varphi_l} \qquad (9.75)$$

and

$$\varphi_k = \varphi(k\mu) = \int_0^\infty e^{-k\mu t}\, dA(t)$$

the Laplace Stieltjes transform of $A(t)$.

Proof: It is clear that the process $\{X_n, n = 1, 2, \ldots\}$ is the imbedded Markov chain associated with the queueing system whose transition probabilities p_{ij} are given by

(a) $\qquad \binom{i+1}{j} \int_0^\infty e^{-j\mu t}(1 - e^{-\mu t})^{i+1-j}\, dA(t) \qquad \text{for } i < m$

(b) $\qquad \binom{m}{j} \int_0^\infty e^{-j\mu t}\left\{\int_0^t \frac{(m\mu u)^{i-m}}{(i-m)!}(e^{-\mu u} - e^{-\mu t})m\mu\, du\right\}A(t)$

$$\text{for } i \geq m \text{ and } j < m \qquad (9.76)$$

(c) $\qquad \int_0^\infty e^{-m\mu t}\frac{(m\mu t)^{i+1-j}}{(i+i-j)!}\, dA(t) \qquad \text{for } i \geq m \text{ and } j \geq m$

If we now assume $m\lambda\mu > 1$, it is easy to prove that the imbedded Markov chain $\{X_n\}$ is ergodic; hence, the limiting probabilities

π_x^*, $x = 0, 1, \ldots$, exist and are independent of the initial distribution of X_1. The π_x^* are uniquely determined by the system of equations

$$\pi_x^* = \sum_{i=x-1}^{\infty} p_{ix}\pi_i^* \qquad (9.77)$$

and satisfy the condition $\sum_{x=0}^{\infty} \pi_x^* = 1$.

We first consider Eq. (9.77) for $x \geqslant m$. From the third expression in (9.76) we have

$$\pi_x^* = \sum_{l=0}^{\infty} \pi_{l+x-1}^* \int_0^{\infty} e^{-m\mu t} \frac{(m\mu t)^l}{l!}\, dA(t) \qquad \text{for } x \geqslant m \qquad (9.78)$$

Now, select a positive real number $\omega \neq 1$ such that

$$\int_0^{\infty} e^{-m\mu(1-\omega)t}\, dA(t) = \omega$$

If $m\lambda\mu > 1$, it can be shown that there exists only one such ω in the interval $(0,1)$.

If we now assume

$$\pi_x^* = H\omega^{x-m} \qquad x \geqslant m \qquad (9.79)$$

we see that Eq. (9.77) is satisfied if $x > m$, and if $x = m$ we obtain

$$\pi_{m-1}^* = H\omega^{-1} \qquad (9.80)$$

Hence, $H = \pi_{m-1}^*\omega$ and therefore

$$\pi_x^* = \pi_{m-1}^*\omega^{x-m+1} \qquad x \geqslant m \qquad (9.81)$$

In order to determine the probabilities π_0^*, π_1^*, $\ldots$, π_{m-1}^*, we utilize the method of generating functions. Let

$$F(s) = \sum_{x=0}^{m-1} \pi_x^* s^x \qquad |s| \leqslant 1$$

From the first two expressions in (9.76) we obtain

$$F(s) = \int_0^{\infty} [1 - e^{-\mu t} + se^{-\mu t}]F(1 - e^{-\mu t} + se^{-\mu t})\, dA(t)$$

$$+ H\int_0^{\infty}\left\{\int_0^t e^{m\mu\omega u}(e^{-\mu u} - e^{-\mu t} + se^{-\mu t})^m m\mu\, du\right\} dA(t) - Hs^m \qquad (9.82)$$

and from (9.79) we obtain

$$F(1) = \sum_{x=0}^{m-1} \pi_x^* = 1 - \sum_{x=m}^{\infty} \pi_x^* = 1 - \frac{H}{1-\omega} \qquad (9.83)$$

Now let

$$U_k = \frac{1}{k!}\left[\frac{d^k F(s)}{ds^k}\right]_{s=1} \qquad k = 0, 1, \ldots, m-1 \qquad (9.84)$$

From (9.83) we obtain

$$U_0 = 1 - \frac{H}{1-\omega} \qquad (9.85)$$

and by differentiating (9.82) k times and putting $s = 1$, we obtain, by using $\varphi(m\mu(1 - \omega)) = \omega$,

$$U_k = \frac{\varphi_k}{1-\varphi_k}U_{k-1} - \frac{H\binom{m}{k}}{1-\varphi_k}\left[\frac{m(1-\varphi_k)-k}{m(1-\omega)-k}\right] \qquad (9.86)$$

for $k = 1, 2, \ldots, m-1$; hence, the U_k satisfy a first-order difference equation with variable coefficients. Now let

$$C_k = 1 \qquad \text{for } k = 0$$

$$= \prod_{i=1}^{k}\frac{\varphi_l}{1-\varphi_l} \quad \text{for } k = 1, 2, \ldots, m-1 \qquad (9.87)$$

By dividing both sides of (9.86) by C_k, we obtain

$$\frac{U_k}{C_k} = \frac{U_{k-1}}{C_{k-1}} - \frac{H\binom{m}{k}}{C_k(1-\varphi_k)}\left[\frac{m(1-\varphi_k)-k}{m(1-\omega)-k}\right] \qquad (9.88)$$

If we now sum this expression for $k = i + 1, \ldots, m-1$ and use the relation $U_{m-1} = \pi_{m-1}^* = H\omega^{-1}$, we obtain

$$\frac{U_i}{C_i} = H\sum_{k=i+1}^{m}\binom{m}{k}\left[\frac{m(1-\varphi_k)-k}{m(1-\omega)-k}\right][C_k(1-\varphi_k)]^{-1} \qquad (9.89)$$

for $i = 0, 1, \ldots, m-1$. If $i = 0$, we obtain, by using (9.85) and (9.87),

$$\frac{U_0}{C_0} = 1 - \frac{H}{1-\omega} \qquad (9.90)$$

On comparing (9.90) with (9.89), we obtain the expression for H as given by (9.74); therefore, the generating function $F(s)$ is uniquely determined. The limiting probabilities can be expressed as

$$\pi_x^* = \frac{1}{x!}\left[\frac{d^x F(s)}{ds^x}\right]_{s=0} \qquad x = 0, 1, \ldots, m-1 \qquad (9.91)$$

The derivatives U_i, $i = 0, 1, \ldots, m - 1$, can be obtained from (9.82), and by utilizing them, we obtain

$$\pi_x^* = \sum_{i=x}^{m-1} (-1)^{i-x} \binom{i}{x} U_i \qquad i = 0, 1, \ldots, m - 1 \qquad (9.92)$$

which together with (9.79) proves (9.72).

We remark that the existence of the limiting distribution $\{\pi_x^*\}$ when $m\lambda\mu > 1$ was proved by Kendall [38], but he did not determine the explicit form of $\{\pi_x^*\}$ as given by (9.72). In the special case when the incoming calls form a Poisson process, the above result yields the limiting probabilities obtained in the last section.

To turn now to the limiting distribution $\{\pi_x\}$, we simply state the result as follows:

Theorem 9.12: If $m\lambda\mu > 1$ and $A(t)$ is not a lattice distribution, the limiting distribution

$$\pi_x = \lim_{t \to \infty} \mathscr{P}\{X(t) = x\} \qquad x = 0, 1, \ldots, 2 \qquad (9.93)$$

exists and is independent of the initial state of the system. The limiting probabilities are given by

$$\pi_x = 1 - \frac{1}{m\lambda\mu} - \frac{1}{\lambda\mu} \sum_{i=1}^{m-1} \pi_{i-1}^* \left(\frac{1}{i} - \frac{1}{m} \right) \qquad \text{for } x = 0$$

$$= \frac{\pi_{x-1}^*}{x\lambda\mu} \qquad \text{for } x = 1, 2, \ldots, m - 1$$

$$= \frac{\pi_{x-1}^*}{m\lambda\mu} \qquad \text{for } x = m, m + 1, \ldots$$

$$(9.94)$$

The next problem we consider is that of determining the probability that a call arriving at the exchange is lost. We now assume that

$$\mathscr{P}\{X_1 = x\} = \pi_x^* \qquad x = 0, 1, \ldots, m$$

that is, the probability that the system is in the state x when the first call arrives is the limiting or stationary probability for the state. In this case the process $\{X_n\}$ is a stationary Markov chain. We denote the process $\{X_n\}$ in this case by $\{X_n^*\}$.

Let p_m denote the *probability that a call is lost*. Palm [66] has shown that[1]

$$p_m = \left\{ \sum_{i=0}^{m} \binom{m}{i} \prod_{j=1}^{i} \left(\frac{1 - \varphi_j}{\varphi_j} \right) \right\}^{-1} \qquad m = 1, 2, \ldots \qquad (9.95)$$

[1] Cf. also Pollaczek [70].

where the empty product means unit, and, as before,

$$\varphi_j = \int_0^\infty e^{-j\mu t}\, dA(t)$$

We give a simple proof of (9.95) which is due to Takács [85] and is based on a method given by Ashcroft [2].[1]

Assume that the channels are numbered $1, 2, \ldots, m$ and that an incoming call realizes a connection through that idle channel which has the lowest serial number. Define $p_i\ (i = 1, 2, \ldots, m)$ as the probability that an incoming call finds channels numbered $1, 2, \ldots, i$ busy. Let us assume that a call has just arrived at the exchange and found channels numbered $1, 2, \ldots, i$ busy. We then denote by e_i the expectation of the random number that tells which of the subsequent calls to arrive at the exchange will be the first to find the channels numbered $1, 2, \ldots, i$ also busy. That is, e_i is the expectation of the number of "events" that take place before the system returns to the state i. Hence, from the theory of Markov chains, we have

$$e_i = \frac{1}{p_i} \qquad i = 1, 2, \ldots, m \tag{9.96}$$

and in particular

$$p_m = \frac{1}{e_m} \tag{9.97}$$

Therefore, in order to obtain p_m, we must determine e_m.

It is clear[2] that for $i = 1, 2, \ldots, m$

$$e_i = 1 + q_{i,1}e_i + q_{i,2}(e_{i-1} + e_i) + \cdots + q_{i,i}(e_1 + \cdots + e_i) \tag{9.98}$$

where

$$q_{i,j} = \binom{i}{j} \int_0^\infty e^{-(i-j)\mu t}(1 - e^{-\mu t})^j\, dA(t) \qquad j = 0, 1, \ldots, i$$

is the probability that, during the interval of time required for the system to return to the state i, exactly j connections $(j = 1, 2, \ldots, i)$ are terminated among the connections taking place in the channels numbered $1, 2, \ldots, i$.

We observe that e_i can be written as

$$e_i = \sum_{j=0}^i \binom{i}{j} . \Delta^j e_0 \tag{9.99}$$

where Δ is the difference operator and $\Delta^0 e_0 = e_0 = 1$. Thus it is sufficient to obtain the unknowns $\Delta^j e_0, j = 1, 2, \ldots, m$. Define $u_0 = 1$,

---

[1] For another proof we refer to Cohen [15].

[2] Obtained by applying the theorem of total expectation.

and $u_i = e_0 + e_1 + \cdots + e_i$ $(i = 0, 1, \ldots, m)$. If we now express (9.98) in terms of the u_i, we have

$$u_i = \sum_{k=0}^{i} q_{i,i-k} u_{k+1} - 1 \tag{9.100}$$

since $\sum_{j=0}^{i} q_{i,j} = 1$. Since $e_0 = \Delta u_0$, we have $\Delta^j e_0 = \Delta^{j+1} u_0$. We also note that $\Delta^j u_0$ can be written as

$$\Delta^j u_0 = \sum_{i=0}^{\infty} (-1)^{j-i} \binom{j}{i} u_i \tag{9.101}$$

If we substitute (9.100) in (9.101) and use the identity

$$\sum_{i=k}^{j} (-1)^{j-i} \binom{j}{i} q_{i,i-k} = (-1)^{j-k} \binom{j}{k} \varphi_j$$

we obtain for $j \geqslant 1$

$$\Delta^j u_0 = \sum_{i=0}^{j} (-1)^{j-i} \binom{j}{i} \sum_{k=0}^{i} q_{i,i-k} u_{k+1}$$

$$= \sum_{k=0}^{j} u_{k+1} \sum_{i=k}^{j} (-1)^{j-i} \binom{j}{i} q_{i,i-k}$$

$$= \varphi_j = \sum_{k=0}^{j} (-1)^{j-k} \binom{j}{k} u_{k+1} = \varphi_j [\Delta^{j+1} u_0 + \Delta^j u_0]$$

Therefore

$$\Delta^{j+1} u_0 = \left(\frac{1 - \varphi_j}{\varphi_j} \right) \Delta^j u_0 \qquad j = 1, 2, \ldots, m$$

Repeated application of this formula, together with the relation $\Delta u_0 = e_0 = 1$, yields

$$\Delta^j e_0 = \Delta^{j+1} u_0 = \prod_{i=1}^{j} \left(\frac{1 - \varphi_i}{\varphi_i} \right) \qquad j = 1, 2, \ldots, m \tag{9.102}$$

By substituting (9.102) in (9.99), we have

$$e_i = \sum_{j=0}^{i} \binom{i}{j} \prod_{k=1}^{j} \left(\frac{1 - \varphi_i}{\varphi_i} \right) \qquad i = 1, 2, \ldots, m \tag{9.103}$$

Since $p_i = 1/e_i$, this proves (9.97).

In the special case when the arrival times t_n form a Poisson process with parameter λ, we obtain from (9.97)

$$p_m = \frac{\dfrac{1}{m!} \left(\dfrac{\lambda}{\mu} \right)^m}{\displaystyle\sum_{j=0}^{m} \dfrac{1}{j!} \left(\dfrac{\lambda}{\mu} \right)^j} \tag{9.104}$$

which is the well-known *Erlang loss formula*.[1] It has been pointed out by several authors that (9.104) is valid in the case when $\{t_n\}$ forms a Poisson process with parameter λ and the distribution of holding times $A(t)$ is an arbitrary distribution function with mean $1/\mu$.

Three other results are of interest. The first gives the *expectation of the number of busy channels at an arbitrary time when a call occurs.* For the stationary process $\{X_n^*\}$ we have

$$\mathscr{E}\{X_n\} = \sum_{i=0}^{m} i\pi_i^* = (1 - p_m)\left(\frac{1 - \varphi_1}{\varphi_1}\right) \tag{9.105}$$

This general expression can be written explicitly in terms of the parameters λ and μ once the distribution function $A(t)$ is stipulated.

The second result, due to Homma [33], gives the *expectation of the number of lost calls* in the interval $(0,t)$ when the incoming calls obey a Poisson process with parameter λ. Let $X(t)$ denote the number of busy channels at time t, and let $Y(t)$ denote the number of lost calls in $(0,t)$. Under the assumption that

$$\mathscr{P}\{X(0) = x\} = p_x \qquad \text{for } x = 0, 1, \ldots, m$$
$$= 0 \qquad \text{for } x > m$$

Homma has shown that

$$\mathscr{E}\{Y(t)\} = \lambda p_m t \tag{9.106}$$

Finally, we determine the *distribution function of the waiting time* for an arbitrary call. The result we state holds for the stationary process, which is defined as follows: Let the random variable $Y(t)$ denote the length of the interval of time between time t and the time of arrival of the next call. If we assume that $m\lambda\mu > 1$ and that the initial distribution of the process $\{X(t), Y(t), t \geqslant 0\}$ is given by

$$\mathscr{P}\{X(0) = x\} = \pi_x$$

and $$\mathscr{P}\{Y(0) \leqslant \xi \mid X(0) = x\} = B_x^*(\xi) \qquad x = 0, 1, \ldots$$

where $$B_x^*(\xi) = \lim_{t \to \infty} \mathscr{P}\{Y(t) \leqslant \xi \mid X(t) = x\}$$

we obtain the stationary process.

The result is given by the following theorem.

Theorem 9.13: If the process is stationary and if the connections at the exchange are performed in the order of arrival of the calls, the distribution function of the waiting time of an arbitrary call is

$$W(t) = 1 - \frac{He^{-m\mu(1-\omega)t}}{1 - \omega} \qquad t \geqslant 0 \tag{9.107}$$

[1] For an exact proof of this formula we refer to Sevastyanov [78].

Proof: We observe that, if a call takes place, the system is in the state i with probability π_i^*. If $i < m$, there is no waiting time; however, if $i \geq m$, the call must wait for $i + 1 - m$ successive terminations of connections, with the times at which these terminations take place forming a Poisson process with parameter $m\mu$. Therefore

$$W(t) = \sum_{i=1}^{m-1} \pi_i^* + \sum_{i=m}^{\infty} \pi_i^* \int_0^t e^{-m\mu u} \frac{(m\mu u)^{i-m}}{(i-m)!} m\mu \, du$$

The result now follows from (9.79) and (9.83).

From (9.107) we have that the *expectation of the waiting time* is

$$\mathscr{E}\{W(t)\} = \int_0^{\infty} [1 - W(t)] \, dt = \frac{H}{m\mu(1-\omega)^2} \tag{9.108}$$

9.4 Applications to the Servicing of Machines

A. Introduction. In the study of many industrial processes the following queueing problem is encountered: A set of m machines which break down from time to time are served by r repairmen. Each time a machine stops, a repairman has to do a certain amount of work in order to put it in an operative state. In the case of a single repairman, if another machine breaks down while the repairman is engaged, the second machine will remain inoperative until the service on the first machine is completed. In the case of several repairmen, if the number of machines inoperative at any given time exceeds the number of repairmen, then the excess number of machines will have to wait until repairmen are available. The loss of production due to the period the machines have to wait for service is termed *machine interference*. The general problem of industrial importance is to estimate the effect of interference on production, given (1) the rate at which breakdowns occur, (2) the time taken to repair the inoperative machine, and (3) the number of machines assigned to each repairman.

In the nomenclature of queueing theory the machines that break down are the customers, the server is the repairman, and the customer's waiting time is the period of time the machine is inoperative before a repairman is available. In the queueing processes associated with the servicing of machines it is necessary to take into consideration two factors. The first is concerned with the cause of the "breakdown." The machines may become inoperative because some phase of the industrial process is completed or because of some defect in the machines or in the raw material. In the first case it is possible to solve the problem of interference by staggering the running of the machines so that not more than one machine will be inoperative during a service period. In the second case, however, the breakdowns occur at random intervals. Hence the queueing problem involved is, in the main, concerned with studying the distribution of the waiting time and determining the characteristics of the queueing system that can be controlled in order to minimize interference. The second factor is concerned with the "arrival of customers." Suppose that at some time all the machines under the care of the repairman are inoperative. In this case there is no possibility of another machine stopping at the time; i.e., no further customer can join the queue. For the queueing processes associated with machine interference, unlike some of the simpler queues studied previously, the probability of an additional customer joining the queue at any time depends on how many customers are already waiting. Hence, the probability of a new customer joining the queue depends on the state of the queue.

In this section we shall discuss several queueing processes associated with the machine interference problem. The first process we consider is based on the now classical work of Palm [67]. The second process we consider is due to Takács [83], who provides a rigorous treatment of several earlier studies. The third process we study is due to Runnenburg [75]. This process is of interest because it considers the case of a patrolling repairman.

B. The Machine Interference Problem. Palm's Model.
Palm [67] developed a model for the machine interference problem based on the following assumptions:

1. All machines are similar with respect to the average number of breakdowns which each experiences in its unit working or running time.

2. All repairmen are similar with respect to skill and aptitude in servicing the machines, and all machines are similar with respect to the skill needed to restore them to working condition.

3. Uninterrupted running time of a machine is an exponentially distributed nonnegative random variable; i.e., the distribution function of the running time is

$$A(t) = 1 - e^{-\lambda t} \qquad \text{for } \lambda > 0, t \geqslant 0$$
$$= 0 \qquad\qquad \text{for } t < 0$$

4. Service or repair time of a machine is an exponentially distributed positive random variable, i.e., the distribution function of the service time is

$$B(\xi) = 1 - e^{-\mu \xi} \qquad \text{for } \mu > 0, \xi \geqslant 0$$
$$= 0 \qquad\qquad \text{for } \xi < 0$$

5. All random variables are independently distributed.

6. The queueing system (machines and repairmen) is in a state of statistical equilibrium.

We first consider the Palm model based on assumptions (1) to (5), i.e., we consider the nonequilibrium behavior of the queueing system. Let the random variable $X(t)$ denote the number of machines not working at time t and let $P_x(t) = \mathscr{P}\{X(t) = x\}$, $x = 0, 1, \ldots, m$. If at time t there are x machines not working, we will say that the queueing system is in state x. We observe that a transition from the state x to state $x + 1$ is caused by the breakdown of one among the $m - x$ working machines, while a transition from the state x to state $x - 1$ occurs when a machine being serviced is repaired and returned to a working state. In view of assumptions (3) and (4) and the above transitions, the queueing system postulated by Palm is of type $M/M/r$ and can, therefore, be described by a Markov process of the birth-and-death type. The formulation of the machine interference problem as a birth-and-death process is due to Feller [22]. We also observe that in case $x \leqslant r$ the queueing system being in the state x means that x machines are being serviced and $r - x$ repairmen are idle, while in case $x > r$ the state x means that r machines are being serviced and $x - r$ machines are waiting to be serviced.

In order to derive the Kolmogorov equations which describe the queueing process, we begin, as usual, by considering the events that can take place in an arbitrary interval of time and cause the system to change from one state to another. From our earlier remarks we have:

1. If at time t a machine is working, the probability that in the interval $(t, t + \Delta t)$ it will break down and require service is $\lambda\, \Delta t + o(\Delta t)$.

2. If at time t a machine is being repaired, the probability that in the interval $(t, t + \Delta t)$ it will be repaired and returned to a working state is $\mu\, \Delta t + o(\Delta t)$.

Let us now assume that the system is in the state x. In this case we have

$$\mathscr{P}\{x \to x + 1\} = \lambda_x\, \Delta t + o(\Delta t)$$
$$\mathscr{P}\{x \to x - 1\} = \mu_x\, \Delta t + o(\Delta t)$$
$$\mathscr{P}\{x \to x\} = 1 - (\lambda_x + \mu_x)\, \Delta t + o(\Delta t)$$

where the birth rates λ_x and death rates μ_x are given by

$$\lambda_x = m\lambda \qquad \text{for } x = 0$$
$$= (m - x)\lambda \qquad \text{for } 1 \leqslant x \leqslant r,\, r \leqslant x \leqslant m$$

and
$$\mu_x = 0 \qquad \text{for } m = 0$$
$$= x\mu \qquad \text{for } 1 \leqslant x \leqslant r$$
$$= r\mu \qquad \text{for } r \leqslant x \leqslant m$$

where λ and μ are positive constants. Therefore, the Kolmogorov equations describing the nonequilibrium behavior of the queueing system are

$$\frac{dP_0(t)}{dt} = -m\lambda P_0(t) + \mu P_1(t)$$

$$\frac{dP_x(t)}{dt} = (m - x + 1)\lambda P_{x-1}(t) - [(m - x)\lambda + x\mu]P_x(t) + (x + 1)\mu P_{x+1}(t)$$

$$1 \leqslant x < r \quad (9.109)$$

$$\frac{dP_x(t)}{dt} = (m - x + 1)\lambda P_{x-1}(t) - [(m - x)\lambda + r\mu]P_x(t) + r\mu P_{x+1}(t)$$

$$r \leqslant x \leqslant m$$

Therefore, the equilibrium behavior of the queueing system, which is of interest in applications, is given by the solutions of the difference equations

$$m\lambda\pi_0 = \mu\pi_1$$

$$[(m - x)\lambda - x\mu]\pi_x = (m - x + 1)\lambda\pi_{x-1} + (x + 1)\mu\pi_{x+1} \qquad 1 \leqslant x < r$$

$$[(m - x)\lambda + r\mu]\pi_x = (m - x + 1)\lambda\pi_{x-1} + r\mu\pi_{x+1} \qquad r \leqslant x \leqslant m$$

$$(9.110)$$

In the above we have put $\pi_x = \lim_{t \to \infty} P_x(t)$. From the first of the above equations we obtain the ratio π_1/π_0. From the second and third equations induction yields

$$(x + 1)\mu\pi_{x+1} = (m - x)\lambda\pi_x \qquad x < r \qquad (9.111)$$

and

$$r\mu\pi_{x+1} = (m - x)\lambda\pi_x \qquad r \leqslant x \leqslant m \qquad (9.112)$$

From (9.111) and (9.112) the ratios π_x/π_0 can be calculated successively, and π_0 can be obtained from the condition $\sum_{x=0}^{m} \pi_x = 1$.

Following Naor [63], we give the solution of (9.111) and (9.112) in terms of the noncumulative and cumulative Poisson distribution functions[1]

$$p(k,\xi) = \frac{\xi^k}{k!} e^{-\xi} \qquad (9.113)$$

and

$$P(k,\xi) = \sum_{i=k}^{\infty} p(i,\xi) \qquad (9.114)$$

---

[1] These functions have been tabulated by E. C. Molina, "Poisson's Exponential Binomial Limit," D. Van Nostrand Company, Inc., Princeton, N.J., 1942.

Naor has shown that the limiting probabilities are

$$\pi_x = \frac{(r^x/x!)p(m - x, r\mu/\lambda)}{S(m, r, \mu/\lambda)} \qquad x < r \qquad (9.115)$$

$$= \frac{[r^{r-1}/(r - 1)!]p(m - x, r\mu/\lambda)}{S(m, r, r\mu/\lambda)} \qquad r \leqslant x \leqslant m \qquad (9.116)$$

where $S\left(m, r, \dfrac{\mu}{\lambda}\right) = \sum_{i=0}^{r-1} \dfrac{r^i}{i!} p\left(m - i, \dfrac{r\mu}{\lambda}\right)$

$$+ \frac{r^{r-1}}{(r - 1)!} \left[1 - P\left(m - r + 1, \frac{r\mu}{\lambda}\right)\right] \qquad (9.117)$$

It is of interest to note that, when only one repairman serves the m machines (i.e., $r = 1$),

$$S\left(m, 1, \frac{\mu}{\lambda}\right) = 1 - P\left(m + 1, \frac{\mu}{\lambda}\right) \qquad (9.118)$$

and the limiting probabilities are given by[1]

$$\pi_x = \frac{p(m - x, \mu/\lambda)}{1 - P(m + 1, \mu/\lambda)} \qquad (9.119)$$

We now consider expressions for several quantities that are of importance in applications. Let a, b, and w denote the average number of machines working, being serviced, and waiting to be serviced, respectively. We have the following identities:

$$a + b + w = m \qquad (9.120)$$

$$\frac{a}{b} = \frac{\mu}{\lambda} \qquad (9.121)$$

$$b = \sum_{x=0}^{r-1} x\pi_x + x \sum_{x=r}^{m} \pi_x = r - \sum_{x=0}^{r-1} (r - x)\pi_x \qquad (9.122)$$

Equation (9.120) is simply an expression of the fact that a machine has to be in one of these states. In (9.121) the left-hand side is the ratio of the average number of machines working and the average number of machines being serviced, while the right-hand side is the ratio of the average uninterrupted servicing time of a machine and the average repair time. Equality is obvious, since these last two quantities are $1/\lambda$ and $1/\mu$, respectively. Equation (9.122) relates to the equality of the number of engaged repairmen and the number of machines being serviced.

[1] Cf. also Benson and Cox [10].

From (9.120) to (9.122) we obtain

$$w = m - b - a = m - b\left(1 + \frac{\mu}{\lambda}\right)$$

$$= m - \frac{\lambda + \mu}{\lambda}\left[r - \sum_{x=0}^{r-1}(r - x)\pi_x\right] \qquad (9.123)$$

For $r = 1$, the case of one repairman, (9.123) becomes

$$w = m - \frac{\lambda + \mu}{\lambda}(1 - \pi_0) \qquad (9.124)$$

These two expressions give the *average number of machines in the waiting line;* hence, they express the *interference loss.* It is of interest to note that, in order to obtain (9.123) and (9.124), only the three basic relations (9.120) to (9.122) were used; in turn, these relations are based on assumptions (1), (2), and (4). Palm and Feller, in order to derive the same expressions for interference loss, used all six assumptions. It has been pointed out by Naor that (9.123) and (9.124) remain valid in the case when some systematic or semisystematic method of breakdowns and servicing machines replaces the randomness assumptions expressed by (3) to (5). In this case, however, it is necessary to modify assumption (6). We can therefore conclude that (9.123) and (9.124) are valid whenever the limiting probability distribution $\{\pi_x\}$ is well defined either as the fraction of time spent in the state x or as probabilities under arbitrary assumptions.

From (9.122), (9.115), and (9.116) we obtain

$$b = r\frac{S(m - 1, r, \mu/\lambda)}{S(m, r, \mu/\lambda)} \qquad (9.125)$$

which gives the *average number of machines being repaired.* By using (9.120) and (9.124), we find that the *average number of machines working is*

$$a = \frac{r\mu}{\lambda}\frac{S(m - 1, r, \mu/\lambda)}{S(m, r, \mu/\lambda)} \qquad (9.126)$$

Similarly, from (9.123) and (9.125), the *interference loss,* i.e., the *average number of machines in the waiting line,* is given by

$$w = m - \left(1 + \frac{\mu}{\lambda}\right)r\frac{S(m - 1, r, \mu/\lambda)}{S(m, r, \mu/\lambda)} \qquad (9.127)$$

Other quantities of interest are the *average number of idle repairmen* given by

$$r - b = r\left\{1 - \frac{S(m - 1, r, \mu/\lambda)}{S(m, r, \mu/\lambda)}\right\} \qquad (9.128)$$

and the *coefficient of loss for repairmen*, which is

$$\frac{r-b}{r} = 1 - \frac{S(m-1,\,r,\,\mu/\lambda)}{S(m,\,r,\,\mu/\lambda)} \qquad (9.129)$$

The *operative efficiency* is defined as the ratio of the number of machines waiting to be serviced to the number of repairmen; hence,

$$\frac{b}{r} = \frac{S(m-1,\,r,\,\mu/\lambda)}{S(m,\,r,\,\mu/\lambda)} \qquad (9.130)$$

which is one minus the coefficient of loss for repairmen.

For machines, three coefficients of loss are defined. The *coefficient of normal loss due to repairs* is given by

$$\frac{b}{m} = \left(\frac{r}{m}\right)\frac{S(m-1,\,r,\,\mu/\lambda)}{S(m,\,r,\,\mu/\lambda)} \qquad (9.131)$$

The *coefficient of loss due to machine interference* is equal to

$$\frac{w}{m} = 1 - \left[1 - \frac{\mu}{\lambda}\right]\left(\frac{r}{m}\right)\frac{S(m-1,\,r,\,\mu/\lambda)}{S(m,\,r,\,\mu/\lambda)} \qquad (9.132)$$

The *combined coefficient of loss* equals

$$\frac{b+w}{m} = 1 - \left(\frac{\mu}{\lambda}\right)\frac{rS(m-1,\,r,\,\mu/\lambda)}{mS(m,\,r,\,\mu/\lambda)} \qquad (9.133)$$

Finally, the *machine efficiency* (or *machine availability*) *of the system* is given by

$$\frac{a}{m} = \left(\frac{\mu}{\lambda}\right)\frac{rS(m-1,\,r,\,\mu/\lambda)}{mS(m,\,r,\,\mu/\lambda)} \qquad (9.134)$$

which is one minus the combined coefficient of loss.

All of the quantities given above are expressed in terms of the S function as defined by (9.117) and can therefore be computed from Molina's tables. However, in the case when the number of repairmen is large or the *servicing factor* λ/μ is small, the parameter $r\mu/\lambda$, which appears in the limiting probabilities π_x, may be greater than 100. Since Molina's tables do not go this high, Naor [64] has derived approximate expressions for the limiting probabilities by using the normal distribution.

In the next section we consider another stochastic process applicable to the machine interference problem which permits a generalization of the Palm model and extends some results obtained by other workers. For additional studies on the machine interference problem we refer to the papers of Ashcroft [2] and Cox [18], as well as to numerous studies listed in Ref. 19.

C.   The Machine Interference Problem.   Takác's Model. We now consider another approach to the machine interference problem which, in the case of a single repairman, generalizes the results obtained by Palm. The model here considered is based on the following assumptions:

1. The queueing system consists of m machines that work independently and a single repairman.

2. If at time t a machine is in a working state, the probability that it will break down and call for service in the interval $(t, t + \Delta t)$ is $\lambda \Delta t + o(\Delta t)$. This probability is the same for each of the m machines. Hence, the distribution function of the running time of a machine is the negative exponential

$$A(t) = 1 - e^{-\lambda t} \qquad \text{for } \lambda \geqslant 0, t \geqslant 0$$

$$= 0 \qquad\qquad \text{for } \lambda < 0$$

3. If a machine breaks down, it will be serviced immediately unless the repairman is servicing another machine, in which case a queue of machines waiting to be serviced is formed.

4. Machines are serviced in the order of their breakdowns.

5. The service time is a positive random variable with arbitrary distribution function $B(\xi)$, each machine having the same service-time distribution. It is also assumed that the mean and variance, given by

$$\mu = \int_0^\infty \xi \, dB(\xi) \qquad \text{and} \qquad \sigma^2 = \int_0^\infty (\xi - \mu)^2 \, dB(\xi)$$

exist and may be infinite.

In view of (1), (2), and (5), the queueing system we consider is of type $M/G/1$.

Let the random variable $X(t)$ denote the number of machines working at time t. The system is in state x ($x = 0, 1, \ldots, m$) if x machines are working simultaneously. Now let $t_1, t_2, \ldots, t_n, \ldots$ denote the end points of the consecutive service-time periods, and let $X(t_n - 0) = X_n$. The limiting probability distributions associated with $X(t)$ and X_n are as follows:

$$\pi_x = \lim_{t \to \infty} \mathscr{P}\{X(t) = x\} \qquad x = 0, 1, \ldots, m$$

and $$\pi_x^* = \lim_{n \to \infty} \mathscr{P}\{X_n = x\} \qquad x = 0, 1, \ldots, m - 1$$

We also introduce the random variable $Y(t)$, which denotes the period of time required from time t until the servicing of a machine is completed. Finally, we denote by $B_x^*(\xi)$ the conditional limiting distribution of $Y(t)$:

$$B_x^*(\xi) = \lim_{t \to \infty} P\{Y(t) \leqslant \xi \mid X(t) = x\} \qquad x = 0, 1, \ldots, m-1$$

The stochastic process described is in general non-Markovian; however, it may be considered as a Markov process if the state of the system at time t is defined in terms of the pair of random variables $(X(t), Y(t))$. We also remark that this process can be applied to the study of telephone traffic in the case of a finite number of trunk lines. In this case we refer to calls and holding times instead of breakdowns and service times, respectively.

We first consider the limiting distributions $\{\pi_x^*\}$ and $\{\pi_x\}$. The results concerning these distributions are stated in the form of theorems, which we do not prove because the method of proof is similar to that employed in Sec. 9.3C. We next consider some properties of the stationary process associated with the above stochastic process. The stationary process is defined by assuming that the distribution of $X(0)$ is π_x and the distribution function of $Y(0)$ under the condition $X(0) = x$ ($x = 0, 1, \ldots, m-1$) is $B_x^*(\xi)$.

We state the two following theorems.

Theorem 9.14: The limiting distribution $\{\pi_x^*\}$, which exists independently of the initial distribution of $X(0)$, is given by

$$\pi_x^* = \sum_{i=x}^{m-1} (-1)^{i-x} \binom{i}{x} B_i^* \tag{9.135}$$

which satisfies $\sum_{x=0}^{m-1} \pi_x^* = 1$, where B_i^*, the ith binomial moment of $\{\pi_x^*\}$, is given by

$$B_i^* = C_i \frac{\sum_{k=i}^{m-1} \binom{m-1}{k}(1/C_k)}{\sum_{k=0}^{m-1} \binom{m-1}{k}(1/C_k)} \tag{9.136}$$

where $\qquad C_i = 1 \qquad\qquad$ for $i = 0$

$$= \prod_{k=1}^{i} \frac{\varphi(k\lambda)}{1 - \varphi(k\lambda)} \qquad \text{for } i = 1, 2, \ldots, m-1 \tag{9.137}$$

and $\qquad\qquad \varphi(s) = \int_0^\infty e^{-s\xi} \, dB(\xi) \qquad \mathscr{R}(s) \geqslant 0$

is the Laplace-Stieltjes transform of $B(\xi)$.

Theorem 9.15: The limiting distribution $\{\pi_x\}$, which exists independently of the initial distribution of $X(0)$, is given by

$$\pi_x = \sum_{i=x}^{m} (-1)^{i-x} \binom{i}{x} B_i \qquad (9.138)$$

which satisfies $\sum_{x=0}^{m} \pi_x = 1$, where B_i, the ith binomial moment of π_x, is given by

$$
\begin{aligned}
B_i &= 1 && \text{for } i = 0 \\[2mm]
&= \frac{mC_{i-1}}{i} \frac{\sum\limits_{k=i-1}^{m-1} \binom{m-1}{k} (1/C_k)}{(1 + m\mu\lambda) \sum\limits_{k=0}^{m-1} \binom{m-1}{k} (1/C_k)} && \text{for } i = 1, 2, \ldots, m
\end{aligned}
$$

$$(9.139)$$

and C_k is defined by (9.137).

Theorems 9.14 and 9.15 give the exact form of the limiting distributions $\{\pi_x^*\}$ and $\{\pi_x\}$. These distributions have been considered by several authors. In particular, Khintchine (cf. [41]) derived a system of linear equations for the π_x^* but did not give the solution. In the case when $B(\xi)$ is the negative exponential, Fry [25] and Palm [67] obtained the distribution $\{\pi_x\}$, and without knowing $\{\pi_x\}$ Ashcroft [2] derived an expression for the expectation of $\{\pi_x\}$.

The next result, which gives the limiting distribution of $Y(t)$, is expressed by the following theorem.

Theorem 9.16: If the expectation of the service-time distribution is finite, then the limiting distribution

$$\lim_{t \to \infty} P\{Y(t) \leqslant \xi \mid X(t) = x\} = B_x^*(\xi)$$

exists, with

$$B_x^*(\xi) = C_x \sum_{i=x}^{m-1} \binom{i+1}{x} \pi_i^* \int_0^\infty e^{-\lambda x \tau} (1 - e^{-\lambda \tau})^{i+1-x} [B(\tau + \xi) - B(\tau)] \, d\tau$$

$$\text{for } x = 0, 1, \ldots, m - 2 \quad (9.140)$$

$$B_{n-1}^*(\xi) = C_{m-1} \pi_{m-1}^* \int_0^\infty e^{-\lambda(m-1)\tau} [B(\tau + \xi) - B(\tau)] \, d\tau$$

where

$$
\begin{aligned}
C_x &= \frac{m\mu/\pi_x^*}{m\alpha\mu + \pi_{m-1}^*} = \frac{x\mu}{\pi_{x-1}^*} && \text{for } x = 1, 2, \ldots, m - 1 \\[3mm]
&= \frac{m\mu\pi_0^*}{m\alpha\mu + \pi_{m-1}^*} && \text{for } x = 0
\end{aligned}
$$

We now consider in some detail the stationary process associated with the basic stochastic process. It is defined by assuming that

1. $\mathscr{P}\{X(0) = x\} = \pi_x$

and

2. $\mathscr{P}\{Y(0) \leqslant \xi \mid X(0) = x\} = B_x^*(\xi)$ $\qquad x = 0, 1, \ldots, m - 1$

Under these assumptions the distribution of $X(t)$ [respectively $Y(t)$] is identical for all t with the initial distribution of $X(0)$ [respectively $Y(0)$]. Furthermore, renewal-theoretic arguments show that the expected number of transitions $(x \to x + 1)$ in the interval $(0,t)$ is

$$\frac{m\lambda\pi_x^*}{m\mu\lambda + \pi_{x-1}^*} t$$

and that the expected number of transitions $(x \to x - 1)$ in the interval $(0,t)$ is

$$\frac{m\lambda\pi_{x-1}^*}{m\mu\lambda + \pi_{x-1}^*} t$$

We next prove three theorems for the stationary process defined above which are of interest in applications.

Theorem 9.17: The *expectation of the production times* of the machines in the interval $(0, T)$ is

$$\mathscr{E}\left\{\int_0^T X(t)\, dt\right\} = \frac{m\sum_{i=1}^{m-1}\binom{m-1}{i}\frac{1}{C_i}}{1 + m\mu\lambda\sum_{i=0}^{m-1}\binom{m-1}{i}\frac{1}{C_i}} T \qquad (9.141)$$

Proof: We have only to note that

$$\mathscr{E}\left\{\int_0^T X(t)\, dt\right\} = \int_0^T \mathscr{E}\{X(t)\}\, dt = \int_0^T B_1\, dt$$

where B_1 is given by (9.139).

Theorem 9.18: Let the random variable $Z(t)$ be defined as follows:

$$Z(t) = 1 \qquad \text{for } X(t) < m$$

$$= 0 \qquad \text{for } X(t) = m$$

The *expectation of the busy time of the repairman* in the interval $(0, T)$ is

$$\mathscr{E}\left\{\int_0^T Z(t)\, dt\right\} = \frac{m\mu\lambda \sum_{i=0}^{m-1} \binom{m-1}{i} \frac{1}{C_i}}{1 + m\mu\lambda \sum_{i=0}^{m-1} \binom{m-1}{i} \frac{1}{C_i}} T \qquad (9.142)$$

Proof: As in the proof of the previous theorem, we have

$$\mathscr{E}\left\{\int_0^T Z(t)\, dt\right\} = \int_0^T \mathscr{E}\{Z(t)\}\, dt$$

$$= \int_0^T \mathscr{P}\{Z(t) = 1\}\, dt = \int_0^T \mathscr{P}\{X(t) < m\}\, dt = \int_0^T (1 - \pi_m)\, dt$$

where π_m, from (9.138), is given by

$$\pi_m = \frac{1}{1 + m\mu\lambda \sum_{i=0}^{m-1} \binom{m-1}{i} \frac{1}{C_i}} \qquad (9.143)$$

Theorem 9.19: Let $W_m(\xi)$ denote the distribution function of the waiting time of a machine and let w_m denote the expected waiting time. We have

$$W_m(\xi) = \sum_{i=1}^{m-1} \pi_{x-1}^* B_{i-1}^*(\xi) * B_{m-i-1}(\xi) + \pi_{m-1}^* B_0 \qquad (9.144)$$

where $B_n(\xi)$ is the n-fold convolution of $B(\xi)$ with itself, with $B_0 = 1$ for $\xi \geqslant 0$ and $B_0 = 0$ for $\xi < 0$.

Furthermore, if $\mu < \infty$, then

$$w_m = (m-1)\mu + \left[1 - \frac{1}{\sum_{i=0}^{m-1} \binom{m-1}{i} \frac{1}{C_i}} \left\{\frac{\sigma^2 + \mu^2}{2\mu} - \frac{\mu}{1 - \varphi(\lambda)}\right\}\right]$$

$$(9.145)$$

Proof: To obtain (9.144), we first note that, when the system is in equilibrium, the probability of a breakdown, i.e., a transition $(x \to x - 1)$, is

$$\frac{x\pi_x}{\sum_{x=0}^{m} x\pi_x} = \frac{x\pi_x}{B_1} = \pi_{x-1}^*$$

If $x = m$, it is clear that the waiting time is zero; however, if $x = 1, 2, \ldots, m - 1$, the waiting time is made up of the period of time until the termination of actual service [which has the distribution $B_{x-1}^*(\xi)$] and the service time of the $m - x - 1$ machines that are in the queue, each machine having the distribution $B(\xi)$. This establishes (9.144).

To obtain (9.145), we observe that

$$\int_0^\infty \xi \, dB^*(\xi) = \frac{\sigma^2 + \mu^2}{2\mu} \tag{9.146}$$

where
$$B^*(\xi) = \frac{1}{\mu} \int_0^\xi [1 - B(u)] \, du$$

From (9.144) and (9.146) we have

$$w_m = \int_0^\infty \xi \, dW_m(\xi) = \sum_{x=1}^{m-1} \pi_{x-1}^* \left[\frac{\sigma^2 + \mu^2}{2\mu} + (m - x - 1)\mu \right]$$

By using (9.135) and (9.136) for π_{x-1}^* and B_1^*, we obtain (9.145).

D. Servicing of Machines by a Patrolling Repairman. We now consider the case when n machines, numbered 1 to n, are placed along a *circular route* which is patrolled by a single repairman walking in a fixed direction. It is assumed that (1) the machines are numbered in such a way that the repairman finds the other machines in natural order along his route and (2) the time required to walk from machine i to machine $i + 1$ is a constant, say c_i (taken modulo n), which is the same for each inspection tour. Hence

$$c = \sum_{i=1}^n c_i$$

is *total walking time* needed to complete an inspection tour.

We assume that at any given time any individual machine is either working (position 0) or not working (position 1). Let $q(u_1, \ldots, u_n)$ denote the probability that at time $t = 0$ machine i ($i = 1, 2, \ldots, n$) is in position u_i, where u_i is either 0 or 1. These probabilities, which are assumed to be given, satisfy the condition

$$\sum_{(u_1, \ldots, u_n)} q(u_1, \ldots, u_n) = 1$$

where $\sum_{(u_1, \ldots, u_n)}$ indicates summation over all possible combinations of 0 and 1 for each of the u_i.

The procedure to be followed by the repairman in servicing the n machines under his supervision is expressed by the following set of rules:

Rule 1. At time $t = 0$ start the inspection tour in front of machine 1 and apply Rule 2.

Rule 2. Notice whether the machine in front of you is in position 0 (working) or 1 (not working), and then apply Rule 3.

Rule 3. If the machine in front of you is in the position 0, walk along your route, stop in front of the next machine, and then apply Rule 2. If the machine in front of you is in position 1, put it in working order again, restart it, and then walk to the next machine, stop in front of it, and apply Rule 2.

With these rules the repairman conducts an indefinite inspection tour.

It is now assumed (1) that the service time, i.e., the total time needed to service a stopped machine and restart it, is a positive random variable with arbitrary distribution function $B(\xi)$, each machine having the same service-time distribution, and (2) that the running time, defined as the time from the moment the machine is restarted to the next stoppage, is a nonnegative random variable with distribution function

$$A(t) = 1 - e^{-\lambda t} \qquad \text{for } \lambda > 0, t \geqslant 0$$
$$= 0 \qquad \text{for } t < 0$$

In view of these assumptions the queueing system here considered is of the type $M/G/1$.

In analyzing this queueing system, we restrict our attention to two random variables X_m and Y_m, which are defined as follows: X_m is the *duration of the complete inspection tour* starting from the time the serviceman leaves the mth machine and terminating at the time when the serviceman leaves the $(m + n)$th machine; Y_m is the duration of time the $(m + n)$th machine has been inoperative when the serviceman reaches that machine. Hence Y_m denotes the *waiting time of the $(m + n)$th machine*. In both cases m is the number of machines the repairman has passed since he started patrolling.

Let the mth inspection, $m = 1, 2, \ldots$, which is performed by the repairman, consist in noting the positions of the first n machines he passes successively, at the moment he reaches them, starting with the mth machine. The result of this inspection may be denoted by a vector $Z_m = (z_m, \ldots, z_{m+n-1})$, where the component z_{m+i} $(i = 1, \ldots, n)$ is 0 if the $(m + i)$th machine is found to be working and 1 if it is not working.

Upon completion of an inspection tour, the repairman may find any one of 2^n vectors $U = (u_1, \ldots, u_n)$, where u_i is either 0 or 1 for all i. These 2^n vectors are the possible *states* that the system of machines may be in when inspection is terminated. From the definition of the state vectors, it is clear that the last $n - 1$ components of the vector specifying the mth state are the first $n - 1$ components of the vector specifying the $(m + 1)$st state.

Before determining the transition probabilities associated with this queueing system, we determine the absolute probabilities for the first state of the system. We denote these probabilities by $p_1(u_1, \ldots, u_n)$, and in general we denote by $p_m(u_1, \ldots, u_n)$ the absolute probabilities of finding the system in state $(u_1, \ldots, u_n)$ at the mth inspection.

Let us now assume that the mth machine has associated with it a *potential service time* ξ_m, which is a nonnegative random variable with distribution function $B(\xi)$. Therefore, when the repairman reaches the mth machine, the *actual service time* will be equal to 0 if the mth machine is working ($z_m = 0$) or equal to ξ_m if the mth machine is not working ($z_m = 1$).

We define

$$\rho(v_i, u_i) = u_i + (1 - 2u_i)(1 - v_i) \exp \left\{ -\lambda \sum_{j=1}^{i-1} c_j + \sum_{j=1}^{i-1} u_j \xi_j \right\} \quad (9.147)$$

where the exponent is zero for $i = 1$. By using this definition, we see that the probability of finding the positions $v_1, \ldots, v_n$ at time zero changed to positions $u_1, \ldots, u_n$ at the first inspection, under the condition that ξ_i is the potential service time of the ith machine, is given by

$$\mathscr{P}\{z_j = u_j, 0 \leqslant j \leqslant n \mid v_j, \xi_j, 1 \leqslant j \leqslant n\}$$

$$= \prod_{i=1}^{n} \mathscr{P}\{z_i = u_i \mid v_k, 1 \leqslant k \leqslant i; \xi_l, 1 \leqslant l \leqslant i - 1\}$$

$$= \prod_{i=1}^{n} \rho(v_i, u_i) \quad (9.148)$$

Hence, we have

$$p_1(u_1, \ldots, u_n)$$

$$= \sum_{(v_1, \ldots, v_n)} q(v_1, \ldots, v_n) \int_0^\infty \cdots \int_0^\infty \prod_{i=1}^{n} \rho(v_i, u_i) \, dB(\xi_1) \cdots dB(\xi_n) \quad (9.149)$$

In order to determine the transition probabilities and the equilibrium distribution of the queueing process, it is necessary to take into consideration the distribution of the running time of a machine just after the repairman has left that machine. There are two cases which must be considered. First, in case the repairman has just serviced the machine, the running time is independent of the time the repairman will be patrolling before he reaches the machine again. In this case $A(t)$, as defined earlier, is the distribution function of the running time. Second, in case the machine was in a working state, we consider the interval of time that has elapsed since the machine was last serviced. This interval of time, which has been spent patrolling and servicing other machines, is independent of, and less than, the present running time. Therefore, the remaining running time of this machine also has the distribution function $A(t)$. That this is true can be demonstrated as follows: Let T denote the running time. Then $\mathscr{P}\{T \leqslant t\} = A(t)$ implies $\mathscr{P}\{T - \tau \leqslant t \mid T > \tau\} = A(t)$ for $t \geqslant 0$, where τ is a constant.

In view of the above, we have for $m \geqslant 1$

$$\mathscr{P}\{z_{m+n} = u_{m+n} \mid z_1 = u_1, \dots, z_{m+n-1} = u_{m+n-1}\}$$
$$= \mathscr{P}\{z_{m+n} = u_{m+n} \mid z_{m+1} = u_{m+1}, \dots, z_{m+n-1} = u_{m+n-1}\} \quad (9.150)$$

Hence the last state determines the transition probabilities to any future state, and this means that our queueing system can be described by a Markov chain. From (9.150) we have

$$\mathscr{P}\{U_{m+1} = (u_1, \dots, u_n) \mid U_m = (u_1', \dots, u_n')\}$$
$$= \mathscr{P}\{z_{m+1} = u_1, \dots, z_{m+n} = u_n \mid z_m = u_1', \dots, z_{m+n-1} = u_n'\}$$
$$= \mathscr{P}\{z_{m+n} = u_{m+n} \mid z_{m+1} = u_2', \dots, z_{m+n-1} = u_n'\}$$
$$\qquad \text{if } u_i = u_{i+1}' \text{ for } 1 \leqslant i \leqslant n - 1$$
$$= 0 \qquad \text{otherwise} \quad (9.151)$$

From (9.151) we observe that the transition probability does not depend on m; hence, the Markov chain is stationary.

For the case $m \geqslant 0$ we have

$$\mathscr{P}\{z_{m+n+1} = u_{m+n+1} \mid z_{m+2} = u_{m+2}, \dots, z_{m+n} = u_{m+n}\}$$
$$= \int_0^\infty \cdots \int_0^\infty \left[(1 - u_{m+n+1}) \exp\left\{ -\lambda\left(c + \sum_{i=2}^n u_{m+i}\xi_{m+i}\right) \right\} \right.$$
$$+ u_{m+n+1}\left(1 - \exp\left\{ -\lambda\left(c + \sum_{i=2}^n u_{m+i}\xi_{m+i}\right) \right\}\right) \right]$$
$$\times dB(\xi_{m+2}) \cdots dB(\xi_{m+n})$$
$$= u_{m+n+1} + (1 - 2u_{m+n+1})e^{-\lambda c}I^{u_{m+2}+\cdots+u_{m+n}} \quad (9.152)$$

where

$$I = \int_0^\infty e^{-\lambda\xi}\, dB(\xi)$$

From the form of the transition probabilities given by (9.151) we see that the Markov chain we are considering is irreducible and aperiodic. Therefore, all states belong to the same class, and since there are a finite number of states, namely, 2^n, they must all be ergodic. This means that the limiting probabilities

$$\pi(u_1, \dots, u_n) = \lim_{m \to \infty} p_m(u_1, \dots, u_n)$$

exist and are given by the solution of the system of equations

$$\pi(u_1, \dots, u_n) = \sum_{(u_1', \dots, u_n')} \pi(u_1', \dots, u_n')\mathscr{P}\{(u_1, \dots, u_n) \mid (u_1', \dots, u_n')\}$$
$$(9.153)$$

where $\mathscr{P}\{(u_1, \ldots, u_n) \mid (u_1', \ldots, u_n')\}$ denotes the transition probability from the state $(u_1, \ldots, u_n)$ to the state $(u_1', \ldots, u_n')$. Equation (9.153) satisfies the condition

$$\sum_{(u_1, \ldots, u_n)} \pi(u_1, \ldots, u_n) = 1 \tag{9.154}$$

By introducing (9.152) into (9.153), we obtain

$$\pi(u_1, \ldots, u_n) = \{\pi(0, u_1, \ldots, u_{n-1}) + \pi(1, u_1, \ldots, u_{n-1})\}$$

$$\times [u_n + (1 - 2u_n)e^{-\lambda c}I^{u_1 + \cdots + u_n}]$$

This relation is satisfied by

$$\pi(u_1, \ldots, u_n) = K \prod_{j=0}^{u-1} (e^{\lambda c}I^{-j} - 1) \tag{9.155}$$

where $u = \sum_{i=1}^{n} u_i$ and K is a constant to be determined by condition (9.154). Hence,

$$K^{-1} = \sum_{(u_1, \ldots, u_n)} \prod_{j=0}^{u-1} (e^{\lambda c}I^{-j} - 1)$$

$$= \sum_{u=0}^{n} \binom{n}{u} \prod_{j=0}^{u-1} (e^{\lambda c}I^{-j} - 1)$$

Let us now consider the limiting distributions of the random variables X_m and Y_m. Let

$$G_m(x) = \mathscr{P}\{X_m \leqslant x\} \qquad \text{and} \qquad H_m(y) = \mathscr{P}\{Y_m \leqslant y\}$$

We define another random variable

$$R_m = c + \sum_{i=1}^{n-1} z_{m+i}\xi_{m+i}$$

which denotes the time required for the repairman to return to the mth machine. From the definitions of X_m and Y_m we have

$$X_m = c + \sum_{i=1}^{n} z_{m+i}\xi_{m+i} = R_m + z_{m+n}\xi_{m+n} \tag{9.156}$$

and
$$Y_m = R_m - T \qquad \text{if } R_m - T > 0$$

$$= 0 \qquad \text{otherwise} \tag{9.157}$$

where the random variable T denotes the running time of the $(m + n)$th machine. Therefore,

$$G_m(x) = \mathscr{P}\{X_m \leqslant x\} = \mathscr{P}\left\{c + \sum_{i=1}^{n} z_{m+i}\xi_{m+i} \leqslant x\right\}$$

$$= \sum_{k=0}^{n} \mathscr{P}\left\{c + \sum_{i=1}^{n} z_{m+i}\xi_{m+i} \leqslant x \,\middle|\, \sum_{i=1}^{n} z_{m+i} = k\right\} \mathscr{P}\left\{\sum_{i=1}^{n} z_{m+i} = k\right\} \tag{9.158}$$

Since the ξ_{m+i} and z_{m+i} are independent, the conditional probability in (9.158) becomes

$$\mathscr{P}\left\{c + \sum_{i=1}^{n} z_{m+i}\xi_{m+i} \leqslant x \,\middle|\, \sum_{i=1}^{n} z_{m+i} = k\right\}$$

$$= \mathscr{P}\left\{c + \sum_{i=1}^{k} \xi_i \leqslant x \,\middle|\, \sum_{i=1}^{m} z_{m+i} = k\right\} = \mathscr{P}\left\{c + \sum_{i=1}^{k} \xi_i \leqslant x\right\}$$

where the service times $\xi_1, \ldots, \xi_k$ are independent and identically distributed random variables, each with distribution function $B(\xi)$. Therefore,

$$G_m(x) = \sum_{k=0}^{n} \mathscr{P}\left\{c + \sum_{i=1}^{k} \xi_i \leqslant x\right\} \mathscr{P}\left\{\sum_{i=1}^{n} z_{m+i} = k\right\} \tag{9.159}$$

On passing to the limit, and using (9.155), we have

$$G(x) = \lim_{m \to \infty} \mathscr{P}\{X_m \leqslant x\}$$

$$= \sum_{k=0}^{n} \mathscr{P}\left\{c + \sum_{i=1}^{k} \xi_i \leqslant x\right\} \lim_{m \to \infty} \mathscr{P}\left\{\sum_{i=1}^{n} z_{m+i} = k\right\}$$

$$= \sum_{k=0}^{n} \mathscr{P}\left\{c + \sum_{i=1}^{k} \xi_i \leqslant x\right\} \binom{n}{k} K \prod_{j=0}^{k-1} [e^{\lambda c}I^{-j} - 1] \tag{9.160}$$

Let

$$X = c + \sum_{i=1}^{m} v_i\xi_i \qquad \text{and} \qquad R = c + \sum_{i=1}^{n-1} f_i\xi_i$$

where f_i, for all i, is either 0 or 1, and let

$$\mathscr{P}\{f_1 = \alpha_1, \ldots, f_n = \alpha_n\} = \pi(\alpha_1, \ldots, \alpha_n)$$

Since the $\xi_1, \ldots, \xi_n$ are independent random variables and each of them is independent of $f_1, \ldots, f_n$, we obtain the following result:

$$\lim_{m \to \infty} \mathscr{P}\{X_m \leqslant x\} = \mathscr{P}\{X \leqslant x\} \tag{9.161}$$

Another result, concerning the moments of X_m and X, is easily obtained. From (9.159) and (9.161) we have

$$\lim_{m \to \infty} \mathscr{E}\{X_m^k\} = \mathscr{E}\{X^k\} \tag{9.162}$$

for all $k \geqslant 0$.

A similar analysis leads to the following result for the limiting form of the waiting-time distribution:

$$\lim_{m \to \infty} H_m(y) = H(y) = \lim_{m \to \infty} \mathscr{P}\{Y_m \leqslant y\}$$

$$= \mathscr{P}\{R - T \leqslant y\}$$

$$= \mathscr{P}\{Y \leqslant y\} \tag{9.163}$$

if $Y = R - T$ if $R - T > 0$

                    $= 0$ otherwise

Also $\lim_{m \to \infty} \mathscr{E}\{Y^k_m\} = \mathscr{E}\{Y^k\}$ for all $k \geqslant 0$ (9.164)

9.5 Some Special Queueing Processes

A. Introduction. In this final section we shall consider several queueing processes that are of interest both in the general theory of queues and in various areas of applications. The first process we consider is termed a *queueing process with bulk service*. In all the queueing situations considered thus far, customers in the queue were served individually; however, in the case of bulk service the customers are served in *batches*, the size of each batch having a fixed maximum. Bulk-service queues arise in the study of hospital outpatient departments and in the study of transport processes (elevators or lifts, buses, etc.). The results we present on queues of this type are due to Bailey [4] and Downton [20,21]. For additional results in the theory of bulk-service queues we refer to the paper of Miller [59].

The second process we study is a *queueing process with balking*. In most of the queueing processes we have studied queue stability or equilibrium was obtained by assuming that the demand for service did not overload the service mechanism; i.e., we assumed that $\rho = \lambda/\mu < 1$, where λ is the average number of departures per unit time. However, Kawata [36] has shown that queue stability can also be obtained by assuming that, although arrivals occur more frequently than departures, *some* arrivals balk or choose not to join the queue. The results presented in this section on queueing processes with balking are due to Haight [29].

Other queueing processes that are of interest are those which consider priorities [40,60], special service [45], and multiple operations [74]. For additional examples and references we refer to Ref. 19.

B. Queueing Processes with Bulk Service. The queueing process we consider can be described as follows: Customers arrive at random at the counter, and then they join a queue to be served at intervals in *batches*, the size of each batch having a fixed maximum. Let us assume that the customers are served in batches of size not greater than m. Hence, each epoch of service represents the withdrawal from the queue of at most m customers.

We now assume (1) that the intervals of time, say of length ξ, between successive occasions of service are independent of each other and have the same probability distribution $B(\xi)$, $0 \leqslant \xi < \infty$, and (2) that the distribution of the number of new arrivals at the counter, say r, in any given interval of time ξ is Poisson; that is,

$$\frac{(\lambda\xi)^r}{r!}\, e^{-\lambda\xi} \qquad r = 0, 1, \ldots \tag{9.165}$$

Now let
$$h = \lambda\mathscr{E}\{\xi\} \tag{9.166}$$

where h is the average number of arrivals per interval and $\mathscr{E}\{\xi\}$ is the expected value of ξ. Hence, the distribution of r in a random interval is given by

$$k_r = \frac{1}{r!} \int_0^\infty (\lambda\xi)^r e^{-\lambda r}\, dB(\xi) \qquad r = 0, 1, 2, \ldots \tag{9.167}$$

If $K(s)$ denotes the generating function of the distribution of the number of arrivals in any interval, we have from (9.167)

$$K(s) = \sum_{r=0}^\infty k_r s^r = \varphi(\lambda(1 - s)) \qquad |s| \leqslant 1 \tag{9.168}$$

where $\varphi(\cdot)$ is the Laplace-Stieltjes transform of $B(\xi)$.

Let the random variable X denote the length of the queue and let X' and X'' denote the lengths of the queue after two consecutive batches of customers have been served. In this case, the transition matrix of the imbedded Markov chain generated by the length of the queue is given by

$$P = (p_{ij}) = \begin{bmatrix} k_0 & k_1 & k_2 & k_3 & k_4 & \cdots \\ k_0 & k_1 & k_2 & k_3 & k_4 & \cdots \\ \cdot & \cdot & \cdot & \cdot & \cdot & \\ \cdot & \cdot & \cdot & \cdot & \cdot & \\ \cdot & \cdot & \cdot & \cdot & \cdot & \\ k_0 & k_1 & k_2 & k_3 & k_4 & \cdots \\ 0 & k_0 & k_1 & k_2 & k_3 & \cdots \\ 0 & 0 & k_0 & k_1 & k_2 & \cdots \end{bmatrix}$$

where the first $m + 1$ rows are *identical*. It is clear from the characteristics of the bulk-service queue, and from the form of the transition matrix, that for the case $m = 1$ (customers are served individually) the bulk-service queue is identical with the system $M/G/1$.

The first problem we wish to consider is that of determining the *equilibrium distribution of queue length*. If we now assume that the Poisson parameter λ is equal to unity, the condition that the queue be in equilibrium is that the traffic intensity $\rho < 1$, where, in the case of the bulk-service queue, $\rho = \mathscr{E}\{\xi\}/m$. Hence, for equilibrium we must have $\mathscr{E}\{\xi\} < m$.

From the structure of the transition matrix, we see that the imbedded Markov chain is irreducible and aperiodic; hence, we have that (1) every state is transient or null-recurrent, and $\lim_{n \to \infty} p_{ij}^{(n)} = 0$ for all i and j, or (2) every state is ergodic and $\lim_{n \to \infty} p_{ij}^{(n)} = \pi_j^* > 0$ for all i and j, and such that $\sum_{j=1}^{\infty} \pi_j^* = 1$.

Let $\Pi^*(s)$ denote the generating function of the limit probabilities π_j^*. Since the π_j^* satisfy the linear system

$$\pi_j^* = \sum_{i=0}^{\infty} \pi_i^* p_{ij} \qquad j = 0, 1, 2, \ldots \tag{9.169}$$

we see that $\Pi(s)$ is given by

$$\Pi(s) = \frac{\sum_{j=0}^{m-1} \pi_j^* (s^m - s^i)}{s^m / [K(s) - 1]} \tag{9.170}$$

In order to obtain the limit probabilities π_j^* ($0 \leqslant j \leqslant m - 1$), it is necessary to consider the zeros of modulus less than or equal to unity of the denominator on the right-hand side of (9.170). If the function $K(s)$ is analytic in the region $\mathscr{R} = \{s \colon |s| < 1 + \delta, \delta > 0\}$ and if the first m π_j^* can be so chosen that the zeros of the numerator and denominator in $\mathscr{R}$ coincide, then $\Pi^*(s)$ will be analytic in $\mathscr{R}$ and will possess a power series expansion, absolutely convergent at $s = 1$, whose coefficients satisfy (9.169).

As an example, we consider a queue characterized by Poisson arrivals and with the service intervals distributed as χ^2 with $2n$ degrees of freedom. Hence

$$dB(\xi) = \frac{\mu^n}{\Gamma(n)} \xi^{n-1} e^{-\mu \xi} d\xi$$

In this case h, the mean of arrivals per service interval, is given by

$$h = \frac{\lambda n}{\mu} = \frac{n}{\mu}$$

since $\lambda = 1$. From (9.168) we have

$$K(s) = \left[1 + \frac{h(1 - s)}{n} \right]^{-n} \tag{9.171}$$

In order for the system to be nondissipative, i.e., $\sum_{j=0}^{\infty} \pi_j^* = 1$, we must have $\Pi^*(1 - 0) = 1$. By applying this condition to (9.170), we obtain the relation

$$\sum_{j=0}^{m-1} (m - j)\pi_j^* = m - h$$

This relation states that, in order for the system to be nondissipative, the average number of customers arriving during a service interval must equal the average batch size actually served.

Analysis of the equation[1]

$$s^m \left[1 + \frac{h(1-s)}{n} \right]^n - 1 = 0 \qquad (9.172)$$

shows that, in addition to the simple zero at $s = 1$, there are just $m - 1$ simple zeros, say $s = s_i$ $(i = 1, 2, \ldots, m - 1)$, with modulus greater than unity. Hence, there are $m + n$ zeros in all, and of this number m have modulus less than or equal to unity. By using the condition $\Pi^*(1) = 1$, it can then be shown that the generating function $\Pi^*(s)$ is of the form

$$\Pi^*(s) = \prod_{j=m}^{m+n-1} \left(\frac{s_j - 1}{s_j - s} \right) \qquad (9.173)$$

From (9.173) the form of the equilibrium distribution in particular cases can be obtained by utilizing the partial-fraction resolution[2] of $\Pi^*(s)$.

The moments of X, the length of the queue, can easily be obtained by utilizing the cumulant generating function $G(u)$. From (9.173) we have

$$G(u) = \sum_{j=m}^{m+n-1} \log (s_j - 1) - \sum_{j=m}^{m+n-1} \log (s_j - e^u)$$

Hence

$$\mathscr{E}\{X\} = \sum_{j=m}^{m+n-1} (s_j - 1)^{-1} \qquad (9.174)$$

and

$$\mathscr{D}^2\{X\} = \sum_{j=m}^{m+n-1} s_j(s_j - 1)^{-2} \qquad (9.175)$$

The second problem we consider is that of determining the *waiting-time distribution for a bulk-service queue*. Let the random variable W denote the waiting time. Downton [20] has shown that in the case of a queue with characteristics as described above the Laplace transform of the equilibrium distribution of the waiting time is given by

$$\gamma(t) = \frac{\Pi^*(1 - t)}{\mathscr{E}\{\xi\}t} \left[\frac{1}{\varphi(t)} - 1 \right] \qquad (9.176)$$

Hence the Laplace transform can be obtained from the generating function of the equilibrium distribution of queue length. Since the

---

[1] The roots of Eq. (9.172) can be found by the root-squaring method (cf. Kunz [50]).

[2] Cf. Appendix A.

queue is characterized by a χ^2 distribution for service intervals, (9.176) becomes

$$\gamma(t) = \frac{\mu[(1 + ht/n)^n - 1]}{nt} \prod_{j=m}^{m+n-1} \left(\frac{s_j - 1}{s_j - 1 + t} \right) \tag{9.177}$$

where, as before, the s_j are the zeros, with modulus greater than unity, of (9.172). Given the s_j, the equilibrium distribution of W can be obtained by inversion. However, it is easy to obtain the moments of W. For example, the mean and variance of the waiting time are given by

$$\mathscr{E}\{W\} = \sum_{j=m}^{m+n-1} (s_j - 1)^{-1} - \frac{n-1}{2\mu} \tag{9.178}$$

$$\mathscr{D}^2\{W\} = \sum_{j=m}^{m+n-1} (s_j - 1)^{-2} + \frac{(n-1)(n-s)}{12\mu^2} \tag{9.179}$$

In another study, Downton [21] has considered the limiting behavior of functions derived from $\Pi^*(s)$ and $\gamma(t)$ as m tends to infinity.

C. Queueing Processes with Balking. Let the random variable $X(t)$ denote the length of the queue at time t. We now assume that an individual arriving at the queue at time t' employs the following rule in deciding whether to join the queue: If $X(t') \leqslant K$, he will join the queue; if $X(t') > K$, he will not join the queue. Here, the (positive) integer K denotes the greatest queue length the individual will join. Since it is reasonable to assume the different individuals will employ different values of K, the values of K may be regarded as being randomly selected from certain distribution. This distribution is termed the *balking distribution.*

Let us now assume that the queueing system is of type $M/M/1$; that is, both arrivals and departure are governed by Poisson processes with, say, parameters λ and μ, respectively. Since the queueing system is of type $M/M/1$, the process $\{X(t), t \geqslant 0\}$ is Markovian, and we can derive the Kolmogorov equations describing the queueing process. We first introduce the following notation:

$$P_x(t) = \mathscr{P}\{X(t) \leqslant x\} \tag{9.180}$$

$$p_x(t) = P_x(t) - P_{x-1}(t) \tag{9.181}$$

$$Q_x(t) = 1 - P_x(t) \tag{9.182}$$

$$H(x) = \mathscr{P}\{K \leqslant x\} \qquad \text{(the balking distribution)} \tag{9.183}$$

$$h(x) = H(x) - H(x - 1) \tag{9.184}$$

$$J(x) = 1 - H(x) \tag{9.185}$$

From the above we see that $p_0(t) = P_0(t) \neq 0$ is the probability that no queue exists at time t; $h(0) = H(0) \neq 0$ is the probability that an individual is absolutely queue-resistant; and, given a queue of length x, the probability that an arrival joins the queue is equal to the probability that *his K* is greater than or equal to x, which in turn is equal to $J(x - 1)$.

In order to derive the Kolmogorov equations describing this queueing process, we consider the following mutually exclusive events that can take place in the interval $(t, t + \Delta t)$:

1. One individual arrives in the interval $(t, t + \Delta t)$ and joins the queue, the probability of this event being $\lambda p_{x-1}(t)J(x - 2) \Delta t + o(\Delta t)$.

2. One individual arrives in the interval $(t, t + \Delta t)$ and does not join the queue, i.e., he balks, the probability of this event being $\lambda p_x(t)H(x - 1) \Delta t + o(\Delta t)$.

3. One individual leaves the queue in the interval $(t, t + \Delta t)$, the probability of this event being $\mu p_{x+1}(t) \Delta t + o(\Delta t)$.

4. There are no arrivals or departures in the interval $(t, t + \Delta t)$, the probability of this event being $[1 - (\lambda + \mu)p_x(t)] \Delta t + o(\Delta t)$.

5. There is more than one arrival or departure in the interval $(t, t + \Delta t)$, the probability of this event being $o(\Delta t)$.

The above considerations lead to the differential-difference equation

$$\frac{dp_x(t)}{dt} = \mu p_{x+1}(t) + [(H(x - 1) - 1) - \mu]p_x(t) + \lambda J(x - 2)p_{x-1}(t)$$

$$x = 0, 1, 2, \ldots \quad (9.186)$$

where, if $x = 0$, we delete μ in the coefficient of $p_x(t)$ and $H(-1) = H(-2) = 0$. If we now sum over all x, we obtain

$$\frac{dP_x(t)}{dt} = \mu p_{x+1}(t) - \lambda J(x - 1)p_x(t) \quad (9.187)$$

If we put $J(x) = \lambda_x$ and use (9.181), Eq. (9.187) becomes

$$\frac{dP_x(t)}{dt} = \mu P_{x+1}(t) - (\mu + \lambda_{x-1})P_x(t) + \lambda_{x-1}P_{x-1}(t) \quad (9.188)$$

which is an equation of the birth-and-death type where the death rate is constant and the birth rate is a function of the state variable x. A method of computing solutions of Eq. (9.186) has been indicated by Haight. This method employs the Laplace transformation and the theory of continued fractions.

We now consider the problem of determining the conditions which must be satisfied in order for equilibrium distributions of the queue length to exist. Let

$$\pi_x = \lim_{t \to \infty} p_x(t)$$

denote these equilibrium probabilities, should they exist. If we put the left-hand side of (9.187) equal to zero, we see that the equilibrium equations are

$$\pi_1 = \rho \pi_0$$

$$\pi_{x+1} = \rho J(x-1)\pi_x \qquad x = 1, 2, \ldots \qquad (9.189)$$

If we let

$$c_0 = c_1 = 1$$

$$c_x = \prod_{i=0}^{x-2} J(i) \qquad x = 2, 3, \ldots$$

(9.189) yields

$$\pi_x = \pi_0 c_x \rho^x \qquad (9.190)$$

In order for $\{\pi_x\}$ to be an honest probability distribution, we must have

$$\sum_{x=0}^{\infty} \pi_x = 1 = \pi_0 \sum_{x=0}^{\infty} c_x \rho^x \qquad (9.191)$$

By utilizing a theorem due to Kawata [36], we find that a necessary and sufficient condition for the queueing process to be ergodic is that

$$\sum_{x=0}^{\infty} c_x \rho^x < \infty$$

Hence π_0, the asymptotic probability that there is no queue, must be greater than zero.

From the above we see that the tails of the balking distribution are proportional to the ratio of ordinates of the queue-length distribution; hence, if ρ is known, either may be obtained from the other.

Also, from (9.184), (9.185), and (9.189), we have (if π_x, $\pi_{x+1} \neq 0$)

$$h(x) = J(x-1) - J(x) = \frac{\pi_0}{\pi_1}\left[\frac{\pi_{x+1}}{\pi_x} - \frac{\pi_{x+2}}{\pi_{x+1}}\right] \qquad (9.192)$$

hence in order for $h(x)$ to be nonnegative, we must have

$$\pi_{x+1}^2 \geqslant \pi_x \pi_{x+2} \qquad (9.193)$$

If for some value of n, $\pi_n \neq 0$ but $\pi_{n+1} = 0$, it follows from (9.189) that $J(n-1) = 0$, which implies that $\pi_x = 0$ for all $x > n$. In this case there is complete balking for queues of length equal to or greater than n.

An examination of (9.192) and (9.193) shows that, associated with any equilibrium distribution of queue length satisfying (9.193), there is a balking distribution. It is of interest to note that there will be a positive probability that K is infinite, that is, $H(\infty) \neq 1$, unless

$$\lim_{x \to \infty} \frac{\pi_{x+1}}{\pi_x} = 0$$

If this condition is not satisfied, there is an upper limit to the traffic intensity ρ for equilibrium to be attained. It is easily verified that ρ must be such that

$$\rho < \lim_{x \to \infty} \frac{\pi_x}{\pi_{x+1}}$$

Let us now obtain expressions for the mean and variance of the equilibrium queue length. We introduce the generating functions

$$F(s,t) = \sum_{x=0}^{\infty} p_x(t)s^x = (1 - s)\sum_{x=0}^{\infty} P_x(t)s^x \qquad |s| \leqslant 1$$

$$G(s,t) = \sum_{x=0}^{\infty} p_{x+1}(t)H(x)s^x \qquad\qquad |s| \leqslant 1$$

Hence, Eq. (9.187) becomes

$$\frac{1}{1 - s}\frac{\partial F(s,t)}{\partial t} = -\lambda p_0(t) + \left(\frac{\mu}{s} - \lambda\right)[F(s,t) - p_0(t)] + \lambda s G(s,t) \quad (9.194)$$

If we let $t \to \infty$, (9.194) yields the equilibrium equation

$$F(s) - \rho s F(s) - p_0 + \rho s^2 G(s) = 0 \tag{9.195}$$

If we put $s = 1$, we obtain, since $F(1) = 1$,

$$\pi_0 = 1 - \rho + \rho G(1) \tag{9.196}$$

and since $0 < \pi_0 \leqslant 1$, we have

$$\frac{\rho - 1}{\rho} < G(1) \leqslant 1 \tag{9.197}$$

Since $G(1)$ is the asymptotic probability that an individual balks, (9.197) gives bounds for this probability, and as one would expect, the lower bound is a function of ρ.[1]

From (9.190), (9.191), and the definition of $F(s,t)$, we have

$$F(s) = \pi_0 \sum_{x=0}^{\infty} c_x \rho^x = \frac{\displaystyle\sum_{x=0}^{\infty} c_x(\rho s)^x}{\displaystyle\sum_{x=0}^{\infty} c_x \rho^x} = \frac{\pi_0}{\pi_0(s)} \tag{9.198}$$

---

[1] It is important to point out that in the derivation of the Kolmogorov equations for this queueing process we did not distinguish between individuals joining and not joining the queue; hence, ρ was calculated for all arrivals. An *effective traffic intensity*, computed only from those joining the queue, say ρ', can be written as $\rho' = \rho(1 - G(1))$.

where $\pi_0(s)$ denotes π_0 with ρ replaced by ρs, and from (9.195) and (9.198) we have

$$G(s) = \frac{\pi_0}{\rho s^2}\left[1 - \frac{1 - \rho s}{\pi_0(s)}\right] \tag{9.199}$$

Hence the equilibrium queue length X and the balking distribution are uniquely determined if π_0, as a function of ρ, is known.

If we differentiate (9.196) and put $s = 1$, we obtain

$$m = \mathscr{E}\{X\} = \frac{\rho}{1 - \rho}\,[1 - G'(1) - 2G(1)] \tag{9.200}$$

By differentiating twice and putting $s = 1$, we obtain

$$\mathscr{D}^2\{X\} = \frac{\rho}{(1 - \rho)^2}\,\{1 - \rho[(G'(1))^2 + 4G^2(1) + 4G(1)G'(1)]$$
$$- (1 - \rho)[G''(1) + 5G'(1) + 4G(1)]\} \tag{9.201}$$

The above expressions give the *mean and variance of the equilibrium queue length*.

The equilibrium distribution $\{\pi_x\}$ can be obtained from the mean length as follows: From (9.190) we have

$$\rho\,\frac{\partial}{\partial\rho}\,\log\pi_x = \rho\,\frac{\partial}{\partial\rho}\,\left\{\log c_x + x\,\log\rho - \log\sum_{x=0}^{\infty}c_x\rho^x\right\}$$

$$= x - m\rho$$

Hence π_x, and in particular π_0, is determined when the mean is known.

As an example we consider the case when

$$\pi_x = \binom{n}{x}u^x(1 - u)^{n-x} \qquad x = 0, 1, 2, \ldots, n \tag{9.202}$$

that is, we assume the equilibrium distribution is binomial. From (9.189) we find the traffic intensity

$$\rho = \frac{nu}{1 - u} \qquad 0 < \rho < \infty \tag{9.203}$$

and

$$J(x) = \frac{n - x - 1}{n(x + 2)} \qquad \text{for } 0 \leqslant x \leqslant n - 1$$
$$= 0 \qquad \text{for } x \geqslant n \tag{9.204}$$

Hence

$$h(x) = \frac{n + 1}{n}\,\frac{1}{(x + 1)(x + 2)} \qquad \text{for } 0 \leqslant x \leqslant n - 1$$
$$= 0 \qquad \text{for } x \geqslant n \tag{9.205}$$

The following expressions are also easily determined:

$$m = \mathscr{E}\{X\} = \frac{n\rho}{n + \rho} \tag{9.206}$$

$$\mathscr{D}^2\{X\} = \rho\left(\frac{n}{n + \rho}\right)^2 \tag{9.207}$$

$$\pi_0 = \left(\frac{n}{n + \rho}\right)^n \tag{9.208}$$

$$G(s) = \left(\frac{n + \rho s}{n + \rho}\right)^n \tag{9.209}$$

Bibliography

1 Arrow, K., S. Karlin, and H. Scarf: "Studies in the Mathematical Theory of Inventory and Production," Stanford University Press, Stanford, Calif., 1958.

2 Ashcroft, H.: The Productivity of Several Machines under the Care of One Operator, *J. Roy. Statist. Soc.*, ser. B, vol. 12, pp. 145–151, 1950.

3 Bailey, N. T. J.: A Study of Queues and Appointment Systems in Outpatient Departments with Special Reference to Waiting Times, *J. Roy. Statist. Soc.*, ser. B, vol. 14, pp. 185–199, 1952.

4 Bailey, N. T. J.: On Queueing Processes with Bulk Service, *J. Roy. Statist. Soc.*, ser. B, vol. 16, pp. 80–87, 1954.

5 Bailey, N. T. J.: A Continuous Time Treatment of a Simple Queue Using Generating Functions, *J. Roy. Statist. Soc.*, ser. B, vol. 16, pp. 288–291, 1954.

6 Bailey, N. T. J.: Some Further Results in the Non-equilibrium Theory of a Simple Queue, *J. Roy. Statist. Soc.*, ser. B, vol. 19, pp. 326–333, 1957.

7 Bartlett, M. S.: "An Introduction to Stochastic Processes," Cambridge University Press, New York, 1955.

8 Beneš, V. E.: On Queues with Poisson Arrivals, *Ann. Math. Statist.*, vol. 28, pp. 670–677, 1957.

9 Benson, F.: Further Notes on the Productivity of Machines Requiring Attendance at Random Intervals, *J. Roy. Statist. Soc.*, ser. B, vol. 14, pp. 200–210, 1952.

10 Benson, F., and D. R. Cox: The Productivity of Machines Requiring Attention at Random Intervals, *J. Roy. Statist. Soc.*, ser. B, vol. 13, pp. 65–82, 1951.

11 Brockmayer, E., H. L. Halstrøm, and A. Jensen: "The Life and Works of A. K. Erlang," Copenhagen Telephone Company, Copenhagen, 1948.

12 Champernowne, D. G.: An Elementary Method of the Solution of the Queueing Problem with a Single Server and Constant Parameter, *J. Roy. Statist. Soc.*, ser. B, vol. 18, pp. 125–128, 1956.

13 Churchman, C. W., R. L. Ackoff, and E. L. Arnoff: "Introduction to Operations Research," John Wiley & Sons, Inc., New York, 1957.

14 Clarke, A. B.: A Waiting Line Process of Markov Type, *Ann. Math. Statist.*, vol. 27, pp. 452–459, 1956.

15 Cohen, J. W.: The Full Availability Group of Trunks with an Arbitrary Distribution of the Inter-arrival Times and a Negative Exponential Holding Time Distribution, *Simon Stevin*, vol. 31, pp. 169–181, 1957.

16 Cohen, J. W.: Certain Delay Problems for a Full Availability Trunk Group Loaded by Two Traffic Sources, *Communication News*, vol. 16, pp. 105–113, 1956.

17 Cohen, J. W.: A Survey of Queueing Problems Occurring in Telephone and Telegraph Traffic Theory, *Proc. First Intern. Conf. on Operations Research*, pp. 138–146, 1953.

18 Cox, D. R.: The Statistical Analysis of Congestion, *J. Roy. Statist. Soc.*, ser. A, vol. 118, pp. 324–335, 1955.

19 Doig, A.: A Bibliography on the Theory of Queues, *Biometrika*, vol. 44, pp. 490–514, 1957.

20 Downton, F.: Waiting Times in Bulk Service Queues, *J. Roy. Statist. Soc.*, ser. B, vol. 18, pp. 265–274, 1956.

21 Downton, F.: On Limiting Distributions in Bulk Service Queues, *J. Roy. Statist. Soc.*, ser. B, vol. 18, pp. 265–274, 1956.

22 Feller, W.: "An Introduction to Probability Theory and Its Applications," vol. 1, 2d ed., John Wiley & Sons, Inc., New York, 1957.

23 Fortet, R.: Random Distributions with an Application to Telephone Engineering, *Proc. Third Berkeley Symposium on Math. Statistics and Probability*, vol. 2, pp. 81–88, 1956.

24 Foster, F. G.: On the Stochastic Matrices Associated with Certain Queueing Problems, *Ann. Math. Statist.*, vol. 24, pp. 355–360, 1953.

25 Fry, T. C.: "Probability and Its Engineering Uses," D. Van Nostrand Company, Inc., Princeton, N.J., 1928.

26 Gani, J.: Problems in the Probability Theory of Storage Systems, *J. Roy. Statist. Soc.*, ser. B, vol. 19, pp. 181–206, 1957.

27 Girault, M.: Files d'attente. Loi de survie d'un intervalle à partir d'un instant quelconque, *Compt. rend. acad. sci. Paris*, vol. 246, pp. 2838–2839, 1958.

28 Gurk, H. M.: Single Server, Time-limited Queues (abstract), *Bull. Am. Math. Soc.*, vol. 63, p. 400, 1957.

29 Haight, F. A.: Queueing with Balking, *Biometrika*, vol. 44, pp. 360–369, 1957.

30 Homma, T.: On a Certain Queueing Process, *Repts. Statist. Appl. Research, Union Japan. Scientists and Engrs.*, vol. 4, pp. 14–32, 1955.

31 Homma, T.: On the Many Server Queueing Process with a Particular Type of Queue Discipline, *Repts. Statist. Appl. Research, Union Japan. Scientists and Engrs.*, vol. 4, pp. 90–101, 1956.

32 Homma, T.: On the Theory of Queues with Some Types of Queue Discipline, *Yokohama Math. J.*, vol. 4, pp. 56–64, 1956.

33 Homma, T.: On Some Fundamental Traffic Problems, *Yokohama Math. J.*, vol. 5, pp. 99–114, 1957.

34 Jackson, R. R. P., and D. G. Nickols: Some Equilibrium Results for the Queueing Process $E_k/M/1$, *J. Roy. Statist. Soc.*, ser. B, vol. 18, pp. 275–279, 1956.

35 Karlin, S., and J. McGregor: Many Server Queueing Processes with Poisson Input and Exponential Service Times, *Pacific J. Math.*, vol. 8, pp. 87–118, 1958.

36 Kawata, T.: A Problem in the Theory of Queues, *Repts. Statist. Appl. Research, Union Japan. Scientists and Engrs.*, vol. 3, pp. 122–129, 1955.

37   Kendall, D. G.: Some Problems in the Theory of Queues, *J. Roy. Statist. Soc.*, ser. B, vol. 3, pp. 151–185, 1952.
38   Kendall, D. G.: Stochastic Processes Occurring in the Theory of Queues and Their Analysis by Means of the Imbedded Markov Chain, *Ann. Math. Statist.*, vol. 24, pp. 338–354, 1954.
39   Kendall, D. G.: Some Problems in the Theory of Dams, *J. Roy. Statist. Soc.*, ser. B, vol. 19, pp. 207–212, 1957.
40   Kesten, H., and J. Th. Runnenburg: Purity in Waiting Line Problems: I, II, *Ned. Akad. Wetensch. Proc.*, ser. A, vol. 60, pp. 312–324, pp. 325–336, 1957.
41   Khintchine, A.: Mathematical Methods of the Theory of Mass Service (in Russian), *Trudy Mat. Inst. im V.A. Steklova*, no. 49, 1955.
42   Kiefer, J., and J. Wolfowitz: On the Theory of Queues with Many Servers, *Trans. Am. Math. Soc.*, vol. 78, pp. 1–18, 1955.
43   Kiefer, J., and J. Wolfowitz: On the Characteristics of the General Queueing Process, with Applications to Random Walk, *Ann. Math. Statist.*, vol. 27, pp. 147–161, 1956.
44   Koenigsberg, E.: Birth, Death and Waiting in Line (abstract), *Am. Math. Monthly*, vol. 62, p. 543, 1952.
45   Koenigsberg, E.: Queueing with Special Service, *Operations Research*, vol. 4, pp. 213–220, 1956.
46   Koenigsberg, E.: Cyclic Queues, *Operations Research Quart.*, vol. 9, pp. 22–35, 1958.
47   Kolmogorov, A. N.: Sur le problème d'attente, *Mat. Sbornik*, vol. 38, pp. 101–106, 1931.
48   Kronig, R.: On Time Losses in Machinery Undergoing Interruptions: I, *Physica*, vol. 10, pp. 215–224, 1943.
49   Kronig, R., and H. Mondria: On Time Losses in Machinery Undergoing Interruptions: II, *Physica*, vol. 10, pp. 331–336, 1943.
50   Kunz, K. S.: "Numerical Analysis," McGraw-Hill Book Company, Inc., New York, 1957.
51   Ledermann, W., and G. E. H. Reuter: Spectral Theory for the Differential Equations of Simple Birth and Death Processes, *Phil. Trans. Roy. Soc. London*, ser. A, vol. 246, pp. 321–369, 1954.
52   Lindley, D. V.: The Theory of Queues with a Single Server, *Proc. Cambridge Phil. Soc.*, vol. 48, pp. 277–289, 1952.
53   Luchak, G.: The Solution of the Single Channel Queueing Equation Characterized by a Time-dependent Poisson-distributed Arrival Rate and a General Class of Holding Times, *Operations Research*, vol. 4, pp. 711–732, 1956.
54   Luchak, G.: The Distribution of Time Required to Reduce to Some Pre-assigned Level a Single-channel Queue Characterized by a Time-dependent Poisson-distributed Arrival Rate and a General Class of Holding Times, *Operations Research*, vol. 5, pp. 205–209, 1957.
55   Marshall, B. O.: Queueing Theory, in "Operations Research for Management," vol. 1, pp. 134–148, Johns Hopkins Press, Baltimore, 1954.
56   McCloskey, J., and J. Coppinger (eds.): "Operations Research for Management," vol. 1, Johns Hopkins Press, Baltimore, 1954.
57   McCloskey, J., and F. Trefethen (eds.): "Operations Research for Management," vol. 2, Johns Hopkins Press, Baltimore, 1956.
58   McMillan, B., and J. Riordan: A Moving Single Server Problem, *Ann. Math. Statist.*, vol. 28, pp. 471–478, 1957.

59 Miller, R. G.: A Contribution to the Theory of Bulk Queues, *Stanford Univ. Tech. Rept.*, 1958.

60 Miller, R. G.: Priority Queues, *Stanford Univ. Tech. Rept.*, 1958.

61 Molina, E. C.: Application of the Theory of Probability to Telephone Trunking Problems, *Bell System Tech. J.*, vol. 6, pp. 461–494, 1927.

62 Morse, P. M.: "Queues, Inventories and Maintenance," John Wiley & Sons, Inc., New York, 1958.

63 Naor, P.: On Machine Interference, *J. Roy. Statist. Soc.*, ser. B, vol. 18, pp. 280–287, 1956.

64 Naor, P.: Normal Approximation to Machine Interference with Many Repairmen, *J. Roy. Statist. Soc.*, ser. B, vol. 19, pp. 334–341, 1957.

65 Naor, P.: Some Problems in Machine Interference, *Proc. First Intern. Conf. on Operations Research*, pp. 147–165, 1958.

66 Palm, C.: Intensitätschwankungen im Fernsprechverkehr, *Ericsson Technics*, no. 44, pp. 1–189, 1943.

67 Palm, C.: The Distribution of Repairmen in Serving Automatic Machines (in Swedish), *Ind. Norden*, vol. 75, pp. 75–80, 90–94, 119–123, 1947.

68 Pollaczek, F.: Über eine Aufgabe der Wahrscheinlichkeitstheorie: I, II, *Math. Zeit.*, vol. 32, pp. 64–100, 729–750, 1930.

69 Pollaczek, F.: Sur l'application de la théorie des fonctions au calcul de certaines probabilités continues utilisées dans la théorie des réseaux téléphoniques, *Ann. Inst. H. Poincaré*, vol. 10, pp. 1–54, 1946.

70 Pollaczek, F.: Généralisation de la théorie probabiliste des systèmes téléphoniques sans dispositif d'attente, *Compt. rend. acad. sci. Paris*, vol. 236, pp. 1469–1470, 1953.

71 Pollaczek, F.: Problèmes stochastiques posés par le phénomène de formation d'une queue d'attente à une quichet et par des phénomènes apparentés, *Mem. sci. math.*, no. 136, Gauthier-Villars, Paris, 1957.

72 Reich, E.: On the Integrodifferential Equation of Takács: I, *Ann. Math. Statist.*, vol. 29, pp. 563–570, 1958.

73 Riley, V.: Bibliography on Queueing Theory, in "Operations Research for Management," vol. 2, pp. 541–556, Johns Hopkins Press, Baltimore, 1956.

74 Riordan, J., and R. E. Fagen: Queueing Systems for Single and Multiple Operations, *J. Soc. Ind. Appl. Math.*, vol. 3, pp. 73–79, 1955.

75 Runnenburg, J. Th.: Machines Served by a Patrolling Operator, July, 1957. (Prepublication copy.)

76 Saaty, T. L.: Resumé of Useful Formulas of Queueing Theory, *Operations Research*, vol. 5, pp. 161–200, 1957.

77 Saaty, T. L.: "Mathematical Methods of Operations Research," McGraw-Hill Book Company, Inc., New York, 1959.

78 Sevastyanov, B. A.: An Ergodic Theorem for Markov Processes and Its Application to Telephone Systems with Refusals (in Russian), *Teor. Veroyatnost. i Primenen*, vol. 2, pp. 106–116, 1957.

79 Smith, W. L.: On the Distribution of Queueing Times, *Proc. Cambridge Phil. Soc.*, vol. 49, pp. 449–461, 1953.

80 Syski, R.: "Introduction to Congestion Theory in Telephone Systems," in press.

81 Takács, L.: Investigation of Waiting Time Problems by Reduction to Markov Processes, *Acta. Math. Acad. Sci. Hung.*, vol. 6, pp. 101–109, 1955.

82 Takács, L.: On the Generalization of Erlang's Formula, *Acta Math. Acad. Sci. Hung.*, vol. 7, pp. 419–433, 1957.

83 Takács, L.: On a Stochastic Process Concerning Some Waiting Time Problems, *Teor. Veroyatnost. i Primenen*, vol. 2, pp. 92–105, 1957.

84 Takács, L.: On Certain Sojourn Time Problems in the Theory of Stochastic Processes, *Acta Math. Acad. Sci. Hung.*, vol. 8, pp. 169–191, 1957.

85 Takács, L.: On a Probability Problem Concerning Telephone Traffic, *Acta Math. Acad. Sci. Hung.*, vol. 8, pp. 319–324, 1957.

86 Takács, L.: On a Queueing Problem in Telephone Traffic, *Acta Math. Acad. Sci. Hung.*, vol. 8, pp. 325–335, 1957.

87 Udagawa, K., and G. Nakamura: On a Certain Queueing System, *Kodai Math. Sem. Rept.*, vol. 8, pp. 117–124, 1956.

88 Volberg, O.: Problème de la queue stationnaire et non-stationnaire, *Compt. rend. acad. sci. U.R.S.S.*, vol. 24, pp. 657–661, 1939.

89 Wishart, D. M. G.: A Queueing System with χ^2 Service-time Distribution, *Ann. Math. Statist.*, vol. 27, pp. 768–779, 1956.

APPENDIX A

Generating Functions

The method of generating functions is a very important tool in the study of stochastic processes with a discrete state space. In this appendix we have collected a number of theorems and remarks concerning generating functions that the reader might find useful. For a more detailed discussion of generating functions, and for the proofs of some of the theorems, Ref. 1 should be consulted.

A. Definition. Consider the sequence of real numbers φ_0, φ_1, φ_2, ...

The power series

$$F(s) = \sum_{i=0}^{\infty} \varphi_i s^i \tag{A.1}$$

can be regarded as a transformation which carries the sequence $\{\varphi_i\}$ into the function $F(s)$. If the series converges in some interval $-s^* < s < s^*$, then the function $F(s)$ is called the *generating function* of the sequence $\{\varphi_i\}$.

Now consider the stochastic process $\{X(t),\ t \geqslant 0\}$ and the sequence of probabilities $P_x(t)$, where $P_x(t) = \mathscr{P}\{X(t) = x\}$. The time parameter t may be discrete or continuous. Let

$$F(s,t) = \sum_{x=0}^{\infty} P_x(t)s^x \tag{A.2}$$

be the generating function of the probabilities $P_x(t)$, $x = 0, 1, 2, \ldots$. Since the $P_x(t)$ are bounded for all x and t and $\sum_{x=0}^{\infty} P_x(t) = 1$ for all $t \geqslant 0$, a comparison with the geometric series shows that the above series converges uniformly in t at least for $|s| < 1$.

439

B. Some Theorems and Properties

Theorem A.1: The expected value of the random variable $X(t)$ is given by

$$\mathscr{E}\{X(t)\} = \frac{dF(s,t)}{ds}\bigg]_{s=1} \qquad (A.3)$$

Proof: By definition, the expected value of $X(t)$ is

$$\mathscr{E}\{X(t)\} = \sum_{x=0}^{\infty} x P_x(t)$$

for $\mathscr{E}\{X(t)\} < \infty$. By differentiating (A.3), we have

$$\frac{dF}{ds} = \sum_{x=0}^{\infty} x P_x(t) s^{x-1}$$

By putting $s = 1$, we obtain the relation

$$\mathscr{E}\{X(t)\} = \sum_{x=0}^{\infty} x P_x(t) = \frac{dF}{ds}\bigg]_{s=1}$$

Theorem A.2: If $\mathscr{E}\{X^2(t)\} = \sum_{x=0}^{\infty} x^2 P_x(t) < \infty$, then

$$\mathscr{E}\{X^2(t)\} = \left[\frac{d^2F}{ds^2} + \frac{dF}{ds}\right]_{s=1} \qquad (A.4)$$

and by the definition of the variance of a random variable, i.e., $\mathscr{D}^2\{X(t)\} = \mathscr{E}\{X^2(t)\} - (\mathscr{E}\{X(t)\})^2$,

$$\mathscr{D}^2\{X(t)\} = \left[\frac{d^2F}{ds^2} + \frac{dF}{ds} - \left(\frac{dF}{ds}\right)^2\right]_{s=1} \qquad (A.5)$$

Proof: By definition,

$$E\{X^2(t)\} = \sum_{x=0}^{\infty} x^2 P_x(t)$$

$$= \sum_{x=0}^{\infty} x(x-1) P_x(t) + \sum_{x=0}^{\infty} x P_x(t)$$

Differentiating $F(s,t)$ and putting $s = 1$ gives (A.4). From Theorem A.1 the result (A.5) follows.

Theorem A.3: Let $X_1(t)$ and $X_2(t)$ be two independent random variables with associated probability distributions $P_x(t)$ and $Q_x(t)$ and generating functions $F_1(s,t)$ and $F_2(s,t)$, respectively. Let $Y(t) = X_1(t) + X_2(t)$ and let $G(s,t)$ be the generating function of the probabilities $R_y(t) = \mathscr{P}\{Y(t) = y\}$; then $G(s,t)$ is given by

$$G(s,t) = F_1(s,t) F_2(s,t) \qquad (A.6)$$

Proof: The distribution of $Y(t)$ is the *convolution* of $P_x(t)$ and $Q_x(t)$, that is,

$$R_y(t) = P_x(t)*Q_x(t) = \sum_{i=0}^{y} P_i(t)Q_{y-i}(t) \qquad (A.7)$$

Termwise multiplication of the power series for $F_1(s,t)$ and $F_2(s,t)$ gives the product $G(s,t)$. On collecting terms, we see that the coefficient $R_y(t)$ of s^y in the expansion of $F_1(s,t)F_2(s,t)$ is given by (A.7).

We remark that the above theorem can easily be extended to the case of N independent random variables. In this case we obtain the result: *The generating function of the sum of N independent random variables is the N-fold product of the generating function associated with each random variable.*

Compound Distributions. Let $\{X_i(t)\}$ be a sequence of mutually independent random variables with common distribution $P_x(t) = \mathscr{P}\{X_i(t) = x\}, x = 0, 1, 2, \ldots$. Consider the sum $S_N = \sum_{i=1}^{N} X_i(t)$, where N is a random variable with distribution $q(n) = \mathscr{P}\{N = n\}$. The distribution of S_N is given by

$$\mathscr{P}\{S_N = x\} = \sum_{n=0}^{\infty} \mathscr{P}\{N = n\}\mathscr{P}\{S_n = x\}$$

$$= \sum_{n=0}^{\infty} q(n)\{P_x(t)\}^{n*} \qquad (A.8)$$

where $\{P_x(t)\}^{n*}$ is the n-fold convolution of $P_x(t)$ with itself. Distributions of the above type are termed *compound distributions*.

Theorem A.4: Let $F_1(s,t)$ and $F_2(s)$ be the generating functions of $P_x(t)$ and $q(n)$, respectively. Then the generating function of the compound distribution (A.8) is the functional $F_2[F_1(s,t)]$.

Proof: Consider n to be fixed. Then by Theorem A.3 the generating function of $\{P_x(t)\}^{n*}$ is $[F_1(s,t)]^n$. Hence, the generating function of $\mathscr{P}\{S_N = x\}$ is given by

$$\sum_{n=0}^{\infty} q(n)[F_1(s,t)]^n \qquad (A.9)$$

which is $F_2(s)$ with the argument s replaced by $F_1(s,t)$.

Maclaurin's Expansion of Generating Functions. It is of interest to state a simple result which is useful in many problems in which generating functions are used: If $F(s,t)$ is analytic at $s = 0$, then we can expand F in the Maclaurin series

$$F(s,t) = F(0,t) + \sum_{x=1}^{\infty} \frac{F^{(x)}(0,t)}{x!} s^x \qquad (A.10)$$

Hence, by comparison with the series $F(s,t) = \sum_{x=0}^{\infty} P_x(t)s^x$, we see that

$$P_0(t) = F(0,t)$$

$$P_x(t) = \frac{F^{(x)}(0,t)}{x!} \qquad x = 1, 2, \ldots \tag{A.11}$$

It follows, therefore, that, if $F(s,t) = F(-s,t)$ (an even function of s), then $P_x(t) = 0$ when x is odd and that, if $F(s,t) = -F(-s,t)$ (an odd function of s), then $P_x(t) = 0$ when x is even.

Partial-fraction Resolution: Asymptotic Approximation of Probabilities

Theorem A.5: Let $F(s,t)$ be a rational function of s, that is,

$$F(s,t) = \frac{G(s,t)}{H(s,t)} \tag{A.12}$$

where G and H are polynomials in s, the degree of G being less than that of H. Then $F(s,t)$ admits a partial-fraction resolution

$$F(s,t) = \sum_{i=1}^{n} \frac{A_i(t)}{s - \xi_i(t)} \tag{A.13}$$

where the $\xi_i(t)$ are the roots of H (real or complex) and the $A_i(t)$ are functions of time. If the roots are distinct with ξ_1 simple, and $|\xi_1| < \xi_i$, $i = 2, 3, \ldots, n$, then putting

$$\sigma_m = \left[-\frac{G(\xi_m,t)}{dH/ds} \right]_{s=\xi_m} \qquad m \leqslant n$$

we have

$$\lim_{x \to \infty} P_x(t) \sim \sigma_1 [\xi_1(t)]^{-(1+x)} \tag{A.14}$$

Bibliography

1 Feller, W.: "An Introduction to Probability Theory and Its Applications," 2d ed., vol. 1, John Wiley & Sons, Inc., New York, 1957.
2 Malécot, G.: Les processus stochastiques et la méthode des fonctions génératrices ou caractéristiques, *Publ. inst. statist. univ. Paris*, vol. 1, pp. 1–25, 1952.
3 Seal, H. L.: The Historical Development of the Use of Generating Functions in Probability Theory, *Bull. assoc. actu. suisses*, vol. 49, pp. 209–228, 1949.

APPENDIX B

The Laplace and Mellin Transforms

A. Let $F(t)$ be a known function of t, defined for all $t > 0$, and let $K(\lambda,t)$ be a known function of the two variables λ and t. If the integral

$$\varphi(\lambda) = \int_0^\infty K(\lambda,t)F(t)\,dt \tag{B.1}$$

is convergent, then (B.1) defines a function of the variable λ. This function is called the *integral transform* of the function $F(t)$ with the kernel $K(\lambda,t)$. A simple example of such a kernel is

$$K(\lambda,t) = e^{-\lambda t}$$

which leads to the *Laplace transform*

$$\varphi(\lambda) = \mathscr{L}\{F(t)\} = \int_0^\infty e^{-\lambda t}F(t)\,dt \tag{B.2}$$

Another kernel is

$$K(\lambda,t) = t^{\lambda-1}$$

which leads to the *Mellin transform*

$$\varphi(\lambda) = \mathscr{M}\{F(t)\} = \int_0^\infty t^{\lambda-1}F(t)\,dt \tag{B.3}$$

Other special integral transforms arise when the kernel $K(\lambda,t)$ assumes other forms.

Integral transforms are indispensable tools in the study of linear functional equations, and they are especially useful in the study of the differential, differential-difference, integral, and integrodifferential equations which arise in the theory of Markov processes. In this appendix we state some theorems and properties of the Laplace and Mellin transforms which are used or referred to in the text. For proofs of the theorems, we refer to Refs. 1 to 8.

B. The Laplace-Stieltjes Transform and the Laplace Transform. 1. *The Laplace-Stieltjes Transform.* If $F(t)$ is a complex function of the real variable t for $0 \leqslant t \leqslant \infty$ and if $F(t)$ is of bounded variation in any closed interval $[0,u]$ of the positive real axis, then the *Stieltjes integral*

$$\int_0^u e^{-\lambda t}\, dF(t) \tag{B.4}$$

exists for all complex values of λ and all real positive values of u. If the limit

$$\lim_{u \to \infty} \int_0^u e^{-\lambda t}\, dF(t) = \int_0^\infty e^{-\lambda t}\, dF(t) \tag{B.5}$$

exists for a certain value of λ, say λ^*, we say that the integral (B.4) converges for $\lambda = \lambda^*$. It can be shown that there exists an abscissa of convergence λ_c with the property that the integral (B.5) converges if $\mathscr{R}(\lambda) > \lambda_c$ and diverges if $\mathscr{R}(\lambda) < \lambda_c$. When the integral (B.5) converges, it defines a function of λ, say $\varphi(\lambda)$, called the *Laplace-Stieltjes transform* of the function $F(t)$, which is analytic in its region of convergence.

2. *The Laplace Transform.* If, in the definition of the Laplace-Stieltjes transform, we can write

$$F(t) = \int_s^t f(\tau)\, d\tau \qquad t > 0 \tag{B.6}$$

where s is an arbitrary positive constant, then (B.5) becomes

$$\varphi(\lambda) = \mathscr{L}\{f(t)\} = \int_0^\infty e^{-\lambda t} f(t)\, dt \tag{B.7}$$

The function $\varphi(\lambda)$ defined by (B.7) is called the (one-sided) *Laplace transform* of the function $f(t)$. In (B.7) the number λ (real or complex) is such that its real part is sufficiently large to make the integral convergent. If $\varphi(\lambda)$ converges for some λ_c, then it is analytic for all λ such that $\mathscr{R}(\lambda) > \mathscr{R}(\lambda_c)$.

Given the Laplace transform $\varphi(\lambda)$ of the function $f(t)$, it is possible to obtain $f(t)$ by application of the following theorem.

Theorem B.1 (*Inversion Theorem for the Laplace Transform*): Let $f(t)$ have a continuous derivative and let $|f(t)| < Me^{\alpha t}$, where M and α are positive constants. Let $\varphi(\lambda) = \mathscr{L}\{f(t)\}$, $\mathscr{R}(\lambda) > \alpha$. Then

$$f(t) = \mathscr{L}^{-1}\{\varphi(\lambda)\} = \lim_{\gamma \to \infty} \frac{1}{2\pi i} \int_{\beta - i\gamma}^{\beta + i\gamma} e^{\lambda t} \varphi(\lambda)\, d\lambda \tag{B.8}$$

where $\beta > \alpha$.

Some Theorems and Properties

Theorem B.2 (*Laplace Transform of Derivatives*): Let the function $f(t)$ be continuous with a sectionally continuous derivative $f'(t)$ in every finite interval $0 \leqslant t \leqslant \tau$. Also, let $f(t)$ be of $0(e^{\alpha t})$ as $t \to \infty$. Then for $\lambda > \alpha$ the Laplace transform of $f'(t)$ exists and is given by

$$\mathscr{L}\{f'(t)\} = \lambda\varphi(\lambda) - f(0+) \qquad (B.9)$$

Also, if $f(t)$ has a continuous derivative $f^{(n-1)}(t)$ of order $n - 1$ and a sectionally continuous derivative $f^{(n)}(t)$ in every finite interval $0 \leqslant t \leqslant \tau$, and if $f(t), f'(t), \ldots, f^{(n-1)}(t)$ are of $0(e^{\alpha t})$ as $t \to \infty$, then the Laplace transform of $f^{(n)}(t)$ exists when $\lambda > \alpha$ and is given by

$$L\{f^{(n)}(t)\} = \lambda^n\varphi(\lambda) - \lambda^{n-1}f(0+) - \cdots - f^{(n-1)}(0+) \qquad (B.10)$$

Theorem B.3: If $\varphi(\lambda) = \mathscr{L}\{f(t)\}$ and k is a positive constant, then

$$\frac{1}{k}\varphi\left(\frac{\lambda}{k}\right) = \mathscr{L}\{f(kt)\} \qquad (B.11)$$

This result is useful when one is interested in putting a problem in dimensionless form.

Theorem B.4 (*Convolution Property*): If $\varphi_1(\lambda) = \mathscr{L}\{f_1(t)\}$ and $\varphi_2(\lambda) = \mathscr{L}\{f_2(t)\}$, then

$$\varphi_1(\lambda)\varphi_2(\lambda) = \mathscr{L}\left\{\int_0^t f_1(\tau)f_2(t - \tau)\,d\tau\right\}$$

$$= \mathscr{L}\left\{\int_0^t f_1(t - \tau)f_2(\tau)\,d\tau\right\}$$

$$= \mathscr{L}\{f_1(t)*f_2(t)\} \qquad (B.12)$$

Theorem B.5 (*Asymptotic Behavior*): If $\varphi(\lambda) = \mathscr{L}\{f(t)\}$, then

$$\lim_{t \to \infty} f(t) = \lim_{\lambda \to 0} \lambda\varphi(\lambda) \qquad (B.13)$$

Theorem B.6 (*Partial-fraction Resolution of the Laplace Transform*): Let $\varphi(\lambda)$ be a rational function of λ, that is,

$$\varphi(\lambda) = \frac{g(\lambda)}{h(\lambda)}$$

where g and h are polynomials in λ, the degree of g being less than that of h. If $h(\lambda)$ is of degree n with roots $\xi_1, \xi_2, \ldots, \xi_n$ which are distinct, then

$$\varphi(\lambda) = \frac{g(\lambda)}{h(\lambda)} = \sum_{i=1}^{n} \frac{g(\xi_i)}{(\lambda - \xi_i)h(\xi_i)}$$

$$= \sum_{i=1}^{n} \frac{1}{\lambda - \xi_i} \left[\frac{(\lambda - \xi_i)g(\xi_i)}{h(\lambda)} \right]_{\lambda = \xi_i} \tag{B.14}$$

and

$$f(t) = \sum_{i=1}^{n} \frac{g(\xi_i)}{h'(\xi_i)} e^{\xi_i t} \tag{B.15}$$

Theorem B.7 (*Laplace Transform Solution for Small Values of Time*): If $\varphi(\lambda) = \mathscr{L}\{f(t)\}$ admits the power series representation

$$\varphi(\lambda) = \sum_{n=1}^{\infty} \alpha_n \lambda^{-n} \tag{B.16}$$

which has a radius of convergence $r < \infty$, then

$$f(t) = \sum_{n=1}^{\infty} \frac{\alpha_n t^{n-1}}{(n-1)!} \tag{B.17}$$

It follows from (B.16) and (B.17) that

$$\lim_{t \to 0} f(t) = \lim_{\lambda \to \infty} \lambda \varphi(\lambda) \tag{B.18}$$

Asymptotic Expansion for Large Values of Time. If $\varphi(\lambda) = \mathscr{L}\{f(t)\}$ is known and $f(t) = \mathscr{L}^{-1}\{\varphi(\lambda)\}$ can be obtained by an application of the inversion theorem

$$f(t) = \lim_{\gamma \to \infty} \frac{1}{2\pi i} \int_{\beta - i\gamma}^{\beta + i\gamma} e^{\lambda t} \varphi(\lambda) \, d\lambda$$

then it is possible to give conditions under which an asymptotic expansion of $f(t)$ for large values of t can be obtained from the behavior of $\varphi(\lambda)$ in the neighborhood of its singularity with the largest real part.

Suppose the singularity to be at $\lambda_0 = \sigma + i\sigma$, $\sigma < \beta$; and let $\lambda = \mu + i\eta$. Assume (1) $f(\lambda)$ is analytic for $\mu \geqslant \sigma - \epsilon$ ($\epsilon > 0$), except at λ_0; (2) as $\eta \to \pm\infty$, $\varphi(\lambda)$ approaches zero uniformly for $\mu \in [\sigma - \epsilon, \beta]$, and in this closed interval

$$\int |\varphi(\lambda)| \, d\eta < \infty$$

at $\eta = \pm\infty$.

Theorem B.8: If conditions (1) and (2) are satisfied and $\varphi(\lambda)$ can be expanded near λ_0 in the form

$$\varphi(\lambda) = \sum_{n=0}^{\infty} A_n (\lambda - \lambda_0)^{n-1} + (\lambda - \lambda_0)^\rho \sum_{n=0}^{\infty} B_n (\lambda - \lambda_0)^n \tag{B.19}$$

$0 < \rho < 1$, and if the series are convergent for $|\lambda - \lambda_0| \leqslant K$, $f(t)$ has the asymptotic expansion for large t (real)

$$f(t) \sim e^{\lambda_0 t}\left\{A_0 + \frac{\sin \pi\rho}{\pi} \sum_{n=0}^{\infty} (-1)^n B_n \Gamma(\rho + n)t^{-\rho-n}\right\} \quad \text{(B.20)}$$

C. The Mellin Transform. From (B.3) the *Mellin transform* of a function $f(t)$ is defined by the equation

$$\varphi(\lambda) = \mathcal{M}\{f(t)\} = \int_0^{\infty} t^{\lambda-1} f(t)\, dt \quad \text{(B.21)}$$

whenever the integral exists. As in the case of the Laplace transform, given the Mellin transform $\varphi(\lambda)$ of a function $f(t)$, it is possible to obtain $f(t)$ by application of the inversion theorem. If we put $t = e^{\xi}$ in (B.21), we obtain

$$\varphi(\lambda) = \int_{-\infty}^{\infty} e^{\lambda\xi} f(e^{\xi})\, d\xi$$

$$= \int_{-\infty}^{\infty} e^{\beta\xi} e^{i\omega\xi} f(e^{\xi})\, d\xi \quad \text{(B.22)}$$

where $\lambda = \beta + i\omega$. Equation (B.22) is the *Fourier transform* of $e^{\beta\xi}f(e^{\xi})$, and the inversion formula for the Fourier transform gives

$$e^{\beta\xi}f(e^{\xi}) = \frac{1}{2\pi} \int_{-\infty}^{\infty} e^{i\omega\xi}\varphi(\lambda)\, d\omega$$

$$= \lim_{\gamma \to \infty} \frac{i}{2\pi} \int_{\beta-i\gamma}^{\beta+i\gamma} e^{-\xi(\lambda-\beta)}\varphi(\lambda)\, d\lambda$$

since $i\omega = \lambda - \beta$. In terms of t, the *inversion formula for the Mellin transform* is

$$f(t) = \mathcal{M}^{-1}\{\varphi(\lambda)\} = \lim_{\gamma \to \infty} \frac{1}{2\pi i} \int_{\beta-i\gamma}^{\beta+i\gamma} t^{-\lambda}\varphi(\lambda)\, d\lambda \quad t > 0 \quad \text{(B.23)}$$

where β has to be so chosen that the above integral exists.

Bibliography

1   Carlslaw, H. S., and J. C. Jaeger: "Operational Methods in Applied Mathematics," Oxford University Press, New York, 1948.
2   Churchill, R. V.: "Modern Operational Mathematics in Engineering," 2d ed., McGraw-Hill Book Company, Inc., New York, 1958.
3   Doetsch, G.: "Theorie und Anwendung der Laplace-Transformation," Dover Publications, New York, 1943.
4   Doetsch, G.: "Handbuch der Laplace-Transformation," 3 vols., Birkhäuser Verlag, Basel and Stuttgart, 1950, 1955, 1956.

5　Erdélyi, A. (ed.): "Tables of Integral Transforms," 2 vols., McGraw-Hill Book Company, Inc., New York, 1954.
6　McLachlan, N. W.: "Modern Operational Calculus with Applications," Macmillan and Company, London, 1949.
7　Sneddon, I. N.: "Fourier Transforms," McGraw-Hill Book Company, Inc., New York, 1951.
8　Tranter, C. J.: "Integral Transforms in Mathematical Physics," Methuen & Co., Ltd., London, 1951.
9　Widder, D. V.: "The Laplace Transform," Princeton University Press, Princeton, N.J., 1941.

APPENDIX C

Monte Carlo Methods
in the Study of Stochastic Processes

A. Introduction. Although the theory of stochastic processes has found many fruitful applications in the sciences and engineering, it is clear to most workers in applied probability that the formulation of more realistic stochastic models leads to processes that are usually non-Markovian, nonstationary, or both. While it is usually not difficult to write down the functional equations describing these processes, a complete, or even partial, analytic treatment of the problem is often wanting. In recent years a very old, but seemingly new, method has been used to treat many problems in applied probability for which analytical methods have not been developed. The method we refer to is called the Monte Carlo method, and it can be described as *the representation of a mathematical or physical system by a sampling procedure which satisfies the same probability laws.* Hence, the Monte Carlo method is concerned with the generation of an artificial realization of a stochastic process by a sampling procedure, the particular procedure being determined by the underlying probability structure of the stochastic process.

The purpose of this appendix is to point out some of the applications of Monte Carlo methods in the study of stochastic processes. No attempt is made to give a complete discussion of this very important area. In the next section we demonstrate the generation of an artificial realization of a birth-and-death process. This is followed by a brief discussion of some applications in various applied fields.

For a general discussion of Monte Carlo methods we refer to the following: Churchman et al. [9], Crone [10], Hammersley [12], Householder [15], Householder et al. [16], Kahn [17], Metropolis and Ulam [21], and Meyer [22].

B.   Generation of an Artificial Realization of a Birth-and-Death Process. In order to demonstrate the generation of an artificial realization, we consider a simple birth-and-death process in which the birth rate λ and the death rate μ are assumed to be equal. This case has been treated by Kendall [18], and the procedure we use, while slightly different, follows the method given in Kendall's paper.

The equations for the birth-and-death process have been given in Sec. 2.4. For the case $\lambda = \mu$ we know that

$$m(t) = \mathscr{E}\{X(t)\} = X(0) = x_0 \tag{C.1}$$

and
$$\mathscr{D}^2\{X(t)\} = 2\lambda t x_0 \tag{C.2}$$

In addition, the probability that the population will die out before time t is

$$\left(\frac{\lambda t}{1 + \lambda t}\right)^{x_0} \tag{C.3}$$

Now let us assume that the system has just arrived in the state $x = N$ and let τ, $0 < \tau < \infty$, denote the length of the interval of time before the next event; that is, τ is the generation time. We now introduce two variables u and v, such that

$$u = N\tau \tag{C.4}$$

and
$$\begin{aligned}v &= 1 \quad &&\text{if the next event is a birth}\\ &= 0 \quad &&\text{if the next event is a death}\end{aligned} \tag{C.5}$$

From Chap. 2 we know that the distribution of the generation time τ for the birth-and-death process is

$$dG(\tau) = N(\lambda + \mu)e^{-(\lambda+\mu)N\tau} d\tau \tag{C.6}$$

while the probability of a birth in the interval $(\tau, \tau + d\tau)$ is

$$N\lambda e^{-(\lambda+\mu)N\tau} d\tau \tag{C.7}$$

Therefore, u has the χ^2 distribution

$$\frac{1}{2(\lambda + \mu)} \chi_2^2 \tag{C.8}$$

and v, which assumes only the values 0 and 1, has mean value $\lambda/(\lambda + \mu)$. Now, u and v are independent variables, and they are also independent of all previous u and v values.

From the above we see that the simple birth-and-death process is equivalent to the following sequential sampling procedure: We draw random variables from the u and v distributions in the following order,

$$u_1, v_1, u_2, v_2, \ldots, u_n, v_n, \ldots$$

and after each drawing calculate the quantities

$$t_n = t_{n-1} + \frac{u_n}{N_{n-1}} \tag{C.9}$$

where
$$N_n = N_{n-1} + 2v_n - 1 \tag{C.10}$$

In the above we assume $t_0 = 0$ and $N_0 = x_0$. We now select a time $T > 0$, which is called the cutoff time, and we stop sampling when either of the following conditions is satisfied:

(a) $\qquad\qquad\qquad\qquad N_n = 0$

(b) $\qquad\qquad\qquad\qquad t_n > T$

Therefore, the function defined by the rule

$$X(t) \equiv N_n \qquad \text{for } t_n \leqslant t < t_{n+1}, n = 0, 1, 2, \dots \tag{C.11}$$

is equivalent to one describing the growth, over the interval $[0, T]$, of a population starting with x_0 individuals and developing according to the birth-and-death process.

To demonstrate the above sampling procedure, we assume

$$\lambda = \mu = 0.5 \qquad x_0 = 5 \qquad T = 10$$

We can obtain the v values by using dice and summing the displayed values such that for

Even sums: $\quad v = 0 \quad (\equiv \text{death—D})$

Odd sums: $\quad v = 1 \quad (\equiv \text{birth—B})$

To obtain the sample of u variables we use the transformation

$$z = e^{-u}$$

hence the z variables are now uniformly distributed $(0,1)$ and can be read from a table of random numbers. In taking the z values from a table of random numbers the digit group 1867, for example, is translated to $z = 0.1867$. The variable u can now be determined by the table of exponential functions of the National Bureau of Standards. In some cases it is necessary to select a new digit group in order to obtain the same number of decimals in u. For example, we might extend 1867 to 18676102, giving $z = 0.18676102$, which in turn gives $u = 1.867$.

In order to check the randomness, the first 18 pairs of the 37 v variables were distributed as follows:

Type	BB	BD	DB	DD
Occurred	3	8	3	4
Expected	4.5	4.5	4.5	4.5

This gives a χ_3^2 equal to 3.8, which is satisfactory. Similarly, the 37 u variables showed the following distribution:

Interval	0.0–0.5	0.5–1.0	1.0–1.5	1.5–2.0	2.0
Occurred	15	7	4	3	8
Expected	14.6	8.8	5.4	3.2	5

This gives a $\chi_4^2 = 2.6$, which is satisfactory.

The table and graph show that the probability of extinction before time $T = 10$ is given by

$$(\tfrac{5}{6})^5 = 0.402$$

and as we see from the graph, extinction did not occur.

C. Remarks on Some Applications. In this section we shall give a brief review of some of the studies dealing with the use of Monte Carlo methods in applied fields. We consider, in particular, some applications in the theory of population growth, epidemic theory, cascade theory, queueing theory, and the theory of high polymers.

1. *Biology.* In the field of biology Monte Carlo methods have been used to study stochastic processes associated with population growth and the spread of epidemics. As we remarked in Chap. 4, a realistic treatment of population growth leads to non-Markovian processes, since the negative exponential distribution of generation times is replaced by a more complicated distribution. Since non-Markovian processes lead to functional equations that are difficult to handle analytically, the use of Monte Carlo methods is indicated. In particular, Monte Carlo methods have been employed by Bartlett[6] and Leslie[20] to generate artificial realizations for the predator-prey interaction, the logistic process, and related processes.

In a very interesting study Hoffman, Metropolis, and Gardiner[14] generated artificial birth processes when various assumptions were made concerning the distribution of the generation time. The first distribution considered was the familiar negative exponential. The second distribution considered was also a negative exponential; however, in this case a "rest period" following the formation of a new cell was introduced; during the period, division could not take place. At the end of this "rest period," the time before division was exponentially distributed. The third distribution was based on an experimental study of generation times.

In each of the above cases the computer was given the probability distribution for the generation time and, starting with one cell, an

t	τ	v	N	u
0.000				
	0.336		5	1.678
0.336		B		
	0.433		6	2.598
0.769		B		
	0.241		7	1.690
1.010		D		
	0.165		6	0.991
1.175		D		
	0.507		5	2.533
1.682		B		
	0.007		6	0.042
1.689		D		
	0.061		5	0.304
1.750		B		
	0.067		6	0.401
1.817		D		
	0.125		5	0.623
1.942		D		
	0.006		4	0.022
1.948		B		
	0.003		5	0.015
1.951		B		
	0.050		6	0.297
2.001		B		
	0.412		7	2.884
2.413		D		
	0.131		6	0.788
2.544		D		
	0.208		5	1.038
2.752		D		
	0.204		4	0.816
2.956		B		
	0.675		5	3.375
3.631		D		
	0.029		4	0.115
3.660		D		
	0.777		3	2.331
4.437		B		
	0.364		4	1.457
4.801		D		
	0.918		3	2.753
5.719		B		
	0.152		4	0.606
5.871		D		
	0.036		3	0.107
5.907		B		
	2.06		4	0.825

t	τ	v	N	u
6.113		D		
	0.151		3	0.452
6.264		B		
	0.337		4	1.349
6.601		D		
	0.125		3	0.374
6.726		B		
	0.395		4	1.579
7.121		D		
	0.046		3	0.137
7.167		B		
	0.019		4	0.076
7.186		D		
	1.189		3	3.567
8.375		D		
	0.233		2	0.465
8.608		B		
	0.028		3	0.084
8.636		B		
	0.179		4	0.717
8.815		B		
	0.041		5	0.205
8.856		D		
	0.276		4	1.104
9.132		D		
	1.002		3	3.005
(10.134)				

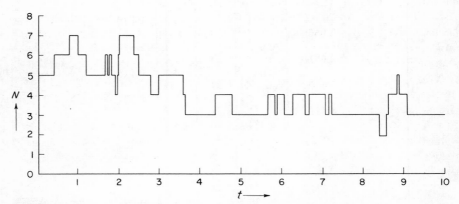

Figure C.1   Development of an artificial birth-and-death process.

artificial population was generated. Following the division of a cell, each daughter cell had the same probability distribution for its generation time. In this way hypothetical populations were generated for each of the assumed distributions.

In the stochastic theory of epidemics Monte Carlo methods have been used to generate artificial epidemics. The epidemic situation is far more complicated than population growth (except perhaps in the case of mutations), since we must take into consideration the numbers of infected and susceptible individuals, death or removal, recovery, immunity, etc. One of the first Monte Carlo studies in epidemiology, by Bartlett [2,3,5], was concerned with the generation of a series of artificial measles epidemics. These epidemics were generated under conditions simulating a partially isolated group such as a boarding school. Recently, Bartlett [4] and Kendall [19] have used Monte Carlo methods to generate artificial realizations of the McKendrick-Bartlett stochastic process.

2. *Physics.* In physics, Monte Carlo methods have been used to study problems arising in cascade theory, the diffusion of gamma rays, and nuclear shielding. In particular, Wilson[1] has studied showers produced by low-energy electrons and photons and the range and straggling of individual high-energy electrons. Wilson [24] has also studied electron- and photon-initiated showers in lead. This study took into consideration the effects of multiple scattering.

The use of Monte Carlo methods in the study of gamma-ray diffusion has been considered by M. J. Berger[2] and by L. A. Beach and R. B. Theus.[3] For a detailed discussion of the Monte Carlo method in cascade theory calculations, with particular reference to the electron-photon showers, we refer to the paper of Butcher and Messel [7].

3. *Chemistry.* In the field of chemistry Monte Carlo methods have been used by King[4] to study the stochastic structure of high polymers and the excluded volume effect in polymer chains. Monte Carlo methods should also play a role in the development of a stochastic theory of reaction kinetics. In particular, Bartholomay [1] has used Monte Carlo methods to generate artificial concentration curves for chemical reactions.

4. *Operations Research.* As in the other applied areas, Monte Carlo methods have been utilized in queueing theory. Foster [11] has used a procedure similar to the one used by Kendall to construct artificial realizations of two queueing systems. In this study one system, say S_1,

[1] Cf. Householder et al. [16].
[2] Cf. [22].
[3] *Ibid.*
[4] Cf. [16].

was given a constant service time of one unit, while the service times for S_2 were drawn from a negative-exponential distribution with a mean value of one unit. Both systems used the same sample of arrival times drawn from a negative-exponential distribution with a mean value of $\frac{3}{2}$. It is of interest to note that S_1 is a stationary non-Markovian process, while S_2 is a Markov process.

In Ref. 9 Monte Carlo methods are applied to a specific queueing situation. The procedure used is explained very clearly, and it can serve as a guide for other studies of similar situations.

Bibliography

1 Bartholomay, A. F.: A Stochastic Approach to Chemical Reaction Kinetics, Harvard University thesis, 1957.
2 Bartlett, M. S.: Stochastic Processes or the Statistics of Change, *Appl. Statist.*, vol. 2, pp. 44–64, 1953.
3 Bartlett, M. S.: "An Introduction to Stochastic Processes," Cambridge University Press, New York, 1955.
4 Bartlett, M. S.: Deterministic and Stochastic Models for Recurrent Epidemics, *Proc. Third Berkeley Symposium on Math. Statistics and Probability*, vol. 4, pp. 81–109, 1956.
5 Bartlett, M. S.: Measles Periodicity and Community Size, *J. Roy. Statist. Soc.*, ser. A, vol. 120, pp. 48–70, 1957.
6 Bartlett, M. S.: On Theoretical Models for Competitive and Predatory Biological Systems, *Biometrika*, vol. 44, pp. 27–42, 1957.
7 Butcher, J. C., and H. Messel: Electron Number Distribution in Electron-Photon Showers, *Phys. Rev.*, vol. 112, pp. 2096–2106, 1958.
8 Camp, G. D.: Bounding the Solutions of Practical Queueing Problems by Analytic Methods, in J. F. McCloskey and J. M. Coppinger (eds.): "Operations Research for Management," vol. 2, pp. 307–339, Johns Hopkins Press, Baltimore, 1956.
9 Churchman, C. W., R. L. Ackoff, and E. L. Arnoff: "Introduction to Operation Research," John Wiley & Sons, Inc., New York, 1957.
10 Crone, I.: Einige Anwendungsmöglichkeiten der Monte-Carlo-Methoden, *Fortschr. Physik*, vol. 3, pp. 97–132, 1955.
11 Foster, F. G.: Discussion of D. G. Kendall's paper, Some Problems in the Theory of Queues, *J. Roy. Statist. Soc.*, ser. B, vol. 13, 151–185, 1951.
12 Hammersley, J. M.: Electronic Computors and the Analysis of Stochastic Processes, *Math. Tables and Aids to Computation*, vol. 4, pp. 56–57, 1950.
13 Hammersley, J. M., and K. W. Morton: Transposed Branching Processes, *J. Roy. Statist. Soc.*, ser. B, vol. 16, pp. 76–79, 1954.
14 Hoffman, J. G., N. Metropolis, and V. Gardiner: Study of Tumor Cell Populations by Monte Carlo Methods, *Science*, vol. 122, pp. 465–466, 1955.
15 Householder, A. S.: "Principles of Numerical Analysis," McGraw-Hill Book Company, Inc., New York, 1953.
16 Householder, A. S., G. E. Forsythe, and H. H. Germond (eds.): "Monte Carlo Method," National Bureau of Standards, 1951.
17 Kahn, H.: "Applications of Monte Carlo," *RAND Research Memorandum* RM-1237-AEC, 1956.

18   Kendall, D. G.: An Artificial Realization of a Simple Birth-and-Death Process, *J. Roy. Statist. Soc.*, ser. B, vol. 12, pp. 116–119, 1950.
19   Kendall, D. G.: Deterministic and Stochastic Epidemics in Closed Populations, *Proc. Third Berkeley Symposium on Math. Statistics and Probability*, vol. 4, pp. 149–165, 1956.
20   Leslie, P. H.: A Stochastic Model for Studying the Properties of Certain Biological Systems by Numerical Methods, *Biometrika*, vol. 45, pp. 16–31, 1958.
21   Metropolis, N., and S. Ulam: The Monte Carlo Method, *J. Am. Statist. Assoc.*, vol. 44, pp. 338–341, 1949.
22   Meyer, H. A. (ed.): "Symposium on Monte Carlo Methods," John Wiley & Sons, Inc., New York, 1956.
23   Wilson, R. R.: The Range and Straggling of High Energy Electrons, *Phys. Rev.*, vol. 84, pp. 100–103, 1951.
24   Wilson, R. R.: Monte Carlo Study of Shower Production, *Phys. Rev.*, vol. 86, pp. 261–269, 1952.

Name Index

Subject Index

R

F